T0181869

Communications in Computer and Information Science 1565

More information about this series at https://link.springer.com/bookseries/7899

Linqiang Pan · Zhihua Cui · Jianghui Cai ·
Lianghao Li (Eds.)

Bio-Inspired Computing: Theories and Applications

16th International Conference, BIC-TA 2021
Taiyuan, China, December 17–19, 2021
Revised Selected Papers, Part I

Editors
Linqiang Pan
Huazhong University of Science and Technology
Wuhan, China

Zhihua Cui
Taiyuan University of Science and Technology
Taiyuan, China

Jianghui Cai
Taiyuan University of Science and Technology
Taiyuan, China

Lianghao Li
Huazhong University of Science and Technology
Wuhan, China

ISSN 1865-0929 ISSN 1865-0937 (electronic)
Communications in Computer and Information Science
ISBN 978-981-19-1255-9 ISBN 978-981-19-1256-6 (eBook)
https://doi.org/10.1007/978-981-19-1256-6

This Springer imprint is published by the registered company Springer Nature Singapore Pte Ltd.
The registered company address is: 152 Beach Road, #21-01/04 Gateway East, Singapore 189721, Singapore

Preface

Bio-inspired computing is a field of study that abstracts computing ideas (data structures, operations with data, ways to control operations, computing models, artificial intelligence, multisource data-driven methods and analysis, etc.) from living phenomena or biological systems such as cells, tissue, the brain, neural networks, the immune system, ant colonies, evolution, etc. The areas of bio-inspired computing include neural networks, brain-inspired computing, neuromorphic computing and architectures, cellular automata and cellular neural networks, evolutionary computing, swarm intelligence, fuzzy logic and systems, DNA and molecular computing, membrane computing, and artificial intelligence and its application in other disciplines such as machine learning, deep learning, image processing, computer science, cybernetics, etc. The Bio-Inspired Computing: Theories and Applications (BIC-TA) conference series aims to bring together researchers working in the main areas of bio-inspired computing, to present their recent results, exchange ideas, and foster cooperation in a friendly framework.

Since 2006, the conference has taken place in Wuhan (2006), Zhengzhou (2007), Adelaide (2008), Beijing (2009), Liverpool and Changsha (2010), Penang (2011), Gwalior (2012), Anhui (2013), Wuhan (2014), Anhui (2015), Xi'an (2016), Harbin (2017), Beijing (2018), Zhengzhou (2019), and Qingdao (2020). Following the success of previous editions, the 16th International Conference on Bio-Inspired Computing: Theories and Applications (BIC-TA 2021) was held in Taiyuan, China, during December 17–19, 2021, which was organized by the Taiyuan University of Science and Technology with the support of the Operations Research Society of Hubei.

We would like to thank the keynote speakers for their excellent presentations: Mingyong Han (Tianjin University, China), Chengde Mao (Purdue University, USA), Ling Wang (Tsinghua University, China), Rui Wang (National University of Defense Technology, China), and Wensheng Zhang (Chinese Academy of Sciences, China). Thanks are also given to the tutorial speakers for their informative presentations: Weigang Chen (Tianjin University, China), Cheng He (Southern University of Science and Technology, China), Tingfang Wu (Soochow University, China), and Gexiang Zhang (Chengdu University of Information Technology, China).

A special mention is given to Honorable Chair Gang Xie for his guidance and support to the conference.

We gratefully thank Xingjuan Cai, Yihao Cao, Guotong Chen, Weigang Chen, Tian Fan, Wanwan Guo, Yang Lan, Zhuoxuan Lan, Jie Wen, Lijie Xie, Linxia Yan, Huan Zhang, Jingbo Zhang, Zhixia Zhang, and Lihong Zhao for their contribution in organizing the conference. We also gratefully thank Shi Cheng, Weian Guo, Yinan Guo, Chaoli Sun, and Hui Wang for hosting the meetings.

Although BIC-TA 2021 was affected by COVID-19, we still received 211 submissions on various aspects of bio-inspired computing, and 67 papers were selected for this volume of Communications in Computer and Information Science. We are grateful to all the authors for submitting their interesting research work. The warmest thanks should be given to the external referees for their careful and efficient work in the reviewing process.

We thank Jianqing Lin and Guotong Chen for their help in collecting the final files of the papers and editing the volume. We thank Lianghao Li and Lianlang Duan for their contribution in maintaining the website of BIC-TA 2021 (http://2021.bicta.org/). We also thank all the other volunteers, whose efforts ensured the smooth running of the conference.

Special thanks are due to Springer for their skilled cooperation in the timely production of these volumes.

January 2022

Linqiang Pan
Zhihua Cui
Jianghui Cai
Lianghao Li

Organization

Steering Committee

Xiaochun Cheng	Middlesex University London, UK
Guangzhao Cui	Zhengzhou University of Light Industry, China
Kalyanmoy Deb	Michigan State University, USA
Miki Hirabayashi	National Institute of Information and Communications Technology, Japan
Joshua Knowles	University of Manchester, UK
Thom LaBean	North Carolina State University, USA
Jiuyong Li	University of South Australia, Australia
Kenli Li	University of Hunan, China
Giancarlo Mauri	Università di Milano-Bicocca, Italy
Yongli Mi	Hong Kong University of Science and Technology, Hong Kong
Atulya K. Nagar	Liverpool Hope University, UK
Linqiang Pan (Chair)	Huazhong University of Science and Technology, China
Gheorghe Paun	Romanian Academy, Romania
Mario J. Perez-Jimenez	University of Seville, Spain
K. G. Subramanian	Liverpool Hope University, UK
Robinson Thamburaj	Madras Christian College, India
Jin Xu	Peking University, China
Hao Yan	Arizona State University, USA

Honorable Chairs

Zhiguo Gui	Taiyuan University, China
Jiye Liang	Shanxi University, China
Gang Xie	Taiyuan University of Science and Technology, China
Jianchao Zeng	North University of China, China

General Chair

Jianghui Cai	North University of China, China

Program Committee Chairs

Zhihua Cui — Taiyuan University of Science and Technology, China
Linqiang Pan — Huazhong University of Science and Technology, China

Special Session Chair

Yan Qiang — Taiyuan University of Technology, China

Tutorial Chair

Weigang Chen — Tianjin University, China

Publication Chairs

Lianghao Li — Huazhong University of Science and Technology, China
Gaige Wang — Ocean University of China, China
Qingshan Zhao — Xinzhou Teachers University, China

Publicity Chair

Haifeng Yang — Taiyuan University of Science and Technology, China

Local Chair

Chaoli Sun — Taiyuan University of Science and Technology, China

Registration Chair

Libo Yang — Taiyuan University, China

Program Committee

Muhammad Abulaish — South Asian University, India
Andy Adamatzky — University of the West of England, UK
Chang Wook Ahn — Gwangju Institute of Science and Technology, South Korea
Adel Al-Jumaily — University of Technology Sydney, Australia
Bin Cao — Hebei University of Technology, China

Xiaobo Liu	China University of Geosciences, China
Wenjian Luo	University of Science and Technology of China, China
Lianbo Ma	Northeastern University, China
Wanli Ma	University of Canberra, Australia
Xiaoliang Ma	Shenzhen University, China
Francesco Marcelloni	University of Pisa, Italy
Efrén Mezura-Montes	University of Veracruz, Mexico
Hongwei Mo	Harbin Engineering University, China
Chilukuri Mohan	Syracuse University, USA
Abdulqader Mohsen	University of Science and Technology Yemen, Yemen
Holger Morgenstern	Albstadt-Sigmaringen University, Germany
Andres Muñoz	Universidad Católica San Antonio de Murcia, Spain
G. R. S. Murthy	Lendi Institute of Engineering and Technology, India
Akila Muthuramalingam	KPR Institute of Engineering and Technology, India
Yusuke Nojima	Osaka Prefecture University, Japan
Linqiang Pan	Huazhong University of Science and Technology, China
Andrei Paun	University of Bucharest, Romania
Gheorghe Paun	Romanian Academy, Romania
Xingguang Peng	Northwestern Polytechnical University, China
Chao Qian	University of Science and Technology of China, China
Balwinder Raj	NITTTR, India
Rawya Rizk	Port Said University, Egypt
Rajesh Sanghvi	G. H. Patel College of Engineering and Technology, India
Ronghua Shang	Xidian University, China
Zhigang Shang	Zhengzhou University, China
Ravi Shankar	Florida Atlantic University, USA
V. Ravi Sankar	GITAM University, India
Bosheng Song	Hunan University, China
Tao Song	China University of Petroleum, China
Jianyong Sun	University of Nottingham, UK
Yifei Sun	Shaanxi Normal University, China
Handing Wang	Xidian University, China
Yong Wang	Central South University, China
Hui Wang	Nanchang Institute of Technology, China
Hui Wang	South China Agricultural University, China

Contents – Part I

DNA and Molecular Computing

Contents – Part II

Machine Learning and Computer Vision

Evolutionary Computation and Swarm Intelligence

An Optimization Task Scheduling Model for Multi-robot Systems in Intelligent Warehouses

Xuechun Jing and Zhihua Cui(✉)

Taiyuan University of Science and Technology, Taiyuan, China
zhihua.cui@hotmail.com

Abstract. Task allocation and path planning are commonly taken into consideration when utilizing multi-robot systems in intelligent warehouses. However, the importance of picking stations, which affects the efficiency of the system, is often ignored. To tackle this problem, this study has designed a novel scheduling model to improve the efficiency of the multi-robot system. Unlike the original scheduling model, the queuing time is taken into account in the newly designed scheduling model. Additionally, two approaches are applied in order to improve the designed model. Specifically, the balanced heuristic mechanism (BHM) is used to choose the optimal picking station to shorten the queuing time. The method of task reordering based on task correlation (TRBTC) is also adopted to reduce the travel cost. To verify the efficiency of this method, the proposed method is applied to the different task allocation schemes of intelligent warehouse systems and compared with the original model. Simulation results show that overall superior performance is achieved in the warehouse system.

Keywords: Multi-robot systems · Intelligent warehouse · Picking station · Balanced heuristic mechanism · Reordering and combining tasks

1 Introduction

With the development and maturity of artificial intelligence technology, the robot system has received widespread attention. To apply the robot system to solve practical problems, Many achievements in robotic systems have been developed in recent years. However, a single robot will not solve practical problems as complexity increases. Therefore, multi-robot systems have been developed [16]. Culmer et al. [8] proposed a dual robot system to treat the upper limb during voluntary reaching exercises. Matsuo et al. [17] proposed a tree formation multi-robot system to search for survivors in a damaged building after an earthquake

Supported by the Key R&D program of Shanxi Province (International Cooperation) under Grant No. 201903D421048.

L. Pan et al. (Eds.): BIC-TA 2021, CCIS 1565, pp. 3–17, 2022.
https://doi.org/10.1007/978-981-19-1256-6_1

or other natural disaster. Ahmadi et al. [1] examined the problem of multi-robot continuous area sweeping and proposed a negotiation-based approach to solve this problem. Turduev et al. [23] described the implementation of various bio-inspired algorithms, including decentralized and asynchronous particle swarm optimization (DAPSO), bacterial foraging optimization (BFO) and ant colony optimization (ACO), to obtain a chemical gas concentration map in an environment filled using the multi-robot system. In addition, warehousing with a large scale and variety of goods is required in the logistics field under the background of e-commerce [2]. Therefore, intelligent warehousing needs to be specially mentioned in multi-robot systems. Li et al. [15] proposed a RFID-based intelligent warehouse management system (RFID-IWMS) to perform inventory control and improve operational efficiency. Tee et al. [22] presented a novel electronic system for container tracking and stock control in a storage warehouse environment.

Specifically speaking, there are three general elements that can be abstracted in the intelligent warehousing system of large-scale logistics robots: logistics robots, tasks and picking stations. The relationship between robots and tasks is commonly considered for multi-robot systems in intelligent warehouses, but there is not much attention given to picking stations. In addition, the cost factor is often not considered because of differences in task sequences. To tackle the problem, a new scheduling model considering queue time at the picking station is described in this paper. To improve the efficiency of the system, the balanced heuristic mechanism (BHM) is used to choose the optimal picking station for shortening queuing time. The TRBTC method is adopted to reduce travel costs.

The rest of this paper is organized as follows: Related work about the development of intelligent warehousing is reviewed in Sect. 2. Detailed descriptions of intelligent warehousing are discussed in Sect. 3. The designed model and methods are explained in Sect. 4. A simulated experiment is performed in Sect. 5. Section 6 is a conclusion.

2 Related Work

In the past few decades, multi-robot systems of intelligent warehouses have attracted the attention of many researchers. One of the most important issues in the research of multi-robot systems is how to achieve coordination and cooperation among robots. Specifically speaking, multi-robot task assignment refers to the assignment of a series of tasks to a series of robots to achieve a set goal under certain constraints. Therefore, the multi-robot task assignment (MRTA) usually considers optimization grouping, linear programming, scheduling optimization and the multi-trip salesman problem (MTSP). Nam et al. [20] regarded multi-robot task allocation as an optimization grouping problem, and proposed an exact algorithm that was used for the general problem and polynomial-time algorithms that were adopted for other problems. Crandall et al. [5] stated that model predictions should compute and guide how operators should allocate their attention. Goyal er al. [13] investigated the general NP-hard problem and instances where interference results in linear or convex penalization functions. Melo et al.

[18] thought of multi-robot task allocation as scheduling optimization problems and proposed a complete methodology that encompasses all related stages of the problem. Faigl et al. [12] regarded multi-robot task allocation as a traveling salesman problem and proposed a novel exploration strategy to the formulation. Given the problem of MRTA, scholars have done a lot of research and proposed a variety of task assignment methods. Multi-robot task assignment methods can be divided into many different types of research. Market mechanism methods [25] and swarm intelligence algorithms [4,21] are often used to assign tasks. The complex dynamic MRTA can be solved using Swarm intelligence algorithms involving Particle Swarm optimization [9], the Bat algorithm [6,24], the pigeon-inspired optimization algorithm [7], or the cuckoo search [26]. However, the parameters of these algorithms are difficult to determine.

Some researchers have invested a lot of time and effort into more easily solving these problems. Nagarajan et al. [19] employed heuristic algorithms to solve multi-robot task assignment problems from a different point of view. Elango et al. [11] solved the dual-objective task assignment model that took into account both total driving distance and robot efficiency by using the k-means clustering method and auction mechanism. Based on the above models, Zhou et al. [27] proposed the balanced heuristic mechanism (BHM) to evaluate different strategies by adopting the travel time (TT), total travel cost (TTC) and coefficient of variation (CV) indicators. Since then, the BHM has been proved to achieve good performance. Dou et al. [10] also introduced the idea of reinforcement learning and optimization algorithm in order to find the optimal path and treat it as the task cost. Bolu et al. [3] proposed an adaptive heuristic approach to assign generated tasks to robots, considering system dynamics. Unlike the above methods, Lamballais et al. [14] built a queuing model to measure the utility of logistics robots and workers. However, the robots are only considered as serving one workstation in the approach, which is not in line with the fact that the robot may have more choices of picking stations with their heavy loads.

The above analysis shows that most of the research done on this topic focuses on task allocation and path planning. Unlike previous research, the new model takes the workloads of picking stations into account. In other words, our work focuses on improving overall speed by balancing the workload of each picking station and reordering the execution sequence of tasks according to the relevance between the tasks.

3 Intelligent Warehouses

3.1 Problem Statement

Intelligent warehouses consist of storage shelves, picking stations and multi-robot systems. In addition, each storage shelf consists of several inventory pods that are delivered to the picking station. Pods are composed of several layers with many order boxes. It is worth noting that these components are usually dispatched uniformly by the central console. Here, the Kiva robots will be employed to illustrate an example. For clear visualization, a sketch of the intelligent warehouse is shown in Fig. 1.

In Fig. 1, the initial position of the storage shelves, picking stations and robots are set. To complete the scheduling process, the central console is responsible for assigning tasks to robots. The tasks contain information the delivery pod position with order boxes and the destination picking station. After receiving the task, the robot will go to the position contained in the pod task. By the time the robot has arrived position, it will lift the whole pod and head for the specific picking station. The robot will get back the original location of the pod after the right box is picked and packed. Then if the robot had completed its tasks, it will return to its initial position. Otherwise, it will move to the pod containing the next task.

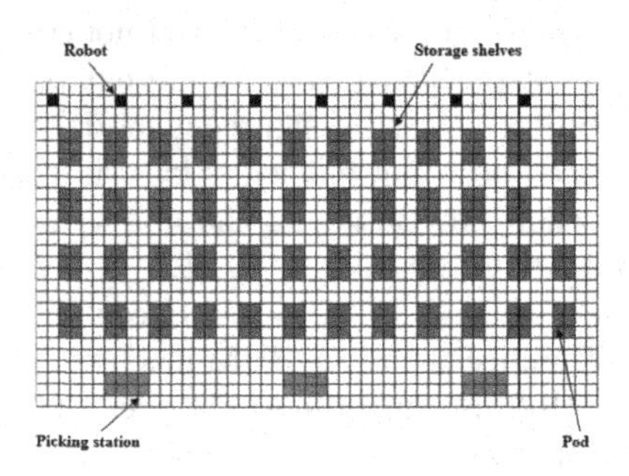

Fig. 1. Sketch of the intelligent warehouse system.

To better explain the problem of the intelligent warehouse, a formula description needs to be carried out. Assuming that a set of tasks $T = t_1, t_2, \ldots, t_n$ needs to be accomplished by the robots, let $R = r_1, r_2, \ldots, r_n$. And the twice distance from the pod to the station is defined as the cost of task t_i and denoted as ω_i. In addition, when the r_i has k tasks $T = t_{i1}, t_{i2}, \ldots, t_{ik}$, the total cost of r_i will be $W(r_i) = \omega_{i1} + \omega_{i2} + \cdots + \omega_{ik}$, and $C_{ij}(i,j) \in R \cup T$ can be defined as transition cost, where i is the robot or the task, and j is the task. $C_{r_i t_j}$ represents the travel cost from the initial position of r_i to the corresponding pod t_j; $C_{t_i t_j}$ represents the travel cost from the current position of t_i to the position of t_j. Therefore, the intelligent warehouse system is described mathematically in Fig. 2.

$ITC(r_i, T_i)$ represents the total travel cost of the individual robot r_i when finishing the task of T_i, and can be computed as Eq. (1).

$$ITC(r_i, T_i) = C_{r_i t_{i1}} + \sum_{j=1}^{k-1} C_{t_{ij} t_{i(j+1)}} + W(r_i) \tag{1}$$

where $W(r_i) = \omega_{i1} + \omega_{i2} + \cdots + \omega_{ik}$, and $C_{r_i t_{i1}}$ represents the distance cost of the robot arriving at the inventory pod assigned in the first task $t_i 1$.

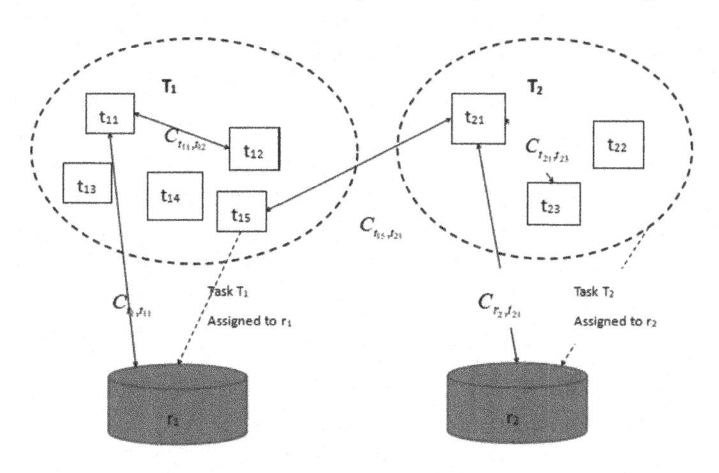

Fig. 2. Mathematical intelligent warehouse.

3.2 BHM

To finish the work as quickly as possible, MRTA and path planning need to be taken into account. The MRTA problem entails how to assign robots to optimize the warehouse system given a set of task sequences, and how to keep the balance between individual travel costs of each robot simultaneously are the key problems that need to be solved. To tackle these problems, the balanced heuristic mechanism (BHM) is proposed to evaluate the performance of different strategies by adopting the travel time (TT), total travel cost (TTC) and coefficient of variation (CV) indicators. In the BHM, the classic single-item auction and the K-means clustering method are introduced to balance the MRTA problems. Meanwhile, the travel costs of different robots would be calculated when the assignment of tasks is completed. For travel costs, the distance from the current position of robot to the target pod and the current workload of the robot are the main cost factors. Therefore, using the BHM, a robot can have both a lower cost and lower workload.

To minimize the distance cost and balance the workforce between different robots, the Eq. (2) will be used:

$$C(t_j, r_i) = \alpha \times dist(t_j, r_i) + (1 - \alpha) \times CW(r_i) \tag{2}$$

where $C(t_j, r_i)$ is the total travel cost of r_i when finishing the tasks of t_j; α is the weight factor, which is usually set within the interval [0.7, 0.9] [27]; $dist(t_j, r_i)$ is the path length from r_i to t_j and $CW(r_i)$ is the current travel cost of r_i.

4 Improved Intelligent Warehouses

4.1 New Model

Based on the above research, time spent waiting in line at the picking station has not been adequately studied. The new design brought forward in this paper has been designed to make up for this deficiency and improve the efficiency of intelligent warehouses systems. Some constraints need to be enforced in this study. Sensor detection devices should be installed in every robot to avoid collision with other robots. The time spent on obstacle avoidance is generally ignored under the circumstance of conflict with other robots. In addition, delivery tasks are independent of each other, and the time of fulfilling the order in picking station is same for all workers.

Assume that δ_i is the sum of queue time and picking time in finishing the task t_i. $Q(r_i)$ is the total queue time of tasks in r_i, and $Q(r_i) = \delta_1 + \delta_2 + \cdots + \delta_n$. Therefore, $ITC(r_i, T_i)$ will be recalculated according to the Eq. (3).

$$ITC(r_i, T_i) = C_{r_i t_{i1}} + \sum_{j=1}^{k-1} C_{t_{ij} t_{i(j+1)}} + W(r_i) + Q(r_i) \tag{3}$$

In the above model, the queuing time factor is taken into account in the scheduling model. In general, the queuing time will continue to lengthen as the number of orders increases.

4.2 New Model

To improve the new scheduling model and make the intelligent warehouse system more efficient, BHM is employed to choose the optimal picking station for shortening queuing time. The TRBTC method is adopted to reduce travel costs. To minimize total costs, the order allocation and cost calculations are performed by the central console. The whole workflow is shown in Fig. 3.

In Fig. 3, the overall control structure includes two parts: the central console and multi-robot execution system. Specifically, when an instruction is sent from the central console to the robots, the robots will execute the given task. After finishing the task, the robots will send back feedback information. Typically, the central console will choose an appropriate robot to complete tasks after receiving orders. Unlike in the previous process, the central console will reorder the tasks before allocating them. For a given robot, the optimal picking station for this task will be determined by comparing the travel costs for different picking stations. Therefore, a better allocation scheme is found to shorten work time, which encourages robots to select an optimal path after receiving instructions.

Modified BHM. Shortening queuing and traveling time is an effective way to reduce the value of $ITC(r_i, T_i)$. To reduce the cost of travel from the pod position to the picking station, the fixed picking station can be replaced with the optimal picking station. Therefore, queue time and distance will be taken

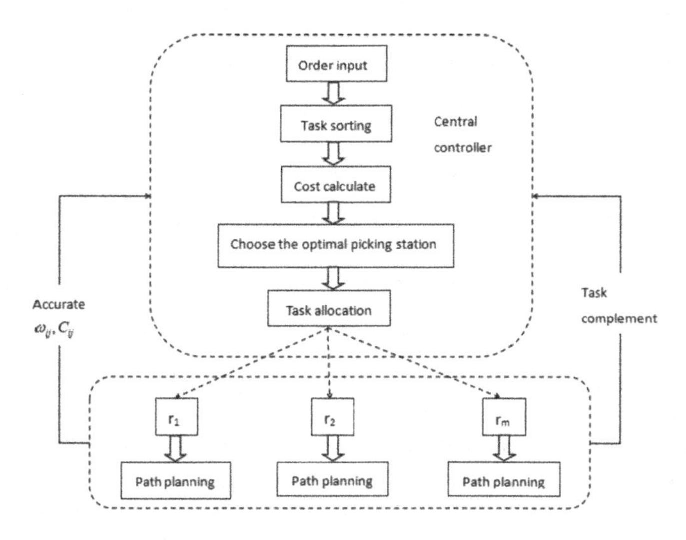

Fig. 3. Overall control structure.

into account when choosing the optimal picking station. Specifically speaking, because the picking station with the shortest queue has the lowest workload, the picking station with the shortest queue can be selected to save queue time. Also, the nearest picking station can be selected to save additional travel time.

In addition, improving the BHM approach will allow for reasonably allocated workloads and minimal travels costs. In the modified BHM approach, a weight factor is used to balance the travel time and queue time for robots. Therefore, the optimal picking station that meets the requirements can be selected by employing the modified BHM. In this way, the Eq. (4) can be obtained:

$$C(t_i, p_j) = \beta \times dist(t_i, p_j) + (1 - \beta) \times CQ(p_j) \tag{4}$$

where $C(t_i, p_j)$ is the degree of preference when selecting a picking station. $dist(t_i, p_j)$ represents the distance from the t_i task to the picking station p_j. $CQ(p_j)$ represents the workload of the picking station p_j. β is a weight factor to balance the distance and workload, which helps assign tasks to the picking station in the shortest time.

Received tasks $T = t_1, t_2, \ldots, t_m$, will be assigned an appropriate picking station in turn. The assignment process can be described mathematically, as shown in Fig. 4.

For Fig. 4, the process of picking station selection is described as an $m \times n$ matrix to be processed. The m and n represent the number of tasks and picking stations, respectively. Therefore, the 0 elements will be replaced with the value of $C(t_i, p_j)$. Specifically speaking $C(t_1, p_1)$, $C(t_1, p_2)$, and $C(t_1, p_3)$ will be calculated to fill the first row. The optimal picking station for the first task can be determined by comparing and selecting the minimum values of $C(t_1, p_1)$, $C(t_1, p_2)$, and $C(t_1, p_3)$. Thereby, every aspect will be updated by following the

$$\begin{bmatrix} C(t_1,p_1) & C(t_1,p_2) & \cdots & C(t_1,p_n) \\ 0 & 0 & \cdots & 0 \\ \vdots & \ddots & & \vdots \\ \vdots & & & \vdots \\ \vdots & & \ddots & \vdots \\ 0 & 0 & \cdots & 0 \end{bmatrix} \Rightarrow \begin{bmatrix} C(t_1,p_1) & C(t_1,p_2) & \cdots & C(t_1,p_n) \\ C(t_2,p_1) & C(t_2,p_2) & \cdots & C(t_2,p_n) \\ 0 & 0 & \cdots & 0 \\ \vdots & \ddots & & \vdots \\ \vdots & & \ddots & \vdots \\ 0 & 0 & \cdots & 0 \end{bmatrix} \Rightarrow \cdots \Rightarrow \begin{bmatrix} C(t_1,p_1) & C(t_1,p_2) & \cdots & C(t_1,p_n) \\ C(t_2,p_1) & C(t_2,p_2) & \cdots & C(t_2,p_n) \\ \vdots & \ddots & & \vdots \\ \vdots & & & \vdots \\ \vdots & & \ddots & \vdots \\ C(t_m,p_1) & C(t_m,p_2) & \cdots & C(t_m,p_n) \end{bmatrix}$$

Fig. 4. The process of picking station selection.

above method in the matrix and the optimal picking stations will be obtained for each task.

After the picking station of the task is determined, the next problem is how to assign tasks to appropriate robots. To express the whole allocation process more intuitively, the robot selection process can be described mathematically as shown in Fig. 5. The process of robot selection is similar to the process of picking station selection shown in Fig. 4. Therefore, the appropriate robot can be determined by comparing and selecting the minimum value of $C(t_j, r_i)$ for each task.

$$\begin{bmatrix} C(t_1,r_1) & C(t_1,r_2) & \cdots & C(t_1,r_n) \\ 0 & 0 & \cdots & 0 \\ \vdots & \ddots & & \vdots \\ \vdots & & & \vdots \\ \vdots & & \ddots & \vdots \\ 0 & 0 & \cdots & 0 \end{bmatrix} \Rightarrow \begin{bmatrix} C(t_1,r_1) & C(t_1,r_2) & \cdots & C(t_1,r_n) \\ C(t_2,r_1) & C(t_2,r_2) & \cdots & C(t_2,r_n) \\ 0 & 0 & \cdots & 0 \\ \vdots & \ddots & & \vdots \\ \vdots & & \ddots & \vdots \\ 0 & 0 & \cdots & 0 \end{bmatrix} \Rightarrow \cdots \Rightarrow \begin{bmatrix} C(t_1,r_1) & C(t_1,r_2) & \cdots & C(t_1,r_n) \\ C(t_2,r_1) & C(t_2,r_2) & \cdots & C(t_2,r_n) \\ \vdots & \ddots & & \vdots \\ \vdots & & & \vdots \\ \vdots & & \ddots & \vdots \\ C(t_m,r_1) & C(t_m,r_2) & \cdots & C(t_m,r_n) \end{bmatrix}$$

Fig. 5. The process of robot selection.

Task Reordering Based Tasks Correlation (TRBTC). In previous research, the independence of each task was highlighted, which facilitates the calculation of certain complex situations, but also brought about some problems. For example, when the adjacent tasks are in the same pod, the robot will return to the same place to execute the next task once the prior task is completed. However, it is unnecessary to return to the picking station after having picked once, because two tasks are in the same pod. Therefore, a new computing method considering task correlation is proposed to solve the problem. Assume that $E_{l,l+1}$ is the adjacent task time cost of l and $l+1$, and $B_{t_{il},t_{i(l+1)}}$ is the relationship of adjacent tasks. In addition, $B_{t_{il},t_{i(l+1)}} = 1$ represents adjacent tasks in the same pod. Otherwise, $B_{t_{il},t_{i(l+1)}} = 0$. Therefore, $E_{l,l+1}$ can be depicted in Eq. (5).

$$E_{l,l+1} = \begin{cases} C_{t_{i(l-1)},t_{il}} + \omega_{il} + \delta_{il} + C_{t_{il},t_{i(l+1)}} \\ +\omega_{i(l+1)} + \delta_{i(l+1)}, \\ C_{t_{i(l-1)},t_{il}} + \omega_{il} + \delta_{il} + T_{l+1}. \end{cases} \tag{5}$$

where T_{l+1} represents the picking time of task $l+1$. δ_{il} refers to the sum of the queue and the picking time of robot i for completing task l. Therefore, T_l must be less than δ_{il}. ω_{il} is twice the distance from the pod to the station in task l. $C_{t_{i(l-1)},t_{il}}$ represents the cost of the former task $l-1$ position to the corresponding pod of the next task l.

From Eq. (5), one can see that system efficiency improves results from $T_{l+1} < \delta_{i(l+1)} < C_{t_{il},t_{i(l+1)}} + \omega_{i(l+1)} + \delta_{i(l+1)}$ when $B_{t_{il},t_{i(l+1)}} = 1$. In other words, task execution order plays an important role in affecting system efficiency. Unfortunately, it is usually neglected in the concrete calculation. To solve the problem, the operations should be performed by TRBTC in the tasks pool before task allocation and before putting sequences of tasks together in the same pod. To describe the TRBTC more intuitively, a task sketch is described in Fig. 6. In addition, it is worth mentioning that, in Fig. 6, tasks of the same color are in the same pod. To take the task r_1 as an example, the other tasks will be judged whether or not they are in the same pod when the task r_1 is assigned. As we can see from Fig. 6, r_1 and r_{10} have the same pod. Therefore, the tasks r_{10} and r_1 should be put together, and other tasks will be automatically postponed.

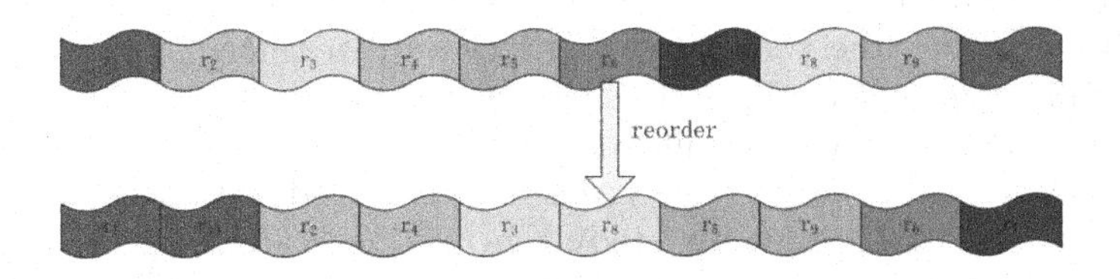

Fig. 6. Tasks reorder.

5 Simulated Experiments

5.1 Performance Metrics

To evaluate the efficiency of the method in the new scheduling model, indicators are needed. In general, the work efficiency of intelligent warehouses can be measured using robot travel time because the work in the warehouse is done by robots. The corresponding travel time can be calculated for each robot. Therefore, the time for the entire task can be obtained by calculating the sum of the traveling time for all robots. Additionally, the system completion time is determined using the longest robot traveling time. The cost and time spent completing tasks also indirectly measures the efficiency of the intelligent warehouse. To guarantee the normal operation of the system, the workload should be reasonable for the robots. An ideal load balancing situation is one in which the longest and the shortest traveling time of the robot are close. The evaluation indicators can be summarized as Eq. (6).

$$TT = \max_i ITC(r_i, T_i), \qquad TTC = \sum_{i=1}^{m} ITC(r_i, T_i),$$
$$STC = T - T_0, \qquad RBU = \frac{\min_i ITC(r_i, T_i)}{\max_i ITC(r_i, T_i)}. \tag{6}$$

Where: TT represents the longest traveling time of the robots; TTC represents the total travel time of robots; STC is the cost time of completing tasks assignment; T and T_0 are the end time and start time of system, respectively; RBU is the workload balancing rate. For TT, TTC, and STC, lower values indicate that the efficiency of the system is high. A high RBU indicates that the workload of robots is balanced.

5.2 Parameter Settings

To achieve a fair comparison, the parameters of the experiment need to be described. The performance of our method was tested in a simulated intelligent warehouse system similar to [10]. For the no-optimization model, the relevant parameters will be set according to the original literature. In the proposed new optimization model, the value of β needs to be determined. To consider the impact of different β values on system efficiency, a discussion of the relationship between the value of β and the different indicators is required.

The traveling time cost of any given robot has a direct effect on TT and STC. Therefore, one can simply study the relationship between β and TTC because TTC is intimately linked to the traveling time cost of each robot. In addition, RBU reflects the balancing of robot workload, which is related to the process of allocating tasks to robots, but not related to choosing appropriate picking stations. Therefore, the relationship between RBU and STC does not need to be researched.

To describe the relationship between β and TTC, a set of parameter experimental results with different numbers of tasks is presented in Fig. 7. The experiment will be repeated three times for each task situation. It can be observed from Fig. 7 that different β values have different effects on the value of TTC. Especially, the TTC value achieves better results for the β values within [0.6, 0.8] when the number of tasks (T) is 50. For other task situations, a similar result can be found within the range of error. Therefore, the β value is set to 0.6 for convenience in the following experiments.

5.3 Analysis of Results

The results of the non-optimized model (NOM) and the optimized model (OM) in different task allocation schema are shown in Table 1. For the non-optimized model, only the queue time is considered. Unlike in the non-optimized model, the queue time is not only considered, but optimal picking stations replace fixed picking stations by modified BHM in the optimized model. Specifically speaking, the results of sequential task allocation and task reorder are shown in Table 1.

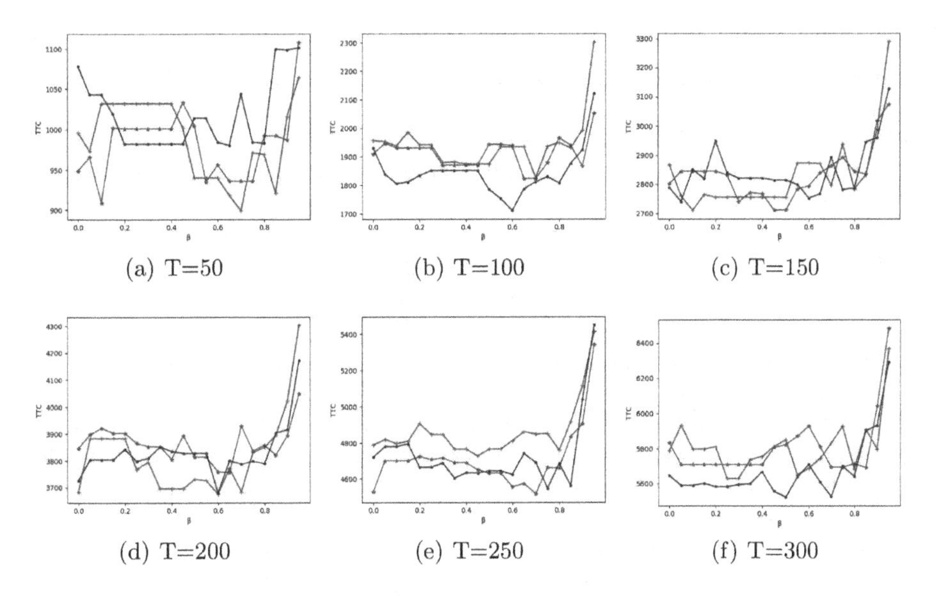

(a) T=50 (b) T=100 (c) T=150

(d) T=200 (e) T=250 (f) T=300

Fig. 7. Relationship between β and TTC

In addition, k-means and Genetic Algorithm (GA) are always applied to solve the problem of tasking allocation in the warehouse system. Usually, k-means can cluster tasks according to their attributes when obtaining a batch of tasks. Then these clusters of tasks are distributed to different picking stations. Similarly, GA can seek a good allocation by heuristic methods [28]. Therefore, we compare with both methods, which are shown in Fig. 8. Meanwhile, the simulation experiments are performed on different size tasks in our experiment.

Looking at Table 1, we can draw some important conclusions:

- There is not much difference in RBU between two models in sequential allocation. RBU shows how balanced the workload of robots is. It is observed that the workload of robots using the optimized model will be more balanced than when using the non-optimized model when the task number is 50 or 100. For other situations, the balance of workload of robots in the optimized and non-optimized models is similar, which means that the experiments have no effect on optimizing for MRTA. In summary, it is ensured that an optimal robot will be chosen to complete the task.
- Table 1 shows the comparison of both models in STC using sequential task allocation. STC is measured in time cost, which represents the overall efficiency of the multi-robot system in the intelligent warehouse. Therefore, the lower the STC, the higher the efficiency of the multi-robot system. In addition, it is obvious that, for the optimized and non-optimized model, the time cost grows as the number of tasks increases. As a result of an increase in the number of tasks, the execution time will also increase. At the same time, it is seen that the optimized model is much more efficient than the non-optimized

model is. Meanwhile, the STC in the optimized model is usually lower than in the non-optimized model. Therefore, the proposed approach improves the efficiency of the system.

- Table 1 shows the difference between the non-optimized model and the optimized model in TT in sequential allocation. Similarly, Table 1 shows the difference in TTC in sequential allocation. TTC represents the sum cost of each robot and TT represents the maximum cost. It is worth mentioning that TTC is distinct from STC. TTC is the sum cost of each robot, while STC is time cost when each robot is working simultaneously. TTC has practical significance on physical cost though robots work in parallel at the warehouse. The results show that the TT and TTC values in the optimized model are almost always lower than in the non-optimized model, no matter the number of tasks, because robot workload is reduced. As a result, the system efficiency is improved. In particular, when the number of tasks is 250, the significant differences between the TT and TTC values for the optimized and non-optimized models are presented in Table 1.

Looking at Fig. 8, we can draw some important conclusions:

- To describe the extensive feasibility of the model, research should be conducted comparing different task allocations. Specifically speaking, Fig. 8(a) shows the comparison of different task allocations in terms of RBU. In addition, the simulation results indicate that there is not much difference in RBU between the optimized and non-optimized models. The RBU is highest for sequential task allocation, which shows that BHM used to allocate tasks helps control robot workload balance.
- Also, the STC, an important indicator, directly reflects system efficiency. As Fig. 8(b) indicates, the value of STC in the optimized model is always lower than that of the non-optimized model. In particular, the STC of GA and k-means are higher than the sequential task allocation. Meanwhile, the optimum solution, which allocates strong correlation tasks to the same robot, is not found in the above algorithm.
- Similarly, as Fig. 8(c) shows, the TT of the optimized model is always lower than the non-optimized model. Additionally, sequential task allocation has a big advantage over k-means and GA. Meanwhile, the results of TT are similar to those of STC, which shows that STC indicates the longest work time among robots.
- For Fig. 8(d), the TTC of the optimized model is always lower than that of the non-optimized model, which is unaffected by the number of tasks, regardless of the allocation scheme. However, the robot workload balancing lead that sequential task allocation has results in higher TTC than other methods do. Specifically, the workload of robots is closer, which means that the TTC is about n times as much as the workload of each robot if there are n robots. Therefore, it is worthwhile to increase the TTC so that the workload will become more balanced, and the efficiency of the system will be improved. In summary, superior performance is achieved using our proposed model of the warehouse system.

Table 1. Comparison of both models in the sequential allocation

Criteria	TT		TTC		STC		RBU (%)	
	NOM	OM	NOM	OM	NOM	OM	NOM	OM
T = 50	1170	949	8386	7004	1174	952	78.72	84.93
T = 100	2279	1824	16947	14112	2284	1833	86.57	93.37
T = 150	3141	2786	24322	21671	3142	2792	94.40	94.51
T = 200	4346	3651	34195	28747	4351	3653	96.57	96.82
T = 250	5489	4661	43120	36727	5496	4675	96.72	97.32
T=300	6166	5649	48786	44741	6173	5652	97.58	97.52
T = 350	7270	6762	57215	53391	7276	6763	97.54	97.71
T = 400	8360	7705	66345	61139	8362	7724	98.58	98.35

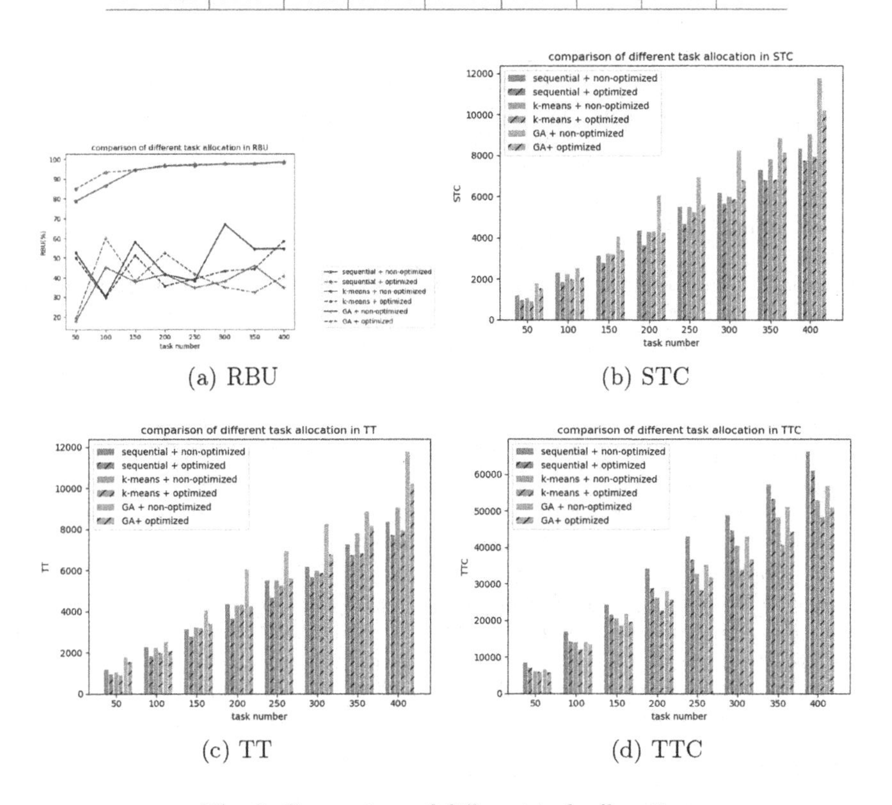

(a) RBU (b) STC

(c) TT (d) TTC

Fig. 8. Comparison of different task allocation

6 Conclusion

The main contribution of this paper is a new task scheduling model considering queuing times. For this new model, two methods are proposed: i) selecting optimal picking station considering travel cost and queue times ii) TRBTC. In this paper, the most important contribution of the optimized model is that a specific picking station is replaced with the optimal station. Meanwhile, queue time is considered in the model of the warehouse system. Simulation results show that *STC*, *TT* and *TTC* are significantly reduced and that the system would be more efficient than traditional methods by using the improved model. At the same time, sequential task allocation has better performance in terms of *RBU*, *TTC*, and *STC*. As a further improvement, it is recommended to focus on optimizing the routings of large-scale multi-robot and individualization of β in different cases.

References

1. Ahmadi, M., Stone, P.: A multi-robot system for continuous area sweeping tasks. In: Proceedings 2006 IEEE International Conference on Robotics and Automation, 2006. ICRA 2006, pp. 1724–1729. IEEE (2006)
2. Bogue, R.: Growth in e-commerce boosts innovation in the warehouse robot market. Ind. Robot Int. J. **43**, 583–587 (2016)
3. Bolu, A., Korçak, O.: Adaptive task planning for multi-robot smart warehouse. IEEE Access **9**, 27346–27358 (2021)
4. Chetty, S., Adewumi, A.O.: Comparison study of swarm intelligence techniques for the annual crop planning problem. IEEE Trans. Evol. Comput. **18**(2), 258–268 (2013)
5. Crandall, J.W., Cummings, M.L., Della Penna, M., De Jong, P.M.: Computing the effects of operator attention allocation in human control of multiple robots. IEEE Trans. Syst. Man Cybern.-Part A Syst. Hum. **41**(3), 385–397 (2010)
6. Cui, Z., Li, F., Zhang, W.: Bat algorithm with principal component analysis. Int. J. Mach. Learn. Cybern. **10**(3), 603–622 (2018). https://doi.org/10.1007/s13042-018-0888-4
7. Cui, Z., Zhang, J., Wang, Y., Cao, Y., Cai, X., Zhang, W., Chen, J.: A pigeon-inspired optimization algorithm for many-objective optimization problems. Sci. China Inf. Sci. **62**(7), 1–3 (2019). https://doi.org/10.1007/s11432-018-9729-5
8. Culmer, P.R., Jackson, A.E., Makower, S., Richardson, R., Cozens, J.A., Levesley, M.C., Bhakta, B.B.: A control strategy for upper limb robotic rehabilitation with a dual robot system. IEEE/ASME Trans. Mechatron. **15**(4), 575–585 (2009)
9. Das, P.K., Behera, H.S., Panigrahi, B.K.: A hybridization of an improved particle swarm optimization and gravitational search algorithm for multi-robot path planning. Swarm Evol. Comput. **28**, 14–28 (2016)
10. Dou, J., Chen, C., Yang, P.: Genetic scheduling and reinforcement learning in multirobot systems for intelligent warehouses. Mathematical Problems in Engineering 2015 (2015)
11. Elango, M., Nachiappan, S., Tiwari, M.K.: Balancing task allocation in multi-robot systems using k-means clustering and auction based mechanisms. Expert Syst. Appl. **38**(6), 6486–6491 (2011)
12. Faigl, J., Kulich, M., Přeučil, L.: Goal assignment using distance cost in multi-robot exploration. In: 2012 IEEE/RSJ International Conference on Intelligent Robots and Systems, pp. 3741–3746. IEEE (2012)
13. Goyal, V., Levi, R., Segev, D.: Near-optimal algorithms for the assortment planning problem under dynamic substitution and stochastic demand. Oper. Res. **64**(1), 219–235 (2016)

14. Lamballais, T., Roy, D., De Koster, M.: Estimating performance in a robotic mobile fulfillment system. Eur. J. Oper. Res. **256**(3), 976–990 (2017)
15. Li, M., Gu, S., Chen, G., Zhu, Z.: A rfid-based intelligent warehouse management system design and implementation. In: 2011 IEEE 8th International Conference on e-Business Engineering, pp. 178–184. IEEE (2011)
16. Mariottini, G.L., et al.: Vision-based localization for leader-follower formation control. IEEE Trans. Rob. **25**(6), 1431–1438 (2009)
17. Matsuo, Y., Tamura, Y.: Tree formation multi-robot system for victim search in a devastated indoor space. In: 2004 IEEE/RSJ International Conference on Intelligent Robots and Systems (IROS) (IEEE Cat. No. 04CH37566), vol. 2, pp. 1071–1076. IEEE (2004)
18. Melo, R.S., Macharet, D.G., Campos, M.F.M.: Collaborative object transportation using heterogeneous robots. In: Robotics, pp. 172–191. Springer (2016)
19. Nagarajan, T., Thondiyath, A.: Heuristic based task allocation algorithm for multiple robots using agents. Procedia Eng. **64**, 844–853 (2013)
20. Nam, C., Shell, D.A.: Assignment algorithms for modeling resource contention in multirobot task allocation. IEEE Trans. Autom. Sci. Eng. **12**(3), 889–900 (2015)
21. Rubio-Largo, A., Vega-Rodriguez, M.A., Gomez-Pulido, J.A., Sanchez-Perez, J.M.: A comparative study on multiobjective swarm intelligence for the routing and wavelength assignment problem. IEEE Trans. Syst. Man Cybern. Part C (Applications and Reviews) **42**(6), 1644–1655 (2012)
22. Tee, K.X., Chew, M.T., Demidenko, S.: An intelligent warehouse stock management and tracking system based on silicon identification technology and 1-wire network communication. In: 2011 Sixth IEEE International Symposium on Electronic Design, Test and Application, pp. 110–115. IEEE (2011)
23. Turduev, M., Cabrita, G., Kırtay, M., Gazi, V., Marques, L.: Experimental studies on chemical concentration map building by a multi-robot system using bio-inspired algorithms. Auton. Agent. Multi-Agent Syst. **28**(1), 72–100 (2012). https://doi.org/10.1007/s10458-012-9213-x
24. Wang, Y., et al.: A novel bat algorithm with multiple strategies coupling for numerical optimization. Mathematics **7**(2), 135 (2019)
25. Zhang, K., Collins, E.G., Jr., Shi, D.: Centralized and distributed task allocation in multi-robot teams via a stochastic clustering auction. ACM Trans. Autonomous Adaptive Syst. (TAAS) **7**(2), 1–22 (2012)
26. Zhang, M., Wang, H., Cui, Z., Chen, J.: Hybrid multi-objective cuckoo search with dynamical local search. Memetic Comput. **10**(2), 199–208 (2017). https://doi.org/10.1007/s12293-017-0237-2
27. Zhou, L., Shi, Y., Wang, J., Yang, P.: A balanced heuristic mechanism for multirobot task allocation of intelligent warehouses. Mathematical Problems in Engineering 2014 (2014)
28. Zhu, W., Liu, J.: Research on multi-robot scheduling algorithm in intelligent storage system. J. Phys. Conf. Ser. **1738**, 012047 (2021)

A Multi-objective Optimization Algorithm for Wireless Sensor Network Energy Balance Problem in Internet of Things

Jiangjiang Zhang, Zhenhu Ning(✉), Kun Zhang, and Naixin Kang

Faculty of Information Technology, Beijing University of Technology, Beijing 100124, China
nzh41034@163.com

Abstract. The rapid popularity of Internet of things devices makes more stable device energy need to face challenges, which has led to wireless sensor network energy balance problem becoming more and more prominent. To meet this challenge, a multi-objective wireless sensor network energy balance model is described, which comprehensively considers the energy consumption of sensor node and neighbor node, sensor node and base station. Meanwhile, an improved multi-objective optimization algorithm based on NSGA-II is employed to address the described model. In the method, the clustering mechanism is introduced to improve the pressure of selection in the later stage of the algorithm. To verify the performance of the algorithm, a wide simulation is performed by comparing it with other advanced methods. And the experiment results show that our method has a good performance in addressing the model.

Keywords: Internet of things · Energy balance problem · Multi-objective optimization · Wireless sensor network

1 Introduction

Along with the development of Internet of Things (IOT) technology, the wireless sensor network (WSN) is derived [1], which is also an important part of wireless communication. They are valuable for many applications, such as healthcare applications, civilian applications, and military applications. WSN nodes can collect the information of the physical area, then process and send it to the base station. However, many requirements need to be considered in WSNs, such as low latency data transmission, long life cycle system and easy to handle smooth deployment [2]. The sensing nodes can achieve rapid data processing, event detection and data transmission. The batteries in sensing network nodes are difficult to replace in environmental sensing, military monitoring, land defense and other applications. Therefore, how to save the energy consumption of the sensing network nodes is an important issue [3].

With the increase in interest, many scholars have proposed some advanced approach to improve the network life cycle [4]. Tam et al. [5] researched a decomposition method to balance the two contradictory objectives of the maximum node energy consumption to extend the network life and the number of relay nodes used,

L. Pan et al. (Eds.): BIC-TA 2021, CCIS 1565, pp. 18–27, 2022.
https://doi.org/10.1007/978-981-19-1256-6_2

which is effective in reducing the overall energy consumption. Zhu et al. [6] introduced the idea of clustering and described an algorithm employing fast search the peaks and evaluate mechanism to balance the WSN energy extend life cycle. In addition, Zhu et al. [7] describe the energy consumption minimization problem as a constrained combinatorial optimization problem, and then used the deep reinforcement learning technology to effectively learn the trajectory strategy of UAV for minimizing the energy consumption of perception nodes.

Intelligent computing is a popular evolutionary method in optimization field because of its characteristics of fast convergence and simple implementation [8]. Encouraged by its computational advantages, researchers integrate it to solve the WSN energy problem. Nayak et al. [9] applied genetic algorithm to WSN clustering, and analyzed the protocol with the help of a custom simulator of Java, and finally proved that it has better performance than the protocol based on fuzzy logic and traditional routing. Cai et al. [10] proposed an improved fusion curve strategy fast triangle flipping bat algorithm, which improved the local and global search ability of selecting sensor nodes, and extended the network lifetime. However, these researches did not analyze the randomness of sensing node election from multiple perspectives.

To solve this problem, many factors of sensing node election in detail are described in this paper. And the distance from the sensing node to the neighbor and the distance from the sensing node to the base station are comprehensively analyzed. By optimizing two distance objectives at the same time, a multi-objective energy balance model is established to address the problem that sensing node election is too random, which improve the network life cycle. To address the built model, an improved multi-objective optimization algorithm is described. In the described algorithm, the clustering mechanism is introduced into the fast-non-dominated sorting algorithm (NSGA-II), which can effectively overcome the selection pressure drop in the later stage of the algorithm. In addition, a rigorous simulation experiment is performed to verify the effectiveness of the algorithm and the built model. And the result show that our method is effective compared to other algorithms.

The remaining structure of the research paper is organized as follows. Some description of WSN energy balance problem is described in Sect. 2. To address this challenge, a multi-objective optimization algorithm is employed in Sect. 3. And extensive experimental simulations are performed to verify the performance of the algorithm in Sect. 4. Finally, a brief summary is presented.

2 WSN Energy Balance Problem

2.1 Problem Analysis

LEACH is executed in a round cycle, including two parts of information collection and transmission. For the node information collection, multiple neighbor nodes are deployed around an IoT sensor node [11]. Generally, it determines whether the node is a sensing node at random. When it is a sensing node, it broadcasts information and waits for the neighbor node to join. Otherwise, it will always wait for the broadcast of the sensing node [12]. Then, neighbor nodes monitor the IOT information based on the instructions of the sensing node, and send it back to the sensing node within the required time. These are sent to the base station after a series of information processing. All transmissions consume energy, so cluster node election is particularly important [13]. It controls the data transmission of each cluster. A reasonable cluster node can save the data transmission distance and reduce the energy consumption of nodes. To address this problem, a multi-objective energy balance model is established to optimize the election of sensing nodes.

2.2 Problem Model

To test the energy balance model of WSN, the constraint conditions of WSN are set as follows [14].

(1) The scale of WSN is small and medium, and the number of nodes is set to 100. For the sake of recording the energy consumption of each node, each node has a unique ID;
(2) After the formation of network clustering, single hop communication mode is adopted between neighbor nodes and IOT sensing nodes, and between sensing nodes and base station.
(3) The location of base station and sensing node is fixed, the range of node is (0, 100), and the location of base station is (50, 50).
(4) And each node has the same energy as 2J. The sensor node is determined by algorithm optimization. When the node energy consumption is 0, the node dies.

In LEACH protocol, the randomness of sensing node election cannot guarantee the rationality of its location. Unreasonable distribution of sensing nodes will increase the transmission distance from neighbor nodes to sensing nodes and from sensing nodes to base stations [15]. When the distance increases, the energy loss also becomes obvious, which makes the network life cycle possible to shorten. To optimize the selection of sensing nodes, this paper describes two distances from different perspectives. One side is the distance between nodes, the other is the node and base station [16]. They are an important part of the multi-objective energy balance model of WSN.

(1) Sensing node and neighbor node distance D_{nc}

In WSN, neighbor nodes need to send information to sensing nodes. When each neighbor node in the cluster surrounds a sensor node, it means that the distance between the neighbor node and the sensor node is the shortest, and the transmission distance is the shortest. The sensor distance model is expressed as follows:

$$\min D_{nc} = \left(\sum_{m=1}^{M}\left(\sum_{n=1}^{N} d_{ncluster}\right)\right) \tag{1}$$

where, M and N are the number of sensing and neighbor nodes, respectively;$d_{ncluster}$ is the Euclidean distance between the neighbor and sensing node. The value of D_{nc} is smaller, the shorter the sensor distance is gained.

(2) Distance between base station and sensing node D_{cs}

Before the base station receives information, the information needs to be processed at the sensing node before it can be transmitted. Therefore, these energy losses cannot be ignored in the entire network system. And it can be described as follows.

$$\min D_{cs} = \sum_{m=1}^{M} d_{csink} \tag{2}$$

where, d_{csink} is the distance from sensing node to base station. The value of D_{cs} is smaller, the shorter the transmission distance is gained, which means that the smaller the energy consumption is gained.

3 Algorithm Principle

3.1 NSGA-II

NSGA-II algorithm is a typical algorithm to handing with multi-objective optimization problems (MOP), which uses genetic algorithm as the updating method of individual population, and uses fast non dominated sorting and crowding degree to choose the solution [17]. Firstly, the initial population (population number is N) is generated, and the non-dominated solution set is determined by calculating the fitness value. Then, the population is duplicated and the new and old populations are combined for cross mutation operation. N individuals are selected as the next generation population from 2N individuals by employing the fast non-dominated sorting. In addition, the individual experience may be divided into multiple layers $(F_1, F_2, \cdots, F_n)$. If the individuals are chosen from the first level, this phenomenon may occur that $F_1, F_2, \cdots, F_{k-1} < N$, and $F_1, F_2, \cdots, F_k > N$. The remaining individuals need to be selected from F_k until N individuals are filled by employing crowding distance. In this way, the pareto solution set is found.

3.2 Clustering Mechanism

Aiming at dividing the multiple individuals into a small population, the clustering mechanism is adopted. Assuming that the population $P = (\overrightarrow{x}_1, \overrightarrow{x}_2, \ldots, \overrightarrow{x}_N)$ with N individuals, and $f(\overrightarrow{x}_i) = (f_1(\overrightarrow{x}_i), f_2(\overrightarrow{x}_i), \ldots, f_M(\overrightarrow{x}_i))$ is the objective function of individual $\overrightarrow{x}_i$, the maximum $f_{j,\max}$ and smallest values $f_{j,\min}$ of the j - th objective in population P are described as follows.

$$f_{j,\max} = \max\{f_j(\overrightarrow{x}_1), f_j(\overrightarrow{x}_2), \ldots.., f_j(\overrightarrow{x}_N)\} \tag{3}$$

$$f_{j,\min} = \min\{f_j(\overrightarrow{x}_1), f_j(\overrightarrow{x}_2), \ldots.., f_j(\overrightarrow{x}_N)\} \tag{4}$$

And the average distance $f_{j,step}$ of the j-th objective of the population is showed as follows.

$$f_{j,step} = \frac{\alpha \cdot (f_{j,\max} - f_{j,\min})}{2N} \quad (5)$$

where, α represents the individuals number contained in each small population. Due to the average distance is $f_{j,step}$, the objective function $f(\overrightarrow{x}_j)$ should meet $f_{\min}(\overrightarrow{x}_j) \leq f(\overrightarrow{x}_j) \leq f_{\max}(\overrightarrow{x}_j)$ when the individual $\overrightarrow{x}_j = (x_1^j, x_2^j, \ldots, x_D^j)$ is in the same small population, where is the upper bound of all individual objective functions in the small population and the lower bound of all individual objective functions in the small population. They can be obtained according to the following instructions:

$$f_{\max}(\overrightarrow{x}_i) = ((f_1(\overrightarrow{x}_i) + f_{1,step}), (f_2(\overrightarrow{x}_i) + f_{2,step}), \ldots, (f_M(\overrightarrow{x}_i) + f_{M,step})) \quad (6)$$

$$f_{\min}(\overrightarrow{x}_i) = ((f_1(\overrightarrow{x}_i) - f_{1,step}), (f_2(\overrightarrow{x}_i) - f_{2,step}), \ldots, (f_M(\overrightarrow{x}_i) - f_{M,step})) \quad (7)$$

Algorithm 1 describes the small population division process, where individuals $\overrightarrow{x}_i$ is randomly selected. In the initial stage, the average distance of the objective function is determined first. On this basis, the small population is further determined, and the unlabeled individuals in the small population are labeled one by one. It should be noted that in the process of small population division, the average distance and the number of small population division are determined by the population in the process of calculation.

Algorithm 1: Clustering mechanism

Input: Population P_0, Population size N

1. Calculate the maximum and minimum dimension by Eq. (3) and Eq. (4);
2. Calculate the average distance by Eq. (5);
3. $k = 1$;
4. **while** (existing unmarked individuals) *do*
5. Randomly selected unlabeled individuals $\vec{x}_i$;
6. Determine the small population range $\vec{x}_i$ by Eq. (6) and Eq. (7);
7. For the unlabeled individuals within the range of average distance $\vec{x}_i$, mark as k;
8. $k = k + 1$
9. **end**
10. Output: Marked population。

Through the above methods, the whole population can be divided into several small populations. The range of each small population is determined according to the average distance of the population, and each small population is roughly evenly distributed relative to the original whole population, which guarantee that the results are uniformly scattered on the Pareto frontier. At the same time, the limited search space of the small population can also ensure that all individuals in the whole small population can effectively explore the frontier.

3.3 Framework of CMNSGA-II

The main process of NSGA-II based on clustering is shown in Algorithm 2. After initializing the population and related parameters, all individuals are sorted quickly and non-dominated. Then the whole population is split into several small populations based on the average distance clustering diversity index, and the small population of individual is further determined according to the range of small population; A selection operator based on average distance clustering is used to execute the selection operator in each small population. According to the dominance level of individuals and the number of individuals in the small population, the optimal individuals are determined, and the average distance clustering crossover operator is used to implement the crossover operator on the optimal individuals in the small population. Mutation operator means that all individuals are mutated, and then the whole population is updated by environment selection. Repeat the above steps until the termination condition is met.

Algorithm 2 : NSGA-II based on clustering mechanism (CMNSGA-II)

Input: Population P with *N*, Crossover and variation probability Pc , Pm ;

1: Initialization population and related parameters;
2: **while** (End conditions met) *do*
3: Perform fast non dominated sorting;
4: Implementation of clustering based crowding distance;
5: Perform crossover operation;
6: Perform variation operation;
7: Update population P;
8: **end while**
9: **return** P

4 Experimental Simulation

4.1 Parameter Setting

Table 1. WSN environment parameter settings

Parameter	Value
Initial energy	2 J
Number of nodes	100
Maximum number of rounds	2000
Information energy loss of transmission and receiving unit	50 nJ/bit
The energy loss of signal amplification is in the free space of unit information	10 pJ/bit/m2
Energy loss of signal amplification under multipath attenuation of unit information	0.0013 pJ/bit/m4
Energy loss of information processing	5 nJ/bit
Information length	4000 bits
Information processing rate	0.7
Propagation threshold	87.5 m

To measure the performance of CMNSGA-II, it is compared with other advanced MOEAs, including classic LEACH, SPEA2 [18] and NSGA-II. Assuming that the IOT deployment scope is a square, sensing nodes are randomly distributed to the scope [19]. In general, the base station is located at (50, 50). And it is considered to perform in an ideal environment, including unlimited base station energy and communication conflicts and delays will not be considered [20]. In addition, other implementation environment settings can be found in in Table 1.

4.2 Result Analysis

For sake of comparing the performance of each algorithm more intuitively, the optimization results and optimization process are displayed respectively. Table 2 shows the surviving nodes number and residual energy of the four algorithms after 2000 rounds. After 2000 iterations, the number of remaining nodes in our algorithm protocol is more than that of the original LEACH and NSGA-II, which is consistent with the number of remaining nodes in SPEA2. In the matter of the total residual energy of nodes, the total residual energy of our algorithm protocol is basically consistent with that of other leach protocols.

Table 2. Number of nodes surviving and remaining energy of the four algorithms

Name	LEACH	SPEA2	NSGA-II [17]	CMNSGA-II
Number of surviving nodes	9	11	9	11
Residual energy of node (J)	3.99	3.77	3.89	3.97

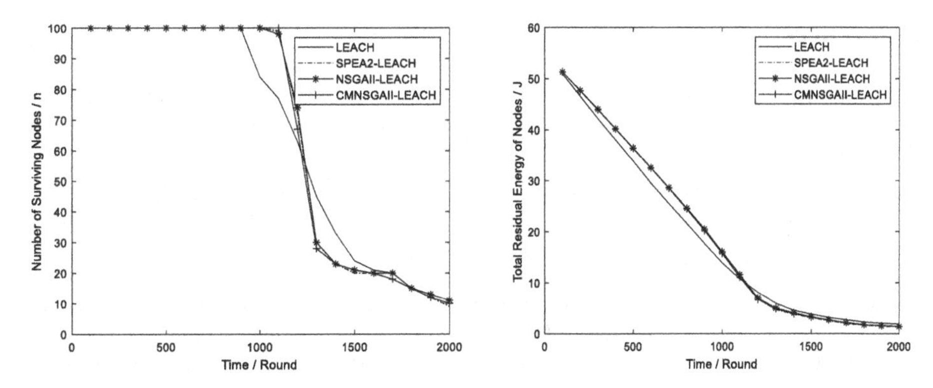

Fig. 1. Number of surviving nodes of four algorithms in different rounds; (b) Residual energy of four algorithms at nodes with different rounds

Figure 1(a) and (b) represent the change of the number of surviving nodes and the residual energy of four different algorithm nodes in the life cycle of the four different algorithms with the iteration of the number of rounds. In Fig. 1, it can be found that the surviving nodes number of our algorithm is obviously more than that of other algorithms from 1200 rounds. And this situation continues until 1300 rounds. Then, it continues with more than other algorithms after 1500 rounds. For the node residual energy, the residual energy of our algorithm is slightly higher than that of LEACH algorithm before 1100 rounds. However, with the increase of the number of rounds, the residual energy of our algorithm is significantly less than that of other algorithms, which is at most the same as that of other algorithms. Therefore, our method has a good performance in addressing the model.

5 Conclusion

In this paper, a multi objective energy balance model of WSN for IoT sensing node election for LEACH. In the described model, the energy loss of neighbor and sensing node, sensing node and base station are comprehensively considered. Meanwhile, the CMNSGA-II is employed to addressing the described model. In the method, the clustering mechanism is introduced to improve the pressure of selection in the later stage of the algorithm. For the sake of verifying the performance of the algorithm, a wide experiment is performed by comparing with other advanced methods. And the simulation results show that our method has a good performance in addressing the model. In the future work, the CMNSGA-II will be applied to handle with other practical problems.

References

1. Ikram, D., Abdennaceur, B., Abdelhakim, B.: A comprehensive survey on LEACH-based clustering routing protocols in Wireless Sensor Networks. Ad Hoc Netw. **114**, 102409 (2021)
2. Kumaresan, P., Prabukumar, M., Subha, S.: Heuristic approach to minimise the energy consumption of sensors in cloud environment for wireless body area network applications. Int. J. Embed. Syst. **12**(4), 475–483 (2020)
3. Li, T., Wang, H., Lian, X., Shi, J., Wang, M.: Improved LEACH-M protocol for processing outlier nodes in aerial sensor networks. IEICE Trans. Commun. **104-B**(5), 497–506 (2021)
4. Rahimifar, A., Kavian, Y., Kaabi, H., Soroosh, M.: Predicting the energy consumption in software defined wireless sensor networks: a probabilistic Markov model approach. J. Ambient Intell. Hum. Comput. **12**(10), 9053–9066 (2021)
5. Tam, N.T., Hung, T.H., Binh, H.T.T., Vinh, L.: A decomposition-based multi-objective optimization approach for balancing the energy consumption of wireless sensor networks. Appl. Soft Comput. **107**, 107365 (2021)
6. Zhu, B., Bedeer, E., Nguyen, H.H., Barton, R., Henry, J.: Improved soft-k-means clustering algorithm for balancing energy consumption in wireless sensor networks. IEEE Internet Things J. **8**(6), 4868–4881 (2021)
7. Zhu, B., Bedeer, E., Nguyen, H.H., Barton, R., Henry, J.: UAV trajectory planning in wireless sensor networks for energy consumption minimization by deep reinforcement learning. IEEE Trans. Veh. Technol. **70**(9), 9540–9554 (2021)
8. Cui, Z.H., Cao, Y., Cai, X.J., Cai, J.H., Chen, J.J.: Optimal LEACH protocol with modified bat algorithm for big data sensing systems in Internet of Things. J. Parallel Distrib. Comput. **132**, 217–229 (2019)
9. Padmalaya Nayak, C., Reddy, P.: Bio-inspired routing protocol for wireless sensor network to minimise the energy consumption. IET Wirel. Sensor Syst. **10**(5), 229–235 (2020)
10. Cai, X.J., Sun, Y.Q., Cui, Z.H., Zhang, W.S., Chen, J.J.: Optimal LEACH protocol with improved bat algorithm in wireless sensor networks. KSII Trans. Internet Inf. Syst. **13**(5), 2469–2490 (2019)
11. Nurgaliyev, M., Saymbetov, A., Yashchyshyn, Y., Kuttybay, N., Tukymbekov, D.: Prediction of energy consumption for LoRa based wireless sensors network. Wirel. Netw. **26** (5), 3507–3520 (2020)
12. Wang, C.: A dynamic evolution model of balanced energy consumption scale-free fault-tolerant topology based on fitness function for wireless sensor networks. Int. J. Secure. Network. **14**(2), 86–94 (2019)
13. Hosseini, R., Mirvaziri, H.: A new clustering-based approach for target tracking to optimize energy consumption in wireless sensor networks. Wirel. Pers. Commun. **114**(4), 3337–3349 (2020)
14. Radhika, M., Sivakumar, P.: Energy optimized micro genetic algorithm based LEACH protocol for WSN. Wirel. Netw. **27**(1), 27–40 (2020). https://doi.org/10.1007/s11276-020-02435-8
15. Jerbi, W., Guermazi, A., Trabelsi, H.: A novel energy consumption approach to extend the lifetime for wireless sensor network. Int. J. High Perform. Comput. Netw. **16**(2/3), 160–169 (2020)
16. Koosheshi, K., Ebadi, S.: Optimization energy consumption with multiple mobile sinks using fuzzy logic in wireless sensor networks. Wirel. Netw. **25**(3), 1215–1234 (2018)
17. Deb, K., Jain, H., Approach, A.-O.-P.-B.: Part I: solving problems with box constraints. IEEE Trans. Evol. Comput. **18**(4), 577–601 (2014)

18. Li, M., Yang, S., Liu, X.: Shift-based density estimation for pareto-based algorithms in many-objective optimization. IEEE Trans. Evol. Comput. **18**(3), 348–365 (2014)
19. Ghaderi, M., Vakili, V., Sheikhan, M.: Compressive sensing-based energy consumption model for data gathering techniques in wireless sensor networks. Telecommun. Syst. **77**(1), 83–108 (2021)
20. Shah, I., Maity, T., Dohare, Y.: Algorithm for energy consumption minimisation in wireless sensor network. IET Commun. **14**(8), 1301–1310 (2020). https://doi.org/10.1049/iet-com.2019.0465

Improved AODV Routing Protocol Based on Multi-objective Simulated Annealing Algorithm

Huijia Wu[1], Wenhong Wei[1(✉)], and Qingxia Li[2]

[1] School of Computer, Dongguan University of Technology, Dongguan 523808, China
weiwh@dgut.edu.cn

[2] School of Computer and Information, Dongguan City College, Dongguan 523419, China

Abstract. Ad Hoc network is a kind of common wireless mobile communication network. Unlike cellular mobile networks and wireless local area networks, Ad Hoc networks do not require preset base stations and are suitable for special scenarios such as rapid networking on the battlefields and post-disaster rescues. The distance vector-based AODV routing protocol is an on-demand routing protocol specially developed for Ad Hoc networks. Due to the simplicity of the routing criteria, problems in path congestion management and node energy consumption still need to be further studied. Based on the AODV routing protocol, this paper proposes an improved protocol MOSA-AODV based on the multi-objective simulated annealing algorithm. In the process of route discovery, a group of optimal routes are selected by considering both the path congestion and the path robustness. The objective of the proposed algorithm is to extend the lifetime of the network while allocating network resources reasonably. This paper uses Matlab to build an Ad hoc network, and carries out simulation experiments in the topology of 25 nodes. In addition, two improved AODV algorithms RA-AODV and IEEAODV are selected to compare with the proposed algorithm. Experimental results show that MOSA-AODV has better performance in indicators such as packet loss rate and path lifetime, which improves the service quality of Ad Hoc network.

Keywords: Ad Hoc network · AODV routing protocol · Multi-objective optimization · Simulated annealing

1 Introduction

The rapid development of switching and communication technology has put forward higher requirements for today's multicast network services. In order to host a large number of multicast sessions, reliable transmission services and multicast mechanisms with QoS parameters are required in multicast application scenarios such as interactive video conference systems and distributed data processing systems. Quality of Service (QoS) means that a network uses various technologies to improve data transmission speed and quality, thus providing better network performance.

L. Pan et al. (Eds.): BIC-TA 2021, CCIS 1565, pp. 28–42, 2022.
https://doi.org/10.1007/978-981-19-1256-6_3

Ad Hoc network is a special wireless self-organizing network. At first, Ad Hoc network was mainly used for military research. Due to its strong environmental adaptability and fast build speed, it has been gradually developed into civil fields such as sensor network, communication in disaster areas and remote distributed equipment dispatch.

In a wireless Ad hoc network, nodes use batteries for power supply. Under normal operation, nodes in the network need to perform tasks such as data receiving and packet forwarding, during which nodes' energy is constantly consumed. When the power of a node is exhausted, the path through the node will fail and affect network transmission. When nodes with many connected links undertake a large number of data forwarding tasks, data packets may be lost due to data congestion, which reduces the working efficiency of nodes and affects the network data transmission performance. Therefore, in order to make full and reasonable use of network resources, certain measures need to be taken for network resource allocation and congestion management.

2 Ad Hoc Network and AODV Protocol

In a mobile Ad Hoc network, data is exchanged between nodes through a multi-hop transmission mechanism. Only adjacent nodes within the communication distance can communicate with each other directly, and then forward packets through a special routing protocol. Unlike traditional wired networks, the rapidly changing network topology, limited wireless channel bandwidth, and exhaustible node battery energy make it no longer suitable for traditional routing protocols.

AODV (Ad Hoc On-Demand Distance Vector Routing) is an on-demand routing protocol with two core mechanisms: route discovery and route maintenance [5]. The former is used to discover new routes, and the latter is used to detect link disconnection and route repair. During the route discovery process, the source node broadcasts a route request (RREQ message) to the neighbor node. When the neighbor node receives the route request, it judges whether the target node is itself, if so, it unicasts the route response (RREP message) to the source node. Otherwise, check whether there is routing information leading to the target node in the local routing table, if so, send a routing response to the source node. If not, continue to forward the routing request to its neighbor nodes until it reaches the destination node or a node with destination node information. When a node receives multiple routing responses, the path length is used as the selection criterion, and only the routing information of the nearest node to the next hop is retained. The intermediate nodes also update their routing table information according to this message.

With the principle of the shortest distance priority, the nodes in the network using the AODV routing protocol [3] only need to store the required routing information. However, in view of different network performance requirements, AODV routing protocol still faces many challenges such as QoS performance issues, node energy consumption issues, and scalability issues [6]. This article proposes an improved routing discovery algorithm based on AODV, focusing on improving the service quality of the AODV routing protocol in Ad Hoc networks and the optimization of node energy consumption.

3 Multi-objective Simulated Annealing

3.1 Single-objective Simulated Annealing

According to the laws of thermodynamics, annealing refers to a physical phenomenon in which an object gradually cools down until it reaches a stable state. The lower the temperature, the lower its energy state. This algorithm [9] simulates the influence of temperature during the annealing process and introduces a random factor in the calculation process, so that the current solution can accept a poor solution with a certain probability to avoid falling into the local optimum. Optimal, so as to achieve the global optimum.

3.2 Multi-objective Simulated Annealing

The multi-objective simulated annealing algorithm [2] deals with the problem of multiple objective functions. The multi-objective simulated annealing algorithm [1] is based on the basic principles of simulated annealing and introduces the concept of amount of domination to calculate the probability of accepting a new solution. The amount of domination of solutions a and b can be expressed by formula (1)

$$\Delta dom_{a,b} = \prod\nolimits_{i=1, f_i(a) \neq f_i(b)}^{M} |f_i(a) - f_i(b)| \tag{1}$$

M is the number of objective functions, and the value of M in this paper is set to 2.

Meanwhile, a certain size of non-dominated solution archive is designed to save the Pareto optimal solution set. The final size of the archive is an integer not exceeding HL, and the user can obtain all alternative solutions from the archive. In the iterative process, the archive capacity can be appropriately increased to *SL* to increase the diversity of non-dominated solutions. When the number of solutions exceeds the specified range, a crowding distance sorting algorithm is used to eliminate some solutions.

The concept of crowding distance comes from NSGA-II [10]. The crowding distance of an individual refers to the distance between it and neighboring individuals i-1 and i + 1. For M objective functions, the calculation steps are as follows.

Initialize the distances of all individuals in the same layer,

$$L(i)_d = 0 \tag{2}$$

Sort the M-th objective function values of all individuals in the same layer in ascending order.

1) Find the two individuals that achieve the maximum value $f_m^{\max}$ and the minimum value $f_m^{\min}$ in the M-th objective function.
2) Calculate the crowding distance for the remaining points, the formula is as formula (3)

$$L(i)_d = L(i)_d + (L(i+1)_m - L(i-1)_m)/(f_m^{\max} - f_m^{\min}) \tag{3}$$

3) Repeat steps 2 to 4 until *M* objective functions are traversed, and the final crowding distance set is obtained.
4) Sort the solution set in ascending order, and select the top *HL* individuals with the highest crowding distance to keep in Archive. At this time, the solutions in Archive can be evenly distributed in the objective function value space to achieve the diversity of solutions.

3.3 Main Process of Multi-target Annealing

In the initialization phase, the parameters need to be set: the starting temperature T_{max}, the ending temperature T_{min}, the annealing rate α, the number of iterations at each temperature iter, the final maximum archive capacity *HL*, and the maximum allowable capacity *SL* before the Archive polymerization.

At the beginning of the iteration, randomly select a solution from the archive as current_pt, and a new solution *new_pt* is obtained through the perturbation function, and the fitness functions of the two are calculated respectively. The dominance relationship between *current_pt* and *new_pt* is divided into the following three situations.

Case 1: When *current_pt* dominates *new_pt*, and k ($k \geq 0$) points in the archive dominates *new_pt*. As shown in Fig. 1, (a) and (b) respectively show the two cases of $k = 0$ and $k \geq 0$. At this time, the probability of *new_pt* being selected as *current_pt* is expressed by formula (4) and formula (5).

$$probability = \frac{1}{\left(1 + \exp\left(\Delta dom_{avg} * temp\right)\right)} \tag{4}$$

$$\Delta dom_{avg} = \left(\left(\sum\nolimits_{i=1}^{k} \Delta dom_{i,new\_pt}\right) + \Delta dom_{current\_pt,new\_pt}\right) \tag{5}$$

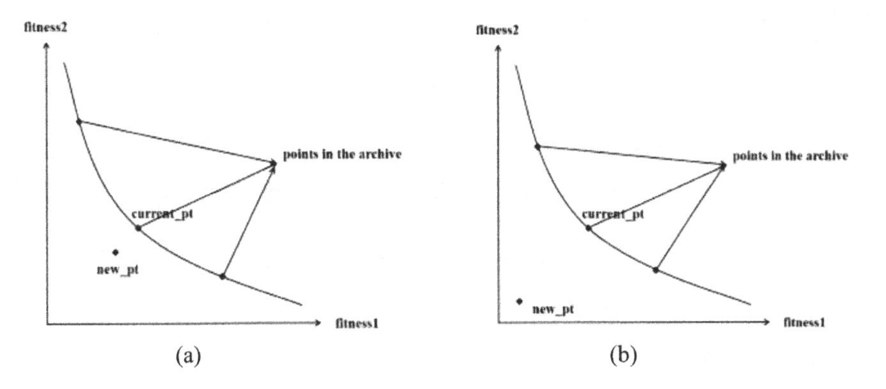

Fig. 1. When current_pt dominates new_pt (a) If the current_pt is in Archive, new_pt and the other points in Archive except current_pt are non-dominated relations (b) Some points in Archive dominate new_pt

Case 2: Current_pt and new_pt are in a non-dominated relationship. According to the comparison results of the points in new_pt and Archive, the following three situations are obtained, as shown in Fig. 2.

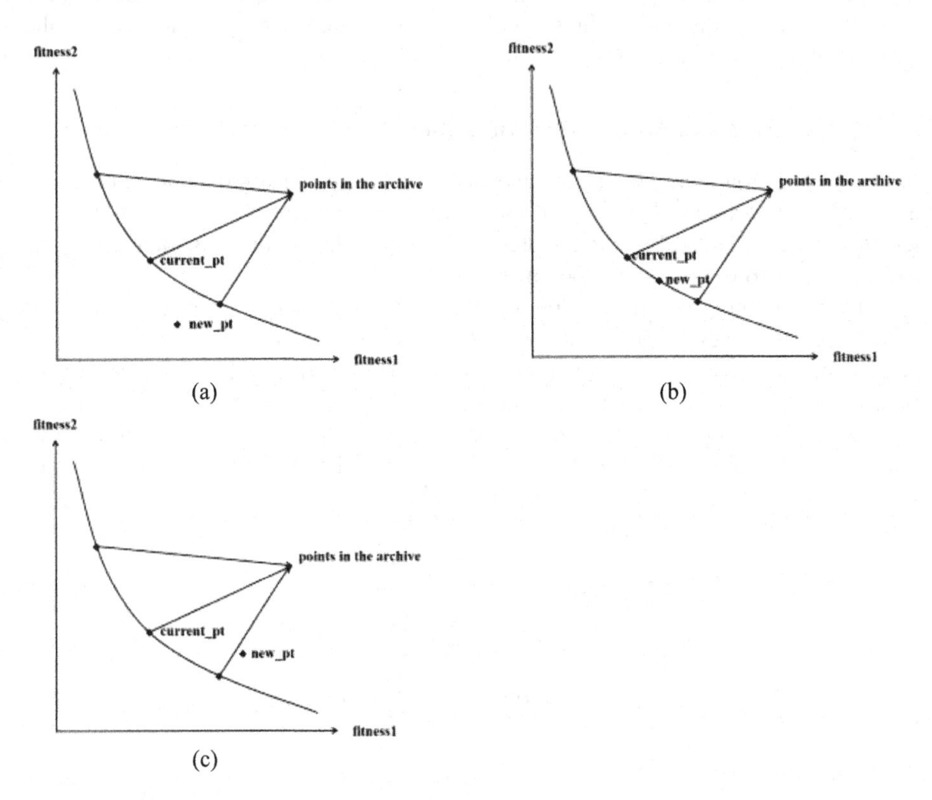

Fig. 2. When the *current_pt* and the *new_pt* are non-dominating to each other, (a) some points in the Archive dominate *new_pt*. (b) all the points in the Archive is non-dominating to *new_pt*. (c) new_pt dominates the $k(k \geq 1)$ points int the Archive

1) The new point is dominated by k points in Archive, of which $k \geq 1$. The probability of *new_pt* being set as *current_pt* is expressed by formula (4) and formula (6).

$$\Delta dom_{avg} = \sum\nolimits_{i=1}^{k} \left(\Delta dom_{i,new\_pt}\right)/k \tag{6}$$

2) All points in Archive is non-dominating to the new point. As shown in (b), the new point and all the points in the Archive are on the same front. *new_pt* is set as current_pt and added to Archive. When the archive size exceeds set range, the archive size is maintained through the aggregation method.
3) The new point dominates $k(k \geq 1)$ points in Archive. *new_pt* is set as current_pt and added to Archive. Meanwhile, remove the points originally dominated by new points in Archive.

Case 3: When *new_pt* dominates *current_pt*, compare the dominance relationship between the new point and the point in Archive. There are three situations as follows, as shown in Fig. 3

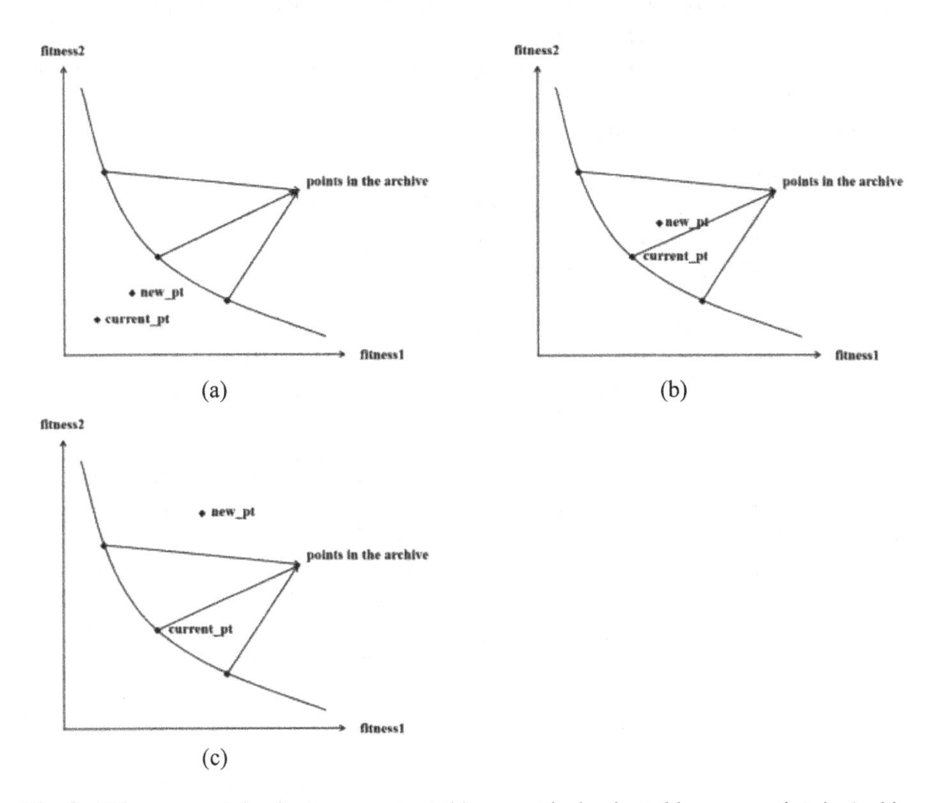

Fig. 3. When new_pt dominates *current_pt*, (a) *new_pt* is dominated by some points in Archive (b) when *current_pt* belongs to Archive, *new_pt* and all points in Archive except *current_pt* is non-dominating to each other, (c) *new_pt* dominate some points in Archive

1) The new point is dominated by some points in the Archive. At this time, the current point does not belong to the Archive. Select the point with the smallest amount of domination between the Archive point and the new point. That is, set the point in the Archive as current_pt with the following probability formula (7). When it is not selected, set new_pt to current_pt.

$$probability = \frac{1}{1 + \exp(-\Delta dom_{\min})} \tag{7}$$

2) When the current point belongs to Archive, the new point and all Archive points except the current point belong to a non-dominated relationship. Set new_pt to current_pt and add it to Archive. Count the number of points in Archive. When the number exceeds the specified size, use the aggregate function to reduce it.
3) When the new point dominates k(k $\geq$ 1) points in the Archive, set new_pt to current_pt and add it to the Archive, and delete those k points in the Archive that are dominated by the new point.

The above process is iterated at each temperature respectively, and the current temperature after each iteration is multiplied by the annealing rate until the minimum temperature Tmin is reached. The final Archive contains all the non-dominated solution sets of this experiment.

4 Multi-objective Optimization AODV Routing Protocol

4.1 Network Model

The Ad Hoc network topology can be abstracted as a graph, and the communication nodes are represented by the node set Nodes. The main attributes of the nodes include: coordinate, adjacent nodes, power and load rate, etc.

Each node maintains a local node routing table routeTable. The Ad Hoc network obtains all path information through the flooding routing table.

The communication distance between each node is represented by Euclidean distance

$$distance_{i,j} = sqrt((nodes(i).x - nodes(j).x)^2 + (nodes(i).y - nodes(j).y)^2) \tag{8}$$

4.2 Fitness Function

In order to optimize the performance and resource utilization of Ad Hoc networks, reduce network congestion and improve the efficiency of node battery use, this paper takes the following two fitness functions [4].

(1) path congestion

Path congestion is used to evaluate the load of a route. In a network topology, the communication range is a fixed value. The number of pairs of all nodes that can communicate with each other constitutes the total number of effective links in the entire network. The greater the bearing pressure of the node. The load rate of a single node is represented by the load rate, which can be expressed as formula (9)

$$\alpha_i = \frac{l_i}{L} \tag{9}$$

L is the total number of links. l_i is the number of links passing through the node α_i. The path congestion of the route can be expressed as formula (10)

$$f_{c_j} = \sum\nolimits_{v_i \in p_j} \alpha_i + cgj \bullet cg \tag{10}$$

f_{c_j} is the path congestion value of path p_j, cgj is the number of congested nodes in path p, cg is the congestion factor, and its value is between 0 and 1. Considering that the greater the number of nodes that have high load rate in the path, the greater the possibility of congestion. This work regards a single node with more than 6 links as a congested node, and the product of the number of congested nodes and the congestion factor is included in the congestion value of the path.

(2) path robustness

Path robustness reflects the survivability of the route in the network topology. In an Ad hoc network, the battery capacity of each node is limited. It should be ensured that the node with strong continuous power supply capability should be selected as the forwarding node while maintaining the normal operation of the network. Among them, the survival time of the route is determined by the node with the smallest remaining power in the path, and the route becomes invalid if the energy of the smallest node is exhausted. In addition, the longer the communication distance, the more energy the routing consumes. The path robustness is determined by the node energy and path length, which can be expressed as formula (11)

$$f_{r_j} = \frac{E(v_i)w_1}{E_{init} \cdot \alpha_i} + \frac{1}{l_{p_j} \cdot (1 - w_1)} \tag{11}$$

f_{r_j} is the path robustness of the path p_j, v_i is the node with the smallest remaining energy in the path, E_{init} is the initial energy of this node, $w_1 \in (0, 1)$ is the robustness weight. The weight of the node's remaining energy status is w_1 while the weight of the path length is (1 – w1). Together, these two factors determine the path robustness of this route. Moreover, the path length is negatively correlated with the path robustness, so the reciprocal is taken in the formula. The more the residual energy of the minimum node takes up the initial ratio, the shorter the path distance, and the stronger the robustness of the path.

4.3 Perturbation Function

The perturbation function is used to obtain a new solution from the current solution. In this paper, the solution space of the optimization problem is all possible paths from the source node to the destination node, and all paths are sorted by the depth-first traversal order. First, select a random floating number from 0 to 1. When the value of the random number is less than or equal to 0.5, select the next solution of the current solution, otherwise select the second solution after the current solution. When the new solution belongs to the Archive skip this solution. When the new solution is out of bounds, select the first solution, and continue the above operation until a new solution that meets the condition is selected.

4.4 Multi-objective Simulated Annealing Optimization Algorithm

The optimization of AODV routing protocol can be regarded as a multi-objective optimization problem. The pseudo code of the algorithm is shown below.

First, select the source node and destination node in the network topology, and send a routing request on the source node. The source node queries the local routing table. If no routing information is found, the source node floods RREQ packets to the neighbor node. If the destination node is reachable, the source node collects all received RREP packets to obtain the flooding routing table. This table retains all the available path, forming the solution space of multi-objective simulated annealing algorithm.

Before the iteration begins, HL solutions are randomly selected and added to the Archive. Then, the current solution *current_pt* is randomly selected from Archive, and the new solution *new_pt* is obtained by perturbation function. Two fitness functions of *current_pt* and *new_pt*, path congestion and path robustness, are calculated respectively. Compare their dominant relationship.

Iterate according to the algorithm process introduced in Sect. 3.3. After the iteration, the final Pareto optimal solution set Archive is obtained. The solutions in the archive are all feasible alternative routes derived from the AODV optimization algorithm.

Algorithm MOSA-AODV

```
Build network topology, set network parameter distance, initialize all nodes
Set the source node and the destination node and query the routing table of the source node
if there is destination node routing information in the source node routing table
    send
else
    form a flood routing table, and sort out all the possible paths as the input of function MOSA

function MOSA(table)
Set the annealing parameters T_max, T_min, iter, α. Initialize the Archive
current_pt=random(Archive)
temp=T_max
    while(temp>T_min)
        for i=1:iter
            new_pt=perturb(current_pt), dm=checkDM(current_pt, new_pt)
            %checkDM: analyze the dominant relationship between them
            switch dm
                case current_pt dominates new_pt
                for j=1:each solution in Archive
                    checkDM(j,new_pt)
                    if solution j dominates new_pt, k++
                  set new_pt using formula (4) and formula (5)
                case current_pt and new_pt is non-dominating to each other
                    for j=1:each solution in Archive
                        checkDM(j, new_pt)
                    if k(k≥1) solutions from Archive dominate new_pt
```

```
                    set new_pt using formula (4) and formula (6)
                  if all solutions in Archive and new_pt are non-dominating
                to each other
                    set new_pt as current_pt and add it to Archive
                    if Archive-size>SL
                      call function cluster
                      %sorting by their crowding distance
                  if new_pt dominates k (k≥1) solutions from Archive
                    set new_pt as current_pt and add it to Archive
                    remove these k solutions from Archive
                case new_pt dominates current_pt
                  for j=1:each solution in Archive
                     checkDM(j,new_pt)
                  if k(k ≥ 1) from Archive dominate new_pt
                     find the minimum amount of domination Δdom_min
                  and the corresponding solution
                    set new_pt using formula (7)
                  if all solutions in Archive and new_pt are non-dominating
                to each other
                     set new_pt as current_pt and add it to Archive
                     if Archive-size>SL
                        call function cluster
                  if new_pt dominates k(k≥1) solutions from Archive
                     set new_pt as current_pt and add it to Archive
                     remove these k solutions from Archive
      end for
      temp=α*temp
    end while
    returns the pareto optimal set
end
```

5 Simulation and Analysis

5.1 Simulation Scenario Description

This paper uses Matlab for simulation experiments. Compare the proposed improved AODV routing protocol with the original AODV protocol, RA-AODV [7] and IEEAODV [8]. According to [4], the number of nodes in the network topology is set to 25, the communication distance is 350 m, and the node energy is a random number not greater than 2000 mAh. The generated network topology is shown in Fig. 4:

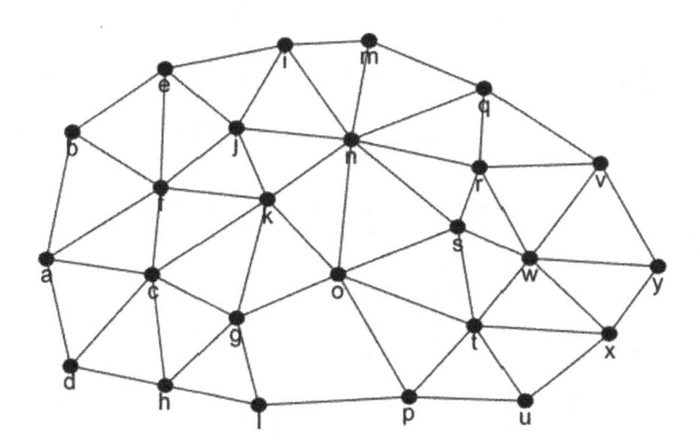

Fig. 4. 25-node network topology

5.2 Simulation Parameter Setting

Considering the size of this network and the original setting of [1], the parameter settings are as follows: T_{max} = 1e10, T_{min} = 1e-5, *Iter* = 2, α = 0.9, *HL* = 8, *SL* = 10.

5.3 Simulation Experiment Results and Comparative Analysis

Table 1 shows the simulation experiment results, according to Table 1 and Fig. 5, in experimental group 1, the packet loss rate of MOSA-AODV is lower than that of AODV and RA-AODV, and is significantly lower than that of IEEAODV. In the second set of experiments, the packet loss rate of MOSA-AODV is very close to that of IEEAODV and both are lower than AODV and RA-AODV. In the third experimental group, MOSA-AODV is significantly lower than the other three algorithms, reaching the lowest packet loss rate. In the fourth set of experiments, it can be seen from the table data that MOSA-AODV and IEEAODV choose the same path, so their packet loss rate is the same, and at the same time lower than AODV and RA-AODV. It can be concluded that the path selected by MOSA-AODV can effectively reduce the packet loss rate.

According to Table 1 and Fig. 6, in experimental group 1 and experimental group 2, the path lifetime of MOSA-AODV is significantly longer than that of AODV, RA-AODV and IEEAODV. In the third set of experiments, the path lifetime of MOSA-AODV ranked third, longer than AODV, and IEEAODV reached the longest path survival time. In the fourth set of experiments, it can be seen from the table data that MOSA-AODV and IEEAODV choose the same path, so the path survival time is the same, and it is significantly longer than AODV and RA-AODV. It can be concluded that MOSA-AODV can effectively select a path with stronger robustness and guarantee a longer lifetime.

Table 1. Comparison of packet loss ratio and path lifetime for the proposed MOSA-AODV, AODV, RA-AODV and IEEAODV for 25 nodes

Group	Algorithm	Path	Packet loss ratio	Path lifetime
group1	MOSA-AODV	[1,3,11,14,17,22,25]	15.46%	51.15625
	AODV	[1,6,11,15,20,24,25]	15.65%	35.73438
	RA-AODV	[1,6,10,14,18,23,25]	15.58%	43.07813
	IEEAODV	[1,3,7,15,20,23,25]	17.56%	47.03125
group2	MOSA-AODV	[1,3,11,14,18,22,25]	16.29%	44.79688
	AODV	[1,6,11,15,20,24,25]	16.87%	10.70313
	RA-AODV	[1,6,10,14,18,23,25]	17.22%	10.17188
	IEEAODV	[1,3,7,15,20,23,25]	16.28%	29.60938
group3	MOSA-AODV	[1,3,11,14,18,22,25]	16.29%	44.79688
	AODV	[1,6,11,15,20,24,25]	16.87%	10.70313
	RA-AODV	[1,6,10,14,18,23,25]	17.22%	10.17188
	IEEAODV	[1,3,7,15,20,23,25]	16.28%	29.60938
group4	MOSA-AODV	[1,3,7,15,20,24,25]	16.17%	44.85938
	AODV	[1,6,11,15,20,24,25]	16.76%	23.26563
	RA-AODV	[1,6,10,14,18,23,25]	16.94%	27.67188
	IEEAODV	[1,3,7,15,20,24,25]	16.17%	44.85938

As can be seen from Table 1, MOSA-AODV simultaneously achieved the lowest packet loss rate and longest path lifetime in experimental group 1 and 2. In experimental group 3, although the path lifetime ranked third, the packet loss rate was the lowest. The IEEAODV in the same group reached the longest path lifetime, but the packet loss rate was the highest. The performance of MOSA-AODV in experimental group 4 was the same as that of IEEAODV, and both of them were better than that of AODV and RA-AODV.

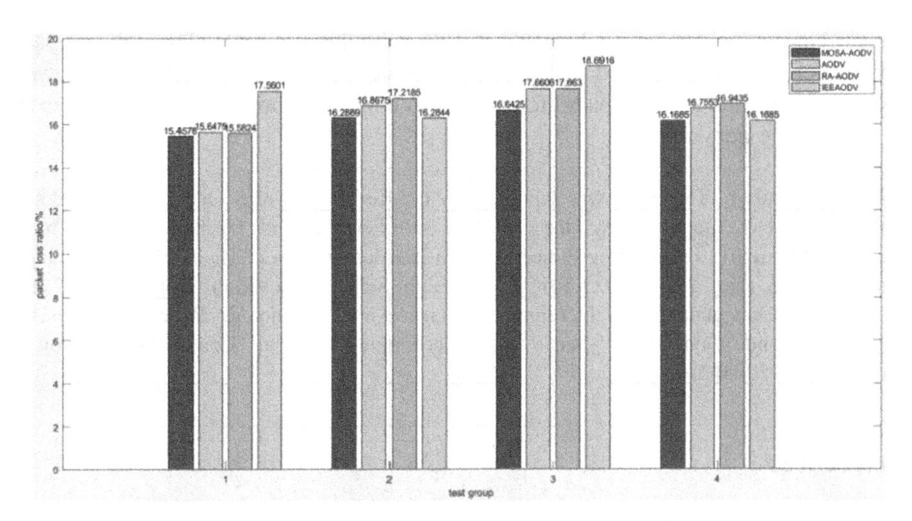

Fig. 5. Comparison of packet loss ratio for MOSA-AODV, AODV, RA-ADOV and IEEAODV in histogram

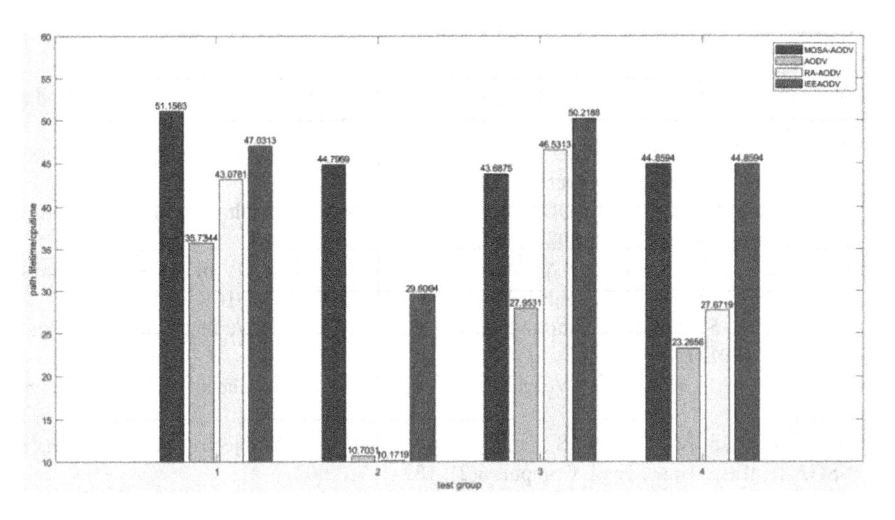

Fig. 6. Comparison of path lifetime for MOSA-AODV, AODV, RA-ADOV and IEEAODV in histogram

6 Conclusion

In the case of limited node working ability and battery energy, how to improve network efficiency and ensure the endurance of the path is an important issue. We propose a multi-objective optimized simulated annealing idea to improve the AODV protocol. The protocol takes the distance between nodes, node load rate and node battery energy into consideration at the same time in the routing method. Set the two fitness functions of path congestion and path robustness to select the optimal route. Through comparative experiments, it is found that the path selected by MOSA-AODV has a lower packet loss rate and a longer path lifetime, which effectively improves the performance of the Ad Hoc network.

Acknowledgement. This work was supported by the Key Project of Science and Technology Innovation 2030 supported by the Ministry of Science and Technology of China (No. 2018AAA0101301), the Key Projects of Artificial Intelligence of High School in Guangdong Province (No. 2019KZDZX1011), The High School innovation Project (No. 2018 KTSCX222), Dongguan Social Development Science and Technology Project (No. 20211 800904722) and Dongguan Science and Technology Special Commissioner Project (No. 20201800500442).

References

1. Bandyopadhyay, S., Saha, S., Maulik, U., et al.: A simulated annealing-based multiobjective optimization algorithm: AMOSA. IEEE Trans. Evol. Comput. **2**(3), 269–283 (2008)
2. Amine, K.: Multiobjective simulated annealing: principles and algorithm variants. Adv. Oper. Res. **2019** (2019)
3. Rajesh, M., Gnanasekar, J.M.: Congestion control using AODV protocol scheme for wireless AD-Hoc network. Adv. Comput. Sci. Eng. **16**(1–2), 19–37 (2016)
4. Wang, Y., Fan, Q., Dai, H., Li, W.: A multi-path routing discovery method based on whale optimization algorithm. J. Hunan Inst. Sci. Technol. (Nat. Sci.) **34**(2), 24–27 (2021)
5. Chang, X.: Research and Simulation of Ad hoc Network Routing Protocols Based on NS-3, Master's thesis of Harbin Institute of Technology (2010)
6. Bi, X., Yang, B.: Improved AODV routing protocol based on multi-objective optimization. Comput. Eng. Design **38**(4), 898–902 (2017)
7. Tyagi, S., Som, S., Rana, Q.P.: A reliability based variant of AODV in MANETs: proposal, analysis and comparison. Procedia Comput. Sci. **79**, 903–911 (2016)
8. Shukla, R.N., Shukla, R.K.: Improve energy efficiency in AODV. Int. J. Sci. Eng. Res. **4** (12), 1812 (2013)
9. Kirkpatrick, S., Gelatt, C.D., Vecchi, A.: Optimization by Simulated Annealing. Science (1983)
10. Deb, K., Pratap, A., Agarwal, S., et al.: A fast and elitist multiobjective genetic algorithm: NSGA-II. IEEE Trans. Evol. Comput. **6**(2), 182–197 (2002)

Solving Satellite Range Scheduling Problem with Learning-Based Artificial Bee Colony Algorithm

Yanjie Song[1(✉)], Luona Wei[1], Lining Xing[3], Yi Fang[2], Zhongshan Zhang[1], and Yingwu Chen[1]

[1] National University of Defense Technology, Changsha 410073, Hunan, China
songyj_2017@163.com, {zszhang,ywchen}@nudt.edu.cn
[2] Northern Institute of Electronics, Beijing 100089, China
[3] School of Electronic Engineering, Xidian University, Xi'an 710126, China

Abstract. Satellite range scheduling problem (SRSP) is a critical and challenging scheduling problem due to the oversubscribed and sequence dependency characteristics. The artificial bee colony algorithm (ABC) is one of the popular evolutionary algorithms to solve large-scale scheduling problems. A new artificial bee colony algorithm is proposed, named the learning-based artificial bee colony algorithm (LB-ABC). We proposed two new learning strategies, named error-based learning strategy and position-based learning strategy, to improve traditional ABC's exploration and exploitation performance. Error-based learning strategy uses differences of each individual in the population to improve the population structure. Position-based learning strategy through the experience of scheduling tasks guide the generation process of new bees. Experiments and algorithm analysis show that the learning strategy we proposed is of great help to planning tasks, and the performance of the new algorithm is better than state-of-the-art algorithms.

Keywords: Satellite range scheduling problem · Artificial bee colony algorithm · Learning-based · Satellite

1 Introduction

Satellite remote-tracking refers to establish a communication link between a satellite and ground station antenna when a satellite passes over the ground station [1]. Several commands will be uploaded from ground station to satellite and various statuses of the satellite will be downloaded. Since satellites fly in fixed orbits, ground stations are also fixed on the ground, the time for medium and low orbit satellites to fly overground stations is quite limited. This time range during which a satellite flies over a ground station and is visible to each other is called a visible time window [2]. A challengeable type of scheduling problem is called the satellite range scheduling problem (SRSP). A task means that one ground station gives a command to one satellite. The goal of this scheduling problem is to determine the execution queue and start time for task sequences. The start time of each task must be within a required time range which is known in advance and the time window of the task. Due to the large-scale, sequence

L. Pan et al. (Eds.): BIC-TA 2021, CCIS 1565, pp. 43–57, 2022.
https://doi.org/10.1007/978-981-19-1256-6_4

dependence characteristics of the SRSP problem, the time window which can be used are quite limited and it is difficult to find a satisfactory solution [3]. The main purpose of scheduling is to improve the efficiency of satellite remote-tracking systems [4].

Solving satellite scheduling problems generally uses heuristic algorithms, meta-heuristic algorithms, or hyper-heuristic algorithms [5]. Algorithms by accurate methods are difficult to solve large-scale problems. A good heuristic algorithm often requires certain characteristics of the problem. Since the SRSP problem is a large-scale problem in practical applications, discipline is often not obvious. The artificial bee colony algorithm is an evolutionary algorithm based on imitating the behavior of honey bees [6]. Compared with other evolutionary algorithms (genetic algorithm, ant colony algorithm, particle swarm algorithm, e. g.), ABC algorithm parameters are simple (number of iterations, parameters of the termination conditions are also in other evolutionary algorithms) [7]. The reduced pressure of repeatedly adjusting the optimal parameter configuration allows us to have more energy to concentrate on the design of the algorithm itself. Researchers try to change the design of population size (expand or shrink), design search strategies, and hybrid some new strategies [8–10]. The search performance of the artificial bee colony algorithm has been greatly improved in these ways.

Early satellite range scheduling research mainly focused on AFSCN scheduling [11], and through a variety of algorithm comparisons, it was found that a genetic algorithm is more suitable to be used to solve this scheduling problem [12]. An obvious advantage of evolutionary algorithms in practical application is that they rely less on the characteristics of the problem, and can quickly solve a complex problem through a combination of strategies or a hybrid of algorithms. Existing evolutionary algorithms to solve satellite scheduling problems are mainly genetic algorithms, tabu-based algorithms, and ant colony algorithm [12–14]. These three algorithms have strong global search capabilities, and the setting of algorithm parameters has a relatively large impact on planning results. Learning mechanism is an important method to improve performance. Allowing algorithms to improve search strategies through information obtained in the optimization process has been applied in many aspects.

Involving some optimization strategies into an evolutionary algorithm can significantly improve the algorithm's exploration and exploitation ability. We try to use several learning strategies to let the ABC algorithm generate a new individual after learning according to the various positions' information of the individual in the optimization process. Meanwhile, an error-based learning strategy is used to reduce the occurrence of invalid solutions and make the search process more targeted. This improved artificial bee colony algorithm is called a learning-based artificial bee colony algorithm (LB-ABC).

The main contributions of this article are as follows:

1. To solve the SRSP problem effectively, we improved the traditional artificial bee colony algorithm and introduced learning strategies into the algorithm process. These strategies let the ABC algorithm can learn by itself.
2. We propose a learning strategy based on position selection, which can instructively generate a brand-new individual based on population individuals' performance for each position of task sequence in a certain optimization stage.

3. We designed an error-based learning strategy, which uses the error between individuals in the population to effectively control the direction of the population optimization and reduce the possibility of invalid searches.

The second part of this paper introduces the mathematical model of the SRSP problem. The third part introduces the ABC algorithm based on learning. The fourth part takes several experimental verifications, and the last part gives the main research conclusions of this paper.

2 Problem Description

Consider a set of tasks *Task*, with a total of M tasks, and a time window set TW, with a total of N time windows. Each task $task_i$ has a given earliest allowable start time est_i, latest allowable end time let_i, task duration dur_i, and task profit p_i. For the j-th time window of task i tw_j^i, it has a start time evt_j^i and an end time lvt_j^i. Limited by the abilities of satellites and ground stations, the minimum mission interval Δ must be met between the two missions. Each task needs to be started and completed within the given time range of the task, and also needs to be started and completed within the time window of ground station. Let x_j^i be a binary 0, 1 decision variable representing whether the i-th task is scheduled in the j-th time window, and let st_i be another decision variable, which represents the start time of the i-th task.

$$f(x) = \sum\nolimits_{i=1}^{M} \sum\nolimits_{j=1}^{N} p_i \times x_j^i \tag{1}$$

$$x_j^i = \begin{cases} 1 & task\, i\, is\, assigned\, to\, time\, window\, j \\ 0 & else \end{cases} \tag{2}$$

Subject to:

$$est_i \leq st_i, i \in M \tag{3}$$

$$st_i + dur_i \leq let_i, i \in M \tag{4}$$

$$evt_j^i \leq st_i, i \in M, j \in N \tag{5}$$

$$st_i + dur_i \leq lvt_j^i, i \in M, j \in N \tag{6}$$

$$st_i + dur_i + \Delta \leq st_j, i,j \in M, i<j \tag{7}$$

The objective function (1) maximizes the profit of whole task sequence. Constraint (2) defines the value range of decision variable. Constraints (3)–(4) define the task started and ended within the required time range $[est_i, let_i]$. Constraints (5)–(6) define the task started and ended within the time window $\left[evt_j^i, lvt_j^i\right] (i \in M, j \in N)$. Constraint

(7) defines the time interval between two tasks that each execution needs to ensure sufficient task conversion time.

Since SRSP problem is an NP-complete problem [15], limited by its complexity, it is difficult to prove that the result obtained for a D dimensional SRSP problem is an optimal solution. A better task execution plan can only be found by designing a smart algorithm.

3 Learning-Based Artificial Bee Colony Algorithm

This section gives a concise introduction to the traditional artificial bee colony algorithm. After that, the overall process and learning strategies of LB-ABC are introduced in detail.

3.1 Traditional Artificial Bee Colony Algorithm

Artificial bee colony algorithm is a popular evolutionary algorithm which consists of three types of bee individuals, employed bee, onlooker bee, and scout bee. These three forms a search population to find better food source alternately. Hiring bees can find better food sources (that is, better solutions) by improving the neighborhood structure. After all the hired bees have searched, Onlooker Bee uses roulette to select individuals for search process. When the population size is *SN*, an onlooker bee i with a fitness value fit_i. Then it has the probability of being selected is $p_i = fit_i / \sum_{i=1}^{SN/2} fit_i$. Scout bee uses a random method to generate new individuals, and determines whether to accept the new scout bee according to the quality of new solution. When the search process without improvement for several generations, a new scout bee will replace the employed bee with worst performance.

In our scheduling problem, a complete task sequence represents an individual bee, and real number coding is used to form the task sequence. Neighborhood improvement operation we used is to exchange two tasks' positions in the task sequence.

Traditional artificial bee colony algorithm searches for food sources by imitating the behavior of a bee colony. Traditional ABC algorithm determines evolution direction of population by evaluating the quality of food sources, but does not make full use of information discovered by bees in the search process or feedback from environment. We try to make traditional ABC algorithm robust, let the algorithm learn information obtained by population. We propose a new artificial bee colony algorithm, named learning-based artificial bee colony algorithm (LB-ABC).

3.2 Algorithm Overall Framework

Compared with the traditional ABC algorithm, a position-based learning strategy is introduced into the LB-ABC algorithm framework, and is used to generate a new individual based on the position information get from the optimization process. What's more, another error-based learning strategy is used to control the search space to avoid search area too large and invalid search. Pseudo code of LB-ABC algorithm is shown in Algorithm 1.

Algorithm1: Learning-based artificial bee colony algorithm (LB-ABC)
1: **Input**: Task set $Task$, Time window set TW, Max generation Gen, Set parameters SN, $Threshold$, Fluctuation threshold Φ, Control parameter η, growth rate λ
2: **Output**: Solution $Solution$
3: $SN_0 \leftarrow$ Generate an initial population; 4: $count = 0; l\_best = 0; g\_best = 0; t\_count = 1;$ $std\_upper = inf;$ //Initialization parameters 5: **For** $gen = 1$ to Gen 6: **For** each employed bee **do** 7: Update employed bee; 8: Calculate fitness; 9: $employed\ bee \leftarrow$ Accept or reject new solution; 10: **End For** 11: **For** each onlooker bee do 12: Do roulette selection for onlooker bee; 13: Update onlooker bee; 14: Calculate fitness; 15: $l_best \leftarrow$Accept or reject new solution; 16: **End For** 17: Update PT_i; 18: $std \leftarrow$Calculate population standard deviation 19: $l_best \leftarrow$Find $local\ best$ in SN; 20: **If** $l\_best > g\_best$ **do** 21: $g\_best \leftarrow l\_best;$ 22: **Else** 23: $count = count + 1;$ 24: **End If** 25: $count, SN', AT_i \leftarrow$Use the position-based strategy; 26: $std_upper', SN', t_count \leftarrow$Use error-based strategy; 27: Generate a new scout bee; 28: $scout\ bee \leftarrow$Accept or reject new solution; 29: **End For** 30: **Return** $Solution$

Two learning strategies (in Line 25, Line 26) are used in LB-ABC algorithm to make algorithm smarter. In the optimization process, internal information of population and external information get from evaluation criteria are used to make the optimization process achieve balance between random and guided search. Detailed process of position-based learning strategy is shown in Algorithm 2, and detailed process of error-based learning strategy is shown in Algorithm 3.

3.3 Position-Based Learning Strategy

If information obtained in the search process of ABC algorithm can be effectively used, it will help to obtain better performance in the subsequent optimization process. One type of important information is which tasks are more suitable for a specific time window. Therefore, we propose a learning strategy based on position selection.

The learning strategy based on location selection judges which task has the highest occurrence probability in each time window according to the prior knowledge obtained in a certain optimization process, and arranges the measurement and control task with the highest occurrence probability at this location. Judging from the initial position of each individual sequence, after obtaining the best matching task at that position, the next position is judged to be the most suitable task for scheduling, and so on until the complete sequence is arranged. In this way, we get a new bee with knowledge, this bee is better than randomly generated individual bees in many cases.

First, introduce a position-task correlation matrix PT_i, which represents the situation where tasks are arranged and successfully executed at each position of generation i. The element at the position (j, k) in the matrix represents task arrangement result of the j-th task in population at the k-th position. For example, if $y_{jk} = c$, it means that the number of times with task j successfully executed at position k is c.

$$PT_i = \begin{bmatrix} y_{11} & \cdots & y_{1n} \\ \vdots & y_{jk} & \vdots \\ y_{m1} & \cdots & y_{mn} \end{bmatrix} \tag{8}$$

When tasks of all individuals in generation i are arranged, add the matrix PT_i and the cumulative task scheduling information matrix $AT_{(i-1)}(i \geq 2)$ to get the cumulative task arrangement information matrix AT_i to current generation, the specific calculation method is shown in formula 9.

$$AT_i = AT_{(i-1)} + PT_i, i \geq 2 \tag{9}$$

In formula 9, $AT_{(i-1)}(i \geq 2)$ represents the cumulative task schedule information before $i - 1$ generation, and AT_i represents the cumulative task schedule to i generation information. After that, the matrix AT_i is transformed into a probability matrix P_i. The element p_{jk} in the probability matrix P_i is determined by the formula 10.

$$p_{jk} = at_{jk} / \sum\nolimits_{j=1}^{m} at_{jk}, at_{jk} \in AT_i \tag{10}$$

In formula 10, at_{jk} represents the value of j, k element in the cumulative task schedule matrix AT_i.

When the optimization algebra reaches a certain threshold *Threshold*, select the row j and column k according to the probability from large to small in the probability matrix

P_i, and arrange the j-th task at the k-th position in a new individual. Subsequent selecting process will not consider the row j and the column k any more. Repeat probability-position match process until all tasks are arranged in a new individual. By using a position learning strategy, the information from the previous optimization process can be effectively used. Pseudo code of position-based learning strategy is shown in Algorithm 2.

Algorithm 2: Position-based Learning Strategy
1: **Input**: $count, SN, gen, Threshold, AT_{(i-1)}$
2: **Output**: $count, SN', AT_i$
3: Update AT_i;
4: **If** $count == Threshold$ **do**
5: Calculate P_i;
6: **For** each row in P_i **do**
7: $j, k \leftarrow argmax\{p_{jk}\}, \forall j, k \notin S$;
8: $S \leftarrow$Update the set of selected positions;
9: $individual' \leftarrow$let the location k be j;
10: **End For**
11: $count = 0$; //reset parameter
12: $SN' \leftarrow$Update the worst individual in SN with new individual $individual'$;
13: **Else**
14: Break; // Don't meet learning conditions jump out
15: **End If**
16: **Return** $count, SN', AT_i$

3.4 Error-Based Learning Strategy

As an evolutionary algorithm that imitates the behavior of bee populations, ABC algorithm obtains excellent solutions through random and instructive population evolution. Random inevitably will cause some invalid areas to be searched, as shown in Fig. 1. For SRSP problem, it is impossible to obtain the structure of search space in advance, and search effect is evaluated according to the performance in optimization process, so that search is carried out in the direction of obtaining a higher quality solution. We propose an error-based learning strategy in LB-ABC. This strategy evaluates the effect of current optimization process by calculating the standard deviation of each generation of the population. Formula 11 gives the method of calculating standard deviation.

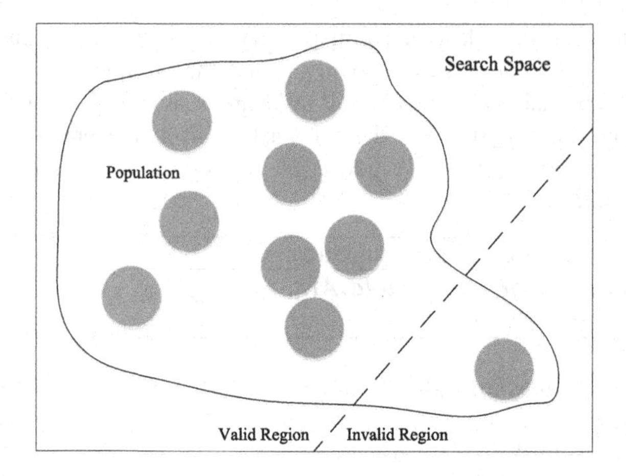

Fig. 1. Schematic diagram of invalid region found during artificial bee colony algorithm search

$$std = \sqrt{\frac{1}{SN}\sum\nolimits_{i=1}^{SN}\left(fit_i - \overline{fit}\right)^2} \tag{11}$$

$$\overline{fit} = \frac{1}{SN}\sum\nolimits_{i=1}^{SN} fit_i \tag{12}$$

In formula 11, fit_i represents the individual profit value, $\overline{fit}$ represents the average value, SN represents population size, and the calculation method of $\overline{fit}$ is shown in formula 12.

Error-based learning strategy compares the standard deviation of contemporary population with the population fluctuation threshold Φ_i to determine whether to trigger this learning strategy. When standard deviation is greater than the fluctuation threshold, the error-based learning strategy will be used to adjust the population to make search more effective.

After using the error-based learning strategy, in order to prevent population exploration ability from reducing significantly, a control parameter η, initial value set 1, and growth rate λ are introduced. When error-based learning strategy is triggered for the i-th time, Control parameters are updated according to formula 13.

$$\eta_{(i+1)} = \eta_i + \lambda, i \geq 1 \tag{13}$$

According to the new standard deviation and control parameters, population fluctuation threshold Φ_i is determined together. Population is controlled according to the standard deviation for the i time, and fluctuation threshold is updated according to formula 14 when new individuals are generated.

$$\Phi_{(i+1)} = \text{std\_upper} \times \eta_{(i+1)} \tag{14}$$

Pseudo-code of error-based learning strategy is shown in Algorithm 3.

Algorithm 3: Error-based Learning Strategy
1: **Input**: $std\_upper, std, SN, \eta, \Phi, \lambda, t\_count$
2: **Output**: std_upper', SN', t_count
3: **If** $std > std\_upper$ **do**
4: $worst\_individual$ ←Find worst individual in SN;
5: $worst\_individual \leftarrow gobal\_best\_individual$;
6: $std\_upper' \leftarrow std\_upper$;
7: Update η; //Update control parameter
8: Update Φ; //Update fluctuation threshold
9: $t\_count = t\_count + 1$;
10: **Else**
11: Break;
12: **End If**
13: **Return** std_upper', SN', t_count

3.5 Parameter Analysis

We choose 300 task-scale scenarios to analyze the influence of threshold *Threshold* and the growth rate of control parameter λ in the error-based learning strategy. According to experimental results, we can select the optimal parameter combination. For threshold parameter, we selected four values of 5, 10, 15 and 20. Under each threshold condition, the growth rate takes 0.01, 0.02, 0.1 and 0.2 respectively. Results are shown in Table 1.

Table 1. Parameter comparison result table. This table lists the maximum value (Max), average value (Mean) and rate of profit (RP). RP = optimization result revenue value/total revenue value of all tasks. We run LB-ABC algorithm 10 times under each parameter condition.

Threshold	λ	Max	Mean	RP
5	0.01	707	699.8	0.7952
	0.02	707	699	0.7943
	0.1	706	699	0.7943
	0.2	701	689.1	0.7831
10√	0.01	709	698.5	0.7938
	0.02√	**709**	**703.3**	**0.7992**
	0.1	708	700.2	0.7957
	0.2	697	687	0.7807
15	0.01	708	698.5	0.7938
	0.02	706	696.3	0.7913
	0.1	699	689	0.7830
	0.2	699	688.9	0.7828
20	0.01	708	696.8	0.7918
	0.02	707	698.4	0.7936
	0.1	700	689.2	0.7832
	0.2	698	684.1	0.7774

It can be seen from experimental results that when the threshold parameter *Threshold* sets 10, and growth rate of control parameter λ sets 0.02, a better optimization result can be obtained. Although the maximum value obtained by this group parameter setting has achieved the same maximum value as other groups, performance of other indicators is better than results obtained by other parameter configurations. This group parameter setting will be used in the subsequent experimental process. It also can be seen from the ideal parameter value setting that parameters that are too large or too small are not reasonable enough. Strategic intervention in optimization process needs to be moderate.

4 Experiment Analysis

4.1 Analysis of Algorithm

We added two new strategies to the traditional ABC algorithm, a learning strategy based on position and a learning strategy based on error. To verify the effectiveness of the strategy effectively, we will separately verify the optimization effect obtained by the LB-ABC algorithm using all strategies with algorithms after removing one strategy. The algorithm removing the position-based learning strategy is named (LB-ABC-WI1), and the algorithm removing the error-based learning strategy is named (LB-ABC-WI2). The scale of our experimental example is set to two groups with 500 and 1000 tasks. Eight experimental scenarios are randomly generated for each scale, and each scenario runs 10 times and the average value is taken as a result. Algorithm strategies' comparison experiment results are shown in Figs. 2, 3, 4 and 5.

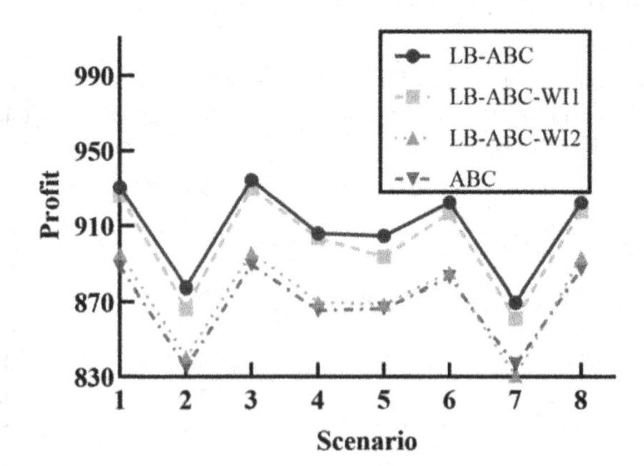

Fig. 2. Profit value comparison under 500 task scale

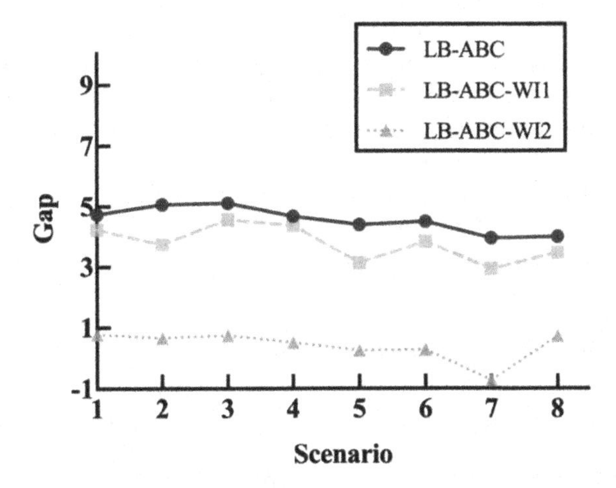

Fig. 3. Gap of different strategies under 500 task scale

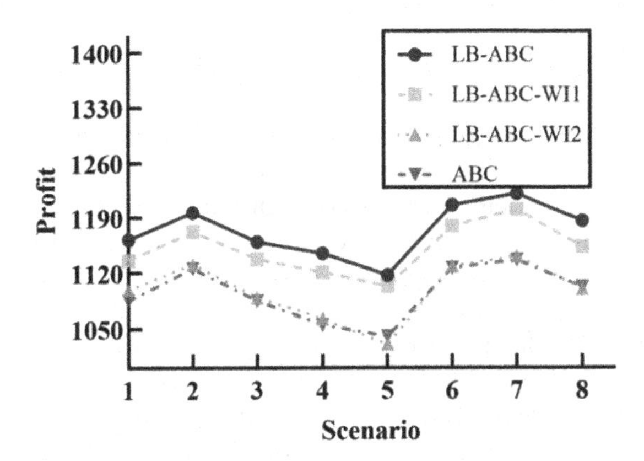

Fig. 4. Profit value comparison under 1000 task scale

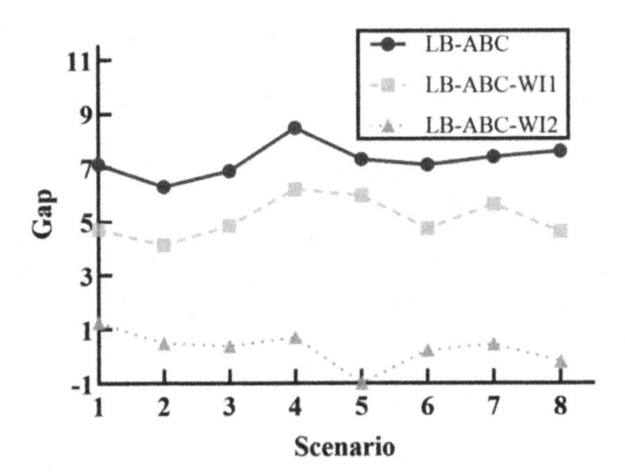

Fig. 5. Gap of different strategies under 1000 task scale

The abscissa of Figs. 2, 3, 4 and 5 represents the id of experimental scenarios, and the ordinate represents the profit value or algorithms' gap. It can be seen from experimental results that LB-ABC using all strategies is better than LB-ABC algorithm without one strategy (position-based learning strategy or error-based learning strategy) and ABC algorithm that does not use any strategy. Using a position-based learning strategy at the 500 task scale is not as effective as at the 1000 task scale. If the position-based learning strategy is used alone, the profit obtained may decrease. If it is used cooperated with the error-based learning strategy, there will improvement in planning performance. Error-based learning strategy can improve planning ability of LB-ABC algorithm by 3%-6%. Combining position-based learning strategy with error control-based learning strategy can further improve performance by 1–2% based on results obtained by ABC algorithm with an error control strategy.

It can also be seen from Fig. 2, 3, 4 and 5 that although the profit obtained is different in each scenario subject to various constraints, the percentage difference between the profit of LB-ABC algorithm and ABC algorithm is always at a stable level. This phenomenon shows that using learning strategies has a weak dependence on characteristics of experimental scenarios.

4.2 Comparison to State-Of-The-Art Algorithms

We compared optimization results of the LB-ABC algorithm with four state-of-the-art algorithms, a knowledge-based genetic algorithm [16], an improved adaptive large neighborhood search algorithm [17], a local search algorithm [18], and a CP Optimizer. The first three comparison algorithms have successfully solved the satellite scheduling problem. To enhance the persuasiveness, we use ILOG CPLEX[1] 12.6, a professional solution tool for MIP problems, also as a benchmark algorithm for comparison. We take the result of the CP Optimizer running for 60 s. All algorithms are implemented by Matlab2017a on a desktop with Core I7–7700 3.6 GHz CPU, 8 GB memory, and Windows 7 operating system. LB-ABC algorithm uses all learning strategies proposed. We set the population size of each optimization algorithm to 10, and the maximum number of generations is 500.

Evaluation indexes of experimental results select the maximum value, the average value, the task profit rate, and the gap between algorithms (GAP) of 10 runs. GAP is calculated by: [(average profit of its algorithm - average profit value of the worst-performing random search algorithm in current scenario)/profit value of the worst-performing random search algorithm in current scenario] * 100%. Because the CP Optimizer is too far from the results of other algorithms except for few scenarios, it is of little significance to use it as a comparison benchmark.

It can be seen from Table 2 that LB-ABC has the best planning results among the four state-of-the-art algorithms, KBGA algorithm ranked second, ALNS-I algorithm ranked third, LS algorithm ranked fourth and CP optimizer ranked last. The performance of the LB-ABC algorithm in all four indicators evaluated by the experiment is very satisfactory. LB-ABC performs better for 1000-task-scale large-scale scenarios

[1] https://www.ibm.com/products/ilog-cplex-optimization-studio.

than 500-task-scale scenarios, which well reflects the algorithm's stronger adaptability to large-scale scenarios. Compared with the KBGA algorithm, LB-ABC has an average profit increase of 12.9%. LB-ABC has an increase of 14.8% compared to the ALNS-I algorithm, and an increase of 16.8% compared to the LS algorithm. It can be said that the algorithm proposed in this paper can get better quality than state-of-the-art algorithms. In all experiments, CP Optimizer results are only ranked second in the 500–2 scenario, and the profit of other scenarios is very low. Especially in the 1000 task scale scenario, it is difficult for CPO to get a good task execution plan in a limited time due to too many constraints to be processed. It is very difficult to solve large-scale SRSP problems with an accurate solving algorithm. We used the proposed algorithm and the first three state-of-the-art algorithms to perform a paired t-test, and P values obtained by LB-ABC with KBGA, ALNS-I, and LS were $2.80 \times 10^{-10}, 1.59 \times 10^{-9}$, 3.72×10^{-10}, indicating the improvement of the solution quality by LB-ABC is significant.

Table 2. Comparison results table between LB-ABC algorithm and state-of-the-art algorithms. GAP represents the percentage difference between average profit value of the algorithm and the average profit value of the worst-performing random search algorithm in current scenario.

Instances	LB-ABC				KBGA				ALNS-I				LS				CP
	Max	Mean	RP (%)	GAP (%)	Max	Mean	RP (%)	GAP (%)	Max	Mean	RP (%)	GAP (%)	Max	Mean	RP (%)	GAP (%)	
500–1	958	930.4	65.89	13.81	880	843.9	59.77	3.23	853	838.7	59.40	2.59	838	817.5	57.90	–	426
500–2	897	877.3	61.48	16.06	819	785.7	55.06	3.94	792	774.5	54.27	2.46	783	755.9	52.97	–	**786**
500–3	971	934.5	64.05	13.00	865	849.8	58.25	2.76	865	838.9	57.50	1.44	844	827	56.68	–	445
500–4	939	906	64.25	14.54	827	810.1	57.54	2.41	821	809.7	57.51	2.36	812	791	56.18	–	769
500–5	922	904.6	63.39	12.82	858	818.3	57.34	2.06	840	816.2	57.20	1.80	817	801.8	56.19	–	491
500–6	949	922.5	64.38	14.51	871	829.4	57.88	2.95	852	826.7	57.69	2.62	835	805.6	56.22	–	373
500–7	892	869.7	63.44	14.16	835	791.4	57.72	3.89	792	781.3	56.99	2.56	803	761.8	55.57	–	404
500–8	931	922.4	61.66	11.82	861	834.4	55.78	1.15	843	825.5	55.18	0.07	848	824.9	55.14	–	429
1000–1	1218	1162.3	39.79	20.16	1053	1004.9	34.40	3.89	1031	997.3	34.14	3.10	995	967.3	33.12	–	3
1000–2	1239	1196.5	42.00	17.99	1095	1053.1	36.96	3.85	1067	1035.6	36.35	2.12	1039	1014.1	35.59	–	3
1000–3	1201	1159.6	40.36	21.42	1034	1016.4	35.38	6.43	1013	984.4	34.26	3.08	1006	955	33.24	–	6
1000–4	1181	1144.7	40.87	18.41	1020	987	35.24	2.10	996	969.2	34.60	0.26	988	966.7	34.51	–	2
1000–5	1140	1116.9	39.52	21.15	990	964	34.11	4.57	971	944.3	33.41	2.43	953	921.9	32.62	–	3
1000–6	1267	1205.9	42.16	16.90	1071	1057.5	36.98	2.51	1061	1033.1	36.12	0.15	1060	1031.6	36.07	–	2
1000–7	1282	1220.3	41.81	18.41	1101	1076.1	36.87	4.41	1110	1046.7	35.86	1.56	1063	1030.6	35.31	–	6
1000–8	1204	1185.9	40.78	19.86	1065	1031.3	35.46	4.23	1021	992	34.11	0.26	1021	989.4	34.02	–	1

5 Conclusion

To effectively solve the SRSP problem, we propose an improved ABC algorithm with position-based and error-based learning strategies. Position-based learning strategy allows algorithms to generate new individuals guided by the optimization performance, and error-based learning strategy allows optimization to proceed in a direction that is more conducive to obtain higher-quality solutions based on the diversity of the population. It can be found through algorithm analysis:

1. In the LB-ABC algorithm, the threshold condition set as and the control parameter growth rate set as, is an ideal parameter combination. We can obtain better optimization results.
2. Two learning strategies introduced are effective. Through experimental verification, it can be found that the error-based learning strategy has the most obvious impact on improving performance. It can improve 3–5% planning performance in 500 tasks scenario and 4–6% in 1000 tasks scenarios on average by using an error-based learning strategy compared with the traditional ABC algorithm.
3. Compared with state-of-the-art algorithms, LB-ABC is 12.9% higher than the KBGA algorithm, 14.8% higher than the ALNS-I algorithm, and 16.8% higher than the LS algorithm on average. In addition, the solution quality of the LB-ABC algorithm is significantly better than CP Optimizer. In a word, the algorithm proposed in this paper is effective to solve the SRSP problem.

The LB-ABC algorithm we proposed can solve the SRSP problem well and improve the quality of the solution effectively compared with state-of-the-art algorithms. In the future, we will try to deploy the LB-ABC algorithm into the actual application system, and consider designing more efficient heuristics into the algorithm framework.

Acknowledgement. This work was supported by the National Natural Science Foundation of China under Grant 71901213 and 72001212.

References

1. Vazquez, A.J., Erwin, R.S.: On the tractability of satellite range scheduling. Optim. Lett. **9** (2), 311–327 (2014). https://doi.org/10.1007/s11590-014-0744-8
2. Gooley, T.D.: Automating the satellite range scheduling process, Master's Thesis Air Force Institute of Technology (1993)
3. He, L., Weerdt, M.D., Smith, N.Y.: Tabu-based large neighbourhood search for time/sequence-dependent scheduling problems with time windows. In: Proceedings of the Twenty-Ninth International Conference on Automated Planning and Scheduling (ICAPS 2019) (2019)
4. Du, Y.H., Xing, L.N., et al.: MOEA based memetic algorithms for multi-objective satellite range scheduling problem. Swarm Evol. Comput. **50**, 100576 (2019)
5. Barbulescu, L., Howe, A.E., Roberts, M., Whitley, L.D.: Understanding algorithm performance on an oversubscribed scheduling application. J. Artif. Intell. Res. **27**(12), 577–615 (2006)
6. Bahriye, A., Dervis, K.: A modified artificial bee colony algorithm for real-parameter optimization. Inf. Sci. **192**(1), 120–142 (2012)
7. Song, X.Y., Zhao, M., Xing, S.Y.: A multi-strategy fusion artificial bee colony algorithm with small population. Expert Syst. Appl. **142**, 112921 (2020)
8. Gao, H., Shi, Y., Pun, C.M., Kwong, S.: An improved artificial bee colony algorithm with its application. IEEE Trans. Ind. Inf. **15**(4), 1853–1865 (2019)
9. Mustafa, S., Kiran, H., et al.: Artificial bee colony algorithm with variable search strategy for continuous optimization. Inf. Sci. **300**, 140–157 (2015)

10. Karaboga, D., Kaya, E.: An adaptive and hybrid artificial bee colony algorithm (aabc) for ANFIS training. Appl. Soft Comput. **49**, 423–436 (2016)
11. Barbulescu, L., Howe, A., Whitley, D.: Afscn scheduling: how the problem and solution have evolved. Math. Comput. Model. **43**(9–10), 1023–1037 (2006)
12. Xhafa, F., Herrero, X., Barolli, A., Barolli, L., Takizawa, M.: Evaluation of struggle strategy in genetic algorithms for ground stations scheduling problem. J. Comput. Syst. Sci. **79**(7), 1086–1100 (2013)
13. Xhafa, F., Herrero, X., Barolli, A., Takizawa, M.: A simulated annealing algorithm for ground station scheduling problem. In: International Conference on Network-based Information Systems. IEEE Computer Society (2013)
14. Xhafa, F, Herrero, X, Barolli, A., Takizawa, M.: A tabu search algorithm for ground station scheduling problem. In: IEEE International Conference on Advanced Information Networking & Applications. IEEE (2014)
15. Luo, K., Wang, H., Li, Y., Li, Q.: High-performance technique for satellite range scheduling. Comput. Oper. Res. **85**, 12–21 (2017)
16. Song, Y.J., Xing, L.N., et al.: A knowledge-based evolutionary algorithm for relay satellite system mission scheduling problem. Comput. Ind. Eng. **150**, 106830 (2020)
17. Chen, Y., Song, Y., Du, Y., Wang, M., Zong, R., Gong, C.: A knowledge-based scheduling method for multi-satellite range system. In: Li, G., Shen, H.T., Yuan, Y., Wang, X., Liu, H., Zhao, X. (eds.) KSEM 2020. LNCS (LNAI), vol. 12274, pp. 388–396. Springer, Cham (2020). https://doi.org/10.1007/978-3-030-55130-8_34
18. Song, B., Chen, Y., Yao, F., et al.: A hybrid genetic algorithm for satellite image downlink scheduling problem. Disc. Dyn. Nat. Soc. **2018**, 1–11 (2018)

Black Widow Spider Algorithm Based on Differential Evolution and Random Disturbance

Shida Wang, Xuncai Zhang(✉), Yanfeng Wang, and Ying Niu

School of Electrical and Information Engineering, Zhengzhou University of Light Industry, Zhengzhou 450002, China
zhangxuncai@163.com

Abstract. Aiming at the problems of unbalanced exploitation and exploration and falling into local optimization in black widow algorithm. In this paper, differential evolution strategy is introduced to avoid unnecessary exploration. At the same time, the random disturbance factor is used to improve the local exploitation performance of the black widow algorithm, and the memory function is added to the individual to improve the population's reproductive strategy, it further reduces the possibility of falling into local and ensures the balance between exploitation and exploration. By testing benchmark optimization functions and engineering problems, it shows the advantages of the improved algorithm, and has better progress in global search ability and convergence speed.

Keywords: Black widow algorithm · Global optimization oriented · Differential evolution algorithm · Random disturbance

1 Introduction

Swarm intelligence algorithm has the characteristics of simple implementation, easy expansion and strong robustness. When facing the search dilemma, it can be combined with other search strategies to balance the process of global search and local search [1]. For a long time, swarm intelligence algorithm has been concerned by many scholars. Such as, bionic swarm intelligence algorithm inspired by bee behavior in nature, Artificial Bee Colony algorithm(ABC) [2]; simulating the flying behavior of moths around the flame, replacing the individual random walk behavior with spiral search, which can effectively solve the optimization problem(MFO) [3]; according to the migration behavior of emperor butterfly, the population will be distributed in two parts, a new swarm intelligence algorithm, Emperor butterfly algorithm(MBO) [4]; according to the process of ant lion digging funnel-shaped trap to prey on ants, the ant lion algorithm is proposed(ALO) [5]; according to the hunting behavior of the gray wolf population, the gray wolf optimization algorithm approaching the three best individuals of the population(GWO) [6]; different whale optimization algorithms are selected according to the random encirclement or drum net expulsion of prey during whale hunting(WOA) [7]. Because of their good characteristics, these algorithms have been applied to the fields of Power engineering [8], control system [9] and network link [10].

L. Pan et al. (Eds.): BIC-TA 2021, CCIS 1565, pp. 58–70, 2022.
https://doi.org/10.1007/978-981-19-1256-6_5

Black widow algorithm(BWO) is a population intelligent algorithm proposed by Hayyolalam et al. [11] in 2020, which simulates the mating rules of black widow spiders and combines the random search of evolutionary computation. Inspired by the self-addictive behavior, the black widow algorithm continuously removes the inferior solution in the population through the competition and reproduction mechanism to find the global optimal solution. At present, many scholars have further studied the algorithm. Agal et al. [12] proposed the black widow spider algorithm (WLBWO) based on weight and feature flight, which improves the bandwidth processing efficiency and reduces the delay in streaming video over the Internet. Micev et al. [13] proposed an adaptive black widow optimization method combined with sudden short circuit test data on synchronous motor. Premkumar et al. [14] used the BWO algorithm to optimize the proportional and integral gain of PI controller. Compared with the optimization results of PSO and GA, the results proved the superiority of the black widow optimization algorithm. ATH et al. [15] used the BWO algorithm combined with OTS method to process the image with a multi-level threshold, which increased the accuracy and efficiency and reduced the computational complexity. Priyade et al. [16] proposed a black widow optimization algorithm(BW-SMO) based on fuzzy and combined with spider monkey optimization algorithm. In the iteration, the parameters are controlled by fuzzy logic, which effectively reduces the cost and delay. Sheriba et al. [17] proposed a hybrid cuckoo algorithm and an improved black widow optimization algorithm (IHCBW), adopted the direction average mutation strategy to improve the convergence rate of the algorithm, divided the population into multiple subgroups, shared the best individuals by combining subgroups and forming a new population. Punithavathi et al. [18] proposed the combination of the black widow algorithm and improved ant colony algorithm(BWO-IACO) for cluster based routing in WSN, which greatly improves energy efficiency and network life.

In the above improved methods, only the part of black widow algorithm is improved, and the overall performance of the algorithm is not considered. However, there are still some disadvantages, such as insufficient local optimization ability and easy premature convergence. To improve the overall optimization ability of the algorithm, this paper uses the differential evolution strategy to improve the mutation stage of the algorithm, and adds a strategy for global optimization in the process of population reproduction to avoid unnecessary exploration and rapid convergence. In addition, the random perturbation strategy is introduced into the algorithm to enhance the advantages of the algorithm in the process of local exploitation.

The rest is organized as follows: the second part introduces the principle of the black widow algorithm; the third part puts forward the improvement strategy of breeding and mutation stage, random disturbance strategy; the fourth part gives the implementation of the improved black widow spider algorithm; the fifth part is the simulation experiment and analysis; the sixth part is the summary.

2 Black Widow Spider Algorithm

The black widow spider is a female spider. The algorithm is inspired the life style of the black widow. When dealing with the objective function, by simulating the behavior of the black widow spider, the algorithm is divided into four steps: population initialization, reproduction, cannibalism and mutation. In the black widow spider algorithm, each black widow spider represents one candidate solution in the population, assuming that the black widow population size set to N and dimension set to D, the position of the i black widow spider in D-dimensional space expressed as $W_i = \left(w_i^1, w_i^2, \ldots w_i^D\right)$, $i = 1, 2 \cdots, N$. Black widow spider reproduces offspring through Eq. (1):

$$\begin{cases} Y_1 = \alpha \times W_1 + (1 - \alpha) \times W_2 \\ Y_2 = \alpha \times W_2 + (1 - \alpha) \times W_1 \end{cases} \tag{1}$$

Where α is a random number between [0, 1], W_1 and W_2 are randomly selected parents by randomly selected, Y_1 and Y_2 are the offspring of reproduction.

In the algorithm, the method of cannibalism is used to keep the population size unchanged, and mutation is used to improve the randomness of the algorithm. The basic steps of the black widow spider algorithm are as follows:

Step 1: Initialization. Initialize a population randomly and set the initial parameters.

Step 2: Population reproduction and cannibalism. According to the reproduction rate, individuals are randomly selected to form parents for reproduction. After reproduction, individuals compete to limit the size of the population.

Step 3: Mutation. According to the mutation rate and the number of randomly selected individuals, two elements are randomly exchanged from the selected individuals to generate new individuals.

Step 4: Merge the population, sort the population according to the fitness value, remove the redundant individuals and carry out the next iteration.

3 Improved Black Widow Spider Algorithm

3.1 Population Reproduction for Global Optimization

From the population reproduction Eq. (1), it can be seen that the standard black widow spider algorithm only considers the randomness of individual generation in the iterative updating process of population reproduction, and each generation of individuals is randomly generated by the parent generation. Therefore, the population reproduction equation has well exploration ability, ignoring the exploitation ability of the algorithm, resulting in the low operating efficiency of the algorithm, it is easy to fall into local optimization when solving multimodal problems.

To enhance the exploitation ability of the black widow spider algorithm and accelerate the convergence of the algorithm. We add the individual memory function to the black widow spider algorithm to save the individual optimal solution of each generation, so that it can remember the best solution in the process of population evolution. Therefore, this paper designs a modified population reproduction Eq. (2) to replace the original population reproduction Eq. (1):

$$\begin{cases} Y_1 = (1-\beta)[\alpha \times W_1 + (1-\alpha) \times W_2] + \beta \cdot \{W_{i,best} - [\alpha \times W_1 - (1-\alpha) \times W_2]\} \\ Y_2 = (1-\beta)[\alpha \times W_1 + (1-\alpha) \times W_2] + \beta \cdot \{W_{i,best} - [\alpha \times W_1 - (1-\alpha) \times W_2]\} \end{cases} \tag{2}$$

Where α is a random number between [0, 1], β is the optimal coefficient, $W_{i,best}$ represents the optimal individual of the i black widow spider in the population during reproduction, W_1 and W_2 are randomly selected parents by randomly selected, Y_1 and Y_2 are the offspring of reproduction. Through adjustment coefficient β can coordinate the effects of population and individual memory on the search of black widow spider algorithm.

Compared with the original population renewal Eq. (1), Eq. (2) consists of two parts. The first part is random individual generation, which enhances the diversity and global search ability of the population; The second part introduces the individual's optimal information to enhance the exploitation ability of the algorithm and speed up the convergence speed of the algorithm. Through adjustment coefficient β to balance the exploitation and exploration capabilities of the algorithm.

3.2 Population Mutation Based on Differential Evolution Algorithm

In the basic BWO algorithm, to increase the diversity, the two elements in the individual are randomly exchanged in the mutation link. In the local extreme case, the mutation rule is used to avoid falling into the local and improve the convergence level, and it is more important to produce a better convergence rate. Black widow spider algorithm, in the self-addictive stage, the population in the algorithm will have strong convergence, and the basic mutation rules can not better meet these conditions. Inspired by the differential evolution algorithm, we use N real- valued parameter vectors with dimension D as the population of each generation. In the search process, the population will mutate first and realize individual mutation through the differential strategy. Randomly select two different individuals in the population, and their vector difference is scaled to synthesize the vector with the individual to be mutated, which is obtained by Eq. (3).

$$A_i(t) = W_i(t) + F \cdot [W_{\mathrm{m}}(t) - W_{\mathrm{n}}(t)] \tag{3}$$

Where F is the scaling factor, is a random number between [0, 2], $W_i(t)$ represents the i individual of the population of generation t, $W_{\mathrm{m}}(t)$ and $W_{\mathrm{n}}(t)$ is two random individuals in the population, and $i \neq m \neq n$.

For mutated individuals $\{A_i(t)\}$ is cross recombined, and two elements in an individual are randomly exchanged with a certain probability, which is obtained by Eq. (4).

$$W_i(t) = \begin{cases} \left(w_i^1, w_i^2, \ldots w_i^b, \ldots, w_i^a, \ldots w_i^D\right), \text{if rand}(0,1) \le \text{CR} \\ \left(w_i^1, w_i^2, \ldots w_i^a, \ldots, w_i^b, \ldots w_i^D\right), \text{if rand}(0,1) > \text{CR} \end{cases} \tag{4}$$

Where CR is the crossover operator between [0, 1], *a* and *b* are two randomly selected individuals. When the random parameter is less than the crossover probability, the two elements are exchanged; on the contrary, maintain the mutated individual. Figure 1 shows the mutation link of the improved black widow spider algorithm.

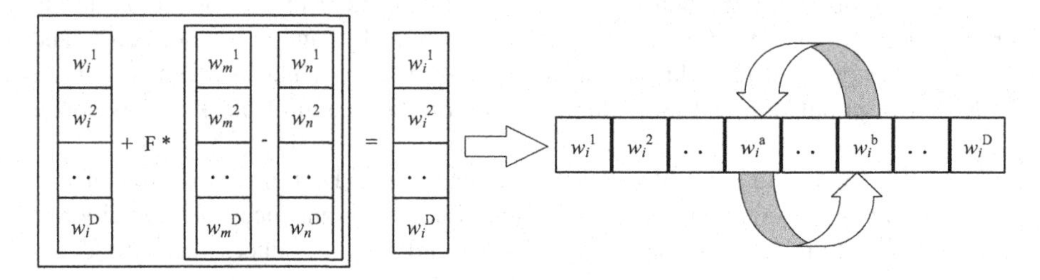

Fig. 1. Mutation link

3.3 Random Disturbance Strategy

The random disturbance search strategy is to find the global optimal solution of the target in the search space combined with the characteristics of probability jump. It can jump out of the local optimal solution with probability and finally tend to the global optimal solution. The random disturbance strategy is represented by Eq. (5).

$$W_i(t+1) = W_i(t) + rand(-1, 1) \tag{5}$$

In this stage, the optimal individual received from the operation of the algorithm will be disturbed, a new solution will be obtained by calculating the fitness function, the two solutions will be compared, and whether to accept the new solution as the current optimal solution will be judged according to the Metropolis criterion, Eq. (6).

$$p = \begin{cases} 1, W_i(t+1)_{fitness} \le W_i(t)_{fitness} \\ e^{-\frac{\Delta f}{T}}, W_i(t+1)_{fitness} > W_i(t)_{fitness} \end{cases} \tag{6}$$

Where, Δf represents the difference of fitness values after disturbance and before disturbance, and T represents control parameters.

At the beginning of the iteration cycle, the transition probability p is large and will accept most of the relatively poor solutions, giving the algorithm more opportunities to explore to jump out of the local optimization. With the operation of the BWO algorithm, the transition probability p decreases gradually, and the different solution is more and more difficult to be accepted. Finally, the optimal solution will tend to be stable.

4 Algorithm Implementation

The main steps of the standard black widow spider algorithm are population initialization, reproduction, cannibalism and mutation. Affected by the algorithm itself, especially when dealing with complex function problems, the algorithm is easy to fall into local optimization, and the exploitation and exploration can not maintain a balance. Therefore, we added the process of global optimal strategy and balanced exploitation and exploration in the breeding process; In the mutation process, the cross recombination method is used to improve the convergence performance of the algorithm; Moreover, the random disturbance strategy is introduced to improve the ability of local exploitation of the algorithm. The overall performance of the algorithm is improved. The basic steps of the improved algorithm are as follows:

Step 1: Initialization. In the BWO algorithm, every individual in the population is a solution to the problem. The variables needed for the problem make up the individuals in the population. To solve the reference function, the structure should be treated as an array. Therefore, population size set to *N* and the dimension set to *D*, the position of the i black widow in dimensional space can be expressed as $W_i = \left(w_i^1, w_i^2, \ldots w_i^D\right)$, $i = 1, 2 \cdots, N$. Finally generate a population with a size of $[N \times D]$ initial population.

Step 2: Reproduction. Black widow spiders produce many individuals in each reproduction, but only a few strong individuals survive in their offspring. In the algorithm, such randomness is copied, an array containing random numbers is created, pairs of parents are randomly selected among individuals, and the optimal solution of each generation is stored by adding the individual memory function, so that it can remember the optimal solution in the evolution process of the population itself, and the reproduction process of the population is simulated by Eq. (2). This process will be repeated *N/2* times. Finally, offspring and parents will be added to an array and arranged in ascending order according to their fitness values. Now, according to the level of cannibalism, the best individuals are added to the newly generated population.

Step 3: Cannibalism. The algorithm introduces a competition mechanism through cannibalism to control the number of populations. In this stage, the algorithm imitates the self addictive behavior of the black widow spider, and there are three cases. The first is cannibalism between parents; The other is cannibalism, which exists in offspring. Strong spiders eat their weaker brothers and sisters; The third kind of cannibalism, young spiders eat their mother, it should be noted that this happens only when the offspring is better than the mother. At this stage, we use the fitness value to determine the quality of the individual.

Step 4: Mutation. This stage is used to avoid falling into local and improve the convergence speed. Two different individuals in the population are randomly selected, and their vector difference is scaled to synthesize with the individual to be mutated from Eq. (3). Get the mutated individual, exchange the elements in the individual randomly, and choose whether to retain it with a certain probability to complete the cross recombination process.

Step 5: Merge populations. The population after reproduction and cannibalism and the population mutated based on differential evolution algorithm are combined, and the combined population is sorted according to the fitness value. This can remove redundant individuals and maintain the population size.

Step 6: Random disturbance. At this stage, for the optimal individual received from the operation of the algorithm, enter the line disturbance through Eq. (5), jump out of the local with probability, and judge whether to accept the solution obtained after the disturbance through Eq. (6), to make the population accept the inferior solution with a certain probability and increase the local exploitation ability of the algorithm.

Step 7: Repeat steps 2 to 6 until the iteration stop condition is met.

The flow chart of the improved algorithm is shown in Fig. 2.

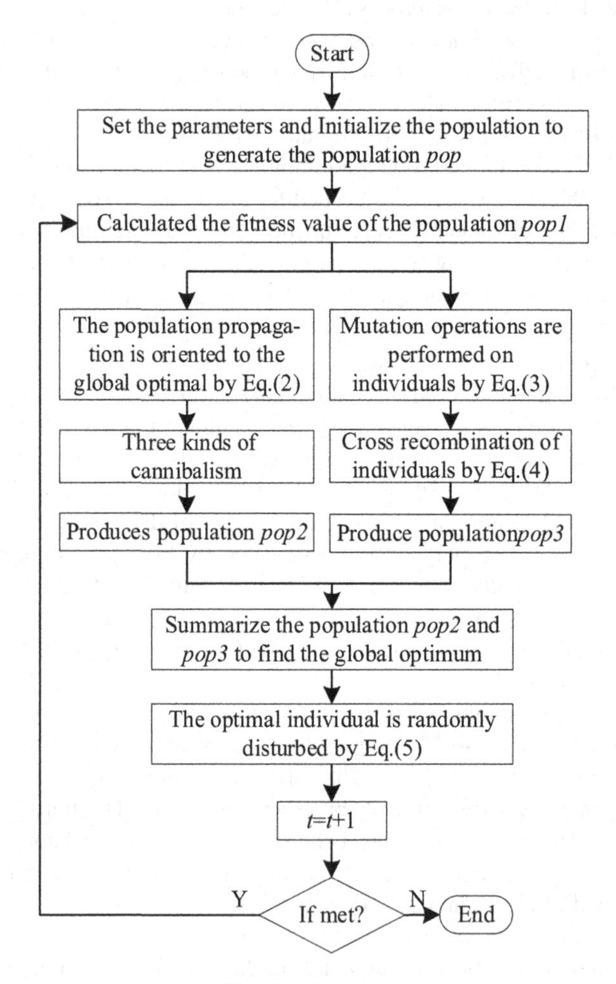

Fig. 2. Flow chart

5 Simulation Experiment

5.1 Test Function

To verify the overall optimization performance of the improved black widow spider algorithm, eight 30 dimensional benchmark functions (as shown in Table 1) in literature [19] are selected for experimental analysis. Of the 8 benchmark functions, $f_1 - f_5$ are unimodal functions, $f_6 - f_8$ are multimodal functions. The control parameters of the improved black widow spider algorithm and the original black widow spider algorithm, particle swarm optimization algorithm [20], differential evolution algorithm [21], the whale algorithm [22] and dragonfly algorithm [23] are shown in Table 2.

Table 1. Benchmark function

Benchmark functions	Range	d	f_{min}
$f_1(x) = \sum_{i=1}^{n} x_i^2$	[−100, 100]	30	0
$f_2(x) = \sum_{i=1}^{n} \lvert x_i \rvert + \prod_{i=1}^{n} \lvert x_i \rvert$	[−100, 100]	30	0
$f_3(x) = \sum_{i=1}^{n} \left(\sum_{j-1}^{i} x_j\right)^2$	[−100, 100]	30	0
$f_4(x) = max_i\{\lvert x_i \rvert, 1 \leq i \leq n\}$	[−100, 100]	30	0
$f_5(x) = \sum_{i=1}^{n} ix_i^4 + random(0, 1)$	[−1.28, 1.28]	30	0
$f_6(x) = \sum_{i=1}^{n} [x_i^2 - 10\cos(2\pi x_i) + 10]$	[−5.12, 5.12]	30	0
$f_7(x) = -20\exp\left(-0.2\sqrt{\frac{1}{n}\sum_{i=1}^{n} x_i^2}\right) - \exp\left(\frac{1}{n}\sum_{i=1}^{n}\cos(2\pi x_i)\right) + 20 + e$	[−32, 32]	30	0
$f_8(x) = \frac{1}{4000}\sum_{i=1}^{n} x_i^2 - \prod_{i=1}^{n}\cos\left(\frac{x_i}{\sqrt{i}}\right) + 1$	[−600, 600]	30	0

For each algorithm, the population size n is set to 50 and the maximum number of iterations t_{max} is set to 1000, run 30 times independently, and take the average value as the final result.

Table 2. Initialization parameters of all algorithms

Algorithm	Parameters	Values
PSO	Learning factor c_1	2
	Learning factor c_2	2
	Inertia coefficient ω	[0.4, 0.9]
DE	The zoom factor F	[0, 2]
BWO	Reproduction rate PR	0.8
	Cannibalism rate PC	0.4
	Mutation rate PM	0.5
IBWO	Reproduction rate PR	0.8
	Cannibalism rate PC	0.4
	Mutation rate PM	0.5
	The optimal coefficient β	0.4
	The zoom factor F	[0, 2]
	Crossover operator CR	[0, 1]
	Disturbance parameters T	1000
GWO	Range control parameter a	2
	The random number C	[0, 2]
DA	The random number $r1$ and $r2$	[0, 1]
	Constant β	1.5

In Table 3, for test functions $f_1 - f_8$. It can be seen that the improved black widow spider algorithm can find satisfactory results.

In unimodal functions $f_1 - f_5$, for f_1 and f_2, it can be seen that the accuracy of the improved black widow spider algorithm has been significantly improved. Compared with the particle swarm optimization algorithm, it does not get a satisfactory answer, which shows that the improved algorithm has strong local convergence performance and good ability to find the global optimal solution.

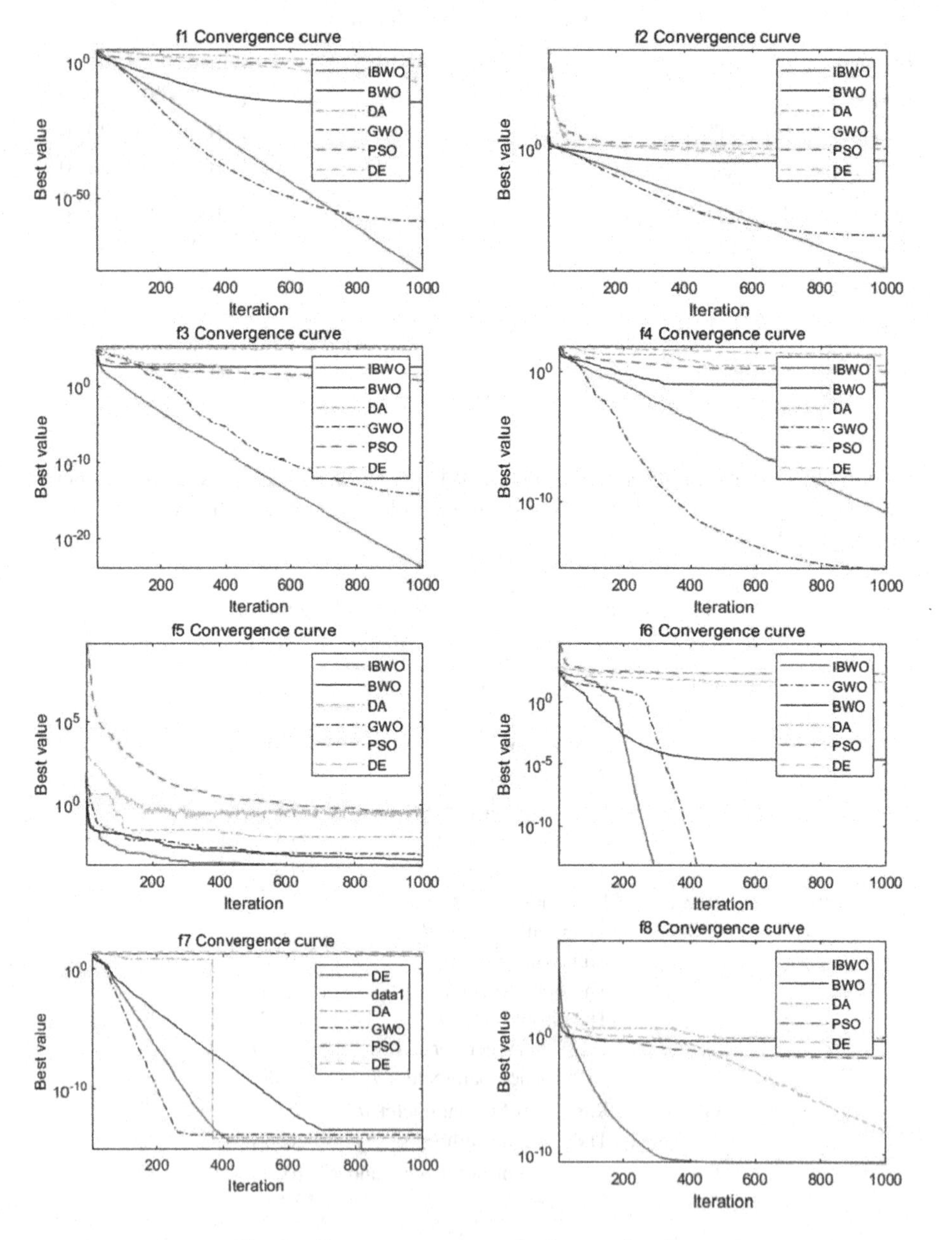

Fig. 3. Convergence curve of reference function

In f_3 and f_4 in the test of functions, the results obtained by the traditional black widow spider algorithm, particle swarm optimization algorithm and differential evolution algorithm are near the optimal value, or the optimal value can not be found, while the improved algorithm is relative to the test functions f_1 and f_2, the accuracy has decreased, but the optimal value can still be found, which shows that the algorithm has been well balanced in exploitation and exploration.

Table 3. Simulation results

Function	PSO	DE	GWO	DA	BWO	IBWO
f_1	0.033115	3.7295E–08	4.938E–59	1.503–05	1.926E–09	**4.336E–76**
f_2	5.3125	1.8963E–05	1.553E–34	0.57897	7.4637E–5	**1.477E–48**
f_3	3.8529	65513.021	1.443E–17	2.4831	260.8533	**1.851E–26**
f_4	1.1234	10.5657	4.648E–15	0.98801	0.10731	**5.492E–10**
f_5	0.41025	0.43739	0.0011637	0.011768	3.483E–04	**1.734E–05**
f_6	159.3897	154.6138	0	31.4286	0.0072464	**0**
f_7	3.225E–06	3.2485	1.509E–14	1.0313	20.1888	**8.881E–16**
f_8	0.029633	2.19E–08	0.008605	0.15563	0.046121	**4.243E–11**

In multimodal function $f_6 - f_8$, test function f_6 is the Rastrigin function, which is often used to test the global search ability of the algorithm. In Table 3, the defects of particle swarm optimization algorithm and differential evolution algorithm fall into the local optimization, while the improved black widow algorithm finds the accurate optimal value. Complex test function f_7 is Ackley function, which is a continuous, rotating and indivisible multimodal function. It is often used to test the ability of the algorithm to jump out of the local extremum. The results also show that the improved black widow spider algorithm has good randomness and can jump out of the local optimization. For the differential evolution algorithm and the traditional black widow spider algorithm, after running many times, they still can not find the global optimization and fall into the local optimization, indicating that the improved algorithm greatly strengthens the ability to resist "premature".

Figure 3 is the convergence curve when solving the reference function. It can be seen from the figure that the improved algorithm has a good improvement in calculation speed, accuracy and convergence performance. In unimodal test function $f_1 - f_5$, the accuracy of the results increases with the increase of the number of iterations. In multimodal function $f_6 - f_8$, which shows that the improved black widow spider optimization algorithm has strong stability and robustness.

5.2 Design Problems of Tension Spring

The purpose of this problem is to minimize the weight of tension or compression spring. As shown in Fig. 4, there are four optimization constraints for this problem, namely, disturbance; shear stress; fluctuation frequency; outside diameter.

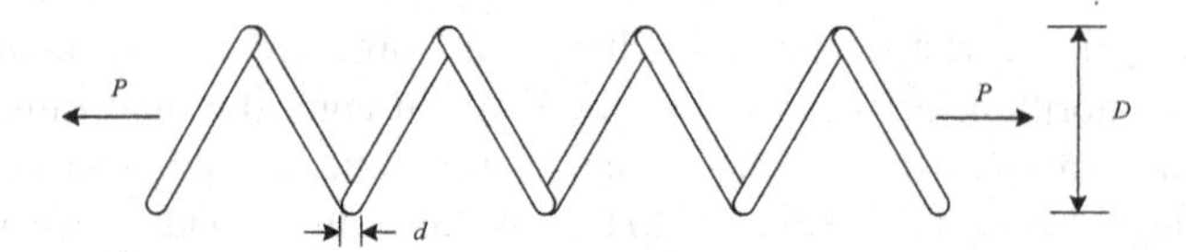

Fig. 4. Design problems of tension spring

There are three design variables, such as wire diameter (*d*), average coil diameter (*D*), and the number of active coils (*P*). The mathematical representation of this problem is described as follows:

Set the individual of the target variable black widow spider population as:

$$Consider\,\vec{z} = [z_1 \quad z_2 \quad z_3] = [d \quad D \quad P] \tag{7}$$

The objective function is:

$$Minimize\,f(\vec{z}) = (z_3 + 2)z_2 z_1^2 \tag{8}$$

The following inequalities need to be satisfied:

$$\text{Disturbance: } g_1(\vec{z}) = 1 - \frac{z_2^3 z_3}{71785 z_1^4} \leq 0 \tag{9}$$

$$\text{Shear stress : } g_2(\vec{z}) = \frac{4z_2^2 - z_1 z_2}{12566\left(z_2 z_1^3 - z_1^4\right)} + \frac{1}{5108 z_1^2} \leq 0 \tag{10}$$

$$\text{Fluctuation frequency: } g_3(\vec{z}) = 1 - \frac{140.45 z_1}{z_2^2 z_3} \leq 0 \tag{11}$$

$$\text{Outside diameter: } g_4(\vec{z}) = \frac{z_1 + z_2}{1.5} \leq 0 \tag{12}$$

Where, $0.05 \leq z_1 \leq 2.0$, $0.25 \leq z_2 \leq 1.3$, $2.0 \leq z_3 \leq 15.0$.

The statistical results of tension spring design are compared in Table 4. The optimal cost of the improved black widow spider optimization algorithm is the smallest of all algorithms, and the control parameters are relatively good. The results show that the improved black widow spider optimization algorithm has higher convergence accuracy in finding the global optimal solution, which not only reduces the cost of spring design problem, but also improves the optimization efficiency of the algorithm.

Table 4. Statistical results

Algorithms	Optimum variables			Optimum cost
	d	*D*	*P*	
IBWO	0.0511887	0.343633	12.1658	**0.012755**
BWO	0.072289	0.78131	7.1068	0.037182
PSO	0.054892	0.43877	7.7151	0.012844
DE	0.059619	0.57919	4.6676	0.013726

6 Summary

The BWO algorithm based on differential evolution and random disturbance in this paper proposed. The differential evolution strategy is introduced in the mutation stage of the algorithm to avoid unnecessary exploration; At the same time, the random disturbance factor is used to improve the local exploitation ability of the algorithm, and the individual memory function is added to improve the reproduction strategy, further reduce the possibility of the algorithm falling into local optimization, and ensure the balance exploitation and exploration. The results show that the improved algorithm is superior to the standard BWO algorithm in terms of convergence speed, global search ability and solution accuracy, overcomes the shortcomings of the standard BWO algorithm.

Acknowledgments. This work was supported in part by the National Natural Science Foundation of China under Grants 62102374, 62072417, and U1804262 and in part by the Henan provincial science and technology research project under Grants 202102210177 and 212102210028.

References

1. Zhou, X., Gao, D.Y., Yang, C., et al.: Discrete state transition algorithm for unconstrained integer optimization problems. Neurocomputing **173**, 864–874 (2016)
2. Song, X., Zhao, M., Yan, Q., et al.: A high-efficiency adaptive artificial bee colony algorithm using two strategies for continuous optimization. Swarm Evol. Comput. **50**, 100549 (2019)
3. Mirjalili, S.: Moth-flame optimization algorithm: a novel nature-inspired heuristic paradigm. Knowl.-Based Syst. **89**, 228–249 (2015)
4. Wang, G.G., Zhao, X., Deb, S.: A novel monarch butterfly optimization with greedy strategy and self-adaptive. In: 2015 Second International Conference on Soft Computing and Machine Intelligence (ISCMI), pp. 45–50. IEEE (2015)
5. Mirjalili, S.: The ant lion optimizer. Adv. Eng. Softw. **83**, 80–98 (2015)
6. Mirjalili, S., Mirjalili, S.M., Lewis, A.: Grey wolf optimizer. Adv. Eng. Softw. **69**, 46–61 (2014)
7. Mirjalili, S., Lewis, A.: The whale optimization algorithm. Adv. Eng. Softw. **95**, 51–67 (2016)

8. Fathy, A., Alharbi, A.G., Alshammari, S., et al.: Archimedes optimization algorithm based maximum power point tracker for wind energy generation system. Ain Shams Eng. J. **13**, 101548 (2021)
9. Dutta, P., Majumder, M., Kumar, A.: An improved grey wolf optimization algorithm for liquid flow control system. IJ Eng. Manuf. **4**, 10–21 (2021)
10. Elsheikh, A.H., Shehabeldeen, T.A., Zhou, J., Showaib, E., Abd Elaziz, M.: Prediction of laser cutting parameters for polymethylmethacrylate sheets using random vector functional link network integrated with equilibrium optimizer. J. Intell. Manuf. **32**(5), 1377–1388 (2020). https://doi.org/10.1007/s10845-020-01617-7
11. Hayyolalam, V., Kazem, A.A.P.: Black widow optimization algorithm: a novel meta-heuristic approach for solving engineering optimization problems. Eng. Appl. Artif. Intell. **87**, 103249 (2020)
12. Agal, S., Gokani, P.K.: An optimized bandwidth estimation for adaptive video streaming systems using WLBWO algorithm. Int. J. Interdisc. Telecommun. Netw. (IJITN) **13**(3), 94–109 (2021)
13. Micev, M., Ćalasan, M., Petrović, D.S., et al.: Field current waveform-based method for estimation of synchronous generator parameters using adaptive black widow optimization algorithm. IEEE Access **8**, 207537–207550 (2020)
14. Premkumar, K., Vishnupriya, M., Sudhakar Babu, T., et al.: Black widow optimization-based optimal pi-controlled wind turbine emulator. Sustainability **12**(24), 10357 (2020)
15. Al-Rahlawee, A.T.H., Rahebi, J.: Multilevel thresholding of images with improved Otsu thresholding by black widow optimization algorithm. Multimedia Tools Appl. **80**(18), 28217–28243 (2021). https://doi.org/10.1007/s11042-021-10860-w
16. Priya, J.S., Bhaskar, N., Prabakeran, S.: Fuzzy with black widow and spider monkey optimization for privacy-preserving-based crowdsourcing system. Soft. Comput. **25**(7), 5831–5846 (2021). https://doi.org/10.1007/s00500-021-05657-w
17. Sheriba, S.T., Hevin, R.D.: Improved hybrid cuckoo black widow optimization with interval type 2 fuzzy logic system for energy-efficient clustering protocol. Int. J. Commun. Syst. **34**(7), e4730 (2021)
18. Punithavathi, R., Kurangi, C., Balamurugan, S.P., et al.: Hybrid BWO-IACO algorithm for cluster based routing in wireless sensor networks. CMC-Comput. Mater. Continua **69**(1), 433–449 (2021)
19. Tang, K., Yáo, X., Suganthan, P.N., et al.: Benchmark functions for the CEC'2008 special session and competition on large scale global optimization. Nat. Insp. Comput. Appl. Lab. USTC China **24**, 1–18 (2007)
20. Bai, Q.: Analysis of particle swarm optimization algorithm. Comput. Inf. Sci. **3**(1), 180 (2010)
21. Qin, A.K., Huang, V.L., Suganthan, P.N.: Differential evolution algorithm with strategy adaptation for global numerical optimization. IEEE Trans. Evol. Comput. **13**(2), 398–417 (2008)
22. Teng, Z.-J., Lv, J.-L., Guo, L.-W.: An improved hybrid grey wolf optimization algorithm. Soft. Comput. **23**(15), 6617–6631 (2018). https://doi.org/10.1007/s00500-018-3310-y
23. Mirjalili, S.: Dragonfly algorithm: a new meta-heuristic optimization technique for solving single-objective, discrete, and multi-objective problems. Neural Comput. Appl. **27**(4), 1053–1073 (2015). https://doi.org/10.1007/s00521-015-1920-1

Attribute Selection Method Based on Artificial Bee Colony Algorithm and Neighborhood Discrimination Matrix Optimization

Yuxuan Ji[1], Jun Ye[1,2](✉), Zhenyu Yang[1], Jiaxin Ao[1], and Lei Wang[1,2]

[1] Nanchang Institute of Technology, Nanchang 330000, Jiangxi, China
2003992646@nit.edu.cn

[2] Key Laboratory of Water Information Cooperative Perception and Intelligent Processing of Jiangxi Province, Nanchang 330000, Jiangxi, China

Abstract. At present, the existing attribute reduction algorithm combining artificial bee colony and neighborhood rough set basically uses the attribute dependence and the number of attribute subsets as parameters to construct the fitness function, while ignoring the role of heuristic information. As a result, the number of bee colony iterations increases, and the convergence speed is slow. Aiming at this kind of problem, an improved method is proposed. First, a discernibility matrix under the neighborhood rough set is defined; secondly, an attribute importance measurement method of the discernibility matrix under the neighborhood decision system is proposed; The attribute importance of the domain discrimination matrix constructs a new fitness function for the heuristic factor; finally, an attribute selection algorithm for artificial bee colony algorithm and neighborhood discrimination matrix importance optimization is designed. Compared with the original algorithm, the new method reduces the number of generations, accelerates the convergence speed, and retains the minimum attribute reduction collected during each iteration, and multiple minimum attribute reductions can be obtained. The experimental results on the UCI data set prove the feasibility and effectiveness of the algorithm.

Keywords: Neighborhood decision system · Discernibility matrix · Artificial bee colony algorithm · Feature selection · Attribute importance

1 Introduction

Among the mathematical tools for dealing with fuzzy and inaccurate problems, rough set theory is widely used in many fields. However, the classic rough set

Supported by the National Natural Science Foundation of China (Nos. 61562061) and Technology Project of Ministry of Education of Jiangxi province of China (No.GJJ211920, GJJ170995).

L. Pan et al. (Eds.): BIC-TA 2021, CCIS 1565, pp. 71–87, 2022.
https://doi.org/10.1007/978-981-19-1256-6_6

model is not suitable for dealing with continuous data problems. For this reason, some scholars have introduced the concept of neighborhood rough set in order to solve such problems. Lin [1] proposed the concept of neighborhood model relation based on topology from the perspective of granularity. Subsequently, Hu et al. [2] proposed a fast reduction algorithm based on the neighborhood rough set model and a hybrid data reduction algorithm based on the neighborhood rough set, which can directly process unprocessed numerical data, thereby extending the classical rough set model. Application areas. In the neighborhood rough model, because the computational workload is much larger than that of discrete data, the attribute reduction and feature dimensionality reduction in the neighborhood rough model have a higher time complexity, especially with the dimensionality of the decision table. The increase in time complexity has increased geometrically. Therefore, reducing time complexity and reducing computational workload are the key research contents of many scholars.

In recent years, swarm intelligence algorithms have achieved good results in solving some optimization problems [4–8]. The researchers introduced the artificial bee colony optimization algorithm [4] into the attribute reduction of the neighborhood rough set, by constructing a moderate function to guide the colony to search for the smallest feature subset to achieve the purpose of reduction. The designed moderate function can reduce the number of iterations and reduce the reduction. Reduce the time complexity, improve the reduction rate of the algorithm, and obtain multiple minimum attribute reductions. For example, literature [5] proposed an information gain-guided bee colony optimization algorithm, which constructs the mutual information between conditional features and decision-making features based on the information entropy of the features, and finally obtains a feature subset. Literature [6] uses a combination of fuzzy rough set and bee colony algorithm for attribute reduction. Through dependence and reduction rate, an objective function that can reflect the size and importance of the attribute set is constructed, and the attribute reduction problem is transformed into an optimization problem. Finally, the goal is The function is an iterative criterion. Literature [7] proposes a new combined bee colony algorithm to solve the MAR problem, in which the lead bee, follow bee and scout bee adopt a search mode based on mutation calculation. Literature [8] proposed a method of combining rough sets in the classic field with artificial bee colonies, and its algorithm can obtain multiple minimum feature subsets.

The neighborhood rough set attribute subset selection algorithms optimized by these artificial bee colony algorithms all use attribute dependence and the number of attribute subsets as parameters to construct fitness functions. Although more satisfactory results have been obtained in the attribute search subset, there are limitations. For this reason, literature [9] introduces attribute importance based on information entropy as heuristic information, and constructs a new fitness function in combination with attribute dependence, thereby proposing an attribute selection algorithm, which improves The speed of attribute reduction reduces the number of iterations. However, taking the importance of information entropy attribute as the fitness function of the heuristic information, the feature subset searched by the bee colony in some decision tables may contain redundant attributes. This paper draws on the fitness function constructed in the literature [9], based on the attribute distinguishing different neighborhood objects in the discernibility matrix, defines a new attribute importance, and replaces the information entropy attribute in the fitness function with the new attribute importance Importance, an improved attribute selection algorithm of artificial bee colony and neighborhood rough set discernibility matrix is proposed. The main advantage of this algorithm is to provide an idea for the neighborhood rough set attribute reduction method, which reduces the number of generations and speeds up convergence speed and retention of the minimum attribute reduction collected during each iteration will make the result of generating attribute reduction more accurate. The rest of this article is organized as follows. Section 2 introduces the knowledge of neighborhood rough set and artificial bee colony algorithm, Sect. 3 gives attribute reduction algorithm, Sect. 4 gives experimental analysis, and Sect. 5 gives conclusions.

2 Related Knowledge

In order to facilitate understanding, the following briefly introduces related knowledge of neighborhood rough set, artificial bee colony optimization and its improved algorithm related to this article. For more detailed knowledge, please refer to literature [1,3].

2.1 Basic Knowledge of Neighborhood Rough Set

In the rough set of neighborhood, the equivalence relationship is extended to the neighborhood relationship, and the information of the original data is retained to the greatest extent.

Definition 1. [2] Given N in the real-dimensional space Ω, $\Delta = R^N \times R \to R$, then called Δ is a metric on R^N. A non-empty finite set $U = \{x_1, x_2, \ldots, x_n\}$ on a given real number space Ω. The neighborhood δ of $\forall \mathbf{x}_i$ is defined $\forall \mathbf{x}_i$:

$$\delta(x_i) = \{x \mid x \in U, \Delta(x, x_i) \leq \delta\} \tag{1}$$

Δ represents the distance, and the Euclidean distance is used in this article, which is p = 2.

$$\Delta_p(x_1, x_2) = \left(\sum_{i=1}^{N} |f(x_1, a_k) - f(x_2, a_k)|^p\right)^{\frac{1}{p}} \tag{2}$$

Definition 2. [2] Neighborhood decision system $NDS = (U, C \cup D, V, f)$, to $B \subseteq C$,$X \subseteq U$, Then the lower approximation and upper approximation of the neighborhood are respectively defined as:

$$\underline{N}_B(X) = \{x_i \mid \delta_B(x_i) \subseteq X, x_i \in U\} \tag{3}$$

$$\bar{N}_B(X) = \{x_i \mid \delta_B(x_i) \cap X \neq \varnothing, x_i \in U\} \tag{4}$$

By Definition 2, the positive domain and negative domain of the neighborhood decision system are:

$$\mathrm{Pos}_B(D) = \underline{N}_B D \tag{5}$$

$$\mathrm{Neg}_B(D) = U - \bar{N}_B D \tag{6}$$

Definition 3. [2] Neighborhood decision system $NDS = (U, C \cup D, V, f)$, The dependence of decision attribute D on conditional attribute subset B is defined as:

$$\gamma_B(D) = \frac{|\mathrm{Pos}_B(D)|}{|U|} \tag{7}$$

Definition 4. [2] Neighborhood decision system $NDS = (U, C \cup D, V, f)$, to $\forall c \in C$, If there is $\gamma_{C-\{c\}}(D) \neq \gamma_C(D)$, then c is called a necessary attribute, otherwise it is an unnecessary attribute, and all necessary attributes form the core attribute set $CORE(C)$.

Definition 5. [2] Neighborhood decision system $NDS = (U, C \cup D, V, f)$, $\forall A \subseteq C$, If the conditional attribute subset A satisfies: (1)$\gamma_A(D) = \gamma_C(D)$ (2)$\forall a \in A, \gamma_A(D) > \gamma_{A-\{a\}}(D)$.

The conditional attribute subset A is said to be a relative reduction of conditional attribute set, and all reduction sets are denoted as $RED(C)$.

2.2 Artificial Bee Colony and Its Improved Algorithm

Karaboga [4] and others proposed the artificial bee colony algorithm (ABC) to simulate the nectar-collecting behavior of bee colonies in nature. The algorithm converts the nectar source location problem into a function to generate the optimal solution. The process of bee colony searching for nectar source is the optimal solution. The process of. The algorithm idea is:

Suppose the dimension of the problem to be solved is D, the position of the honey source i in t iterations is expressed as $X_i^t = [x_{i1}^t, x_{i2}^t, \ldots, x_{iD}^t]$. Represents the current number of iterations. $x_{im} \in (K_m, U_m)$,K_m and U_m respectively represent the upper and lower bounds of the cost parameter, $m = 1, 2, \ldots, D$. Randomly generate the initial position of the desired nectar source i. At the beginning of the search, the hired bee searches around the nectar source i to generate a new nectar source.

When the fitness of the new honey source $W_i = [\mathrm{w}_{i1}, w_{i2}, \ldots w_{id}]$ is better than X_i, the greedy selection method is used to replace X_i with W_i, otherwise X_i is retained. After the hired bee completes the calculation, he flies back to the information exchange area to share the nectar source information. Follow the bee based on the nectar source information shared by the hired bee. The calculated probability is followed.

Then, the follower bee uses the roulette method to select the hired bee. During the search process, if the nectar source X_i reaches the threshold limit after trial iterations and no better nectar source is found, the nectar source

X_i will be abandoned and the corresponding employment The role of the bee is transformed into a scout bee. The scout bee will randomly generate a new nectar source in the search space to replace X_i.

Flow chart of artificial bee colony algorithm (Fig. 1.)

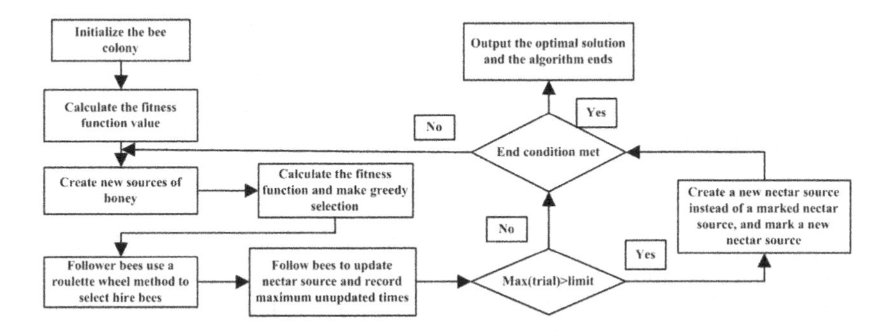

Fig. 1. Flow chart of artificial bee colony algorithm.

Artificial bee colony algorithm has the advantages of strong robustness and strong search ability. However, there are problems such as easy to fall into the local optimum and slow convergence speed in the later stage. Researchers have proposed a variety of improvement methods. Among them, the improved method of literature [10] has been widely used. Gao et al. [10] combined the differential evolution algorithm with the global optimal particle-improved bee colony algorithm, and proposed an algorithm with faster convergence speed. The information of the best solution is introduced into the generation of candidate honey sources, and the position update formula is modified to:

$$v_{ij} = x_{ij} + \varphi_{ij}\left(x_{ij} - x_{kj}\right) + \alpha_{ij}\left(x_{gj} - x_{kj}\right) \tag{8}$$

Among them, x_{ij} is the best nectar source found by the bee colony so far, $a_{ij} \in [0, 1]$. This paper draws on the Formula (8) for updating the location of the honey source constructed by Gao et al. to optimize the algorithm proposed in this paper.

3 Improved Attribute Selection Algorithm for Artificial Bee Colony and Neighborhood Discrimination Matrix

3.1 Definition of Attribute Importance of Neighborhood Discernibility Matrix

A. Skowron [11] first proposed the use of discernibility matrix to express knowledge system, which is defined as follows:

Definition 6. [11] Given a decision-making system $DS = (U, C \cup D, V, f)$, where C is the conditional attribute set, D is the decision attribute set, and the universe is a non-empty finite set of objects $U = \{x_1, x_2, \ldots, x_n\}$, where is $|U| = n$. Then define the discernibility matrix of the decision-making system as:

$$(M_{\mathrm{ij}})_{n \times n} = (c_{ij})_{n \times n} = \begin{bmatrix} c_{11} & c_{12} & \cdots & c_{1n} \\ c_{21} & c_{22} & \cdots & c_{2n} \\ \vdots & \vdots & \ddots & \vdots \\ c_{n1} & c_{n2} & \cdots & c_{nn} \end{bmatrix} \tag{9}$$

Among, $i, j = 1, 2, 3, \ldots, n$

$$C_{ij} = \begin{cases} \{a \mid a \in C \wedge f(x_i) \neq f(y_i)\}, f_D(x_i) \neq f_D(x_i) \\ \varnothing, \quad f_D(x_i) = f_D(x_i) \end{cases} \tag{10}$$

In this paper, the above Definition 7 is extended to the rough set of the neighborhood, and the lower discernibility matrix of the neighborhood relationship is defined. The definition method is as follows:

Definition 7. Neighborhood decision system $NDS = (U, C \cup D, V, f)$, where C is the condition attribute set, D is the decision attribute set, for any x_i, $x_j \in U$ and attributes $a \in C$, the discrimination matrix M_{ij} of the decision system NDS is defined as:

$$\{M_{\mathrm{ij}}\}_{n \times n} = \begin{cases} \{a|a \in C \wedge |f_a(x_i) - f_a(x_j)| > \delta(x_i)\}, f_D(x_i) \neq f_D(x_i) \\ 0, f_D(x_i) \neq f_D(x_i) \wedge f_C(x_i) - f_C(x_i) < \delta(x_i) \\ \varnothing, f_D(x_i) = f_D(x_i) \end{cases} \tag{11}$$

According to Definition 7, M_{ij} can be known that it is an upper triangle or a lower triangle matrix.

Theorem 1. *In the neighborhood discernibility matrix M_{ij}, The set of all single attribute elements is the necessary (core) attribute set of C relative to D, namely* $\mathrm{CORE}_C(D) = \{a \mid a \in C \wedge (\exists c_{ij}, ((c_{ij} \in M) \wedge (c_{ij} = \{a\})))\}$

Through the analysis of Eq. (11), we know that for objects with equal decision attribute values, these objects do not need to be distinguished, so there is $c_{ij} = \varphi$. For objects whose decision attribute values are not equal and belong to the same neighborhood, if these objects are indistinguishable, there is $c_{ij} = 0$. For objects with unequal decision attribute values and belonging to different neighborhoods, these objects can be distinguished, then c_{ij} is composed of conditional attributes in C that can distinguish these objects, which may include the attributes in $CORE(C)$ or include Hor K. All the attributes in c_{ij} play a role in distinguishing objects x_i and x_j in different neighborhoods, but the effects are different. The necessary attributes have the greatest effect, the relative necessary attributes are the second, and the unnecessary attributes contribute the least. In order to accurately measure the role of different types of conditional attributes in distinguishing objects in different neighborhoods, this paper uses the frequency of occurrence of each attribute in c_{ij} and the weight of each attribute to measure its importance, which is defined as follows:

Definition 8. Neighborhood decision-making system $NDS = (U, C \cup D, V, f)$, C is a set of conditional attributes, and D is a set of decision-making attributes. c_{ij} is a set of non-empty elements, K is the total number of sets of non-empty elements, for $\forall c \in C$, the importance of attribute c to D is defined as:

$$\mathrm{Nsig}(c) = \begin{cases} 1, & |c_{ij}| = 1 \\ \frac{\sum_{i,j=1,2,\ldots,n} r_c \cap c_{ij}}{K}, & |c_{ij}| > 1 \end{cases} \tag{12}$$

Among, $r_c \cap c_{ij} = \begin{cases} \frac{1}{|c_{ij}|}, & c \cap c_{ij} \neq \varnothing \\ 0, & c \cap c_{ij} = \varnothing \end{cases} i, j = 1, 2, \ldots, n, |c_{ij}| = 1$ represents the number of attributes in the collection c_{ij}.

It can be seen from the Formula (12) in Definition 8 that when $|c_{ij}| = 1$ is a single attribute, its importance is 1, which means that a single attribute contributes the most when distinguishing objects in different neighborhoods. This is the same as the single attribute in Theorem 2. The attributes match. When $|c_{ij}| > 1$, that is, the set of c_{ij} is composed of multiple attributes, it means that distinguishing objects in different neighborhoods is completed by multiple attributes, which may be necessary attributes, relatively necessary attributes or unnecessary attributes. Each attribute functions as $\frac{1}{|c_{ij}|}$. Obviously, the importance of a nuclear attribute is the sum of its importance as a single attribute 1 and the importance of multiple attributes, namely: $1 + \frac{\sum_{i,j=1,2,\cdots,n} r_c \cap c_{ij}}{K}$, which also reflects that the nuclear attribute has the strongest ability to distinguish neighboring objects. For Definition 8, we can get two properties:

Property 1: For the set c_{ij} of non-empty elements in the discernibility matrix M, if $\forall c \in C$ and $c \in c_{ij}$, then $Nsig(c) > 0$.

It can be seen from Property 1 that Definition 9 can not only obtain the importance of necessary attributes, but also the relative importance of necessary attributes and unnecessary attributes. It shows that the attributes in the set all play a role in distinguishing objects in different neighborhoods, avoiding the situation that the importance of non-core attributes in Definition 6 are all zero.

Property 2: For the non-empty element set c_{ij} in the discernibility matrix M, there is $\forall a, b \in c_{ij}$, if there are $a \in CORE(C)$, $b \in N$, then there is $Nsig(a) > Nsig(b) > 0$. Among them, $N = C - CORE(C)$ represents the non-core attribute set.

Property 2 shows that the importance of nuclear attributes is the largest, and the importance of non-nuclear attributes is less than it. This is consistent with the properties of the rough set of neighborhoods. Therefore, Definition 9 accurately measures the role of attributes in neighborhood resolution.

3.2 Fitness Function Construction

The fitness function directly determines the evolution direction of the colony, the number of iterations, and the pros and cons of the solution. Therefore, the appropriate fitness function plays an important role in the attribute selection method. At present, the attribute selection algorithm optimized by the artificial bee colony algorithm uses the fitness function of the following Formula (13):

$$fit = \alpha \cdot \gamma_B(D) + \beta \cdot \frac{|C| - |B|}{|C|} \tag{13}$$

Among them, $\alpha + \beta = 1$ and $\alpha + \beta \in [0, 1]$, $|C|$ is the total number of feature attributes of the data set, $|B|$ is the length of the currently selected feature subset, and $\gamma_B(D)$ is the attribute dependency. Equation (13) has been widely used in the research of various feature selection methods based on neighborhood rough set attribute dependence, and has achieved good results. However, the fitness function also has some shortcomings; one is that the attribute dependence in the fitness function is only calculated for the selected attributes, and the analysis and evaluation of other attributes in the attribute set are lacking [9]; the other is that the fitness function is only The feature is selected based on the attribute dependence and the number of attributes, ignoring the role of heuristic information that can speed up finding the reduced subset. For this reason, literature [9] introduces attribute importance based on information entropy as heuristic information, and combines attribute dependence to construct a new fitness function, as shown in the following Formula (14):

$$fit^2 = \lambda_1 \cdot \gamma_B(D) + \lambda_2 \cdot \frac{1}{1 + e^{sig(C-B)}} + \lambda_3 \cdot \frac{|C| - |B|}{|C|} \tag{14}$$

Among them, $\lambda_1, \lambda_2, \lambda_3$ are weighting factors, where $\lambda_1 + \lambda_2 + \lambda_3 = 1$ and $\lambda_1, \lambda_2, \lambda_3 \in [0, 1]$; $sig(C - B)$ is the attribute importance of the conditional feature set outside the selected feature subset B;

However, the attribute importance of information entropy is a quantitative measure of the change in the uncertainty domain. The change of any attribute may cause the uncertainty domain to change. Therefore, the information entropy method improves the non-kernel attributes under the rough set algebraic view. When the importance is zero. However, in some decision tables, the information entropy method has the situation that the importance of redundant attributes is greater than the importance of core attributes [16]. The subset of attributes found by the bee colony may contain redundant attributes, resulting in selection accuracy. Decreased significantly.

Aiming at the shortcomings of the fitness function constructed by the conditional entropy attribute importance in the literature [9], this paper replaces the fitness function with the attribute importance of the discernibility matrix defined in Sect. 3.1. The new fitness function is as follows (15).

$$\text{fitness'} = \omega_1 \cdot \gamma_B(D) + \omega_2 \cdot \frac{1}{1 + e^{Nsig(c, C-B, D)}} + \omega_3 \cdot \frac{|C| - |B|}{|C|} \tag{15}$$

Among them, the weight factor $\omega_1, \omega_2, \omega_3$ satisfies the condition $\omega_1 + \omega_2 + \omega_3 = 1$ and $\omega_1, \omega_2, \omega_3 \in [0, 1]$; $Nsig(c, C - B, D)$ is the attribute importance of the conditional feature set outside the selected feature subset B under the neighborhood rough set discernibility matrix. let $C = \{c_1, c_2, \ldots, c_n\}$, $\{0,1\}^n$ denote m-dimensional Boolean space, define $x = (x_1, x_2, \ldots, x_n) \notin \{0,1\}^n$.

3.3 Neighborhood Discernibility Matrix Importance and Artificial Bee Colony Feature Selection Algorithm

According to the fitness function constructed by the new neighborhood discernibility matrix attribute importance, we design an artificial bee colony feature search algorithm (ABCNRT algorithm) based on the discernibility matrix

Algorithm 1: ABCNRT algorithm

Input: Neighborhood decision system $NDS = (U, C \cup D, V, f)$, the population size of the bee colony is nPop, the number of iterations iter and the location of the nectar source is Pop

Output: the best nectar source x_b in history

1 **while** $U \neq \varnothing$ **do**
2 **for** *each* $M_j \in U$ **do**
3 Select the neighborhood for the decision table δ;
4 $POS_i \cup M_j \rightarrow POS_i$
5 **end**
6 The dependence degree $\gamma_B(D)$ is calculated according to Equation (7);
7 The importance $Nsig(c, C - B, D)$ is calculated according to Equation (12)
8 **end**
9 **while** $iter < iter_{max}$ **do**
10 **for** *each* $x_i \in nPop$ **do**
11 $\gamma_B(D)$ and $Nsig(c, C - B, D) \rightarrow \lambda_1 \cdot \gamma_B(D) + \lambda_2 \cdot \frac{1}{1+e^{Ncig(c,C-B,D)}} + \lambda_3 \cdot \frac{|C|-|B|}{|C|}$;
12 **end**
13 **for** *each* $x_i \in C$ **do**
14 L_i=(a*rand([-1,1][1, x_b]))*(Pop [i]- Pop [k])+(a*rand([1, x_b]))*(Pop [i]-x_b);
15 Judging the value of L_i gives W_i ;
16 Use formula (20) to calculate the value of the W_i fitness function;
17 **if** $f(W_i) \leq f(X_i)$ **then**
18 Replace X_i with W_i
19 **else**
20 $trial = trial + 1$
21 **end**
22 **end**
23 **for** *each* $x_i \in nPop$ **do**
24 Execute roulette algorithm, Repeat the above steps to hire bees;
25 Go through the stage of detection bee;
26 Output historical optimal solution x_b;
27 **end**
28 **end**

attribute importance. The main idea is: First, obtain the discernibility matrix and calculate Resolve the importance of each attribute in the matrix; secondly, construct the fitness function; finally, iterate according to the direction of the fitness function value, until the search for the smallest feature set stops the algorithm. The specific steps of the ABCNRT algorithm are as follows. In the experiment part, the fitness function of the algorithm is used to conduct two sets of experiments with Eqs. (13) and (15).

Algorithm Complexity Analysis. The time complexity of the algorithm in this paper mainly depends on the number of bee colonies $nPop$, the variable dimension D, the number of conditional attributes M, and the number of universe objects $|U|$. Among them, the time complexity of calculating the discernibility matrix in the fitness function is $O\left(|M||U|^2\right)$, Suppose the sample is divided into k sets, so the time complexity of importance in the fitness function is $O\left((|M|+N)\frac{|U|^2}{k}\right)$. In addition to the neighborhood rough set calculation, the outer loop also has the number of bee colonies and variable dimensions. Therefore, in general, the time complexity of the ABCNRT algorithm is $O\left((|M|+N)\frac{|U|^2}{k} * k{*}D * nPop\right)$. Among them, it can be calculated in parallel in the process of calculating the degree of dependence, reducing the time complexity as $O\left((|M|+N)\frac{|U|^2}{k} * D * nPop\right)$.

4 Experiment Analysis

In order to verify the superiority and reliability of the proposed algorithm. We select 8 data sets from UCI machine learning library (Table 1) for experiments, and compare the algorithm results obtained from experiments with the neighborhood rough set NRS algorithm proposed in literature [3]. Feature selection FSRSWOA algorithm based on rough set and improved Whale optimization algorithm in literature [14] and feature selection PSORSFS algorithm based on rough set and particle swarm optimization algorithm in literature [15] were performed 20 times in the experiment. The software environment is Win 10 operating system, the programming software is Python3.6, the hardware is Intel I7 1040 3.6 Ghz, and the memory is 8 GB.

Table 1. Data set

Serial number	Data set	Number of samples	Number of attributes	Number of categories
1	Zoo	101	16	7
2	Wine	178	13	3
3	Sonar	208	60	2
4	Ionosphere	351	34	2
5	Heart	270	13	7
6	WDBC	569	30	2
7	German	1000	24	2
8	CMC	1473	9	2

4.1 Selection of δ

In order to remove the influence of dimensions on the data, the data set data is normalized first. Experiment on the data set again, and the experimental results are shown in Fig. 2.

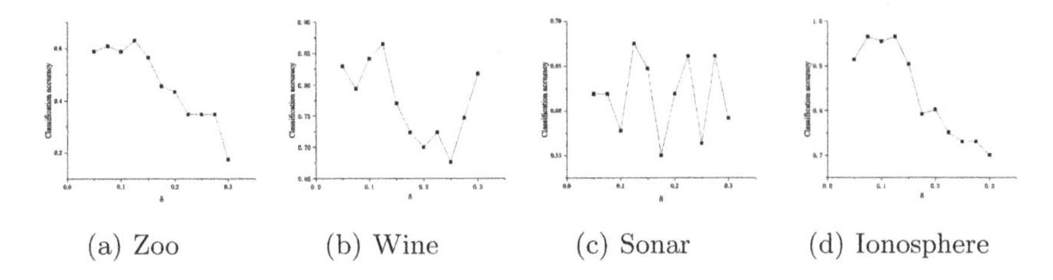

(a) Zoo (b) Wine (c) Sonar (d) Ionosphere

Fig. 2. Classification accuracy varies with threshold

In the experiment, due to the selected threshold δ, the classification accuracy of different data sets will be different. When calculating the classification accuracy, four data sets in the data set are selected, and the CART classifier is used to classify these four data sets to select the best δ value. The classification accuracy changes with the threshold δ were cross-validated and analyzed ten times.

It can be seen from Fig. 2 that the threshold is within the range of $[0.1, 0.15]$, and the classification accuracy of the data set is better. In this paper, $\delta = 0.125$ is selected for the experiment.

4.2 Algorithm Comparison Results

Compare this algorithm with NRS algorithm [3], FSRSWOA algorithm [14] and PSORSFS algorithm [15]. The value of ABCNRT algorithm is: initial colony size N = 20, the maximum number of iterations M = 50. When the values of the neighboring particles δ are all set to 0.125, three algorithms are used to perform attribute reduction on 8 data sets.

Original Fitness Function. The original fitness function $fit = \alpha \cdot \gamma_B(D) + \beta \cdot \frac{|C| - |B|}{|C|}$ is selected, and the minimum attribute reduction problem is transformed into the maximum fitness function value problem. Take $\alpha = 0.7$ and $\beta = 0.3$. First, use the ABCNRT algorithm to reduce the 8 UCI data. Table 2 shows the reduction of ABCNRT algorithm.

Table 2. ABCNRT algorithm reduction result

Serial number	Data set	Number of samples	Number of optimal attribute subsets	Optimal attribute subset	Reduction rate	Optimal classification accuracy
1	Zoo	16	3	c_5, c_7, c_9	81.25	0. 7619
2	Wine	13	4	c_1, c_5, c_7, c_{10}	69.23	0.9444
3	Sonar	60	7	$c_{10}, c_{12}, c_{22}, c_{28}, c_{31}, c_{32}, c_{48}$	88.33	0.7380
4	Ionosphere	34	6	$c_4, c_5, c_7, c_{16}, c_{29}, c_{34}$	82.35	0.9295
5	Heart	13	9	$c_1, c_4, c_5, c_7, c_8, c_9, c_{11}, c_{12}, c_{13}$	30.77	0.6393
6	WDBC	30	4	$c_8, c_{21}, c_{22}, c_{27}$	86.67	0.9649
7	German	24	2	c_4, c_6	91.67	0.6950
8	CMC	9	2	c_4, c_7	77.78	0.4033

According to Table 2, 5 of the 8 UCI data sets have a reduction rate of more than 80%, and 2 of them have a reduction rate of about 70%. Only the Heart data set has a lower reduction rate of about 30%. There are more subsets of reductions. The first six data are compared with the initial classification accuracy of Fig. 1. According to Table 2, it can be seen that all data sets are almost better than the initial classification accuracy. It shows that the above table fully reflects the better reduction results achieved by ABCNRT algorithm reduction.

Table 3, 4 shows the comparative experimental results of PSORSFS algorithm, FSRSWOA algorithm, NRS algorithm, ABCNRT algorithm attribute reduction and the best classification accuracy.

According to Table 3 below, the attribute reduction number analysis of the four algorithms for the eight data sets shows that when comparing the NRS algorithm, it is found that except for the Heart data set, the other data ABCNRT algorithm is better than the NRS algorithm. When comparing the PSORSFS algorithm, except for Wine and Heart, the number of other attribute reductions is more than that of the ABCNRT algorithm. When comparing the FSRSWOA algorithm, except for the Heart, Wine and Ionosphere data sets, the number of reductions in some running results is less than that of the ABCNRT algorithm, and the number of other attribute reductions is more than that of the ABCNRT algorithm.

According to the following Table 4, the classification accuracy analysis of the eight data sets in the four algorithms shows that the classification accuracy of the Heart data set is better than the ABCNRT algorithm. The classification accuracy of the Zoo data set is better than the PSORSFS algorithm and the FSRSWOA algorithm. For the ABCNRT algorithm, the Sonar data set FSRSWOA algorithm is better than the ABCNRT algorithm, and the classification accuracy of the ABCNRT algorithm in other data sets is better than the other three algorithms.

Table 3. Attribute reduction

Data set	PSORSFS	FSRSWOA	NRS	ABCNRT
Zoo	$8^{[10]}, 9^{[10]}$	$7^{[5]}, 8^{[12]}, 9^{[3]}$	5	$\mathbf{3^{[15]}, 4^{[5]}}$
Wine	$3^{[5]}, 4^{[15]}, 5^{[5]}$	$\mathbf{3^{[12]}, 4^{[8]}}$	5	4
Sonar	$11^{[15]}, 12^{[5]}$	$10^{[13]}, 11^{[7]}$	16	$\mathbf{7^{[17]}, 8^{[3]}}$
Ionosphere	$7^{[8]}, 8^{[12]}$	**5**	9	$5^{[4]}, 6^{[16]}$
Heart	$6^{[5]}, 7^{[15]}$	$7^{[5]}, 8^{[15]}$	**6**	9
WDBC	$10^{[13]}, 11^{[7]}$	$9^{[12]}, 10^{[8]}$	6	$\mathbf{4^{[18]}, 5^{[2]}}$
German	$5^{[16]}, 6^{[4]}$	$3^{[13]}, 4^{[7]}$	9	$\mathbf{2^{[10]}, 3^{[10]}}$
CMC	5	$6^{[16]}, 7^{[4]}$	9	$\mathbf{1^{[8]}, 2^{[12]}}$

Table 4. Optimal classification accuracy

Data set	PSORSFS	FSRSWOA	NRS	ABCNRT
Zoo	0.8443	**0.8743**	0.7425	0.7619
Wine	0.8764	0.8967	0.8764	**0.9444**
Sonar	0.7234	**0.7829**	0.6782	0.7380
Ionosphere	0.9072	0.9156	0.8576	**0.9295**
Heart	0.6672	**0.7023**	0.6516	0.6393
WDBC	0.8976	0.9425	0.8796	**0.9649**
German	**0.6950**	0.6909	0.5672	**0.6950**
CMC	0.3782	0.3965	0.3928	**0.4033**

In summary, it can be proved that the ABCNRT algorithm has advantages.

Improve Fitness Function. The fitness function formula fitness' $= \omega_1 \cdot \gamma_B(D) + \omega_2 \cdot \frac{1}{1+e^{Nsig(c,C-B,D)}} + \omega_3 \cdot \frac{|C|-|B|}{|C|}$ improved in this paper is adopted, in which the problem of minimum attribute reduction is transformed into the problem of maximum fitness function value, and $\omega_1 = 0.6, \omega_2 = 0.3, \omega_3 = 0.1$ is taken. Table 4 shows the reduction situation obtained with the improved fitness function formula.

Table 5. ABCNRT algorithm reduction result

Serial number	Data set	Number of samples	Number of optimal attribute subsets	Optimal attribute subset	Reduction rate	Optimal classification accuracy
1	Zoo	16	3	c_5, c_7, c_9	81.25	0.7619
2	Wine	13	4	c_1, c_5, c_7, c_{10}	69.23	0.9444
3	Sonar	60	6	$c_{10}, c_{12}, c_{22}, c_{31}, c_{32}, c_{48}$	90.00	0.7619
4	Ionosphere	34	5	$c_4, c_7, c_{16}, c_{29}, c_{34}$	85.29	0.9437
5	Heart	13	6	$c_1, c_2, c_3, c_5, c_8, c_{10}$	53.84	0.7213
6	WDBC	30	4	$c_8, c_{21}, c_{22}, c_{27}$	86.67	0.9649
7	German	24	2	c_4, c_6	91.67	0.6950
8	CMC	9	1	c_4	88.89	0.4644

It can be seen from Table 5 that the reduction rate of the three data sets Sonar, Ionosphere, Heart, and CMC in the table has been significantly improved. Among them, the reduction rate of 6 data sets has reached more than 80%, and the reduction rate of the Heart data set The reduction rate has also increased to about 50%. In terms of classification accuracy, Sonar, Ionosphere, Heart, and CMC have significantly improved the classification accuracy with the original fitness function based on the improvement of the reduction rate.

This paper improves the predecessor's fitness function formula, changes the predecessor's use of information entropy to calculate the importance of the attribute, and calculates the importance of the attribute more accurately. It also provides a new way to calculate the fitness function. Experiments have also proved that the new fitness function is better than the original fitness function.

In the algorithm comparison, according to Table 6, it can be seen that in addition to the Wine data set, the number of attribute reductions in the FSRSWOA algorithm is slightly less than that of the ABCNRT algorithm. The Heart and Ionosphere data sets are partially equal to other algorithms, which can be clearly seen in other data sets. The ABCNRT algorithm is better than the compared algorithm. It can be seen from Table 7 that the optimal classification accuracy of Zoo and Sonar data sets is better than the ABCNRT algorithm in other algorithms, and the classification accuracy of the ABCNRT algorithm in the remaining data sets is better than the comparison algorithm.

Table 6. Attribute reduction

Data set	PSORSFS	FSRSWOA	NRS	ABCNRT
Zoo	8[10], 9[10]	7[5], 8[12], 9[3]	5	**3[15], 4[5]**
Wine	3[5], 4[15], 5[5]	**3[12], 4[8]**	5	4
Sonar	9[15], 10[5]	8[13], 9[7]	16	**6[17], 7[3]**
Ionosphere	6[8], 7[12]	**5[13], 6[7]**	9	5[15], 6[5]
Heart	**6[15], 7[5]**	**6[9], 7[5], 8[6]**	**6**	**6**
WDBC	10[13], 11[7]	9[12], 10[8]	6	**4[18], 5[2]**
German	5[16], 6[4]	3[13], 4[7]	9	**2[10], 3[10]**
CMC	4[14], 5[6]	5[14], 6[3], 7[3]	9	**1**

Table 7. Optimal classification accuracy

Data set	PSORSFS	FSRSWOA	NRS	ABCNRT
Zoo	0.8447	**0.8746**	0.7425	0.7619
Wine	0.8764	0.8967	0.8764	**0.9444**
Sonar	0.7356	**0.7889**	0.6782	0.7619
Ionosphere	0.9192	0.9256	0.8576	**0.9437**
Heart	0.6972	0.7123	0.6516	**0.7213**
WDBC	0.8776	0.9425	0.8796	**0.9649**
German	**0.6951**	0.6920	0.5672	**0.6951**
CMC	0.3982	0.4173	0.3928	**0.4644**

In summary, it fully illustrates the advantages and reliability of the algorithm in this paper after comparing the other three algorithms.

4.3 Algorithm Performance Comparison

In order to further verify the advantages of the algorithm in this paper, the performance of the above-mentioned various algorithms is compared, and the experiment uses the methods of Friedman test and Nemenyi follow-up test to verify.

First, according to the classification accuracy in Table 4 and Table 7, sort and assign the classification accuracy of the 4 algorithms involved in the experiment on 8 data sets. Tables 8 and 9 show the classification accuracy of each algorithm in each data set. The ranking value of the test performance and its average value.

Table 8. The original function experimental algorithm performance ranking results

Data set	PSORSFS	FSRSWOA	NRS	ABCNRT
Zoo	2	1	4	3
Wine	3.5	2	3.5	1
Sonar	3	1	4	2
Ionosphere	3	2	4	1
Heart	2	1	3	4
WDBC	3	2	4	1
German	1.5	3	4	1
CMC	4	2	3	1
Average	2.75	1.75	3.6875	1.1825

Table 9. Improved function experiment algorithm performance ranking results

Data set	PSORSFS	FSRSWOA	NRS	ABCNRT
Zoo	2	1	4	3
Wine	3.5	2	3.5	1
Sonar	3	1	4	2
Ionosphere	3	2	4	1
Heart	3	2	4	1
WDBC	4	2	3	1
German	1.5	3	4	1.5
CMC	3	2	4	1
Average	2.875	1.875	3.8125	1.4375

Then use Friedman test to determine whether the above algorithms have the same performance. Through the above observations, there is no algorithm with the same performance. Suppose we compare n algorithms on M data sets, the commonly used variable $\tau_F = \frac{(M-1)\tau_X^2}{M(n-1)-\tau_X^2}$, where τ_F obeys the F distribution with $n-1$ and$(n-1)(M-1)$ degrees of freedom.

If the algorithm performance is not the same, use the Nemenyi follow-up test method, use the Nemenyi follow-up test to calculate the critical value range CD, $CD = q_\alpha \sqrt{\frac{n(n+1)}{6M}}$ of the average ordinal difference, where n is the number of algorithms and M is the number of data sets. When $n = 4$ and $\alpha = 0.1$ are given after looking up the table, the value of q_α is 2.291, so the critical value $CD = 1.4789$ is obtained.

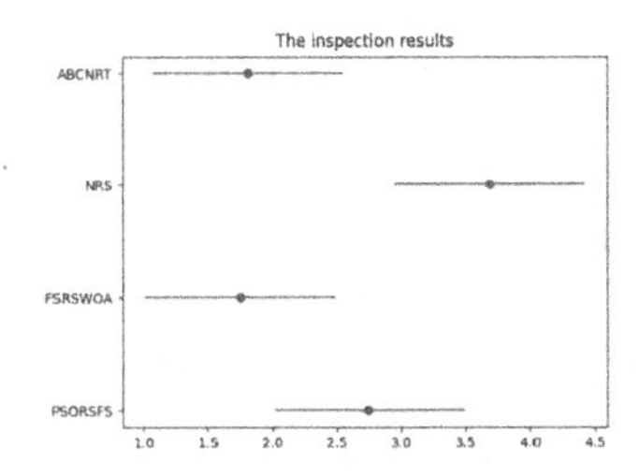

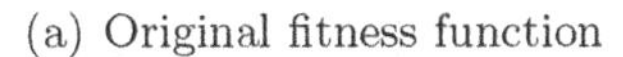

(a) Original fitness function

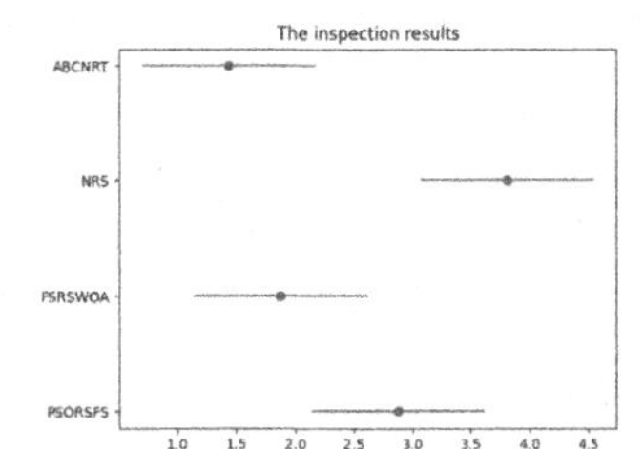

(b) Improve fitness function

Fig. 3. Inspection result graph

According to the result of the test (Fig. 3.), it can be seen that in the original fitness function test result graph, the performance of the FSRSWOA algorithm is slightly better than the ABCNRT algorithm, and the two are similar, but in the improved fitness function test result graph It can be clearly seen that the ABCNRT algorithm is better than the other three algorithms. It fully shows that the performance of the attribute reduction algorithm in this paper has significant advantages compared with other algorithms.

In summary, through the above experiments, it can be seen that the ABC-NRT algorithm proposed in this paper has more advantages in selecting the optimal attribute reduction and obtaining the optimal classification accuracy. In the above experiments, the algorithm in this paper has selected a better attribute reduction and obtained a better classification accuracy, which shows the effectiveness of the algorithm in this paper.

5 Concluding Remarks

This paper proposes an attribute importance measurement method based on discrimination matrix, and introduces the attribute importance as heuristic information into the bee colony algorithm, constructs a new fitness function, and designs based on artificial bee colony and discrimination matrix The attribute selection method of the optimization algorithm, this method avoids the redundant attributes in the searched attribute subset, reduces the number of iterations, and reduces the reduction time complexity. The experimental results of the UCI data set show that this method improves the reduction rate. After testing, it can be found that the attribute selection method in this paper has better performance and better classification accuracy. However, the algorithm in this paper also has shortcomings. First, it is unable to calculate the importance of attributes that do not appear in the neighborhood discrimination matrix; second, although the number of colony iterations is reduced, the importance of the neighborhood discrimination matrix must be calculated, which increases the amount of calculation. To further reduce the computational workload and find the optimal attribute subset is the method for future research.

References

1. Lin, T.Y.: Granular computing on binary relations I: data mining and neighborhood systems. Rough Sets Knowl. Disc. **18**(1), 107–121 (1998)
2. Hu, Q.H., Yu, D.R., Liu. J., et al.: Neighborhood rough set based heterogeneous feature subset selection. Inf. Sci. **178**(18), 3577–3594 (2008)
3. Liu, Y., Huang, W., Jiang, Y., et al.: Quick attribute reduct algorithm for neighborhood rough set model. Inf. Sci. **271**(7), 65–81 (2014)
4. Karaboga, D., Basturk, B.: A comparative study of artificial bee colony algorithm. Appl. Math. Comput. **214**(1), 108–132 (2009)
5. Tao, W., Fan, H.L.: Attribute reduction of rough set based on bee colony optimization. Comput. Meas. Control **20**(01), 193–195 (2012)
6. Wang, S.Q., et al.: Attribute reduction based on fuzzy rough set and bee colony algorithm. J. Central South Univ. (Sci. Technol.) **44**(01), 172–178 (2013)
7. Ye, D.Y., Chen, Z.J.: An efficient combined artificial bee colony algorithm for minimum attribute reduction problem. Acta Electronica Sinica **43**(05), 1014–1020 (2015)

8. Peng, X.R., Liu, Z.R., Gao, Y.: Artificial bee colony decision table reduction algorithm based on neighborhood rough model. J. Qingdao Univ. (Nat. Sci.) **30**(03), 55–59 (2017)
9. Fang, B., Chen, H.M., Wang, S.W.: Feature selection method based on rough set and fruit fly optimization algorithm. Comput. Sci. **46**(07), 157–164 (2019)
10. Gao, W.F., Huang, L.L., Wang, J., et al.: Enhanced artificial bee colony algorithm through differential evolution. Appl. Soft Comput. **48**, 137–150 (2016)
11. Wang, G.Y., Yao, Y.Y., Yu, H.: Chin. J. Comput. **32**(07), 1229–1246 (2009)
12. Wang, H., Wu, Z.J., Rahnamayan, S., et al.: Improving artificial Bee colony algorithm using a new neighborhood selection mechanism. Inf. Sci. **527**, 227–240 (2020)
13. Zhou, X.Y., Lu, J.X., Huang, J.H., Zhong, M.S., Wang, M.W.: Enhancing artificial bee colony algorithm with multi-elite guidance. Inf. Sci. **543**(prepublish), 242–258 (2021)
14. Wang, S.W., Chen, H.M.: Feature selection method based on rough set and improved whale optimization algorithm. Comput. Sci. **47**(02), 44–50 (2020)
15. Wang, X.Y., Yang, J., Teng, X.L., Xia, W.J., Jensen, R.: Feature selection based on rough sets and particle swarm optimization. Patt. Recogn. Lett. **28**(4), 459–471 (2006)
16. Ye, J., Wang, L.: Approach of ascertaining combinatorial attribute weight based on discernibility matrix. Comput. Sci. **41**(11), 273–277 (2014)
17. Cui, Z.H., Cao, Y., Cai, X.J., Cai, J.H., Chen, J.J.: Optimal LEACH protocol with modified bat algorithm for big data sensing systems in internet of things. J. Parallel. Distrib. Comput. **132**, 217–229 (2019)
18. Cui, Z.H., Chang, Y., Zhang, J.J., Cai, X.J., Zhang, W.S.: Improved NSGA-III with selection-and-elimination operator. Swarm. Evol. Comput. **49**, 23–33 (2019)

A Cuckoo Quantum Evolutionary Algorithm for the Graph Coloring Problem

Yongjian Xu and Yu Chen[✉]

School of Science, Wuhan University of Technology, Wuhan 430070, China
ychen@whut.edu.cn

Abstract. A typical combinatorial optimization problem, Graph Coloring Problem (GCP), has a wide range of applications in the fields of science and engineering. A cuckoo quantum evolutionary algorithm (CQEA) is proposed for the GCP, which is based on the framework of quantum-inspired evolutionary algorithm. To reduce iterations for the search of the chromatic number, the initial quantum population is generated with random initialization assisted by inheritance. Moreover, improvement of global exploration is achieved by incorporating the cuckoo search strategy, and a local search operation, as well as a perturbance strategy, is developed to enhance its performance on GCP. Numerical results show that CQEA has strong exploration and exploitation ability, and is competitive compared with the state-of-the-art heuristic algorithms.

Keywords: Graph coloring problem · Quantum-inspired evolutionary algorithm · Cuckoo search algorithm

1 Introduction

Graph Coloring Problem (GCP), one of the most important combinatorial problems, is used extensively in the scientific and engineering fields [1]. Due to its NP-completeness, heuristic algorithms have been widely investigated to develop efficient algorithms for large-scale GCP. GCP's state-of-the-art heuristic algorithms include the cuckoo search algorithm [2–5], the ant colony algorithm [6, 7], the genetic algorithm [8, 9], the memetic algorithm [10, 11] and the tabu search algorithm [12] and so on. Recently, the marriage in honey bees optimization [13] and the DNA algorithm [14] are also developed for GCP.

Most of heuristic algorithms employ population-based search schemes, which enhance their convergence performance, but has the side-effect of heavy complexity. The quantum-inspired evolutionary algorithm (QEA) [15] operates with a small population, and thus, is of less complexity compared to most of population-based heuristic algorithms. Unfortunately, the small population would inevitably reduce its ability of global exploration, and sometimes results in local convergence to problems with complicated landscapes.

L. Pan et al. (Eds.): BIC-TA 2021, CCIS 1565, pp. 88–99, 2022.
https://doi.org/10.1007/978-981-19-1256-6_7

Imitating the behavior of cuckoos, the cuckoo search algorithm (CSA) with a few parameters has been successfully applied in engineering applications [16–18]. Introducing the quantum representation to the framework of CSA, Djelloul et al. [2] proposed a quantum inspired cuckoo search algorithm for GCP. However, this research has shortcomings such as easy to fall into the local optimal solution and long running time [1].

In [19], a new memetic Teaching-Learning-Based Optimization algorithm, which is combined with tabu search algorithm was proposed for GCP. A one-level membrane systems membrane algorithm with dynamic operators for GCP was proposed in [20]. The discrete selfish herd optimizer for GCP was proposed in [21], which has few iterations and high success rate of coloring. Instead of trying to get the chromatic number, these researches try to find a coloring-scheme without conflicting edges for a given number of colors.

A greedy genetic algorithm was proposed in [22], which combines genetic algorithm and greedy sequential algorithm for GCP. A new enhanced binary dragonfly algorithm was proposed in [23] for GCP. However, performance of these algorithms depends on quite a few algorithm parameters, which results in the difficulty of their applications in various complicated GCP.

Recalling that QEA has low complexity arising from its small population size, we propose a novel cuckoo quantum evolutionary algorithm (CQEA) for GCP, where the cuckoo search (CS) schema is introduced to improve global exploration of QEA. Meanwhile, the inherited initialization strategy and special search strategies are developed to improve its performance on GCP.

The rest of this article is organized as follows. The formulation of GCP is presented in Sect. 2, and Sect. 3 elaborates details of CQEA for GCP. Performance of CQEA is validated in Sect. 4 by numerical experiments, and finally, our work is summarized in Sect. 5.

2 Problem Description

In this paper, the GCP refers to the problem of vertex coloring. Given an undirected graph $G(V, E)$, where the vertex set is represented by V and the edge set is represented by E, the GCP is dedicated to minimize the amount of colors to paint adjacent vertices by inconsistent colors. If k different colors are sufficient to coloring adjacent vertices with different colors, $G(V, E)$ is called k-colorable. The smallest value of k such that G is called k-colorable is named as the *chromatic number* of $G(V, E)$.

If not confused, an undirected graph $G(V, E)$ is denoted as G. GCP to get the *chromatic number* of G can be model as a bi-level combinatorial optimization problem:

$$
\min k
$$

$$
s.t.\quad \min f(S) = \sum_{(u,v)\in E} \delta_{uv} = 0,
$$

$$
s.t.\begin{cases} \delta_{uv} = \begin{cases} 1, & \textit{if } i = j, \\ 0, & \textit{if } i \neq j, \end{cases} \quad u \in V_i, v \in V_j; \\ S = (V_1, V_2, \cdots, V_k),\ \bigcup_{i=1}^{k} V_k = V,\ V_i \cap V_j = \emptyset, i \neq j; \\ i,j = 1, 2, \cdots, k. \end{cases} \tag{1}
$$

where $S = (V_1, V_2, \cdots, V_k)$ is a partition of V that represents an assignment of colors of vertices. When $\delta_{uv} = 1$, incident vertices of edge (u, v) are assigned the consistent color, and (u, v) is named as a *conflicting edge*. Then, $f(S)$, the total amount of conflicting edge of S, is the objective function that evaluates qualities of color assignments. To get the *chromatic number* and the corresponding color assignment S, one must iteratively search the graph to minimize $f(S)$ for successively increasing/decreasing k, which is tedious and time consuming.

3 The Cuckoo Quantum Evolutionary Algorithm for GCP

In this section, we first introduce the representation of the solution and outline the framework of the cuckoo quantum evolutionary algorithm. Then, details of population initialization, search strategy, cuckoo search operation, and perturbance strategy are presented.

3.1 Representation of the Solution to the Graph Coloring Problem

For an undirected simple graph G of n vertices, a color assignment with k colors can be represented by a $k \times n$ binary matrix, where each column includes only one "1" bit [2]. Figure 1 illustrates an example that a graph of five vertices is colored by three colors. For a graph included in Fig. 1(a), the assignment of color illustrated in Fig. 1(b) can be represented by the 3×5 binary matrix in Fig. 1(c). When nodes 1 and 4 are colored by the first color (namely yellow), elements at the first row of columns 1 and 4 are set to be "1"; for nodes 2 and 3 painted by the second color (namely green), and node 5 given the third color (namely blue), we know the "1"-bits in columns 2, 3 and 5 are located at rows 2, 2 and 3, respectively.

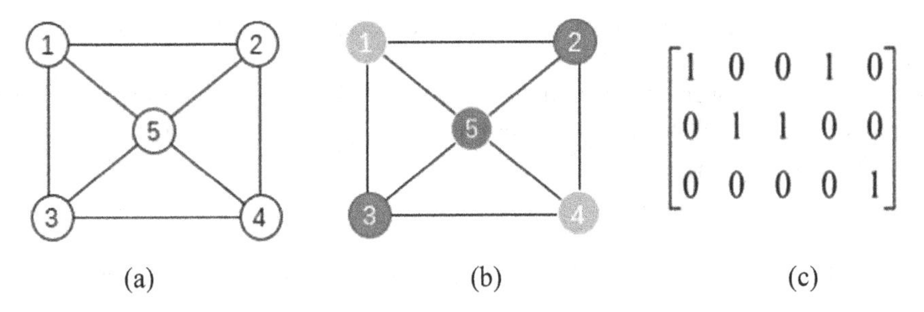

Fig. 1. (a) an undirected simple graph; (b) an assignment of color by 3 colors; (c) binary representation of color assignment.

Since a binary matrix **x** representing an assignment of color is equivalent to a partition S of the vertex set V, its quality can be evaluated by the number of conflicting edges, that is,

$$f(\mathbf{x}) = f(S). \tag{2}$$

3.2 Quantum Matrix in the Cuckoo Quantum Evolutionary Algorithm

In cuckoo quantum evolutionary algorithm (CQEA), individuals are not binary matrices of coloring assignments but quantum matrices that present probability models about distributions of colors. An assignment of K colors to N vertices can be modelled by a $2K \times N$ quantum matrix.

$$\mathbf{q} = \left[\begin{array}{c|c|c|c} \alpha_{11} & \alpha_{12} & \cdots & \alpha_{1N} \\ \beta_{11} & \beta_{12} & \cdots & \beta_{1N} \\ \vdots & \vdots & \vdots & \vdots \\ \alpha_{K1} & \alpha_{K2} & \cdots & \alpha_{KN} \\ \beta_{K1} & \beta_{K2} & \cdots & \beta_{KN} \end{array}\right],$$

where $\alpha_{i,j}^2 + \beta_{i,j}^2 = 1,\ i = 1,\ldots,K, j = 1,\ldots,N.$ $\alpha_{i,j}^2$ denotes the probability to assign the ith color to vertex j, and $\beta_{i,j}^2$ denotes the probability not to assign the ith color to vertex j. By performing the quantum measurement operation [15] on **q**, a binary matrix **x** can be generated. Since there could be more than one "1" in a column of **x**, a repair operation is necessary to transfer the generated **x** to a solution of GCP.

3.3 Framework of the Cuckoo Quantum Evolutionary Algorithm

The CQEA for GCP consists of two nested iterations. The inner loop tries to minimize number of conflicting edges for given color number K, and the external one is performed to minimize color numbers K to get the chromatic number of the graph G. Termination criterions of loops are set as follows.

If iteration number of the external loop reaches a preset value or the chromatic number of G is obtained, termination-condition 1 is satisfied; it quits the inner loop if the maximum number of generations is reached or a color assignment with no conflicting edges is obtained. Framework of the CQEA is presented as follows.

```
Framework of the CQEA
Begin
    gen ← 0;
    Initialize the color number K and set solution archive
B(gen)=∅;
    while (termination-condition 1 is not satisfied) do
        t ← 0;
        Initialize the quantum population Q(t) and the so-
lution population P(t);
        evaluate P(t) and update the solution archive
B(gen);
        perform the local search on B(gen);
        while (termination-condition 2 is not satisfied)
do
            t ← t+1;
            generate P(t) by Q(t−1);
            perform the local search on P(t);
            evaluate P(t) and update B(gen);
            Generate Q(t) by applying the cuckoo search
operation to Q(t−1);
            apply the perturbance strategy to update Q(t)
and B(gen);
        End While
        If f(B(gen)) = 0
            K=K-1;
        End If
        gen ← gen+1;
    End While
End
```

Note that CQEA performs iterations on with its quantum population and the solutions to GCP are binary matrices. Thus, it generates binary matrices by implementing the quantum measurement operation on quantum matrices [15]. In the quantum measurement operation, an infeasible color assignment could be generated with uncolored nodes or nodes painted by more than one colors [2]. Then, a feasible color assignment is generated by the repair operation detailed as follows.

1. For a vertex with multiple colors, randomly choose one to color the vertex;
2. for a vertex that is not assigned with a color, randomly choose a color to paint the vertex.

After generation of the solution population, a binary matrix $\mathbf{x}$ that represents an assignment of color is evaluated by (2). While the solution archive is empty, the best solution with the least number of conflicting edges in $P(t)$ is put into $B(gen)$; if is not empty, the best solution in $P(t) \cup B(gen)$ is kept in $B(gen)$.

3.4 Initialization of the CQEA

Initialization of Color Number K

Since the CQEA is implemented by successive iterations of the outer loop, promising initialization of K is helpful to reduce the number of iterations to get the chromatic number. At the beginning, it sets $K = \Delta(G) + 1$ according to the following theorem [24].

Theorem 1. Given undirected simple graph G, it holds $k \leq \Delta(G) + 1$, where $\Delta(G)$ is the degree of G, and k is the chromatic number of G.

Initialization of the Quantum Population and the Solution Population

There are two different strategies employed to initialize the quantum population. While $gen = 0$, the quantum population is randomly initialized; when $gen > 0$, it is generated by the inheritance initialization strategy.

Random Initialization. At the beginning, the state expressed by each individual in the population is the equal probability superposition of all its possible states [2, 15]. By setting $K = \Delta(G) + 1$, each individual in the initial quantum population is taken as

$$\left[\begin{array}{c|c|c|c} \alpha_{11} & \alpha_{12} & \cdots & \alpha_{1N} \\ \beta_{11} & \beta_{12} & \cdots & \beta_{1N} \\ \vdots & \vdots & \vdots & \vdots \\ \alpha_{K1} & \alpha_{K2} & \cdots & \alpha_{KN} \\ \beta_{K1} & \beta_{K2} & \cdots & \beta_{KN} \end{array}\right] = \left[\begin{array}{c|c|c|c} 1/\sqrt{2} & 1/\sqrt{2} & \cdots & 1/\sqrt{2} \\ 1/\sqrt{2} & 1/\sqrt{2} & \cdots & 1/\sqrt{2} \\ \vdots & \vdots & \vdots & \vdots \\ 1/\sqrt{2} & 1/\sqrt{2} & \cdots & 1/\sqrt{2} \\ 1/\sqrt{2} & 1/\sqrt{2} & \cdots & 1/\sqrt{2} \end{array}\right]. \quad (3)$$

Once the quantum population $Q(0)$ is generated, the solution population $P(0)$ can be generated by perform quantum measurement on each individual of $Q(0)$.

Inheritance Initialization. If CQEA has successfully colored G by K colors, it attempts to color G by $K - 1$ colors. To further improve searching efficiency of the upcoming iteration, the inheritance initialization is performed to generate an initial quantum population, which can be implemented by obtained best binary matrix A and the corresponding quantum matrix Q_A obtained with K feasible colors. Detailed operations of the inheritance initialization are described as follows.

1. Take the best binary matrix A and the corresponding quantum matrix Q_A obtained by $K(K \leq \Delta(G)+1)$ colors. Locate zero rows[1] of the binary matrix A, and denote row indexes of zero rows as $n_j, j = 1, \ldots, row_0$.
2. If $row_0 > 0$, G can be colored by $K - row_0$ colors. Set $K = K - row_0$. Delete zero rows of A and the corresponding rows of Q_A, *i.e.*, rows $2n_j - 1, 2n_j (j = 1, \ldots, row_0)$ of Q_A.
3. If row_0=0, locate a row of A that has the least number of "1" and label it as row n_0. Since row n_0 has the least number of "1", we try to delete the corresponding color to get the initial coloring of iteration g. Then, delete row n_0 of A and rows $2n_0 - 1, 2n_0$ of Q_A, where the number of colors is $K - 1$.

In this way, we can get an updated quantum matrix Q_A and the corresponding binary matrix A. Take them as individuals in $Q(0)$ and $P(0)$, respectively. Other individuals of $Q(0)$ are generated by random initialization. Performing quantum measurement on $Q(0)$, one can get the other members of $P(0)$.

3.5 Local Search in the Solution Space

Inspired by the mutation operator proposed in [2] and the simple decentralized graph coloring (SDGC) algorithm proposed in [25], we introduce a local search performed in the solutions space to improve local exploitation of CQEA. Given a set P of binary matrix, the local search is performed for every $\mathbf{x} \in P$.

1. randomly traverse vertices of graph G; according to the color assignment represented by $\mathbf{x}$, record the collection of alternative colors AC_j for vertex v_j, $j = 1, 2, \ldots, n$;
2. generate a candidate solution $\mathbf{y}$ as follows. For vertex v_j with nonempty AC_j, assign the smallest color index in AC_j to v_j if P is the solution archive $B(gen)$; otherwise, randomly assign a color index in AC_j to v_j;
3. replace $\mathbf{x}$ by $\mathbf{y}$ if $f(\mathbf{y}) < f(\mathbf{x})$.

Note that the local search applies different strategies when adjusting color indexes of vertices. While it is applied to the best solution set $B(gen)$, it is assigned the smallest available color index to attempt to reduce the amount of color number K; however, when searching neighborhoods of $P(t)$, a randomly selected color index is selected to try to improve global convergence by introducing randomness to the result.

The local search is repeated for *Limit* times before a solution without conflicting edges is generated.

3.6 The Cuckoo Search

Global exploration of CQEA is enhanced by introducing the cuckoo search (CS) to the quantum population [16–18]. For a quantum population Q, the cuckoo search to generate new quantum population Q_{new} and associated update of $B(gen)$ are performed for each quantum matrix $\mathbf{q} \in Q$.

[1] Zero rows are rows of a matrix where all elements are equal to "0".

1. perform the Levy flight on $\mathbf{q}$ to generate new $\mathbf{q}_{new}$; generate binary matrix $\mathbf{x}$ and $\mathbf{x}_{new}$ by $\mathbf{q}$ and $\mathbf{q}_{new}$, respectively;
2. evaluate $\mathbf{q}$ and $\mathbf{q}_{new}$ by $f(\mathbf{q}) = f(\mathbf{x})$ and $f(\mathbf{q}_{new}) = f(\mathbf{x}_{new})$, respectively;
3. if $f(\mathbf{q}_{new}) < f(\mathbf{q})$, replace $\mathbf{q}$ and $\mathbf{x}$ by $\mathbf{q}_{new}$ and $\mathbf{x}_{new}$, respectively; otherwise, perform a random walk for $\mathbf{q}_{new}$ and generate a $\mathbf{x}_{new}$ to replace $\mathbf{q}$ and $\mathbf{x}$;
4. if $\mathbf{x}_{new}$ is better than a binary matrix $\mathbf{y} \in B(gen)$, replace y by $\mathbf{x}_{new}$.

3.7 The Perturbance Strategy

The quantum population $Q(t)$ and archive $B(new)$ are further updated by performing the perturbance strategy.

1. According to the color assignment represented by $B(new)$, traverse the graph to locate all conflict vertices of G.
2. Sort quantum matrices $\mathbf{q}$ in $Q(t)$ in ascending order of $f(\mathbf{q})$.
3. For all quantum matrices belonging to the second half of sorted $Q(t)$, resort elements corresponding to the conflicting vertices to $1/\sqrt{2}$.
4. For conflicting vertex v of G, denote the collection of its adjacent vertices as V_v.
5. For $\mathbf{x} \in B(new)$, assign the same color to the non-adjacent vertices in V_v; Record available colors of v by AC_v; If $AC_v \neq \emptyset$, randomly assign a color in AC_v to v when the color of v is not in AC_v.

4 Experimental Results

In order to evaluate the efficiency of the CQEA, it is implemented in MATLAB R2020a. The running environment is Intel(R) Core(TM) i7 CPU 860 @ 2.80 GHz, system memory 8GB and Win7 operating system. Benchmark problems are selected from the graph coloring instance library[2]. Numerical results for 10 independent runs are collected in Table 1, where the obtained color number and iterations are recorded.

Table 1. Numerical results of CQEA on benchmark graph coloring problems

Problems				Population size	Number of colors				Ieterations
Instance	$\|V\|$	$\|E\|$	Best known		Min	Max	Mean	Std	Mean
myciel3	11	20	4	10	4	4	4	0	1
myciel4	23	71	5	10	5	5	5	0	1
myciel5	47	236	6	10	6	6	6	0	1
myciel6	95	755	7	10	7	7	7	0	1
myciel7	191	2360	8	10	8	8	8	0	1
queen5_5	25	160	5	10	5	5	5	0	2.8

(*continued*)

[2] http://mat.gsia.cmu.edu/COLOR/instances.html.

Table 1. (*continued*)

Problems				Population size	Number of colors				Ieterations
Instance	$\|V\|$	$\|E\|$	Best known		Min	Max	Mean	Std	Mean
queen6_6	36	290	7	10	7	8	7.7	0.458	87.5
huck	74	301	11	6	11	11	11	0	1
jean	80	254	10	6	10	10	10	0	1
david	87	406	11	6	11	11	11	0	1.1
games120	120	638	9	6	9	9	9	0	1
miles250	128	387	8	6	8	8	8	0	8.7
miles500	128	1170	20	6	20	21	20.5	0.5	1
anna	138	493	11	6	11	11	11	0	1
fpsol2.i.1	496	11654	65	6	65	65	65	0	1
fpsol2.i.2	451	8691	30	6	30	30	30	0	1
fpsol2.i.3	425	8688	30	6	30	30	30	0	1
mulsol.i.1	197	3925	49	6	49	49	49	0	1
mulsol.i.2	188	3885	31	6	31	31	31	0	1
mulsol.i.3	184	3916	31	6	31	31	31	0	1
mulsol.i.4	185	3946	31	6	31	31	31	0	1
mulsol.i.5	186	3973	31	6	31	31	31	0	1

Numerical results in Table 1 indicate that the CQEA can successfully get the found the known optimal solution of 22 instances. Besides 20 problems that are optimized by CQEA with 100% success rate, the worst results of queen6_6 and miles500 are one greater than the known optimal results.

Competitiveness of CQEA is demonstrated by comparing it with state-of-the-art algorithms [2, 3, 9]. Table 2 tabulates the results on color number of these algorithms. It is shown that CQEA and a new hybrid genetic algorithm in [9] perform better on "queen6_6" than the other two. In this paper, the new hybrid genetic algorithm in [9] is abbreviated as NHGA.

Since both CQEA and NHGA can find the currently known optimal solutions for 22 instances in Table 2, we compare their stability by success rates to get the best known number of colors. Results in Table 3 show that CQEA outperforms NHGA in terms of stability in 12 instances (namely myciel6, myciel7, queen6_6, miles250, anna, fpsol2.i.1, fpsol2.i.2, fpsol2.i.3, mulsol.i.1, mulsol.i.2, mulsol.i.3, mulsol.i.4, mulsol. i.5), however, performs a bit worse than NHGA on instance queen6_6. We conclude that CQEA is competitive to NHGA, and more efficient than MCOA and QICSA.

Table 2. Comparison between CQEA and state-of-the-art algorithms

Instance	*k*			
	CQEA	MCOA[3]	NHGA[9]	QICSA[2]
myciel3	4	4	4	4
myciel4	5	5	5	5
myciel5	6	6	6	–
myciel6	7	7	7	–
myciel7	8	8	8	–
queen5_5	5	5	5	5
queen6_6	**7**	8	**7**	8
huck	11	11	11	11
jean	10	10	10	10
david	11	11	11	11
games120	9	9	9	9
miles250	8	8	8	8
miles500	20	20	20	20
anna	11	11	11	11
fpsol2.i.1	65	65	65	65
fpsol2.i.2	30	30	30	–
fpsol2.i.3	30	30	30	–
mulsol.i.1	49	49	49	–
mulsol.i.2	31	31	31	–
mulsol.i.3	31	31	31	–
mulsol.i.4	31	31	31	–
mulsol.i.5	31	31	31	–

Table 3. Comparison of success rate between CQEA and NHGA[9]

Instance	Success rate	
	CQEA	NHGA [9]
myciel6	**15/15**	11/15
myciel7	**15/15**	9/15
queen6_6	4/15	**15/15**
miles250	**15/15**	11/15
anna	**15/15**	12/15
fpsol2.i.1	**15/15**	2/15
fpsol2.i.2	**15/15**	6/15
fpsol2.i.3	**15/15**	4/15
mulsol.i.1	**15/15**	2/15
mulsol.i.2	**15/15**	7/15
mulsol.i.3	**15/15**	5/15
mulsol.i.4	**15/15**	4/15
mulsol.i.5	**15/15**	7/15

5 Conclusion

Considering that the quantum-inspired evolutionary algorithm can perform well with a small population, this paper develops a cuckoo quantum evolutionary algorithm (CQEA) to efficiently address the graph coloring problem. Enhancement of global exploration is achieved by introduction of general cuckoo search and specific strategies specially designed for the GCP, including an inherited initialization strategy, a local search strategy and a perturbance strategy. Numerical comparison based on benchmark problems demonstrates that the proposed CQEA can outperform most of the state-of-the-art algorithms. The results also indicate that development of quantum-based heuristic algorithms could be a feasible solution to efficient address of large-scale GCP. To further improve the performance of CQEA on large-scale GCP, our future work would focus on designing efficient representation of solution and reduction strategy of graph scale.

References

1. Mostafaie, T., Modarres, F., Navimipour, N.J.: A systematic study on meta-heuristic approaches for solving the graph coloring problem. Comput. Oper. Res. **120**, 104850 (2020)
2. Djelloul, H., Layeb, A., Chikhi, S.: Quantum inspired cuckoo search algorithm for graph colouring problem. Int. J. Bio-Inspired Comput. **7**, 183–194 (2015)
3. Mahmoudi, S., Lotfi, S.: Modified cuckoo optimization algorithm (MCOA) to solve graph coloring problem. Appl. Soft Comput. **33**, 48–64 (2015)
4. Aranha, C., Toda, K., Kanoh, H.: Solving the graph coloring problem using cuckoo search. In: Tan, Y., Takagi, H., Shi, Y. (eds.) ICSI 2017. LNCS, vol. 10385, pp. 552–560. Springer, Cham (2017). https://doi.org/10.1007/978-3-319-61824-1_60
5. Zhou, Y., Zheng, H., Luo, Q., Wu, J., Guangxi, N.: An improved cuckoo search algorithm for solving planar graph coloring problem. Appl. Math. Inf. Sci. **7**, 785–792 (2013)
6. Silva, A.F., Rodriguez, L.G., Filho, J.F.: The improved Colour Ant algorithm: a hybrid algorithm for solving the graph colouring problem. Int. J. Bio-Inspired Comput. **16**, 1–12 (2020)
7. Mohammadnejad, A., Eshghi, K.: An efficient hybrid meta-heuristic ant system for minimum sum colouring problem. Int. J. Oper. Res. **34**, 269–284 (2019)
8. Marappan, R., Sethumadhavan, G.: Solution to graph coloring problem using divide and conquer based genetic method. 2016 International Conference on Information Communication and Embedded Systems (ICICES), 1–5 (2016)
9. Douiri, S.M., Elbernoussi, S.: Solving the graph coloring problem via hybrid genetic algorithms. J. King Saud Univ. Eng. Sci. **27**, 114–118 (2015)
10. Lü, Z., Hao, J.: A memetic algorithm for graph coloring. Eur. J. Oper. Res. **203**, 241–250 (2010)
11. Moalic, L., Gondran, A.: Variations on memetic algorithms for graph coloring problems. J. Heuristics **24**(1), 1–24 (2017). https://doi.org/10.1007/s10732-017-9354-9
12. Hertz, A., Werra, D.: Using tabu search techniques for graph coloring. Computing **39**, 345–351 (2005)
13. Bessedik, M., Toufik, B., Drias, H.: How can bees colour graphs. Int. J. Bio-Inspired Comput. **3**, 67–76 (2011)

14. Wang, Z., Wang, D., Bao, X., Wu, T.: A parallel biological computing algorithm to solve the vertex coloring problem with polynomial time complexity. J. Intell. Fuzzy Syst. **40**, 3957–3967 (2021)
15. Han, K., Kim, J.: Quantum-inspired evolutionary algorithm for a class of combinatorial optimization. IEEE Trans. Evol. Comput. **6**, 580–593 (2002). https://doi.org/10.1109/TEVC.2002.804320
16. Yang, X., Deb, S.: Engineering optimisation by cuckoo search. Int. J. Math. Modell. Numerical Optim. **1**, 330–343 (2010)
17. Santillan, J.H., Tapucar, S., Manliguez, C., Calag, V.: Cuckoo search via Lévy flights for the capacitated vehicle routing problem. J. Ind. Eng. Int. **14**(2), 293–304 (2017). https://doi.org/10.1007/s40092-017-0227-5
18. Yang, X.: Cuckoo search for inverse problems and simulated-driven shape optimization. J. Comput. Methods Sci. Eng. **12**, 129–137 (2012)
19. Dökeroglu, T., Sevinç, E.: Memetic teaching-learning-based optimization algorithms for large graph coloring problems. Eng. Appl. Artif. Intell. **102** 104282 (2021)
20. Andreu-Guzmán, J.A., Valencia-Cabrera, L.: A novel solution for GCP based on an OLMS membrane algorithm with dynamic operators. J. Membrane Comput. **2**(1), 1–13 (2019). https://doi.org/10.1007/s41965-019-00026-x
21. Zhao, R., et al.: Discrete selfish herd optimizer for solving graph coloring problem. Appl. Intell. **50**(5), 1633–1656 (2020). https://doi.org/10.1007/s10489-020-01636-0
22. Basmassi, M.A., Benameur, L., Chentoufi, J.A.: A novel greedy genetic algorithm to solve combinatorial optimization problem. ISPRS - International Archives of the Photogrammetry, Remote Sensing and Spatial Information Sciences, pp. 117–120 (2020)
23. Baiche, K., Meraihi, Y., Hina, M.D., Ramdane-Cherif, A., Mahseur, M.: Solving graph coloring problem using an enhanced binary dragonfly algorithm. Int. J. Swarm Intell. Res. **10**, 23–45 (2019)
24. West, D.B.: Introduction to Graph Theory, 2nd edn. Prentice Hall, Upper Saddle River (2001)
25. Galán, S.F.: Simple decentralized graph coloring. Comput. Optim. Appl. **66**(1), 163–185 (2016). https://doi.org/10.1007/s10589-016-9862-9

Feature Selection Algorithm Based on Discernibility Matrix and Fruit Fly Optimization

Jiaxin Ao[1], Jun Ye[2(✉)], Yuxuan Ji[1], and Zhenyu Yang[1]

[1] College of Information Engineering, Nanchang Institute of Technology, Nanchang 330000, China
[2] Jiangxi Province Key Laboratory of Water Information Cooperative Sensing and Intelligent Processing, Nanchang 330000, China
2003992646@nit.edu.cn

Abstract. Feature selection methods that combine swarm intelligence optimization algorithms with rough sets have been widely used. Normally, this type of method can quickly search for feature subsets, effectively reducing the computational workload; However, in some decision tables of this type of method, there were also problems such as the searched feature subset containing redundant features, which led to a decrease in accuracy. Aiming at such problems in the current feature selection algorithms based on rough sets and fruit fly optimization, improvements have been made from two aspects. Firstly, a method for measuring the importance of discernibility matrix attributes was proposed and used as a heuristic factor to constructed a new fitness function; Secondly, guided by the fitness function, a fruit fly optimized feature selection method was designed, which can avoid redundant attributes in the feature subset. Examples and experimental results on the UCI datasets show that the new feature selection method can effectively search for the smallest feature subset and improve the stability.

Keywords: Discernibility matrix · Fruit fly optimization algorithm · Attribute importance · Feature subset · Fitness function

1 Introduction

Feature selection can achieve the preprocessing of the data, rough set theory [1] has displayed strong ability in data dimensionality reduction, and the feature selection method using attribute importance as heuristic information is the most widely used dimensionality reduction method in rough set. However, this type of method has a huge computational workload when processing massive and

Supported by the National Natural Science Foundation of China (Nos. 61562061) and Technology Project of Ministry of Education of Jiangxi province of China (No.GJJ211920, GJJ170995).

L. Pan et al. (Eds.): BIC-TA 2021, CCIS 1565, pp. 100–115, 2022.
https://doi.org/10.1007/978-981-19-1256-6_8

high-dimensional data, and searched feature subset is inaccurate [2]. Therefore, researchers introduced swarm intelligence optimization algorithms in rough set theory, and proposed some optimized feature selection methods. For example, [3] proposed a combined artificial bee colony algorithm to efficiently solve the minimum feature subset problem. The algorithm used the attribute dependence in rough set theory as heuristic information to avoid invalid local search and save search time. [4] used information entropy in rough set theory to describe importance, and proposed an attribute reduction algorithm based on information entropy and ant colony optimization. [5] used the sensitivity of features to different datasets to find the optimal genes, and used the attribute importance and dependence of rough sets as heuristic information, and proposed a feature selection algorithm based on ant colony optimization. [6] raised a particle swarm algorithm in view of information entropy, which solved the discrete issue in the feature selection strategy by modified the fitness function in the algorithm, so as to find the optimal feature subset. These algorithms improved the search speed and accuracy to varying degrees.

The Fruit Fly Optimization Algorithm (FOA) is an optimized evolutionary algorithm proposed by PAN [7] to simulate the foraging behavior of fruit flies. Compared with swarm intelligence algorithms such as particle swarm algorithm and ant colony algorithm, the fruit fly optimization algorithm has simple operation and fewer parameters, especially has advantages in solving the extreme value of mathematical functions [8,9]. But fruit fly optimization algorithms also have limitations such as low optimization accuracy and easy to fall into local optima, so many researchers have proposed improved algorithms. For example, [10] proposed a hierarchical guidance strategy-assisted fruit fly optimization algorithm with a collaborative learning mechanism, introducing hierarchical guidance strategies for local search, and assigning inferior individuals to different levels of elite individuals for exploration and development. The cooperation between the inferior subgroups and the elite subgroups balanced the algorithm and improved the algorithm's global search capabilities. [11] introduced an improved dual-strategy evolutionary fruit fly optimization algorithm into rough set theory, and proposed a feature selection algorithm based on rough sets and fruit fly optimization. This method used the information entropy attribute importance in rough set theory as the heuristic search factor, constructed a new fitness function, and reduced the number of iterations to search for feature subsets. But this method have redundant attributes in the feature subsets searched in some decision tables, resulting in a significant decrease in search accuracy. Because the importance of information entropy is measured by the change of uncertainty domain from a quantitative point of view, any change in an attribute may cause change of uncertainty domain. Although the information entropy method improves the situation that the importance of non-core attributes is zero in algebraic rough set, there are still situations where the importance of redundant features is greater than the importance of feature kernels [12].

Aiming at the deficiencies in [11], this paper has proposed a method for measuring the importance of attributes of discernibility matrix. And used this to replace the attribute importance of information entropy to construct a new fitness function. On this basis, a feature selection strategy based on discernibility matrix and fruit fly algorithm was designed. The feature subset searched by this method avoids redundant attributes and provides a way of thinking for feature selection methods.

2 Related Knowledge

2.1 Rough Set Theory

For better understanding, the rough set knowledge related to this paper is briefly introduced below. For more detailed knowledge, please refer to [1].

Decision information system $S = (U, C \cup D, L, g)$, Where U is the universe of discourse, C is the conditional attribute set, D is the decision attribute set, and L is the attribute value set; g is an info equation, namely $g : U \times (C \cup D) \rightarrow L$, Attribute subset $K \subseteq (C \cup D)$ determines an indistinguishable relationship $IND(K) = (m,n) \in U \times U | \forall p, q \in K, g(m,p) = g(m,q)$, $IND(K)$ constitutes an equivalence class division of U, abbreviated as U/K

Definition 1 *[1]. (Degree of dependence) Decision information system $S = (U, C \cup D, L, g)$, and the dependency between condition attribute set C and decision attribute set D is defined as:*

$$r(C,D) = \frac{|POS_C(D)|}{|U|} \tag{1}$$

Researchers introduced information theory into rough set, and put forward a viewpoint based on information theory rough set theory [12]. Knowledge can be regard as a random variable in rough set theory, and it is a method defined by importance measurement in rough set of information theory.

Definition 2 *[13]. (Information Entropy) Given knowledge P and its probability distribution, the information entropy of P can be expressed as:*

$$H(P) = -\sum_{i=1}^{t} p(Z_i) log_2 p(Z_i) \tag{2}$$

Definition 3 *[13]. (Conditional Entropy) Given knowledge P and Q and their respective probability distributions and conditional probability distributions, the conditional entropy of Q comparative to P is:*

$$H(Q|P) = -\sum_{i=1}^{t} p(Z_i) \sum_{j=1}^{v} p(W_j|Z_i) log_2 p(W_j|Z_i) \tag{3}$$

Definition 4 *[13]. (Attribute Importance) Decision information system $S = (U, C \cup D, L, g)$, where U is the universe of discourse, C is the conditional attribute set, D is the decision attribute set. There are two methods for calculating the importance of attributes: deletion method and addition method. The equations are:*

$$\forall c \in C, sig(c, C, D) = H(D \mid C - \{c\}) - H(D \mid C) \tag{4}$$

$$\forall c \in C - B, B \subseteq C, sig(c, B, D) = H(D \mid B) - H(D \mid B \cup \{c\}) \tag{5}$$

It can be seen from Definition 4 that the influence of the uncertain classification subset in the universe is a prerequisite for defining the importance of the conditional entropy attribute. By adding or deleting a certain attribute and examining the changes of the uncertainty domain, it is found that the change of any one attribute may change the uncertainty domain. Therefore, the conditional entropy attribute importance measurement method avoids the situation that the importance of the non-core attributes in the rough set of algebraic view are all zero. However, in some incompatible decision tables, there are errors in the attribute importance calculated by this method, that is, the importance of redundant features is greater than the importance of core features. Therefore, the subset of features searched by the fruit fly algorithm may contain redundant features.

2.2 Fruit Fly Optimization Algorithm

According to the different operation contents, fruit fly optimization algorithm [7] can be divided into the following six stages:

(1) Initialization phase: Initialize the population size *pop*, the maximum number of iterations N_{max}, and randomly initialize the initial position (M, N) of the fruit fly.
(2) Searching stage: All fruit fly individuals have random search distance and search direction(V);

$$\begin{cases} M_i = M + V \\ N_i = N + V \end{cases} \tag{6}$$

(3) The length *Len* from individual drosophila to the origin can be calculated by the Pythagorean theorem, and the taste density determination value S and *Len* are the reciprocal of each other;

$$\begin{cases} Len_i = \sqrt{M^2 + N^2} \\ S_i = 1/Dist \end{cases} \tag{7}$$

(4) Substitute S into the fitness function $fitness$ to get the taste concentration $flavor_i = fitness(S_i)$, and record the taste concentration value of the optimal individual ($flavor_{best}$) and the corresponding position ($index$);

$$[flavor_{best}, index] = flavor'_{best} \tag{8}$$

(5) The fruit fly population then moved closer to the best individual $flavor_{best}$;

$$\begin{cases} M = M(index) \\ N = N(index) \end{cases} \tag{9}$$

(6) Iteration phase: Repeat the search phase until the number of iterations $N = N_{max}$.

3 Feature Selection Algorithm Based on Discernibility Matrix and Fruit Fly Optimization

3.1 Method of Defining Attribute Importance Based on Discernibility Matrix

A. Skowron [14] first proposed to use discernibility matrix to express knowledge system, which is more concise and efficient than other knowledge expression systems, which are defined as follows:

Definition 5 *[14]. (Discernibility matrix) Decision information system $S = (U, C \cup D, L, g)$, where C is the conditional attribute set, D is the decision attribute set, domain $U = \{m_1, m_2, \cdots, m_t\}$, and U satisfies the non-empty finite set condition, $|U| = t$, then the discernibility matrix of the knowledge expression system is:*

$$M_{t\times t} = (c_{ij})_{t\times t} = \begin{bmatrix} c_{11} & c_{12} & \cdots & c_{1t} \\ c_{21} & c_{22} & \cdots & c_{2t} \\ \vdots & \vdots & \ddots & \vdots \\ c_{t1} & c_{n2} & \cdots & c_{tt} \end{bmatrix}$$

In the above formula $\forall i, j = 1, 2, \cdots, t$:

$$c_{ij} = \begin{cases} \{p \mid (p \in C) \cap (f(m_i) \neq f(n_j))\}, & f_D(m_i) \neq f_D(m_i) \\ \emptyset, & f_D(m_i) = f_D(m_i) \end{cases} \tag{10}$$

Element c_{ij} is used to distinguish the set composed of m_i and m_j, when $m_i = m_j, c_{ij} = \emptyset$;

Because $c_{ij} = c_{ji}(i, j = 1, 2, \cdots, t)$, the discernibility matrix is the main diagonal symmetric matrix, so the discernibility matrix is usually written as an upper triangular matrix or a lower triangular matrix. Theorem 1 in [15] gave a method to found the core attribute of discernibility matrix. This method is simple and efficient. The theorem is as follows:

Theorem 1 *[15]. Decision information system $S = (U, C \cup D, L, g)$, where C is the conditional attribute set, D is the decision attribute set, the set combined by all single elements in the discernibility matrix is the core attribute set of C relative to D in the decision table, which is expressed as:*

$$CORE_C = \{p \mid (p \in C) \cap (\exists c_{ij}, ((c_{ij} \in M) \cap (c_{ij} = \{p\})))\} \tag{11}$$

Through the analysis of Formula (10), it can be seen that when the elements c_{ij} in the domain of discourse U belong to the same equivalence class, the decision attribute values of these elements are equal and indistinguishable, which can be expressed as $c_{ij} = \emptyset$; When the components in the domain of discourse belong to different equivalence classes, the decision attribute values of these elements are not equal and can be distinguished, expressed as $c_{ij} \neq \emptyset$; Therefore, it can be inferred that c_{ij} is composed of attributes that can distinguish these elements in the conditional attribute set C, which may include core attributes, or ordinary or redundant attributes. All attributes in c_{ij} will play a role in distinguishing objects m_i and m_j in different equivalence classes. Different attributes play different roles, and the nuclear attribute should have the greatest effect. For able to accurately measure the role of each conditional attribute in distinguishing different objects, this paper uses the frequency of these conditional attributes in c_{ij} and the proportion of each time to measure their importance, which is defined as follow:

Definition 6. *(Importance of Discernibility Matrix) Decision information system $S = (U, C \cup D, L, g)$, where C is the conditional attribute set, D is the decision attribute set, the elements in the discernibility matrix M that are not equal to the empty set are denoted by c_{ij}, and the total number of c_{ij} is denoted by K. Then for the conditional attribute subset c, $\forall c \in C$, the importance of c about D is defined as:*

$$Nsig(c, C, D) = \begin{cases} \frac{\sum_{i,j=1,2,\cdots,t} r_c \cap c_{ij}}{K} & |c_{ij}| > 1 \\ 1, & |c_{ij}| = 1 \end{cases} \tag{12}$$

In the above formula:

$$r_c \cap c_{ij} = \begin{cases} \frac{1}{|c_{ij}|}, & c \cap c_{ij} \neq \varnothing, i, j = 1, 2, \cdots, t \\ 0, & c \cap c_{ij} = \varnothing, i, j = 1, 2, \cdots, t \end{cases} \tag{13}$$

$|c_{ij}|$ represents the quantity of attributes in the set c_{ij}, $|c_{ij}| = 1$ indicates that there is only a single attribute in the set, and $|c_{ij}| > 1$ indicates that there are multiple attributes in the set.

It can be seen from Eq. (13) that when $|c_{ij}| = 1$, the importance of C is equal to 1, indicating that a single attribute contributes the most when distinguishing objects of different equivalence classes. This is consistent with the view that all single elements in the discernibility matrix are core attributes in Theorem 1. When $|c_{ij}| > 1$, it means that there are multiple attributes participating in distinguishing different equivalence classes. The attributes may be core attributes or ordinary attributes, and the function of each attribute is $1/c_{ij}$. Obviously, the importance of a core attribute is derived from the sum of its importance as a single attribute and the importance of multiple attributes, namely:

$$1 + \frac{\sum_{i,j=1,2,\cdots,t} r_c \cap c_{ij}}{K} \tag{14}$$

It can be concluded from the Formula (13) of Definition 6:

Conclusion. For the non-empty element set c_{ij} in the discernibility matrix M, $\forall p, q \in c_{ij}$, if $p \in CORE(C), q \in N$, then: $Nsig(p) > Nsig(q) > 0$. $CORE(C)$ represents the feature core element set in the discernibility matrix, and N represents the non-core feature element set.

Proof. If $p \in CORE(C)$ and $p \notin N$, then $Nsig(p, C, D) = 1$. if $p \in CORE(C)$ and $p \in N$, then:

$$N\,\text{sig}(p) = 1 + \frac{\sum\limits_{i,j=1,2,\cdots,t} r_p \cap c_{ij}}{K} > 1$$

If $q \notin CORE(C)$ and $q \in N$, according to Theorem 1, if the quantity of characteristic attributes in the set element c_{ij} is at least one ($|c_{ij}| > 1$), and $q \cap c_{ij} = 1/|c_{ij}| < 1$, then there is $\sum\limits_{i,j=1,2,\cdots,t} r_q \cap c_{ij} < K$, so:

$$Nsig(q) = \frac{\sum\limits_{i,j=1,2,\cdots,t} r_q \cap c_{ij}}{K} < \frac{K}{K} = 1$$

In summary, we can get: $Nsig(p) > Nsig(q) > 0$

From the above conclusions, we know that the importance of core attributes is greater than the importance of non-core attributes, which solves the phenomenon that redundant features are greater than feature cores in the importance of information entropy.

3.2 Rough Set Fitness Function

The fitness function plays a key role in the iterative process of algorithm optimization, and the purpose of optimizing the algorithm can be achieved by improving the fitness function. In the feature selection method combined with rough set, the fitness function [16] often used is:

$$Fitness = \alpha \cdot \gamma_B(D) + \beta \cdot \frac{|C| - |B|}{|C|} \tag{15}$$

This fitness function has achieved good results in feature selection, but it cannot evaluate unselected feature attributes. In response to this problem, [11] introduced the importance of information entropy attributes in the dual-strategy evolutionary fruit fly optimization algorithm, improved the fitness function in the original algorithm, and quickly searched out the feature subset. The improved fitness function is:

$$fitness = \eta_1 \cdot \gamma_B(D) + \eta_2 \cdot \frac{1}{1 + e^{SGF(C-B)}} + \eta_3 \cdot \frac{|C| - |B|}{|C|} \tag{16}$$

The weight factor η_1, η_2, η_3 in the function satisfies the conditions $\eta_1 + \eta_2 + \eta_3 = 1$ and $\eta_1, \eta_2, \eta_3 \in [0, 1]$; $\gamma_B(D)$ is attribute dependency; $SGF(C - B)$ is attribute importance of the conditional feature set except for the feature subset B; $|C|$ refers to the sum of all conditional feature sets; $|B|$ refers to the sum of selected feature subsets. This fitness function improves the shortcomings of the fitness function in the original algorithm, and evaluates all the feature subsets in the feature space by combining the attribute dependence and the attribute importance based on information entropy.

Aiming at the fitness function constructed by the importance of conditional entropy attributes in [11], the feature subset searched by the fruit fly algorithm in some decision tables will contain redundant features and other issues. This paper raises a way for measuring the importance of discernibility matrix attributes, and improves the fitness function based on this, and designs a method for searching feature subsets. In the fitness function, $Ncig(c)$ is used instead of $Sig(c)$, as shown in the following formula:

$$fitness' = \eta_1 \cdot \gamma_B(D) + \eta_2 \cdot \frac{1}{1 + e^{Ncig(c,C-B,D)}} + \eta_3 \cdot \frac{|C| - |B|}{|C|} \tag{17}$$

$Ncig(c, C - B, D)$ is attribute importance of the conditional feature set except for the feature subset B. The fitness function not only evaluates all the feature subsets in the feature space at the same time, but also reduces the redundant attributes in the feature subset. In the experiment, we believe that the feature subset should meet the maximum attribute dependency and the minimum feature length. The purpose of maximizing the attribute dependency is to improve the stability of the algorithm. During the experiment, we found that the smaller the length of the feature subset, the larger the fitness function value. For maximize the fitness function value, set $\eta_1 = 0.7, \eta_2 = 0.2, \eta_3 = 0.1$ in the experiment.

3.3 Feature Selection Algorithm Based on Discernibility Matrix and Improved Fruit Fly Optimization

This paper draws on the improved fruit fly optimization algorithm of Zhao et al. [10], and used Eq. (17) to designed a feature selection algorithm based on discernibility matrix and fruit fly optimization (DMFOAFS). The algorithm divided the fruit fly population into elite subgroups and ordinary subgroups, and dynamically divided these two seed groups. The chaotic variable parameters was introduced into the elite subgroup to improve the local search ability of the algorithm, and the weight factor were introduced into the ordinary subgroup to improve the global search ability of the algorithm. The weight factor should decrease with the increase of the number of iterations. For speed up the convergence speed at the beginning of the iteration, the value of the weight factor should be too large. In the latter part of the iteration, for improve the search accuracy, the value of the weight factor should be too small. According to the different operation contents, DMFOAFS can be divided into the following four stages:

(1) Randomly generate N fruit fly populations whose values of each dimension are in the range of [0, 1]. The position of the individual fruit flies represents the feature number of the datasets;

(2) Use Formula (17) to calculate the flavor concentration of individual fruit flies, and divide the population into elite subgroups and ordinary subgroups according to the taste concentration;
(3) Introduce chaotic variable parameters into the elite subgroups and introduce weight factors into the ordinary subgroups to make the fruit fly population closer to the optimal individual fruit fly;
(4) If it reaches the upper limit of iteration, terminate the algorithm and output the optimal individual position of the fruit fly, otherwise go to step (2).

Combined with [6], the detailed step description of the feature selection method based on improved discriminant matrix proposed in this paper is shown

Algorithm 1: DMFOAFS

Input: Knowledge expression system $DT = (U, C \cup D, L, g)$, population size pop and the iteration upper limit N_{max}, chaotic variable parameter α, weight factor ω_1, ω_2.
Output: Feature subset $C' \subseteq C$.

1 $N \leftarrow 0$;
2 **while** $N < N_{max}$ **do**
3   **for** $\forall p_i \in pop$ **do**
4   compute $\gamma_{p_i}(D)$;
5     $flavo_i \leftarrow \eta_1 \cdot \gamma_{p_i}(D) + \eta_2 \cdot \frac{1}{1+e^{Ncig(c, C-p_i, D)}} + \eta_3 \cdot \frac{|C|-|p_i|}{|C|}$
6   **if** $flavor_i > flavor_{best}$ **then**
7       $flavor_{best} \leftarrow flavor_i$;
8       $S_{best} \leftarrow S_i$
9   **end**
10   **if** $flavor_i < flavor_{best}$ **then**
11       $flavor_{worst} \leftarrow flavor_i$;
12   **end**
13   compute $d_1 \leftarrow \mid flavor_i - flavor_{best} \mid$ and $d_2 \leftarrow \mid flavor_i - flavor_{worst} \mid$;
14   **if** $d_1 \leq exp(-\frac{N}{N_{max}}) \cdot d_2$ **then**
15       $pop_e \leftarrow pop_e \cup p_i$
16   **else**
17       $pop_o \leftarrow pop_o \cup p_i$
18   **end**
19   **end**
20   **for** $\forall p_i \in pop_e$ **do**
21     $S_i \leftarrow S_i + \alpha(S_i - S_{best})$
22   **end**
23   **for** $\forall p_i \in pop_o$ **do**
24     $S_i \leftarrow \omega_1 S_i + 2\omega_2(V - 0.5)$
25   **end**
26   $N \leftarrow N + 1$
27 **end**
28 $C' \leftarrow S_{best}$;
29 *output* C'

in Algorithm 1. Before the iteration starts, first randomly initialize the population $pop = \{p_1, \cdots, p_m\}$ $(1 \leq i \leq m)$, pop_e represent elite subgroup, pop_o represent ordinary subgroup, and the initial position $S = \{S_1, \cdots, S_m\}, S_i \subseteq C(1 \leq i \leq m)$, the flavor concentration is expressed as $flavor$.

3.4 Algorithm Time Complexity

The time complexity of the algorithm in this paper can be split into two sections: the complexity of the dual-population fruit fly algorithm and the complexity of the importance of the discernibility matrix. The intricacy of the dual-population algorithm mainly determined by the dimensionality of the population *pop*, the iteration upper limit N_{max}, and the time T required for each individual fruit fly iteration. Although the fruit fly population is compartmentalized into double parts, the sum of the individual fruit flies of the two populations after each iteration is still *pop*, so the complexity of the dual-strategy fruit fly algorithm is the same order of magnitude as the classic fruit fly algorithm, which is $O\left(Sizepop \cdot N_{max} \cdot T\right)$; The time intricacy of the importance of the discernibility matrix is $O(\alpha|U|)$, where α is the sum of all elements in the discernibility matrix that are not equal to the empty set, and $|U|$ is the amount of domain of discourse objects. So the time intricacy of the algorithm in this paper is $O\left(Sizepop \cdot N_{max} \cdot T + \alpha|U|\right)$. The time intricacy of the traditional algorithm for solving feature subsets based on dependency is $O\left(n|U|^2\right)$, The time intricacy of calculating the importance of each attribute in the feature subset is $O\left(n|U|^2\right)$, Therefore, the total time intricacy is $O\left(2n|U|^2\right)$. So contrasted with the traditional feature selection strategy, the time complexity of the algorithm in this paper is better.

4 Experimental Data Analysis

For verify the effectiveness and practicability of the feature selection algorithm based on the discernibility matrix and fruit fly optimization proposed in this paper, we analyzed and compared the experimental consequences from two aspects of decision table instance verification and UCI datasets experiment.

4.1 Instance Verification

The decision table of the cited [17] is shown in Table 1. Using [11] and the feature selection method of this article to searched for feature subsets and analyzed the results.

Existence knowledge expression system $DT = (U, C \cup D, L, g)$, conditional attribute set $C = \{c_1, c_2, c_3, c_4, c_5\}$, decision attribute $D = \{d\}$, universe $U = \{x_1, x_2, \cdots, x_{10}\}$ is a non-empty finite set, $|U| = 10$, The feature kernel is c_5, and the minimum feature subset is $\{c_1, c_4, c_5\}$.

Table 1. Decision table.

U	c_l	c_2	c_3	c_4	c_5	D	U	c_l	c_2	c_3	c_4	c_5	D
x_1	0	0	0	0	1	0	x_6	1	1	0	1	0	1
x_2	0	1	1	1	0	1	x_7	0	1	1	1	1	1
x_3	1	1	0	1	1	1	x_8	1	1	1	0	1	1
x_4	0	1	1	1	0	0	x_9	1	1	0	1	1	0
x_5	0	0	1	0	1	0	x_{10}	0	1	1	1	1	0

(1) [11] method:
Initially, calculated the importance of the feature, the operation is as follows: $sig\,(c_1, C, D) = H\,(D \mid C - c_1) - H(D \mid C) = 0.6 - 0.4 = 0.2$. The same calculation method can be obtained $sig\,(c_2, C, D) = sig\,(c_3, C, D) = sig(c_4, C, D) = 0.2$, $sig\,(c_5, C, D) = 0.275$.
Then, fruit flies searched feature subsets:
Substituting the importance of conditional features into $fitness$ of equation (16) in turn, the subset of features found by fruit fly according to the optimization algorithm is $\{c_2, c_3, c_4, c_5\}$. Clearly, this consequence is inconsistent with the correct consequence $\{c_1, c_4, c_5\}$, and the feature subset obtained by this method contains redundant feature c_3

(2) The method of this paper:
Firstly, in line with Definition 5, the discernibility matrix of decision Table 1 is obtained as:

$$\begin{bmatrix} \emptyset \\ c_2c_3c_4c_5 & \emptyset \\ c_1c_2c_4 & \emptyset & \emptyset \\ \emptyset & \emptyset & c_1c_3c_5 & \emptyset \\ \emptyset & c_2c_4c_5 & c_1c_2c_3c_4 & \emptyset & \emptyset \\ c_1c_2c_4c_5 & \emptyset & \emptyset & c_1c_3 & c_1c_2c_3c_4c_5 & \emptyset \\ c_2c_3c_4 & \emptyset & \emptyset & c_5 & c_2c_4 & \emptyset & \emptyset \\ c_1c_2c_3 & \emptyset & \emptyset & c_1c_4c_5 & c_1c_2 & \emptyset & \emptyset & \emptyset \\ \emptyset & c_1c_3c_5 & \emptyset & \emptyset & \emptyset & c_5 & c_1c_3 & c_3c_4 & \emptyset \\ \emptyset & c_5 & c_1c_3 & \emptyset & \emptyset & c_1c_3c_5 & \emptyset & c_1c_4 & \emptyset & \emptyset \end{bmatrix}$$

Secondly, calculated the importance of features by Definition 6 as:

$$Nsig\,(c_5) = 1 + \frac{\frac{1}{4} + \frac{1}{4} + \frac{1}{3} + \frac{1}{3} + \frac{1}{3} + \frac{1}{3} + \frac{1}{5} + \frac{1}{3}}{22} = 1.108$$

The same calculation method can be obtained $Nsig\,(c_1) = 0.236$, $Nsig\,(c_2) = 0.149$, $Nsig\,(c_3) = 0.172$, $Nsig\,(c_4) = 0.198$.

Finally, fruit flies searched for a subset of features:

Substituting the importance values of the features into $fitness'$ in Eq. (17) in turn, the smallest feature subset finally searched by fruit flies is $\{c_1, c_4, c_5\}$. The result is completely consistent with the correct result, which shows the feasibility of the feature selection strategy raised in this paper.

From the above calculation results can know that $Nsig\left(c_5\right) > Nsig\left(c_1\right) > Nsig\left(c_4\right) > Nsig\left(c_3\right) > Nsig\left(c_2\right) > 0$. It shows that the importance of core attribute c_5 is greater than that of non-core attributes, and the importance of all attributes is greater than zero, which is consistent with the conclusion obtained in Definition 6 of this paper. At the same time, it also shows that Definition 6 in this paper reasonably measures the importance of each attribute.

4.2 UCI Data Sets Experimental Data Analysis

For further verify the effectiveness and reliability of the feature selection strategy in this paper, we selected five datasets such as Lymphography from the UCI classification repository, the data sets information is shown in Table 2. Feature selection strategy software environment is Win10 operating system and application software Python3.8, the hardware is AMD Ryzen 5 4500U with Radeon Graphics 2.38 GHz, and the memory is 16 GB. This paper (DMFOAFS algorithm) will compare the index of feature subset size and fitness function value with [11] (RSFOAFS algorithm), [18] (ABCFS algorithm) and [19] (FSRSWOA algorithm).

Table 2. Experimental data sets information.

Data sets	Number of samples	Feature number	Number of categories
Lymphography	148	18	4
Soybean	307	35	19
Dermatology	366	33	6
Car	1728	6	4
Chess	3196	36	2

Relevant data settings in the experiment: In the iterative process of DMFOAFS and RSFOAFS, the search range of the common population should be changed from large to small, where the weight factor ω_1 controls the individual's historical cognition, and ω_2 adjusts the search step size; The value of the weighting factor is a variable. For achieve excellent performance, the value range of ω_1 is set to [1.0, 1.5], and the value range of ω_2 is set to [0.5, 1.5]; The chaos factor α in the elite population is 0.9; in the fitness function, $\eta_1 = 0.7, \eta_2 = 0.2, \eta_3 = 0.1$; ABCFS is an incremental feature selection strategy. For achieve data dimensionality reduction without reducing accuracy, the parameters in the fitness function are set as: $k = 2K, z = 0.05, \omega = 0.55$; The fitness function used in FSRSWOA is Eq. (15), and the parameter $\alpha = 0.9, \beta = 0.1$;

Through a large number of experiments, it can be known that if the size of the fruit fly population is not set properly, it will lead to a large error in the output result and the algorithm will fall into a local optimum. For reduce the impact of the population size on the output results, set the size of the fruit fly

Table 3. Comparison of feature subset length of algorithms.

Data sets	DMFOAFS	RSFOAFS	ABCFS	FSRSWOA
Lymphography	7	**6**	**6**	7
Soybean	9	10	**7**	10
Dermatology	**8**	**8**	9	10
Car	**3**	4	5	5
Chess	**21**	24	25	27

population to half the number of features in the data set, the maximum number of iterations $N_{max} = 100$, run 10 experiments on 5 data sets, and select the average of the experimental results. The final feature subset length is shown in Table 3:

It can be seen from Table 3 that in the Lymphography data set with a small number of samples and features, DMFOAFS was slightly inferior to RSFOAFS, but the gap has small; In the Soybean data set with a small number of samples but a high number of features, DMFOAFS had better than RSFOAFS and FSRSWOA, but weaker than ABCFS; The sample number and feature number of the Dermatology data set are similar to Soybean, but DMFOAFS got the smallest feature subset, where the length of the feature subset was not much different from that in Soybean, while RSFOAFS and ABCFS were in the data set Dermatology and Soybean and the length of feature subsets was quite different, indicating that DMFOAFS was more stable; In the data set Car with a large number of samples but few features, and in the data set Chess with a large number of samples and a high number of features, DMFOAFS has found the smallest feature subset. This shows that the algorithm in this paper has excellent performance when processing high-dimensional, multi-sample data.

It can be seen from Sect. 3.2 that the purpose of the weighting factor is to maximize the attribute dependence and minimize the length of the feature subset, so as to obtain the maximum fitness function value, so the problem of searching for the minimum feature subset can be transformed into the problem of finding the fitness function value. For further prove the effectiveness of the algorithm in this paper, the following will compare the four algorithms to obtain stable fitness function values and fitness function value variances in different data sets. Run 10 experiments on 5 data sets, and select the average of the experimental results. The final results are shown in Table 4:

It can be seen from Table 3 and Table 4 that in the same data set, the algorithm with a smaller feature subset also has a larger fitness function value, which proves the feasibility of transforming the problem of searching for the minimum feature subset into the problem of seeking the maximum fitness function value. In the Soybean data set, although the fitness function value of DMFOAFS was smaller than ABCFS, the variance of the fitness function value was less than ABCFS, indicating that DMFOAFS is more stable. In the three data sets of Dermatology, Car and Chess, the fitness function value of DMFOAFS was all

Table 4. Comparison of fitness function values and variances between algorithms.

Data sets	Algorithm	Fitness function value	Variance
Lymphography	DMFOAFS	0.8118	3.30E−6
	RSFOAFS	**0.8395**	3.28E−6
	ABCFS	0.8277	**3.2E−6**
	FSRSWOA	0.8104	3.34E−6
Soybean	DMFOAFS	0.8353	**3.14E−6**
	RSFOAFS	0.8297	3.21E−6
	ABCFS	**0.8504**	3.15E−6
	FSRSWOA	0.8146	3.34E−6
Dermatology	DMFOAFS	**0.8345**	**3E−6**
	RSFOAFS	0.8325	3.11E−6
	ABCFS	0.8297	3.17E−6
	FSRSWOA	0.8274	3.36E−6
Car	DMFOAFS	**0.8014**	**2.83E−6**
	RSFOAFS	0.7931	2.86E−6
	ABCFS	0.7708	3.04E−6
	FSRSWOA	0.7420	3.52E−6
Chess	DMFOAFS	**0.7153**	**2.69E−6**
	RSFOAFS	0.6827	2.81E−6
	ABCFS	0.6538	3.11E−6
	FSRSWOA	0.6083	3.64E−6

the smallest, which further proves the stability of the algorithm in this paper. Although in the Lymphography data set, the fitness function value variance of DMFOAFS was only better than FSRSWOA, but the variance value was not much different from the other two algorithms.

To sum up, compared with the other three feature selection strategy, the feature selection strategy proposed in this paper can search for a smaller feature subset and has better stability. When dealing with small data sets, this method is underperformance; But when dealing with high-dimensional features and multi-sample data sets, this method has obvious advantages in terms of feature subset length and stability.

5 Concluding Remarks

Based on previous research results, this paper combines the discernibility matrix to give a new attribute importance definition method, and introduces the attribute importance as a heuristic factor matrix into the hierarchical two-strategy evolutionary fruit fly algorithm. On this basis, a feature selection way in view of discernibility matrix and fruit fly algorithm is designed. This method avoids redundant attributes in the searched feature subset, improves the convergence speed of the feature selection strategy, and has a more obvious effect on high-dimensional features and data sets with large sample sizes. In the follow-up, we will combine the method of determining the dual-population threshold proposed by [20] to further find the optimal feature subset. However, the definition method proposed in this paper also has limitations: One is that in a very small number of decision tables, there will be situations where attributes do not emerge in the discernibility matrix, and the importance of feature attributes that pass off in the discernibility matrix set may also be zero; The Second is for some decision tables without core attributes, the effectiveness of the algorithm needs further research.

References

1. Miao, D., Li, D.: Rough Set Theory, Algorithm and Application. Tsinghua University Press, Beijing (2008)
2. Miao, J., Niu, L.: A survey on feature selection. Procedia Comput. Sci. **91**, 919–926 (2016)
3. Ye, D., Chen, Z.: A new approach to minimum attribute reduction based on discrete artificial bee colony. Soft Comput. **19**(7), 1893–1903 (2014). https://doi.org/10.1007/s00500-014-1371-0
4. Chen, Y., Chen, Y.: Attribute reduction algorithm based on information entropy and ant colony optimization. J. Chin. Comput. Syst. **36**(3), 586–590 (2015)
5. Hou, Y.: Feature gene selection algorithm based on ant colony optimization. J. Sun Yat-sen Univ. **36**(06), 120–123 (2019)
6. Zhai, J., Liu, B.: Feature selection method based on rough set relative classification information entropy and particle swarm optimization. J. Intell. Syst. **12**(03), 397–404 (2017)
7. Pan, W.T.: A new fruit fly optimization algorithm: taking the financial distress model as an example. Knowl.-Based Syst. **26**, 69–74 (2012)
8. Mitić, M., Vuković, N., Petrović, M., Miljković, Z.: Chaotic fruit fly optimization algorithm. Knowl.-Based Syst. **89**, 446–458 (2015)
9. Pan, L.: A Classification-based surrogate-assisted evolutionary algorithm for expensive many-objective optimization. IEEE Trans. Evol. Comput. **23**(1), 74–88 (2019)
10. Zhao, F., Ding, R., Wang, L., Cao, J., Tang, J.: A hierarchical guidance strategy assisted fruit fly optimization algorithm with cooperative learning mechanism. Expert Syst. Appl. **183**, 115342 (2021)
11. Fang, B., Chen, H., Wang, S.: Feature selection method based on rough set and fruit fly optimization algorithm. Comput. Sci. **46**(07), 157–164 (2019)
12. Ye, J., Wang, L.: A method of attribute combination weight construction based on discrimination matrix. Comput. Sci. **41**(11), 273–277 (2014)
13. Wang, G., Yu, H., Yang, D.: Decision table reduction based on conditional information entropy. Chin. J. Comput. **25**(7), 759–766 (2002)
14. Skowron, A., Rauszer, C.: The discernibility matrices and functions in information systems. In: Intelligent Decision Support. Springer, Dordrecht, pp. 331–362 (1992). https://doi.org/10.1007/978-94-015-7975-9_21
15. Ye, J., Zhu, H., Li, M.: An attribute importance definition method and its application in reduction. Appl. Res. Comput. **33**(07), 2075–2078 (2016)

16. Wang, X., Yang, J., Teng, X., Xia, W., Jensen, R.: Feature selection based on rough sets and particle swarm optimization. Patt. Recogn. Lett. **28**(4), 459–471 (2007)
17. Huang, G., Zeng, F., Wen, H.: Conditional information entropy representation of algebraic reduction and its efficient reduction algorithm. Comput. Sci. **41**(07), 236–241+274 (2014)
18. Gao, W., Xie, H.: Dynamic feature selection based on rough set and artificial bee colony algorithm. Comput. Eng. Des. **40**(09), 2697–2703 (2019)
19. Wang, S., Chen, H.: Feature selection method based on rough set and improved whale optimization algorithm. Comput. Sci. **47**(02), 44–50 (2020)
20. Cui, Z., Li, F., Zhang, W.: Bat algorithm with principal component analysis. Int. J. Mach. Learn. Cybern. **10**(3), 603–622 (2018). https://doi.org/10.1007/s13042-018-0888-4

Feature Selection Method Based on Ant Colony Optimization Algorithm and Improved Neighborhood Discernibility Matrix

Zhenyu Yang[1], Jun Ye[1,2](✉), Jiaxin Ao[1], and Yuxuan Ji[1]

[1] College of Information Engineering, Nanchang Institute of Technology, Nanchang 330000, China
2003992646@nit.edu.cn

[2] Jiangxi Province Key Laboratory of Water Information Cooperative Sensing and Intelligent Processing, Nanchang 330000, China

Abstract. Under the condition that the classification ability of the characterization data remains unchanged, selecting the smallest feature subset from the feature set is one of the problems that many researches try to solve. In view of the existing neighborhood rough set and ant colony optimization feature selection methods, the algorithms are premature, and the searched feature subset contains redundancy and other problems. Improvements have been made in three aspects. One is to extend the discrimination matrix based on equivalence to the rough set of neighborhoods, and define a neighborhood discernibility matrix; second, to give a method for the importance of neighborhood discrimination matrix features; The third is to use the importance of the neighborhood feature as the heuristic information factor of the ant colony algorithm, and propose a feature selection method based on the ant colony optimization algorithm and the improvement of the neighborhood discernibility matrix. Numerical examples and data set experiments show that the new feature selection method improves accuracy and reduces the number of iterations.

Keywords: Rough set · Neighborhood rough set · Feature selection · Discernibility matrix · Ant colony algorithm · Feature subset

1 Introduction

With the improvement of technology in the field of data mining and knowledge discovery, feature dimensions have rapidly increased, and feature selection has attracted the attention of a large number of researchers. Feature selection [1]

Supported by the National Natural Science Foundation of China (Nos. 61562061) and Technology Project of Ministry of Education of Jiangxi province of China (No.GJJ211920, GJJ170995).

L. Pan et al. (Eds.): BIC-TA 2021, CCIS 1565, pp. 116–131, 2022.
https://doi.org/10.1007/978-981-19-1256-6_9

refers to selecting a subset of features from all the features, but the classification accuracy of prediction does not decrease. Usually feature selection consists of two parts: evaluation function and subset generator. Evaluation function is a measure to distinguish feature subsets, and subset generator is a process of searching feature subsets using evaluation functions. The neighborhood rough set [2] is widely used in feature selection of continuous data decision table because it can directly process numerical data. For example, literature [3] proposed a neighborhood rough set feature selection algorithm; [4] gave related knowledge concepts such as neighborhood rough mutual information entropy, and proposed a non-monotonic based on neighborhood rough mutual information entropy feature selection algorithm; Literature [5] proposed an incremental feature selection algorithm based on mixed data neighborhood discrimination. However, these algorithms are based on changes in the positive domain. In the neighborhood rough model, the computational workload to obtain the positive domain is much larger than that of discrete data, which makes the feature selection method in the neighborhood rough model more time-complex. In particular, the increase in the dimensions of the decision table is more prominent. Therefore, researchers introduced swarm intelligence algorithm [6–10] to optimize to reduce time complexity and reduce computational workload. Feature importance is an important concept of neighborhood rough set. It is usually used as part of the evaluation function in feature selection. Researchers combine it with swarm intelligence algorithms to design many neighborhood rough sets and swarm intelligence optimization feature selection algorithm; For example, literature [11] combines neighborhood rough set with artificial bee colony algorithm. Put the dependency under the neighborhood into the fitness function, and propose an artificial bee colony reduction algorithm based on the rough neighborhood model; Literature [12] introduces the dependency of neighborhood rough set into the objective function, combine it with the binary firefly algorithm to find reduction; Literature [13] combines neighborhood rough set with fish school intelligence algorithm, and uses the attribute dependency of neighborhood in the evaluation function, and proposes a gene selection method based on fish school intelligence and neighborhood rough set; Literature [14] introduces the attribute dependence of neighborhood rough set into the fitness function of quantum genetic algorithm, and proposes a feature selection method based on quantum genetic algorithm and neighborhood rough set.

Ant colony algorithm is a swarm intelligence algorithm proposed by Dorigo [15] in Italy. It has the characteristics of robustness, distribution, positive feedback and global optimization. Combining it with neighborhood rough set can effectively reduce continuous data, and many scholars have studied it. For example, the literature [16] combined the ant colony optimization algorithm with the neighborhood rough set, and took the attribute importance as the heuristic information, and proposed an attribute reduction algorithm based on the neighborhood rough set and ant colony optimization. However, this method uses the positive domain to measure the importance of attributes, and there is a case where the weight of non-core attributes is zero, which will cause the algorithm to be premature and unable to search for the smallest feature subset; Literature [17] uses the importance of neighborhood information entropy as the enlightening information of the ant colony, and proposes a gene selection algorithm based on neighborhood entropy and ant colony optimization. However, in this method, the importance of non-core attributes may be greater than the importance of core attributes, and the feature subset may contain redundant features, resulting in a significant decrease in the reduction accuracy.

Therefore, the subset generator in this paper uses the ant colony optimization algorithm and gives a new feature importance definition based on the neighborhood discernibility matrix. Combine this importance method as heuristic information with ant colony algorithm to obtain the optimal feature subset, and verify it through examples and experiments.

2 Related Knowledge

In order to facilitate understanding, the following briefly introduces the related knowledge of neighborhood rough set related to this article, and for more detailed knowledge, please refer to literature [17–19].

2.1 Neighborhood Rough Set Theory

Definition 1. *[3] In the neighborhood decision information system $NS = (U, C \cup S, V, f)$, C is the condition attribute, S is the decision attribute. About $R \subseteq C$, The neighborhood of $x_i \in U$ is:*

$$\delta_R(x_i) = \{y \mid y \in U, \Delta(x_i, y) \leq \delta\} \tag{1}$$

In the formula, Δ represents the distance function, This paper uses Euclidean distance, $g = 2$

$$\Delta_g(y_1, y_2) = \left(\sum_{i=1}^{n} \mid f(y_1, a_i) - f(y_2, a_i)^g\right)^{\frac{1}{g}} \tag{2}$$

Definition 2. *[3] In the neighborhood decision information system $NS = (U, C \cup S, V, f)$, About $A \subseteq U, R \subseteq C$, The lower approximation, upper approximation and boundary area of the neighborhood are defined as:*

$$\underline{N}_R(A) = \{a_i \mid \delta_R(a_i) \subseteq A, a_i \in U\} \tag{3}$$

$$\bar{N}_R(A) = \{a_i \mid \delta_R(a_i) \cap A \neq \emptyset, a_i \in U\} \tag{4}$$

$$NBX = \overline{N_R}(A) - \underline{N}_R(A) \tag{5}$$

Definition 3. *[3] In the Neighborhood decision information system $NS = (U, C \cup S, V, f)$, The dependency of decision attribute on conditional attribute subset is defined as:*

$$\gamma_R(S) = \frac{|Pos_R(S)|}{|U|} \tag{6}$$

In the formula, $Pos_R(S)$ is the positive domain of the neighborhood, and $Pos_R(S) = \underline{N}_R(A)$.

Definition 4. *[3] In the Neighborhood decision information system* $NS = (U, C \cup S, V, f)$, $\forall c \in C$, *if* $\gamma_{C-(c)}(S) \neq \gamma_c(S)$ *, then* c *is called a necessary attribute, otherwise it is an unnecessary attribute. All necessary attributes form the core attribute set* $CORE(C)$.

Definition 5. *[3] In the Neighborhood decision information system* $NS = (U, C \cup S, V, f)$, $\forall R \subseteq C$, *If the condition attribute subset* R *satisfies the following conditions: (1)* $\gamma_R(S) = \gamma_C(S)$, $(2)\forall a \in R, \gamma_R(S) > \gamma_{R-\{a\}}(S)$, *Then the conditional attribute subset* R *is a relative reduction of the conditional attribute set* C. *All reduction sets are denoted as* $RED(C)$.

Definition 6. *[3] (Neighborhood Positive Attribute Importance) In the Neighborhood decision information system* $NS = (U, C \cup S, V, f)$, *After deleting* a *from the conditional attribute set* R, *the importance of* a *relative to* S *is:*

$$Sig(a, R, S) = \gamma_R(S) - \gamma_{R-(a)}(S) \tag{7}$$

After adding a *from the conditional attribute set* $R \subseteq C$, *the importance of* a *with respect to* S *is:*

$$Sig(a, R, S) = \gamma_{RUa}(S) - \gamma_R(S) \tag{8}$$

The importance of the neighborhood positive domain attribute in Definition 6 is the same as the classic rough set, which is based on the dependency of the attribute. Investigate changes in the positive domain by adding or deleting an attribute. But the only thing that can change the positive domain is the core attribute.

Definition 7. *[17] In the Neighborhood decision information system* $NS = (U, C \cup S, V, f)$, $R \in C$, *any* $x \in U$, *The neighborhood* δ *of* x *on* R *is* $\delta_R(x_i)$, *Then the neighborhood information entropy of* R *is defined as:*

$$H_\delta(R) = -\sum_{i=1}^{|U|} \left(\frac{|\delta_R(x_i)|}{|U|^2} \log_2 \frac{1}{|\delta_R(x_i)|} \right) \tag{9}$$

Definition 8. *[17] (Neighborhood Information Entropy Attribute Importance)In the Neighborhood decision information system* $NS = (U, C \cup S, V, f)$, $a \in C, R \subseteq C$ *Then the importance of* a *relative to* R *is expressed as:*

$$sig(a, R, S) = \left|M_{RU\{a\}}(S)_\delta\right| - |M_R(S)_\delta| \tag{10}$$

In the formula, $M_R(S)_\delta = \gamma_R(S)_\delta \cdot (1 - H_\delta(S))$

The definition of the importance of the neighborhood information entropy is conditioned on the influence of the attribute on the uncertain classification subset in the universe. Check the uncertain classification changes after adding or deleting an attribute. Therefore, the importance of this attribute is 0 under the definition of the importance of the neighborhood positive domain attribute, and it is not 0 under the definition of the neighborhood information entropy. This avoids the situation where the importance of non-core attributes is all zero. However, in some decision tables, the importance of redundant attributes may be greater than the importance of core attributes.

2.2 Ant Colony Algorithm and Its Optimized Feature Selection Method

Dorigo M proposes an ant colony algorithm for the foraging behavior of real ants in nature. The algorithm simulates the process of ant colony searching for food to find the shortest path [21]. Combination optimization ability and robustness are relatively strong [22]. In a non-static environment, the flexibility and robustness of the ant colony optimization algorithm is also very high.

Jensen first introduced the ant colony algorithm to the rough set, converted the minimum feature subset into the optimal path combination optimization problem, and proposed a feature selection method based on the ant colony optimization algorithm. The following briefly introduces the rough set feature selection method optimized by ant colony algorithm, and refer to the literature [23] for more details.

Problem Presentation. In the literature [23], Jensen combined the ant colony algorithm to transform the feature selection into a complete graph form as shown in Fig. 1. In the neighborhood decision information system $NS = (U, C \cup S, V, f)$, C represents the set of conditional features set, S represents the set of decision-making features set, features are represented by each node in the figure. The edge connecting the two nodes represents the path of the ant choosing the next feature, and the path contains heuristic information. The heuristic information in this paper uses the importance method defined in Sect. 3.1. The ants traverse the whole graph with the heuristic information on the path, and the sign of searching for the optimal feature combination is the shortest path with the least node.

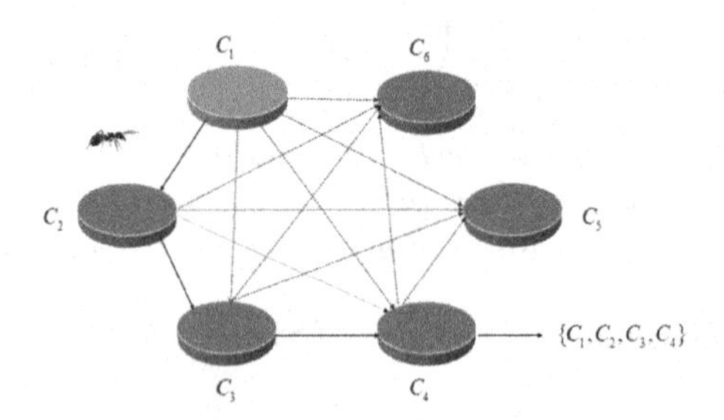

Fig. 1. ACO problem representation for FS

Pheromone Update. After the ants go through a path, the pheromone of the path is updated. After all the ants have found an feature, it represents the completion of an iteration. The pheromone update formula is:

$$\tau_{ij}(t+1) = \rho\tau_{ij}(t) + \Delta\tau_{ij}(t) \tag{11}$$

$\tau_{ij}(t)$ represents the pheromone concentration on the path from feature i to feature j at the t-th iteration, ρ is the volatilization degree of the pheromone, and $\Delta\tau_{ij}(t)$ is the sum of all the pheromone along the path, the formula is:

$$\Delta\tau_{ij}(t) = \begin{cases} \sum \frac{q}{|R(t)|}, & Ants\ go\ through\ path\ (i,j) \\ 0, & other \end{cases} \tag{12}$$

$|R(t)|$ represents the number of features of the smallest feature subset in t iterations, and q is a set constant. Starting from the feature core, the ant selects the next feature according to the probability formula:

$$P_{ij}^{k}(t) = \frac{\tau_{ij}^{\alpha}(t)\eta_{ij}^{\beta}(t)}{\sum_{1 \in allowed_k} \tau_{il}^{\alpha}(t)\eta_{il}^{\beta}(t)}, j \in allowed_k \tag{13}$$

In the formula, $\eta_i(t)$ is the heuristic information on the path ij, this paper uses the importance definition method in Sect. 3.1 to represent. α, β represents its proportion, and $allowed_k$ represents the feature that the k ant has not yet selected. Heuristic information $\eta_{ij}(t)$ is expressed by the importance of feature j, $\eta_{ij}(t)$is defined as:

$$\eta_{ij}(t) = sig(j, C, S) \tag{14}$$

In the above Formula (14), $sig\,(j, C, S)$ is usually expressed by the importance of Definition 6 and Definition 8 in Sect. 2.1. For example, literature [16] uses the attribute importance of the neighborhood positive domain in Definition 6. And the literature [17] uses the attribute importance degree based on the neighborhood information entropy in Definition 8. Since the importance of the positive domain based on the neighborhood is based on the dependence of the attribute, it can only measure the importance of the feature core, but cannot measure the importance of other features. This makes the ant colony unable to accurately find the next feature according to the probability formula when searching for non-core features. The importance of the neighborhood information entropy will get a situation in some decision tables that the importance of the redundant attribute is greater than the importance of the core attribute, which also makes the ant colony unable to accurately find the optimal solution.

When the solution searched by the ant meets the following conditions, the search stops and the ant dies:

(1) $\gamma_{R_k}(S) = \gamma_c(S)$, When the dependency of the local solution R_k is equal to the feature set C, the k ant constructs a local solution.
(2) If the cardinality of the set searched by the k ant is greater than the cardinality of the current global optimal solution, the ant will die.

3 Feature Selection Method Based on ACO and Improved Neighborhood Discernibility Matrix

3.1 The Definition of Attribute Importance of Neighborhood Discernibility Matrix

A. Skowron [24] first proposed the discernibility matrix under the equivalence relation:

Definition 9. *[24] In the decision information system* $TS = (U, C \cup S, V, f)$, *the set of objects is* U, C *is the conditional attribute,* S *is the decision attribute, and the universe is a non-empty finite set of objects,* $U = x_1, x_2, x_3, \ldots x_n$, $|U| = n$, *Then the discernibility matrix of the decision table is:*

$$M_{n \times n} = (c_{ij}) = \begin{bmatrix} c_{11} & c_{12} & \cdots & c_{1n} \\ c_{21} & c_{22} & \cdots & c_{2n} \\ \vdots & \vdots & \ddots & \vdots \\ c_{n1} & c_{n2} & \cdots & c_{nn} \end{bmatrix}$$

In the matrix,$j = 1, 2, \ldots, n$.

$$C_{ij} = \begin{cases} \{a \mid a \in C \wedge f(x_1) \neq f(y_j)\}, f_S(x_1) \neq f_S(x_j) \\ \emptyset, \quad f_S(x_i) = f_S(x_j) \end{cases} \tag{15}$$

Hu gave the method of finding the core attribute set of the discernibility matrix under the equivalence relation in the literature [25], Such as Theorem 1:

Theorem 1. *[25] In the discernibility matrix, the set of all single attribute elements is a relatively necessary (core) attribute set.*

In this paper, the above Definition 9 is extended to the rough set of the neighborhood, and the discernibility matrix of the neighborhood is defined. The definition method is as follows:

$$c_{ij} = \begin{cases} \{a|a \in C \wedge |f_a(x_1) - f_a(x_j)| > \delta(x_i)\}, & |f_a(x_i) - f_a(x_j)|_S > \delta(x_i) \\ \emptyset, & |f_a(x_i) - f_a(x_j)|_S \leq \delta(x_i) \end{cases} \tag{16}$$

For the neighborhood discernibility matrix, we get the following Theorem 2:

Theorem 2. *In the neighborhood discernibility matrix, the set of all single attribute elements is a relatively necessary (core) attribute set:*

$$CORE_C(S) = \{a \mid a \in C \wedge (\exists c_{ij}, ((c_{ij} \in M) \wedge (c_{ij} = \{a\})))\} \tag{17}$$

Proof. It is directly proved by Definition 4.

Through the analysis of Formula (16), it can be known that the objects with the same decision attribute value do not need to be distinguished, then $c_{ij} = \emptyset$. For objects whose decision attribute values are not equal and belong to different neighborhoods, these objects can be distinguished. Then c_{ij} is composed of conditional attributes in C that can distinguish these objects, it may contain core attributes or non-core attributes. All the attributes in c_{ij} play a role in distinguishing objects x_i and x_j in different neighborhoods, but they are different. Core attributes have the greatest effect, while non-core attributes have relatively small effects. In order to accurately measure the role of each conditional attributes in distinguishing objects in different neighborhoods, this paper uses the frequency of appearance of each attribute in c_{ij} and the proportion of each attribute to measure its importance. Defined as follows:

Definition 10. *In the neighborhood decision information system $NS = (U, C \cup S, V, f)$, C is the conditional attribute, Sis the decision attribute. c_{ij} is the non-empty element in the discrimination matrix M, and W is the total number of all non-empty c_{ij} in the discrimination matrix M, $\forall c \in C$, The importance of attribute C to D is:*

$$Nsig(c) = \begin{cases} \frac{\sum\limits_{i,j=1,2,\cdots,n} r_c \cap c_{ij}}{W} & |c_{ij}| > 1 \\ 1, & |c_{ij}| = 1 \end{cases} \tag{18}$$

$$r_c \cap c_{ij} = \begin{cases} \frac{1}{|c_{ij}|} & c \cap c_{ij} \neq \emptyset, i,j = 1,2,\cdots,n \\ 0 & c \cap c_{ij} = \emptyset, i,j = 1,2,\cdots,n \end{cases} \tag{19}$$

According to the Formula (18) in Definition 10, when $|c_{ij}| = 1$, that is, when it is a single attribute, its importance is 1. This means that a single attribute has the greatest contribution to distinguishing objects in different neighborhoods. This is consistent with the single attribute being a core attribute in Theorem 1. When $|c_{ij}| > 1$, that is, the set c_{ij} is composed of multiple attributes, which means that distinguishing objects in different neighborhoods is completed by the joint action of multiple attributes. It may be a necessary attribute, a relatively necessary attribute or an unnecessary attribute, each attribute functions as $\frac{1}{c_{ij}}$. Obviously, the importance of a core attribute is composed of the sum of its importance as a single attribute 1 and the importance of multiple attributes, that is: $1 + \frac{\sum\limits_{i,j=1,2,\cdots,n} r_c \cap c_{ij}}{W}$, This also reflects that the core attribute has the strongest ability to distinguish objects in the neighborhood.

From Definition 10, the following properties can be obtained.

Property 1. For the non-empty element set c_{ij} in the neighborhood discernibility matrix M, $\forall c \in C$ and $c \in c_{ij}$, then $Nsig(c) > 0$.

Proof. $\forall c \in c_{ij}$, If c_{ij} is a single attribute, according to Formula (18): $Nsig(c) = 1 > 0$, $\forall c \in c_{ij}$, if c_{ij} is a set of multiple attributes, that is: $|c_{ij}| > 1$, according to Definition 10: $c \cap c_{ij} = \frac{1}{|c_{ij}|} > 0$, therefore $\frac{\sum\limits_{i,j=1,2,\cdots,n} r_c \cap c_{ij}}{W} > 0$.

Property 2. In the non-empty element set c_{ij} in this matrix M, $\forall a, b \in c_{ij}$, if $a \in CORE(C), b \in N$, then $Nsig(a) > Nsig(b) > 0$.

Among them, $CORE(C)$ represents the core attribute element set in the neighborhood discernibility matrix, and N represents the non-core attribute element set.

Proof. if $a \in CORE(C)$ and $a \in N$, According to Theorem 1, the number of features in set C is at least two or more. $|c_{ij}| > 1$, therefore, $b \cap c_{ij} = \frac{1}{c_{ij}} < 1$.

$$\sum_{i,j=1,2,\cdots,n} r_b \cap c_{ij} < W,\ Nsig(b) = \frac{\sum_{i,j=1,2,\cdots,n} r_b \cap c_{ij}}{W} < \frac{W}{W} = 1.$$

In summary: $Nsig(a) > Nsig(b) > 0$

3.2 Description of Algorithm Steps

Using the importance defined by the discernibility matrix in Sect. 3.1 as the heuristic information, according to the principle of ant colony algorithm, a feature selection algorithm based on neighborhood discernibility matrix and ant colony algorithm is proposed:

Algorithm 1: NMACO

Input: $NS = (U, C \cup D, V, f)$, *maxcycle* and K

Output: the minimum reduction R_{min} and its cardinal number L_{min}

1 $R_{min} \leftarrow C, L_{min} \leftarrow |C|, R_k \leftarrow \emptyset, t \leftarrow 0$;
2 computer $\gamma_C(S)$;
3 Compute $CPRE$ (Use the discernibility matrix);
4 **while** $t < maxcycle$ **do**
5     $R_k \leftarrow R_k \cap c_k$, (According to the probability Formula (13), each ant selects the next characteristic feature separately);
6   **if** $\gamma_{R_k}(S) == \gamma_C(S)$ *or* $|R_k| >= L_k$ **then**
7         $K \leftarrow K - 1$
8   **end**
9   **if** $\gamma_{R_k}(S) == \gamma_C(S)$ *and* $|R_k| < L_k$ **then**
10         $R_{min} \leftarrow R_k,\ L_{min} \leftarrow |R_{min}|$
11   **end**
12     $\tau_{ij}(t+1) \leftarrow \rho\tau_{ij}(t) + \Delta\tau_{ij}(t)$
13   **if** $K = 0$ **then**
14   break
15   **end**
16 **end**
17 *output*R_{min}

3.3 Analysis of Algorithm Time Complexity

Compared with the algorithm in literature [16,17], the time complexity of the algorithm in this paper is that the heuristic information of the probability formula is different, that is, the calculation time of the importance of the feature is different. The time complexity of calculating the discernibility matrix in the algorithm in this paper is: $O\left(n|U|^2\right)$, Use the discernibility matrix to calculate the importance: $O(a|U|)$, Among them, a is the number of non-empty elements in the discrimination matrix, n is the number of features. The time complexity of the heuristic information is: $O\left(n|U|^2\right)+ O(a|U|)$, The time complexity of the traditional importance to find the neighborhood is: $O\left(n|U|^2\right)$, Using the neighborhood to find the importance of each feature: $O\left(n|U|^2\right)$, The total time complexity of its importance is: $O\left(2n|U|^2\right)$, It can be seen that the algorithm proposed in this paper can effectively reduce the computational workload.

4 Analysis of Experimental Data

4.1 Case Analysis

In order to verify the effectiveness of the algorithm in this paper, given a neighborhood decision Table 1. Use three attribute importance methods as heuristic information to find the optimal reduction, and analyze the results. $NS = (U, C \cup D, V, f)$, C is the conditional attribute, S is the decision attribute, the domain $U = \{x_1, x_2, x_3, x_4, x_5, x_6, x_7, x_8, x_9, x_{10}\}$ is a non-empty finite set. $|U| = 10$, The feature core set is:$\{x_5\}$. Minimum feature subset: $\{x_1, x_3, x_5, \}$.

Table 1. Decision table.

U	c_1	c_2	c_3	c_4	c_5	S	U	c_1	c_2	c_3	c_4	c_5	S
x_1	0.13	0.5	0.33	0.23	0.12	0	x_6	0.37	0.89	0.38	0.58	0.46	1
x_2	0.12	0.87	0.78	0.56	0.45	1	x_7	0.15	0.91	0.73	0.55	0.17	1
x_3	0.36	0.92	0.34	0.55	0.15	1	x_8	0.35	0.91	0.77	0.28	0.16	1
x_4	0.17	0.89	0.76	0.53	0.44	0	x_9	0.34	0.89	0.39	0.56	0.12	0
x_5	0.13	0.59	0.37	0.26	0.17	0	x_{10}	0.17	0.88	0.76	0.59	0.19	0

The radius of the neighborhood is set to 0.125, and the distance function uses Euclidean distance.

Calculate importance:

The Importance of the Positive of the Neighborhood Calculated from Definition 6:

$Pos_{C-\{c_1\}}(S) = \{x_1, x_5, x_6, x_8\}$ $Pos_{C-\{c_2\}}(S) = \{x_1, x_5, x_6, x_8\}$
$Pos_{C-\{c_3\}}(S) = \{x_1, x_5, x_6, x_8\}$ $Pos_{C-\{c_0\}}(S) = \{x_1, x_5, x_6, x_8\}$
$PoS_{C-\{c_5\}}(S) = \{x_1, x_5, x_8\}$ $PoS_C(S) = \{x_1, x_5, x_6, x_8\}$

$sig\left(c_5, C, S\right) = \frac{\left|POS_C(S) - POS_{C-\{c_5\}}(S)\right|}{|U|} = 0.1$

The same can be calculated:

$sig\left(c_1, C, S\right) = 0$ $sig\left(c_2, C, S\right) = 0$ $sig\left(c_3, C, S\right) = 0$ $sig\left(c_4, C, S\right) = 0$

Use the Neighborhood Information Entropy Attribute Importance of Definition 8 to calculate:

$sig\left(c_5, C, S\right) = 0.34$ $sig\left(c_1, C, S\right) = 0$ $sig\left(c_2, C, S\right) = 0$ $sig\left(c_3, C, S\right) = 0$ $sig\left(c_4, C, S\right) = 0$

If the above two methods of defining importance are used as heuristic information, and the feature core set is the starting point, because the non-core feature $\eta_{ij}(t) = sig(j, C, D) = 0$, the ant colony cannot accurately select the next feature to be added to the core set according to the probability formula of Eq. (15). As a result, the algorithm converges prematurely and cannot find the optimal feature subset.

Use the Method of this Paper to Calculate the Importance of the Neighborhood Resolution Matrix Attribute:

The discrimination matrix of Table 1 is as follows:

$$\begin{bmatrix}
\emptyset \\
c_2c_3c_4c_5 & \emptyset \\
c_1c_2c_4 & \emptyset & \emptyset \\
\emptyset & \emptyset & c_1c_3c_5 & \emptyset \\
\emptyset & c_2c_3c_4c_5 & c_1c_2c_4 & \emptyset & \emptyset \\
c_1c_2c_4c_5 & \emptyset & \emptyset & c_1c_3 & c_1c_2c_4c_5 & \emptyset \\
c_2c_3c_4 & \emptyset & \emptyset & c_5 & c_2c_3c_4 & \emptyset & \emptyset \\
c_1c_2c_3 & \emptyset & \emptyset & c_1c_4c_5 & c_1c_2c_3 & \emptyset & \emptyset & \emptyset \\
\emptyset & c_1c_3c_5 & \emptyset & \emptyset & \emptyset & c_5 & c_1c_3 & c_3c_4 & \emptyset \\
\emptyset & c_5 & c_1c_3 & \emptyset & \emptyset & c_1c_3c_5 & \emptyset & c_1c_4 & \emptyset & \emptyset
\end{bmatrix}$$

It can be seen from the discrimination matrix:

Number of non-empty elements $W = 22$. According to Theorem 1, the feature core set is $\{c_5\}$

Calculate the importance of features according to the Formula (18) in Definition 10:

The importance of a feature core is composed of the sum of its importance as a single feature and the importance of multiple features:

$Nsig\left(c_5\right) = 1 + \frac{\frac{1}{4} + \frac{1}{4} \cdots + \frac{1}{3}}{22} = 1.1061$

The same can be calculated:
$Nsig(c_1) = 0.236, Nsig(c_2) = 0.136, Nsig(c_3) = 0.219, Nsig(c_4) = 0.166$

According to the calculation results, the ant colony starts from the feature core set $\{c_5\}$. According to the probability Formula (13), the ants will preferentially select the most important feature c_1 and add it to the core set, and then select c_3, and finally the most feature subset searched by the ant is $\{c_1c_3c_5\}$, Consistent with the correct result.

The above calculation results show that the importance of each feature is greater than 0, which is consistent with the Property 1 obtained in this paper. $Nsig(c_5) > Nsig(c_1) > eNsig(c_3) > Nsig(c_4) > Nsig(c_2)$, This result is the same as the Property 2 obtained in this paper.

Based on the above, using the importance of the neighborhood resolution matrix as heuristic information, the ant colony accurately finds the optimal feature subset $\{c_1c_3c_5\}$, This is consistent with the results obtained by the rough set reduction theory of neighborhood. It proves the effectiveness of the algorithm in this paper.

4.2 Analysis of UCI Data Set

In order to further prove the effectiveness of the algorithm in this paper, the UCI data set is selected for experiment, as shown in Table 2. And use the formula $f(x_i) = \frac{x_i - x_{\min}}{x_{max} - x_{min}}$ to normalize. Use the algorithm NMACO in this paper to compare with the NRSACO algorithm in Literature [16] and the NEACOGS algorithm in Literature [17] for comparative experiments. The experimental environment: processor Intel (R) Core (TM) i5-8265U @1.60 GHz @1.80 GHz, RAM 8 GB, Using pycharm64 under Windons10 as the experimental platform, programming language: python.

Experimental parameter settings: $\alpha = 0.9, \beta = 0.1$, The initial pheromone concentration is 1. $\rho = 0.5q = 0.1$. The number of ants k satisfies the condition: $n = 1.5k$, n is the feature dimension. Since the difference in the radius of the neighborhood will affect the experimental results, in order to improve the effectiveness of the algorithm, the classification accuracy changes with the radius of the neighborhood are analyzed. All use CART classification, SVM classification algorithm for experiments. The experimental results are shown in Fig. 2. Among them, the classification accuracy formula is:

$$Classification\ accuracy = \frac{Number\ of\ samples\ corectly\ classified}{Total\ number\ of\ samples} \times 100\% \tag{20}$$

From Fig. 2 below, when the neighborhood radius is within the range of 0.1 0.175, the classification accuracy of the data set is relatively good. In this paper, the neighborhood radius is $\delta = 0.125$.

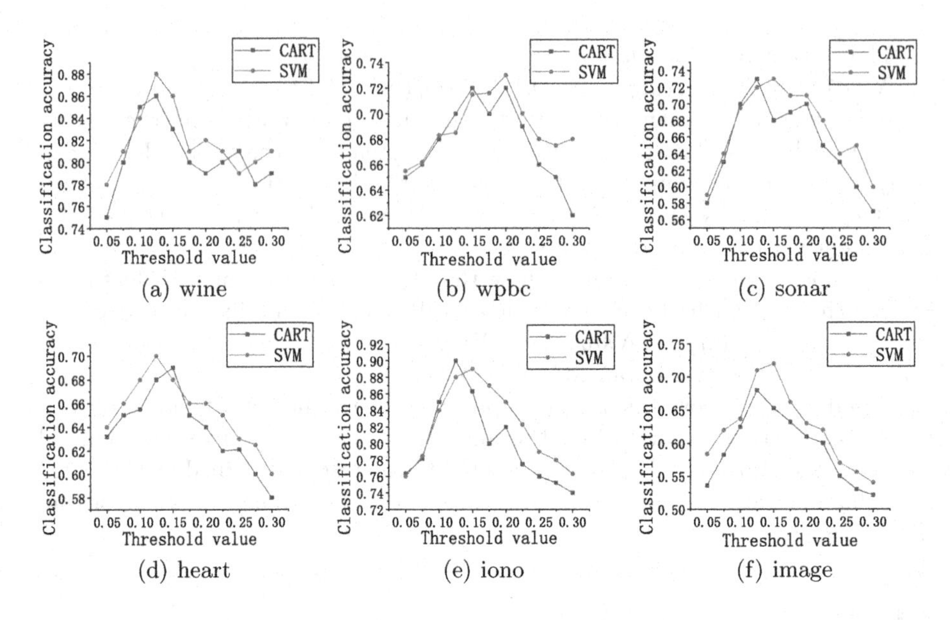

Fig. 2. Variation of classification accuracy with threshold value

Each algorithm was run 20 times, and the average value was taken as the result. The characteristic numbers of the three algorithms on the data set are shown in Table 3.

Table 2. Testing data set

UCI	Feature	Instances
1. wine	13	178
2. wpbc	33	198
3. sonar	60	208
4. heart	13	270
5. iono	34	351
6. image	19	2310

Table 3. Feature reduction.

UCI	NMACO	NEACOGS	NRSACO
1	4.5	5	5.5
2	5.2	7.2	8.5
3	5	7	7.2
4	5.5	6.5	7
5	6	7.5	7.5
6	7	7.2	8

From Table 3 above, it can be seen that the result of using the importance definition method based on the discernibility matrix as the heuristic information proposed in this paper has a better dimensionality reduction effect. It further proves the effectiveness of the algorithm proposed in this paper.

The following analyzes the convergence speed of the algorithm, these 6 data sets are used for testing. Figures 3 shows the curves of the fitness values of these data sets. The calculation formulas for the fitness values are as follows:

$$fitness = \lambda * \frac{|POS_R(S)|}{|U|} + (1-\lambda)\left|\frac{|C|-|R|}{|C|}\right| \tag{21}$$

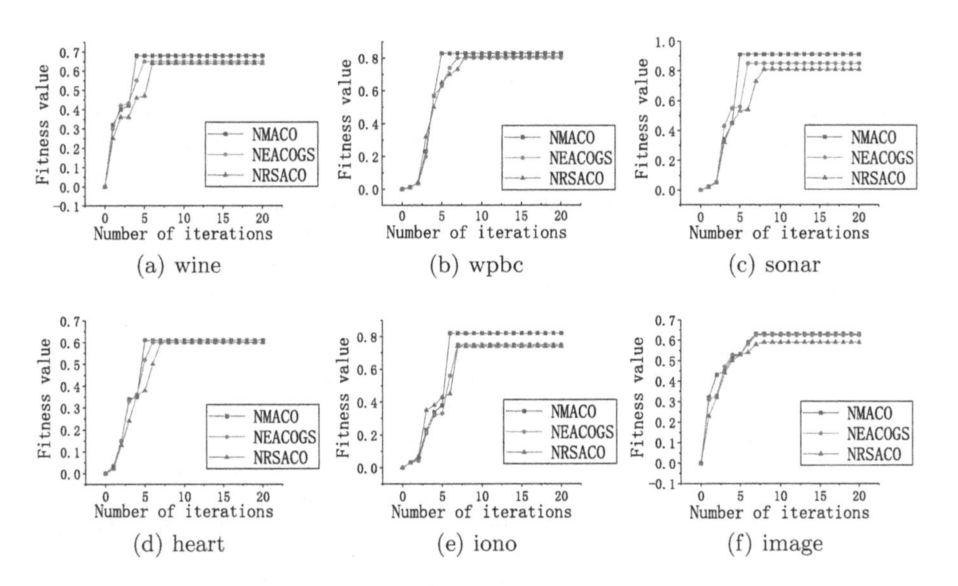

Fig. 3. The fitness value change curve of the data set

It can be seen from Fig. 3 that compared to the other two algorithms, the algorithm NMACO proposed in this paper has a faster convergence speed, reduces the number of iterations, and has a relatively high fitness value.

5 Concluding Remarks

This paper combines the definition of the importance of a neighborhood discernibility matrix with the ant colony optimization algorithm, and applies this importance to the heuristic information of the ant colony algorithm, and propose an feature selection algorithm based on neighborhood discernibility matrix and ant colony optimization. Using this neighborhood discrimination matrix to define the importance degree effectively and reasonably reflects the role played by each feature, so that the ant colony algorithm can accurately find the optimal reduction in the search process. The example analysis and experiment show that the algorithm is effective and feasible. However, the algorithm in this paper also has limitations. For example, in a very small number of neighborhood decision tables, some features will not appear in the neighborhood discrimination matrix, and the importance of these features that do not appear in the neighborhood discrimination matrix cannot be calculated. Making the ant colony unable to obtain heuristic information will also cause the algorithm to mature prematurely.

References

1. Dash, M., Liu, H.: Feature selection for classification. Intell. Data Anal. **1**(1), 131–156 (1997)
2. Lin, T.Y.: Granular computing on binary relations I: data mining and neighborhood systems. Rough Sets Knowl. Disc. **18**(1), 107–121 (1998)
3. Hu, Q.H., Yu, D.R., Xie, Z.X.: Numerical attribute reduction based on neighborhood granulation and rough approximation. J. Softw. **19**(03), 640–649 (2008)
4. Yao, S., Xu, F., Wu, Z.Y., Chen, J., Wang, J., Wang, W.: Nonmonotonic attribute reduction based on neighborhood rough mutual information entropy. Control Dec. **34**(02), 353–361 (2019)
5. Sheng, K., Wang, W., Bian, X.F., et al.: An incremental attribute reduction algorithm based on neighborhood differentiation for mixed data. Acta Electron Sini-ca **48**(4), 682–696 (2020)
6. Qu, B.Y., Li, C., Liang, J., Yan, L., Yu, K.J., Zhu, Y.S.: A self-organized speciation based multi-objective particle swarm optimizer for multimodal multi-objective problems. Appl. Soft Comput. **86**, 105–886 (2020)
7. Pan, L.Q., He, C., Tian, Y., Wang, H.D., Zhang, X.Y., Jin, Y.C.: A classification-based surrogate-assisted evolutionary algorithm for expensive many-objective optimization. IEEE Trans. Evol. Comput. **23**(1), 74–88 (2019)
8. Liu, Z.Z., Wang, B.C., Tang, K.: Handling constrained multiobjective optimization problems via bidirectional coevolution. IEEE Trans. Cybern. (99), 1–14 (2021)
9. Cui, Z.H., Chang, Y., Zhang, W.S., et al.: Improved NSGA-III with selection-and-elimination operator. Swarm Evol. Comput. **49**, 23–33 (2019)
10. Wang, H., Wu, Z.J., et al.: Improving artificial bee colony algorithm using a new neighborhood selection mechanism. Inf. Sci. **527**, 227–240 (2020)
11. Peng, X.R., Liu, Z.R., Gao, Y.: Artificial bee colony decision table reduction algorithm based on neighborhood rough set. J. Qingdao Univ. (Nat. Sci. Edn.) **30**(03), 55–59 (2017)
12. Peng, P., Ni, Z.W., Zhu, X.H., Xia, P.F.: Attribute reduction method based on improved binary glowworm swarm optimization algorithm and neighborhood rough set. Patt. Recogn. Artif. Intell. **33**(02), 95–105 (2020)
13. Chen, Y.M., Zhu, Q.X., et al.: Gene selection method based on neighborhood rough sets and fish swarm intelligence. J. Univ. Electron. Sci. Technol. China **47**(01), 99–104 (2018)
14. Feng, L., Li, C., Shen, L.: Facial expresion feature selection method based on neighborhod rough set theory and quantum genetic algorithm. J. Hefei Univ. Technol. (Nat. Sci. Edn.) **36**(01), 39–42+128 (2013)
15. Dorigo, M., Maniezzo, V., Colorni, A.: Ant system: optimization by a colony of cooperating agents. IEEE Trans. Syst. Man Cybern. Part B: Cybern. **26**(1), 29–41 (1996)
16. Zhang, D.W., Wang, P., Qiu, J.Q.: Approach to feature selection based on neighbourhood rough set and colony optimization. J. Hebei Univ. Sci. Technol. **32**(05), 403–408 (2011)

17. Xu, M., Zheng, L.B., Xie, Y.Q., Chen, Y.M.: Gene selection algorithm based on neighbor-hood entropy and ant colony optimization. J. Fuzhou Univ. (Nat. Sci. Edn.) **45**(06), 815–821 (2017)
18. Hu, Q.H., Yu, D.R., Liu, J., et al.: Neighborhood rough set based heterogeneous feature subset selection. Inf. Sci. **178**(18), 3577–3594 (2008)
19. Liu, Y., Huang, W., Jiang, Y., et al.: Quick attribute reduct algorithm for neighborhood rough set model. Inf. Sci. **271**(7), 65–81 (2014)
20. Lou, C., Liu, Z.R., Guo, G.Z.: Fast reduction algorithm of neighborhood rough set based on block set. Comput. Sci. **41**(S2), 337–339+363 (2014)
21. Sudholt, D., Thyssen, C.: A simple ant colony optimizer for stochastic shortest path problems. Algorithmica **64**(4), 643–672 (2012)
22. Zhang, X.L., Chen, X.F., He, Z.J.: An ACO-based algorithm for parameter optimization of support vector machines. Expert Syst. Appl. **37**(9), 6618–6628 (2010)
23. Jensen, R., Shen, Q.: Finding rough set reducts with ant colony optimization, pp. 15–22. UK, Proc of the UK Workshop on Computational Intelligence. Bristol (2003)
24. Skowron, A., Rauszer, C.: The discernibility matrices and functions in information system . In: Slowinski Red. Intelligent Decision Support Handbook of Applications and Advances of the Rough Sets Theory. Dordrecht: Kluwer Academic Publishers, pp. 331–362 (1992)
25. Hu, X.H., Cercone, N.: Learning in relational databases: a rough set approach. Comput. Intell. **11**(2), 323–337 (1995)

Implementation and Application of NSGA-III Improved Algorithm in Multi-objective Environment

Fei Xue[1], Yuelu Gong[1(✉)], Qiuru Hai[1], Huilin Qin[1], and Tingting Dong[2]

[1] Beijing Wuzi University, Beijing 101149, China
gylxbt@163.com
[2] Beijing University of Technology, Beijing 100124, China

Abstract. Recently, the development of evolutionary multi-objective optimization (EMO) algorithm to deal with multi-objective optimization problems (with four or more objectives) has gradually become a hot spot. NSGA-III algorithm is effective in dealing with evolutionary multi-objective optimization problems. In this paper, we recognize some advantages of the existing NSGA-III algorithm and make some improvements. The improved NSGA-III algorithm has higher adaptability and can provide more dense Pareto-optimal front under the same amount of computation. The improved NSGA-III algorithm is applied to many multi-objective testing problems with 3 to 8 objectives, and its performance is compared with the existing multi-objective evolutionary algorithms. Experimental results show that the improved algorithm can produce satisfactory results for all the problems considered in this study. Among the 28 environments with all values, the improved NSGA-III algorithm has 22 optimal values, accounting for 78.57%. After that, we analyze the results and put forward the future improvement and research direction.

Keywords: Many-objective optimization · Multi-criterion optimization · Large dimension · NSGA-III · Evolutionary computation

1 Introduction

Multi-objective optimization method (EMO) has been proposed since the early 1990s. Its purpose is to find a group of non dominated solutions in two-objective or three-objective problems, so as to make its convergence and diversity better. However, with the development of practical research, more and more problems involve four objectives or even more multi-objective optimization, and sometimes even 15 objectives [1, 2]. These optimization problems put forward more requirements for the optimization algorithm. One of the most important problems is that with the increase of the number of objectives, the proportion of non dominated solutions in the randomly selected objective vector set will increase exponentially. The consequence of this phenomenon is that the non dominated solution will occupy a lot of space in the population, so the evolutionary algorithms retained by the elite lack space to accommodate new solutions. At the same time, the visualization of large-dimensional frontier, such as hyper volume

L. Pan et al. (Eds.): BIC-TA 2021, CCIS 1565, pp. 132–144, 2022.
https://doi.org/10.1007/978-981-19-1256-6_10

measure [3] or other metrics [4, 5], is facing the situation of too high computing cost, or the diversity preserving operators such as crowding distance operator [6] or clustering operator [7] are complex and expensive in computing.

By extending NSGA-II [6] framework, a new multi-objective evolutionary algorithm, NSGA-III [9], is proposed. NSGA-III algorithm converges to Pareto optimal front by predefining a group of reference points, and the species are distributed around the whole front. Next, the algorithm is searched intensively to determine the relevant Pareto optimal solution of each reference point. In this paper, we extend NSGA-III algorithm to deal with more challenging multi-objective constrained optimization problems.

In the next part of this paper, we will introduce and review some problems related to emo algorithm in Sect. 2. In Sect. 3 of this paper, we will introduce our proposed improved NSGA-III program. The comparative experiments with other algorithms and the experimental results will be arranged in Sect. 4 of this paper. The conclusions and future improvement directions will be summarized in Sect. 5.

2 Many-Objective Problems and EMO Methodologies

2.1 Potential Difficulties in Dealing with Multi-objective Problems

Some existing emo algorithms [10] may encounter the following difficulties:

The Proportion of Non Dominated Solutions Is Large. More and more parts of the population will become non dominated groups with the increase of the number of targets [4, 11]. Many EMO algorithms focus on the non dominated solution in the whole, which makes the space of innovative solution occupied, so the search process becomes slow and the implementation efficiency of the algorithm is relatively low.

It Is Difficult to Deal with Large Dimensional Space. In a large dimensional space, if only a small decomposition of a multi-objective problem can be found, the distance between these decomposition may be large. As a key operator in EMO algorithm, the effect of reorganization operator will also be greatly reduced. For example, two parent solutions are far away, which may cause the resulting child solution to be farther away from the parent solution.

Assessing Diversity Requires High Costs. The larger the dimension of a space, the greater the computational cost and the higher the computational cost of identifying the domain when measuring the crowding degree of the solution of the population. If the calculation speed is accelerated by approximation, the final solution may become undesirable in diversity evaluation.

2.2 Two Strategies to Face These Difficulties

In the face of these possible difficulties, we can have the following two ways:

Look for a Special Dominant Principle. We can set a special control principle, which can find the adaptive discrete Pareto-optimal front and a well distributed point set.

For example, using a dominance principle [12, 13], all points of a group of Pareto-optimal points are in the dominant state. In this process, a limited number of Pareto-optimal points will be generated. Other special principles can also achieve this goal [14, 15]. The difficulty in diversity can be solved by SBX [16] with large distribution index. The idea can also be realized by mating restriction or special recombination. For example, Sato et al. [17] proposed to check the dominance of the target subset, and then make different combinations in each generation. Of course, the shortcomings of these solutions are carried out in low dimensional space, and the success of solving multi-objective optimization problems can not be guaranteed.

Predefined Multiple Reference Points. When it is in a large dimensional space, it is very difficult for a single optimization algorithm to achieve the two objectives of simultaneously converging the generated population near the Pareto-optimal frontier and evenly distributing it in the whole frontier. Therefore, we can take some external auxiliary means to realize the problem of diversity maintenance. We can preset multiple predefined target searches, so we don't have to find Pareto-optimal solutions in the whole space. This method effectively solves the problem that most of them are non dominated solutions, because we have found the search best corresponding to each target task. NSGA-III algorithm is also based on this principle, so it will be analyzed in more detail here. The first method is to specify the search direction across the entire Pareto-optimal front in advance, and multiple search tasks can be carried out along each direction. In this method, due to the wide range of search directions, the most advantages obtained in most problems can be distributed on the optimal frontier. For example, MEOAD [18] uses this idea. Another method is to specify multiple predefined reference points and highlight the corresponding points of each reference point, so as to obtain widely distributed Pareto-optimal points. This paper perfects the method of [19].

3 NSGA-III Algorithm and Its Improvement

The basic framework of NSGA-III algorithm is similar to its original version NSGA-II [6], but the selection mechanism is completely different. NSGA-III updates a large number of reference points through adaptive methods. These reference points are well distributed and can ensure the diversity of population members.

3.1 Determines the Reference Point on the Hyperplane

As we mentioned above, NSGA-III obtains the diversity of solutions by predefined reference points. If there is no preference, we can use any solution structure. In this paper, we select a systematic method [1] of DAS and Dennis in [20]. On a normalized hyperplane, if p-division method is adopted along each target, the total number of reference points(H) in M target problems is:

$$H = \binom{M + p - 1}{p} \tag{1}$$

As shown in Fig. 1, we give a three objective problem, that is, M = 3, and establish reference points on triangles with vertex coordinates of (1, 0, 0), (0, 1, 0), (0, 0, 1). In NSGA-III algorithm, on the one hand, we pay attention to the non dominated solution, on the other hand, we will also pay attention to the middle group members related to each reference point. The reference points we created are widely distributed in the normalized hyperplane, so the obtained solutions may also be widely distributed near the Pareto-optimal front. The method of Algorithm 1 is to try to find the approximate Pareto-optimal solution corresponding to the reference point. The specific steps of Algorithm 1 are given below.

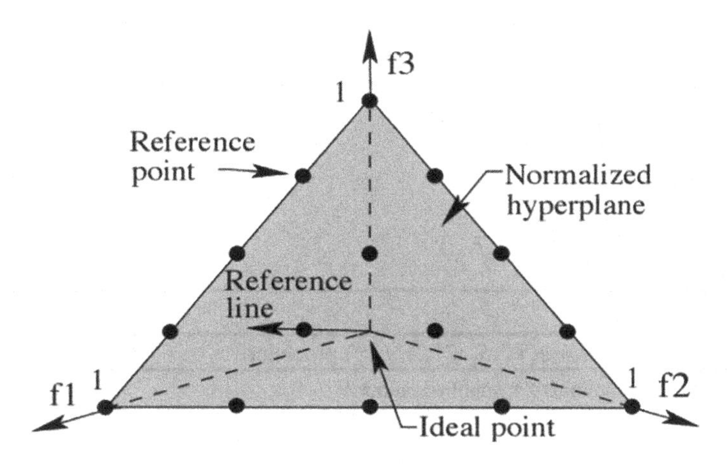

Fig. 1. For the three objective problem with P = 4, the reference point is established in the schematic diagram of the standardized reference plane.

3.2 Normalization of Population Members

As shown in Fig. 2, we give an m-dimensional hyperplane composed of M (M = 3) extreme vectors.

On this hyperplane, we calculate the intercept between the i-th target axis and the linear hyperplane. We use the methods of DAS and Dennis [20]. The calculation method of the objective function is as follows:

$$f_i^n(X) = \frac{f_i^{'}(X)}{a_i - z_i^{min}} = \frac{f_i(X) - z_i^{min}}{a_i - z_i^{min}}, \quad \text{for i} = 1, 2, \ldots, M \tag{2}$$

Algorithm 1 Generation t

Input: H structured reference points Z^s, population P_t
Output: P_{t+1}
1: $S_t = \emptyset$, i = 1
2: Q_t =Recombination+Mutation(P_t)
3: $R_t = P_t \cup Q_t$
4: $(F_1, F_2 \ldots)$ = Non-dominated-sort(R_t)
5: **repeat**
6: $S_t = S_t \cup F_i$ and $i = i + 1$
7: **until** $|S_t| \geq N$
8: $F_l = F_i$
9: **if** $|S_t| = N$ **then**
10: $P_{t+1} = S_t$,break
11: **else**
12: $P_{t+1} = \cup_{j=1}^{l-1} F_j$
13: $F_l: K = N - |P_{t+1}|$
14: $Z^r: Normalize(f^n, S_t, Z^r, Z^s, Z^a)$
15: $[\pi(s), d(s)] = \mathrm{Associate}(S_t, Z^r)$
16: $j \in Z^r: \rho_j = \sum_{\in S_t/F_l}((\pi(s) = j)?1:0)$
17: $P_{t+1}: Niching(K, \rho_j, \pi, d, Z^r, F_l, P_{t+1})$
18: **end if**

Algorithm 2 $Normalize(f^n, S_t, Z^r, Z^s/Z^a)$ procedure

Input: S_t ,structured points Z^s, supplied points Z^a
Output: f^n, Z^r
1: **for** j = 1 **to** M **do**
2: $z_j^{min} = min_{s \in S_t} f_j(s)$
3: $f_j'(s) = f_j(s) - z_j^{min} \quad \forall s \in S_t$
4: $z^{j,max}$ = s: $\mathrm{argmin}_{s \in S_t} \mathrm{ASF}(s, w^j)$,where $w^j = (\epsilon, \ldots, \epsilon)^T$ $\epsilon = 10^{-6}$,and $w_j^j = 1$
5: **end for**
6: Compute intercepts a_j for j =1,...,M
7: Calculation (f^n) using formula 2
8: **if** Z^a is given **then**
9: Save each point in Z^r with formula 2
10: **else**
11: $Z^r = Z^s$
12: **end if**

Select the reference point on the normalized hyperplane, and then map the reference point to the normalized hyperplane using formula 2. In this way, NSGA-III program can adaptively maintain the cross spatial diversity of members of each generation, and NSGA-III algorithm can better solve the problem that the target value has Pareto-optimal front in different proportions. The specific process will be given in Algorithm 2.

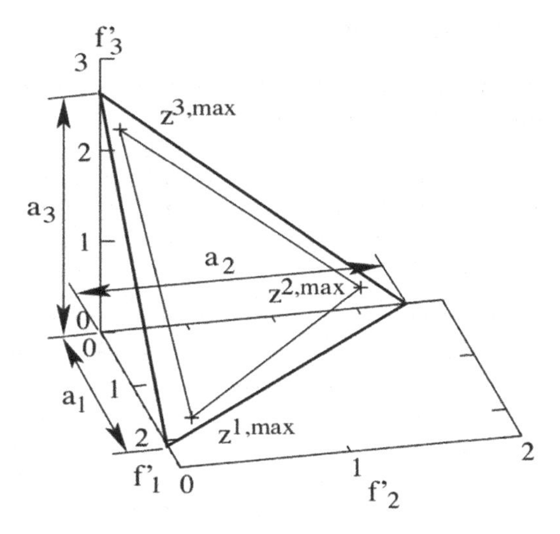

Fig. 2. Calculate the intercept of a three objective problem and form a hyperplane from the pole value.

3.3 Association Operation

Previously, we have carried out adaptive normalization on the member range of the target space. Next, we will associate the member reference points in the individual population. As shown in Fig. 3, we connect the reference point with the origin and define a reference line for each reference point on the hyperplane. Next, we calculate the vertical distance of each population member from the reference line and the reference point closest to the normalized target space, which we call population member association.

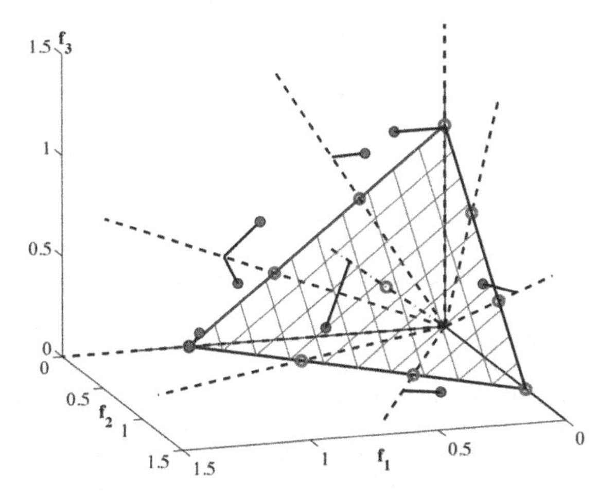

Fig. 3. Associated operation diagram.

3.4 Inheritance of Population Offspring

After the formation of a new generation population, we can use genetic operators to create a new offspring population. In the improved NSGA-III algorithm, we make a new elite selection, and ensure the diversity of the population by finding the reference line closest to each reference point. We attach great importance to every member of the population, so we do not perform explicit operations. However, if you encounter large distribution index values, we recommend using the SBX operator.See Algorithm 3 and Algorithm 4 for the specific process.

Algorithm 3 Associate(S_t, Z^r) procedure

Input: S_t , Z^r
Output: $\pi(s \in S_t), d(s \in S_t)$
1: **for** $z \in Z^r$ **do**
2: $w = z$
3: **end for**
4: **for** each $s \in S_t$ **do**
5: **for** each $w \in Z^r$ **do**
6: $d\perp(s, w) = s - w^T/||w||$
7: **end for**
8: Assign w: $argmin_{w \in Z^r} d \perp (s, w)$
9: Assign $d(s) = d \perp (s, \pi(s))$
10: **end for**

Algorithm 4 $Niching(K, \rho_j, \pi, d, Z^r, F_l, P_{t+1})$ procedure

Input: K, ρ_j,$\pi(s \in S_t)$,Z^r,F_l
Output: P_{t+1}
1: $k = 1$
2: **while** $k \leq K$ **do**
3: $J_{min} = \{j: argmin_{j \in Z^r} \rho_j\}$
4: $\bar{j} = random(J_{min})$
5: $I_{\bar{j}} = \{s: \pi(s) = \bar{j}, s \in F_l\}$
6: **if** $I_{\bar{j}} \neq \emptyset$ **then**
7: **if** $\rho_{\bar{j}} = 0$ **then**
8: $P_{t+1} = P_{t+1} \cup (s: argmin_{s \in I_{\bar{j}}} d(s))$
9: **else**
10: $Z^r = Z^r/\{\bar{j}\}$
11: **end if**
10: **end while**

3.5 Evaluation of Service Performance

The improved NSGA-III algorithm still does not need to set other parameters other than conventional genetic parameters, such as population size, termination factor, mutation probability, crossover probability, or other relevant parameters. The user is free to determine the number of reference points. Therefore, it can be seen that our algorithm is more convenient and fast in use.

Next, we will show some comparison results between the improved NSGA-III algorithm and some other classical algorithms in multi-objective problems.

4 Experiment

We use the improved NSGA-III algorithm and several classical algorithms to simulate the multi-objective optimization problem. The classical algorithms include MOEAD and some of its derivative algorithms [23],which have similar frameworks,but also have their own characteristics.

As a performance measure, we select the general distance (GD) metric [24, 25], which can measure the distance between approximate solution sets on the real Pareto-optimal front. It is necessary to obtain a set of solution sets uniformly sampled on the real Pareto optimal front in advance. Let P* be a set of solutions uniformly sampled on the real Pareto-optimal front, and S be the Pareto optimal front approximate solution obtained by the multi-objective evolutionary algorithm, then GD is defined as follows:

$$\mathrm{GD}(\mathrm{S}, \mathrm{P}^*) = \frac{\sqrt{\sum_{\mathrm{x} \in \mathrm{S}} \mathrm{dist}(\mathrm{x}, \mathrm{P}^*)^2}}{|\mathrm{S}|} \tag{3}$$

The smaller the the value of GD, the better the convergence of S and the closer it can approach the whole Pareto-optimal front.

As a performance measure, we selected the inverse general distance(IGD) metric [24, 25]. This index can provide information about the quality of solutions,including convergence and diversity. We call all Pareto-optimal points in the target space a set. The measurement method of IGD is to calculate the average Euclidean distance of all points in the set and the nearest member of all points in the set. The calculation method is as follows:

$$\mathrm{IGD}(\mathrm{A}, \mathrm{Z}) = \frac{1}{|\mathrm{Z}|} \sum_{\mathrm{i}=1}^{|\mathrm{Z}|} \min_{j=1}^{|A|} d(z_i, a_j) \tag{4}$$

The smaller the value of IGD, the better the quality of the solution.

4.1 Experimental Environment Setting

We test in DTLZ4, WFG4, WFG5, WFG6, WFG7, WFG8 and WFG9. We compare our improved NSGA-III algorithm with MOEAD, MOEADDE, MOEADDRA and MOEAPC [23]. The basic parameters are configured according to the original experiment to maximize the performance of the algorithm as much as possible, and the multi-objective value is between 3–8.

4.2 Analysis of Experimental Results

The following test results will be given in the form of tables. The best value obtained by each group will be expressed in rent increase for analysis and comparison.

When M = 3, the experimental comparison results of GD values are shown in Table 1. From Table 1, we can see that when M = 3, our improved algorithm achieves the best value in five environments of WFG4, WFG6, WFG7, WFG8 and WFG9.

Table 1. When M = 3, the GD value of the improved NSGA-III and several comparison algorithms in the given target test problem.

Problem	MOEADDRA	MOEADDE	NSGA-III	MOEAPC	MOEAD
DTLZ4	5.6454e−3 (5.80e−3)	1.3675e−3 (7.08e−4)	5.9750e−4 (8.84e−5)	9.2617e−2 (2.44e−2)	**3.9643e−4** **(2.33e−4)**
WFG4	1.9557e−2 (2.05e−3)	1.7040e−2 (1.88e−3)	**6.5045e−3** **(5.83e−4)**	1.8485e−2 (1.29e−3)	1.4483e−2 (2.32e−3)
WFG5	8.6292e−3 (5.34e−4)	**8.3374e−3** **(5.40e−4)**	9.2807e−3 (3.40e−4)	8.7141e−3 (6.27e−4)	1.2813e−2 (1.51e−3)
WFG6	2.6055e−2 (2.97e−3)	2.3846e−2 (3.48e−3)	**1.6247e−2** **(3.62e−3)**	2.4645e−2 (3.68e−3)	2.3588e−2 (4.28e−3)
WFG7	1.6142e−2 (3.62e−3)	1.3848e−2 (2.17e−3)	**6.4379e−3** **(7.49e−4)**	1.7374e−2 (1.65e−3)	3.1140e−2 (8.61e−3)
WFG8	5.6854e−2 (1.52e−2)	3.9505e−2 (5.28e−3)	**2.4316e−2** **(1.10e−3)**	3.4428e−2 (1.62e−3)	3.4538e−2 (5.40e−3)
WFG9	1.7533e−2 (7.08e−3)	9.8729e−3 (3.29e−3)	**9.2974e−3** **(1.45e−3)**	2.0745e−2 (7.23e−3)	2.7783e−2 (8.82e−3)

When M = 5, the experimental comparison results of GD values are shown in Table 2. From Table 2, we can see that when M = 3, our improved algorithm achieves the best value in five environments of WFG4, WFG5, WFG6, WFG7 and WFG8.

Table 2. When M = 5, the GD value of the improved NSGA-III and several comparison algorithms in the given target test problem.

Problem	MOEADDRA	MOEADDE	NSGA-III	MOEAPC	MOEAD
DTLZ4	1.9306e−2 (1.03e−2)	9.2509e−3 (3.15e−3)	5.2014e−3 (5.94e−4)	7.4709e−2 (4.37e−3)	**3.6456e−3** **(9.38e−4)**
WFG4	4.1396e−2 (6.46e−3)	4.3131e−2 (5.03e−3)	**3.2820e−2** **(1.42e−3)**	4.6424e−2 (1.63e−3)	3.4725e−2 (3.04e−3)
WFG5	4.2554e−2 (4.77e−3)	3.6631e−2 (3.55e−3)	**3.2063e−2** **(2.54e−3)**	3.5764e−2 (1.76e−3)	3.6388e−2 (1.16e−3)
WFG6	4.7208e−2 (5.76e−3)	4.5131e−2 (4.49e−3)	**4.3210e−2** **(6.92e−3)**	7.0390e−2 (5.05e−3)	4.3411e−2 (3.88e−3)
WFG7	5.4378e−2 (1.45e−2)	6.5739e−2 (1.18e−2)	**3.5081e−2** **(4.28e−3)**	5.0816e−2 (1.29e−3)	4.1365e−2 (3.69e−3)
WFG8	1.2786e−1 (2.87e−2)	1.1493e−1 (1.88e−2)	**6.5634e−2** **(2.39e−3)**	6.9557e−2 (2.58e−3)	9.5939e−2 (1.70e−2)
WFG9	5.8654e−2 (8.26e−3)	4.7755e−2 (4.89e−3)	5.1650e−2 (6.35e−3)	**3.8896e−2** **(3.20e−3)**	4.3408e−2 (7.44e−3)

When M = 3, the experimental comparison results of IGD values are shown in Table 3. From Table 3, we can see that when M = 3, our improved algorithm achieves the best value in seven environments of DTLZ4, WFG4, WFG5, WFG6, WFG7, WFG8 and WFG9.

Table 3. When M = 3, the IGD value of the improved NSGA-III and several comparison algorithms in the given target test problem.

Problem	MOEADDRA	MOEADDE	NSGA-III	MOEAPC	MOEAD
DTLZ4	2.7550e−1 (1.13e−1)	1.8257e−1 (7.32e−2)	**1.5237e−1** **(1.98e−1)**	3.5586e−1 (7.23e−2)	4.2787e−1 (3.44e−1)
WFG4	4.1538e−1 (2.23e−2)	4.0061e−1 (1.64e−2)	**2.3161e−1** **(2.44e−3)**	3.3291e−1 (8.87e−3)	2.9322e−1 (1.12e−2)
WFG5	3.3846e−1 (7.91e−3)	3.3733e−1 (6.07e−3)	**2.3719e−1** **(2.04e−3)**	2.8641e−1 (3.21e−3)	2.7337e−1 (9.71e−3)
WFG6	4.7403e−1 (8.45e−2)	4.4108e-1 (2.20e−2)	**2.7491e−1** **(2.06e−2)**	3.7143e−1 (2.94e−2)	3.3229e−1 (2.21e−2)
WFG7	4.2451e−1 (3.06e−2)	3.8205e−1 (1.32e−2)	**2.3151e−1** **(2.76e−3)**	3.2107e−1 (1.09e−2)	3.9619e−1 (5.07e−2)
WFG8	5.7634e−1 (8.85e−2)	4.8748e−1 (2.73e−2)	**3.2130e−1** **(8.60e−3)**	4.2020e−1 (1.56e−2)	3.8001e−1 (2.80e−2)
WFG9	3.9581e−1 (3.57e−2)	3.5169e−1 (1.63e−2)	**2.3858e−1** **(7.03e−3)**	3.5230e−1 (5.62e−2)	3.5655e−1 (5.22e−2)

When M = 5, the experimental comparison results of IGD values are shown in Table 4. From Table 4, we can see that when M = 5, our improved algorithm performs best in DTLZ4, WFG6, WFG7 and WFG8,and is close to the optimal value in WFG4, WFG5, and WFG9.

Table 4. When M = 5, the IGD value of the improved NSGA-III and several comparison algorithms in the given target test problem.

Problem	MOEADDRA	MOEADDE	NSGA-III	MOEAPC	MOEAD
DTLZ4	6.0085e−1 (8.31e−2)	4.8841e−1 (3.60e−2)	**3.2242e−1** **(1.13e−1)**	6.1287e−1 (2.18e−2)	6.9800e−1 (1.81e−1)
WFG4	2.1201e+0 (1.96e−1)	2.3511e+0 (1.47e−1)	1.2278e+0 (4.70e−3)	**1.2217e+0** **(2.11e−2)**	2.5540e+0 (2.73e−1)
WFG5	2.3886e+0 (1.39e−1)	2.4916e+0 (9.25e−2)	1.2069e+0 (5.05e−3)	**1.1597e+0** **(1.92e−2)**	2.5754e+0 (1.67e−1)
WFG6	2.3124e+0 (2.45e−1)	2.3140e+0 (2.17e−1)	**1.2452e+0** **(9.32e−3)**	1.4267e+0 (5.14e−2)	3.1756e+0 (2.70e−1)
WFG7	2.2909e+0 (2.20e−1)	2.5627e+0 (1.50e−1)	**1.2520e+0** **(9.31e−3)**	1.3035e+0 (2.16e−2)	2.8317e+0 (2.44e−1)
WFG8	2.5187e+0 (2.59e−1)	2.6572e+0 (1.71e−1)	**1.3104e+0** **(7.79e−2)**	1.4955e+0 (2.88e−2)	2.3342e+0 (2.90e−1)
WFG9	2.1996e+0 (2.01e−1)	2.2804e+0 (1.41e−1)	1.2276e+0 (2.94e−2)	**1.1974e+0** **(2.65e−2)**	2.5378e+0 (1.66e−1)

When M = 6, the experimental comparison results of IGD values are shown in Table 5. From Table 5, we can see that when M = 6, our improved algorithm performs best in DTLZ4, WFG6, WFG7 and WFG8, and is close to the optimal value in WFG4, WFG5, and WFG9.

Table 5. When M = 6, the IGD value of the improved NSGA-III and several comparison algorithms in the given target test problem.

Problem	MOEADDRA	MOEADDE	NSGA-III	MOEAPC	MOEAD
DTLZ4	7.3334e−1 (7.47e−2)	6.5108e−1 (5.57e−2)	**3.7693e−1** **(1.04e−1)**	6.5970e−1 (1.84e−2)	7.2475e−1 (1.82e−1)
WFG4	2.9100e+0 (1.74e−1)	3.1579e+0 (1.91e−1)	1.9534e+0 (3.91e−2)	**1.8692e+0** **(3.05e−2)**	4.6898e+0 (3.37e−1)
WFG5	2.9569e+0 (1.53e−1)	3.2573e+0 (1.37e−1)	1.9276e+0 (1.37e−2)	**1.8933e+0** **(3.56e−2)**	4.6056e+0 (1.65e−1)
WFG6	3.2697e+0 (3.84e−1)	3.3586e+0 (3.39e−1)	**1.9689e+0** **(1.88e−2)**	2.1519e+0 (3.64e−2)	5.1034e+0 (1.99e−1)
WFG7	3.0275e+0 (2.40e−1)	3.3668e+0 (2.25e−1)	**1.9748e+0** **(1.79e−2)**	1.9967e+0 (2.48e−2)	4.7863e+0 (3.14e−1)
WFG8	3.3704e+0 (2.14e−1)	3.5533e+0 (2.01e−1)	**2.1254e+0** **(1.30e−1)**	2.2566e+0 (4.41e−2)	3.9381e+0 (4.80e−1)
WFG9	2.8940e+0 (1.95e−1)	3.1225e+0 (1.73e−1)	1.9619e+0 (7.08e−2)	**1.9161e+0** **(3.32e−2)**	4.3642e+0 (1.66e−1)

When M = 8, the experimental comparison results of IGD values are shown in Table 6. From Table 6, we can see that when M = 8, our improved algorithm achieves the best value in seven environments of DTLZ4, WFG4, WFG5, WFG6, WFG7, WFG8 and WFG9.

Table 6. When M = 8, the IGD value of the improved NSGA-III and several comparison algorithms in the given target test problem.

Problem	MOEADDRA	MOEADDE	NSGA-III	MOEAPC	MOEAD
DTLZ4	8.8073e−1 (8.35e−2)	8.3028e−1 (5.56e−2)	**5.0404e−1** **(7.89e−2)**	7.0599e−1 (1.34e−2)	8.3377e−1 (1.59e−1)
WFG4	5.1049e+0 (2.89e−1)	5.2623e+0 (2.38e−1)	**3.6011e+0** **(9.73e−2)**	3.9755e+0 (7.76e−2)	8.5423e+0 (2.37e−1)
WFG5	4.7660e+0 (1.41e−1)	5.3015e+0 (1.95e−1)	**3.5653e+0** **(7.63e−2)**	4.1034e+0 (7.08e−2)	8.0331e+0 (2.73e−1)
WFG6	5.7972e+0 (6.14e−1)	6.0757e+0 (4.35e−1)	**3.6692e+0** **(1.61e−1)**	4.4307e+0 (8.36e−2)	8.5717e+0 (1.70e−1)
WFG7	5.5091e+0 (4.98e−1)	5.4877e+0 (3.44e−1)	**3.6264e+0** **(1.45e−1)**	4.2203e+0 (6.74e−2)	8.5777e+0 (5.39e−1)
WFG8	5.8371e+0 (5.07e−1)	5.6213e+0 (2.71e−1)	**3.8574e+0** **(1.52e−1)**	4.4898e+0 (4.43e−2)	7.2763e+0 (8.31e−1)
WFG9	4.8121e+0 (2.65e−1)	5.1757e+0 (3.14e−1)	**3.6461e+0** **(6.47e−2)**	4.1505e+0 (7.92e−2)	7.9925e+0 (2.84e−1)

Among the 14 environments with all GD values, the improved NSGA-III algorithm has 10 optimal GD values, accounting for 71.42%.

Among the 28 environments with all IGD values, the improved NSGA-III algorithm has 22 optimal IGD values, accounting for 78.57%.

5 Conclusion

In this paper, the improvement of our proposed NSGA-III algorithm is mainly to use the reference point method to optimize the algorithm and improve the quality of the solution. At the same time, we also prove through the experimental data that the improved NSGA-III algorithm has a better effect in maintaining the diversity of solutions. Our algorithm needs less computation and saves the computational cost based on multi-objective optimization and decision-making methods. At the same time, we do not need to set additional parameters, which is also a more convenient side proof of our algorithm.Among the 14 environments with all GD values, the improved NSGA-III algorithm has 10 optimal GD values, accounting for 71.42%. Among the 28 environments with all values, the improved NSGA-III algorithm has 22 optimal IGD values, accounting for 78.57%.

In future research, we will consider combining the improved NSGA-III algorithm with other excellent algorithms, carry forward their respective advantages, deal with the best part of each algorithm, save computing time and computing resources, and further solve multi-objective problems.We can also consider applying it to some new problems and expanding its application scenarios. Of course, if there are more ideal improvements, we will continue to study and follow up.

References

1. Chikumbo, O., Goodman, E., Deb, K.: Approximating a multidimensional pareto front for a land use management problem: a modified MOEA with an epigenetic silencing metaphor. In: Proceedings of Congress on Evolutionary Computation (CEC-2012), pp. 1–8 (2012)
2. Coello, C.A.C., Lamont, G.B.: Applications of Multi-Objective Evolutionary Algorithms. World Scientifiic, Singapore (2004)
3. Zitzler, E., Thiele, L.: Multiobjective evolutionary algorithms: a comparative case study and the strength Pareto approach. IEEE Trans. Evol. Comput. **3**(4), 257–271 (1999)
4. Deb, K.: Multi-objective Optimization Using Evolutionary Algorithms. Wiley, Chichester (2001)
5. Zitzler, E., Thiele, L., Laumanns, M., Fonseca, C.M., Fonseca, V.G.: Performance assessment of multiobjective optimizers: an analysis and review. IEEE Trans. Evol. Comput. **7**(2), 117–132 (2003)
6. Deb, K., Agrawal, S., Pratap, A., Meyarivan, T.: A fast and elitist multi-objective genetic algorithm: NSGA-II. IEEE Trans. Evol. Comput. **6**(2), 182–197 (2002)
7. Zitzler, E., Laumanns, M., Thiele, L.: SPEA2: improving the strength Pareto evolutionary algorithm for multiobjective optimization. In: Giannakoglou, K.C., Tsahalis, D.T., P′eriaux, J., Papailiou, K.D., Fogarty, T. (eds.) Evolutionary Methods for Design Optimization and

Control with Applications to Industrial Problems. Athens, Greece: International Center for Numerical Methods in Engineering (CIMNE), pp. 95–100 (2001)

8. Jain, H., Deb, K.: An evolutionary many-objective optimization algorithm using reference-point based nondominated sorting approach, Part II: handling constraints and extending to an adaptive approach. IEEE Trans. Evol. Comput. **18**(4), 602–622 (2014)
9. Deb, K., Jain, H.: An improved NSGA-II procedure for manyobjective optimization Part I: Problems with box constraints. Indian Institute of Technology Kanpur, Technical report, 2012009 (2012)
10. Chankong, V., Haimes, Y.Y.: Multiobjective Decision Making Theory and Methodology. North-Holland, New York (1983)
11. Garza-Fabre, M., Pulido, G.T., Coello, C.A.C.: Ranking methods for many-objective optimization. In: Aguirre, A.H., Borja, R.M., Garciá, C.A.R. (eds.) MICAI 2009. LNCS (LNAI), vol. 5845, pp. 633–645. Springer, Heidelberg (2009). https://doi.org/10.1007/978-3-642-05258-3_56
12. Laumanns, M., Thiele, L., Deb, K., Zitzler, E.: Combining convergence and diversity in evolutionary multi-objective optimization. Evol. Comput. **10**(3), 263–282 (2002)
13. Hadka, D., Reed, P.: Diagnostic assessment of search controls and failure modes in many-objective evolutionary optimization. Evol. Comput. J., in press
14. Farina, M., Amato, P.: A fuzzy defifinition of 'optimality' for manycriteria decision-making and optimization problems. IEEE Trans. Syst. Man Cybern. **34**(3), 315–326 (2004)
15. Zou, X., Chen, Y., Liu, M., Kang, L.: A new evolutionary algorithm for solving many-objective optimization problems. IEEE Trans. Syst. Man Cybernet. Part B: Cybernet. 38(5), 1402–1412 (2008)
16. Deb, K., Agrawal, R.B.: Simulated binary crossover for continuous search space. Complex Syst. **9**(2), 115–148 (1995)
17. Sato, H., Aguirre, E., Tanaka, K.: Pareto partial dominance moea in many-objective optimization. In: Proceedings of Congress on Computational Intelligence (CEC-2010), 2010, pp. 1–8
18. Zhang, Q., Li, H.: MOEA/D: A multiobjective evolutionary algorithm based on decomposition. IEEE Trans. Evol. Comput. **11**(6), 712–731 (2007)
19. Deb, K., Jain, H.: Handling many-objective problems using an improved NSGA-II procedure. In: Proceedings of World Congress on Computational Intelligence (WCCI-2012), pp. 1–8 (2012)
20. Das, I., Dennis, J.: Normal-boundary intersection: a new method for generating the Pareto surface in nonlinear multicriteria optimization problems. SIAM J. Optim. **8**(3), 631–657 (1998)
21. Xue, F., Wu, D.: NSGA-III algorithm with maximum ranking strategy for many-objective optimisation. Int. J. Bio-Inspired Comput. **15**(1), 14–23 (2020)
22. Essiet, I.O., Sun, Y., Wang, Z.: Improved genetic algorithm based on particle swarm optimization-inspired reference point placement. Eng. Optimizat. **51**(7), 1097–1114 (2019)
23. Zhang, Q.: MOEA/D. http://dces.essex.ac.uk/staff/zhang/webofmoead.htm
24. Veldhuizen, D.V., Lamont, G.B.: Multiobjective evolutionary algorithm research: a history and analysis. Dayton, O.H. (ed.) Department of Electrical and Computer Engineering, Air Force Institute of Technology, Technical report TR-98–03 (1998)
25. Zhang, Q., Zhou, A., Zhao, S.Z., Suganthan, P.N., Liu, W., Tiwari, S.: Multiobjective optimization test instances for the cec-2009 special session and competition. Singapore: Nanyang Technological University, Technical report (2008)

A Differential Evolution Algorithm for Multi-objective Mixed-Variable Optimization Problems

Yupeng Han[1,2], Hu Peng[1,2(✉)], Aiwen Jiang[3], Cong Wang[2], Fanrong Kong[2], and Mengmeng Li[3]

[1] School of Information Management, Jiangxi University of Finance and Economics, Nanchang 330013, People's Republic of China
hu_peng@whu.edu.cn
[2] School of Computer and Big Data Science, Jiujiang University, Jiujiang 332005, People's Republic of China
[3] School of Computer and Information Engineering, Jiangxi Normal University, Nanchang 330022, People's Republic of China
{jiangaiwen,limengmeng}@jxnu.edu.cn

Abstract. Multi-objective mixed-variable optimization problems (MO-MVOPs) are common and complex practical design optimization problems. MO-MVOPs often include multiple complex functions, constraints and mixed types of decision variables. Compared with single objective mixed-variable optimization problems (MVOPs), the decision space of MO-MVOPs presents more complex spatial distribution features. These features of MO-MVOPs make solving such problems face a big challenge. In this paper, fundamental advancements are made to MCDE_{mv} which is previously proposed for single objective MVOPs. This improved version can solve MO-MVOPs, which can be named as $\text{MO-MCDE}_{\text{mv}}$. In $\text{MO-MCDE}_{\text{mv}}$, the best solution in the population is no longer the solution with the best fitness value, but a random solution in the first rank after executing the fast non-dominated sorting approach in NSGA-II. The generation of offsprings is generated by using the selection operator in NSGA-II. In addition, the local search in MCDE_{mv} is utilized to improve the parents. The quality of the newly generated individual depends on the dominance relationship between itself and its parent. The experimental results of two actual MO-MVOPs are obtained by using two advanced multi-objective algorithms, i.e., CMOEA/D and NSGA-II, and the proposed $\text{MO-MCDE}_{\text{mv}}$. The experimental results show that the $\text{MO-MCDE}_{\text{mv}}$ has better performance than the two advanced multi-objective algorithms.

Keywords: Differential evolution · Multi-strategy · Non-dominated sorting · Multi-objective mixed-variable optimization problems

L. Pan et al. (Eds.): BIC-TA 2021, CCIS 1565, pp. 145–159, 2022.
https://doi.org/10.1007/978-981-19-1256-6_11

1 Introduction

In real world, modern design optimization problem is no longer just a simple single-objective optimization problem with decision variables containing only continuous variables, but a complex optimization problem with multi-objective, continuous and discrete variables simultaneously. Multi-objective optimization (MOO) problems are such problems with these features. MOO aims to simultaneously improve multi-objectives and make the best trade-offs between these conflicting objectives in a practical design optimization problem. Unlike single-objective optimization, the target of MOO is to obtain a set of optimal trade-off designs or solutions that can be called Pareto optimal solutions, rather than searching for a single optimal solution in the decision space.

However, in practical design optimization problems, multi-objective optimization problems always have more complex decision variables, i.e., including continuous and discrete variables simultaneously. The multi-objective mixed-variable optimization problems (MO-MVOPs) are such problems with these features. The general formula of MO-MVOPs can be expressed as follows [1]:

$$
\begin{aligned}
&\min_{i\in[1,N]} [f_1(\boldsymbol{X}), f_2(\boldsymbol{X}), f_i(\boldsymbol{X}), \ldots, f_N(\boldsymbol{X})] \\
&\boldsymbol{X} = (x_1^c, x_2^c, \ldots x_m^c, x_1^d, x_2^d, \ldots, x_n^d) \\
\mathbf{s.t.}\quad & \\
&\mathbf{g}_p(\boldsymbol{X}) \leq 0, p = 1, 2, \ldots, P \\
&\mathbf{g}_q(\boldsymbol{X}) \leq 0, q = 1, 2, \ldots, Q
\end{aligned}
\tag{1}
$$

where $f_i(\boldsymbol{x})$ represents the ith objective function; m, n represents the number of continuous and discrete variables (x^c, x^d), respectively; g_p and h_q represent the pth inequality constraint and the qth equality constraint, respectively; P and Q respectively represent the number of inequality and equality constraints. Specifically, MO-MVOPs usually have the following features:

- high non-linearity and multimodality of multiple objective functions;
- high complex constrained decision space;
- continuous and discrete decision variables.

The above features of MO-MVOPs make solving such problems face a big challenge. Hence, it is significative to design an efficient method to cope with this problem. The aim of this paper is to structure a multi-objective implementation of previously proposed multi-strategy co-evolutionary differential evolution algorithm (MCDE$_{mv}$ [2]).

As a simple and effective optimization algorithm, evolutionary algorithms (EAs), do not use the information about functions and slope continuity and are already extensively used to solving many complex practical problems [3–5]. Based on EAs, many multi-objective EAs (MOEAs) have been proposed for solving MOO problems. On the basis of the methods that MOEAs coping with MOO problems, these algorithms designed for MOO problems can be categorized into three categories. The first method of MOEAs to deal with MOO problems is based on dominance. In this category, the most typical example is fast and elitist multi-objective genetic algorithm (NSGA-II) [6]. In, NSGA-II, a selection operator and a fast non-dominated sorting approach is proposed. The experiment results on several functions show that is NSGA-II effective and promising. The second method of MOEAs to deal with MOO problems is based on decomposition. The main idea of this method is to decompose a multi-objective optimization problem into multiple scalar optimization sub-problems. In this category, Zhan et al. proposed a decomposition-based method, named MOEA/D [7], in which three approaches are utilized to transform the PF approximation problem into multiple scalar optimization problems. The third method of MOEAs to deal with MOO problems is based on indicators. In this category, Zitzler et al., proposed an indicator-based algorithm, named IBEA [8], in which the additional diversity protection mechanisms are no longer needed.

As one of an effective and powerful real-parameter evolutionary algorithms (EAs), differential evolution (DE) [9], has been utilized in many multi-objective optimization algorithms, such as [10] etc. These excellent algorithms provide a foundation for coping with MO-MVOPs. Thus, it is feasible to use MOEAs solve MO-MVOPs. However, MOEAs are rarely proposed for dealing with MO-MVOPs in the literature of recent years, due to its complex spatial characteristics and decision variable types. Therefore, in this paper, a new algorithm combining the multi-objective technology in NSGA-II and our previously proposed MCDE_{mv}, named $\text{MO-MCDE}_{\text{mv}}$, is presented to solve MO-MVOPs. In $\text{MO-MCDE}_{\text{mv}}$, the best solution in the population is no longer the solution with the best fitness value, but a random solution in the first rank after executing the fast non-dominated sorting approach in NSGA-II. The generation of offsprings is generated by using the selection operator in NSGA-II. In addition, the local search in MCDE_{mv} is utilized to improve the parents. The quality of the newly generated individual depends on the dominance relationship between itself and its parent.

The structure of this paper will be constructed as follows. The briefly introduce about NSGA-II and MCDE_{mv} are presented in Sect. 2. Section 3 illustrates our proposed $\text{MO-MCDE}_{\text{mv}}$. In Sect. 4, the comparative analysis of experimental results on MO-MVOPs between $\text{MO-MCDE}_{\text{mv}}$, NSGA-II and CMOEA/D are presented. The conclusion is drawn in Sect. 5.

2 Related Work

Compared with continuous optimization problems (CnOPs), discrete optimization problems (DOPs) and single objective mixed-variable optimization problems (MVOPs), MO-MVOPs have more optimization objective functions, which make these algorithms more difficult to solve. However, for MO-MVOPs, some excellent algorithms have been proposed to cope with them.

In order to solve MO-MVOPs, a new approach, named the Pareto set pursuing (PSP) method (MV-PSP), was proposed by Khokhar et al. [20]. In MV-PSP, important modifications have been made so that the PSP method that originally only solved CnOPs can handle MO-MVOPs. Based on a performance comparison strategy, it discusses the performance of the algorithm from many different aspects. When function evaluations is limited, MV-PSP is competitive compared with other methods. In addition, a new scheme for multi-objective constrained portfolio optimization problem, named Compressed Coding Scheme (CCS) was proposed by Yi et al. [21]. This scheme compresses two dependent variables into one variable to deal with the challenge of the dependence among variables, which is the problem faced by the MOEAs' direct coding scheme. Moreover, a new variant of PSO, named multi-objective MDPSO algorithm (MO-MDPSO), was proposed by Tong et al. [1]. Based on the mixed-discrete PSO (MDPSO) algorithm, this algorithm proposed leader selection mechanism and diversity preservation to maintain the convergence performance and diversity of the population.

3 Overview

3.1 Review of NSGA-II

Among many multi-objective optimization algorithms, NSGA-II [6] is the most influential and widely used multi-objective algorithm. After its appearance, due to its simplicity, effectiveness and obvious superiority, this algorithm has become one of the basic algorithms in multi-objective optimization problems. This is why we use this multi-objective technology combined with $MCDE_{mv}$ to cope with MO-MOVPs. The main idea of NSGA-II can be concluded as follows:

(1) A fast non-dominated sorting approach is proposed for reducing the computational complexity.
(2) An elite strategy is proposed for ensuring the convergence performance of algorithm.
(3) A crowded-comparison operator is presented for ensuring the diversity of the population.

Next, we will briefly introduce the three main approaches. For fast non-dominated sorting approach, each individual i in the population has two parameters n_i and S_i. n_i is the number of individuals dominating the solution of individual i in the population, and S_i is the set of individuals dominated by individual i. First, find all individuals with $n_i = 0$ in the population and store them in the non-dominated set Z_1; Second, for each individual j in the non-dominated set Z_1, traverse the individual set S_j dominated by it, and subtract 1 from the n_t of each individual t in the set S_j. Because the individual j that dominates the individual t has been stored in the non-dominated set Z_1. If n_t-1 $= 0$, the individual t will be stored in another set H; Finally, Z_1 is regarded as the first-level non-dominated individual set, so the solution individual in Z_1 is the best. They only dominate individuals, and are not dominated by any other individuals. All individuals in the set are assigned the same non-dominated level $rank_i$, and then the above-mentioned hierarchical operation on the set H is continued, and all individuals will be assigned to the corresponding non-dominated level.

For crowded-comparison operator, if there are the following two situations, then individual i is better than individual j. First, the non-dominated level of individual i is better than individual j, i.e., $rank_i < rank_j$; Secondly, if two individuals in the population have the same rank and the crowded distance of i is smaller than individual j, i.e., $rank_i = rank_j$ and $i_d < j_d$.

For elite strategy, this strategy combines the parent population with its offspring population, and jointly generates the next generation population through competition which ensures the convergence performance of algorithm.

3.2 Review of $MCDE_{mv}$

$MCDE_{mv}$ [2] is a multi-strategy co-evolutionary variant of DE, which is previously proposed by us to deal with MVOPs. In $MCDE_{mv}$, Based on a scheme that considers both two types of variables and in order to adapt to the all-inclusive situation in MVOPs, a multi-strategy co-evolution method is proposed. In addition, in order to ensure the optimization performance of discrete variables and improve the efficiency and flexibility of $MCDE_{mv}$, a statistical-based local search (SBA) is proposed. Next, we will briefly introduce the two methods mentioned above.

For the multi-strategy co-evolutionary approach, six different trial vector generation schemes and a dynamic adaptive selection mechanism are proposed. In this approach, the Gaussian distribution is used to dynamically update the scaling factor F. In the dynamic adaptive selection mechanism, the specific choice of mutation strategy and crossover operator for application is determined by the roulette method. An adaptive shift strategy [11] is utilized to dynamically update the crossover probability CR. The values of the larger CR_l and smaller CR_s values are set to 0.9 and 0.1, respectively. The updated equation of F and CR are given as follows:

$$F = \mathrm{N}(0.9, 0.1) \tag{2}$$

Algorithm 1 Adaptive shift strategy [11]

```
1: if f(Ui) > f(Xi) then
2:     Flag =~Flag;
3:     if Flag == 1 then
4:         CR = N (CRl, 0.1);
5:     else
6:         CR = N (CRs, 0.1);
7:     end if
8: end if
```

For generating the trial vectors, the choice between three mutation strategies (i.e., DE/rand/1, DE/best/1 and DE/target-to-rand/1) and two crossover operators (i.e., multiple exponential crossover [12] and binomial crossover) is determined by the roulette method. It is obvious that there are six different generation schemes. In addition, it is worth noting that the use of the above six trial vector generation schemes cannot generate discrete variables. Therefore, in order to use the above scheme to optimize discrete variables, a mapping method is used on the basis of the relaxation method described in [13]. Before the fitness function is evaluated, the continuous value is rounded to the integer index value of each feasible discrete value. Then, a feasible discrete value is selected according to the integer index value. Through this conversion method, the co-evolution between the different types of variables of the individual can be maintained. As the description of the above, the pseudo-code of the approach can be given in Algorithm 2.

For statistics-based local search, this method is proposed to ensure the optimization performance of discrete variables and improve the efficiency and flexibility of MCDEmv. In SBA, the *b* discrete dimensions of *a* selected individual are adjusted based on probability. Before the start of the iteration, the probability that the feasible values of all discrete variables are selected is initialized to zero. Equation 3 below will be used to update this probability.

Algorithm 2 The multi-strategy co-evolutionary approach

Input: The population P;
Initial crossover probability with CR_i;
The mutation strategy candidate pool: DE/rand/1, DE/best/1 and DE/target-to-rand/1;
The crossover operator candidate pool: binomial crossover and multiple exponential crossover.
1: **for** $i = 1$:NP **do**
2: Update F by using equation 2;
3: Select a scheme from mutation strategy and crossover operator candidate pools according to the roulette method;
4: Generate the trial vector $\mathbf{U}_i$;
5: Evaluate the fitness function value of the trial vector after executing the mapping method;
6: FEs = FEs+1;
7: **if** $f(\mathbf{U}_i) < f(\mathbf{X}_i)$ **then**
8: $\mathbf{X}_i = \mathbf{U}_i$;
9: Update the number of successful updates of the selected mutation strategy and crossover operator;
10: **else**
11: Update CR according to the Algorithm 1;
12: **end if**
13: Update the total number of the selected mutation strategy and cross over operator;
14: **end for**
Output: The optimized population;

$$Prob_i(j) = \frac{count_{i,j}}{count_i} \tag{3}$$

where $count_i$ represents the total number of all feasible values that appear in the ith discrete variable. $count_{i,j}$ represents the number of times that the jth feasible value appears in the ith discrete variable. $Prob_i$ (j) represents the probability that the jth feasible value is assigned to the ith discrete variable.

In SBA, only the first half of the better individuals in the population are used for collecting information for ensuring the convergence of the algorithm. The pseudo-code of SBA is shown in Algorithm 3. In addition, the constraints in MVOPs are coped with feasibility rule [14]. But, in this paper, the used constraint-handling technique is same as the method in NSGA-II.

Algorithm 3 Procedure of the statistics-based local search

Input: The optimized population P, the number of individuals a, the dimensions b and the first c high probabilities to be selected

1: Sort the population from the best to the worst according to the fitness value;
2: **for** each discrete variable $\mathbf{X}_i^{m+n}$, i = 1 to n **do**
3: **for** each feasible value V, j = 1 to V_s **do**
4: **for** the half of the superior individuals, k = 1 to NP/2 **do**
5: **if** $\mathbf{X}_{k,i}$ == $\mathbf{V}_j$ **then**
6: $count_{i,j} = count_{i,j} + 1$;
7: **end if**
8: **end for**
9: $count_i = count_i + count_{i,j}$;
10: **end for**
11: Calculate the probability by using equation 3;
12: **end for**
13: Select a individuals from the top 50% of the population;
14: **for** the select individual $\mathbf{X}_i$, i = 1 to a **do**
15: **for** the select dimension j, j =1 to b **do**
16: Sort from small to large according to the probability;
17: Choose feasible values with the first c high probabilities;
18: **end for**
19: Evaluate fitness value $f(\mathbf{X}_i)$ of new generated individual after executing the mapping method;
20: FEs = FEs+1;
21: **end for**
22: Compare the generated individuals with the randomly selected individuals from the first a superior individuals;

Output: The next generation population;

---

4 The Proposed MO-MCDE$_{mv}$

As described above, many practical design optimization problems can be naturally modeled as MO-MVOPs. These problems often have three complex features, which makes designing an effective method to cope with such problems face a challenge. Nonetheless, it is indispensable and significative to design an efficient algorithm to deal with them. It is known that EAs simple, effective and do not use the information about functions and slope continuity. Therefore, a classic algorithm of EAs, DE, combined with the excellent multi-objective technology in NSGA-II is utilized to cope with MO-MVOPs.

Thus, in the proposed MO-MCDE$_{mv}$, we combine the multi-objective technology in NSGA-II and our previously proposed MCDE$_{mv}$ to coping with MO-MVOPs. This is not a simple joint. In order to make MCDE$_{mv}$ better suitable for multi-objective technology and maintain the characteristics of handling with mixed-variable, important modifications have made in the optimal individual and statistics-based local search.

4.1 Selection of the Optimal Individual

In MCDE$_{mv}$, the best individual information is used to guide the evolution of the population and maintain the convergence performance of the algorithm. In single-objective optimization problems, it is known that the optimal individual refers to the individual with the smallest fitness function value (minimization problem). However, in multi-objective optimization problems, the optimal individual does not exist because its goal is to obtain a set of optimal trade-off designs or solutions with good convergence and distribution. Therefore, in order to make MCDE$_{mv}$ suitable for multi-objective algorithms and maintain its original characteristics, important modification has been made to the selection of the optimal individual.

As described in NSGA-II, all individuals are assigned to the corresponding non-dominated level $rank_i$ after fast non-dominated sorting. In all non-dominated level, Z_1 is regarded as the first-level non-dominated individual set, so the individuals in Z_1 are the best. They only dominate individuals, and are not dominated by any other individuals. Therefore, it is feasible to select the individual in the first dominance level as the optimal individual in MO-MCDE$_{mv}$. It is worth noting that there may be more than one individual at this level. Therefore, an individual is selected at random as the optimal individual.

4.2 The Modifications of Statistics-Based Local Search

As described in above, after the population is generated by the multi-strategy co-evolution method, a local search is specifically proposed to optimize discrete variables. However, in MO-MCDE$_{mv}$, this method is utilized to optimize discrete variables of individual in parent population rather than the optimized population.

In addition, it is not infeasible to learn only the first half of the better individuals in the population. The reason is that the individuals in multi-objective optimization problems are no longer good or bad. Thus, after fast non-dominated sorting, we sort individuals in ascending order according to the serial number of each individual in the corresponding dominance level. The first NP/2 individuals in all dominance levels will be selected for dimensional learning in local search.

Finally, Note that the new individual generated in the local search is compared with the original parent individual. In single goals, the comparison criterion is based on their fitness values. But this situation is no longer applicable in multi-objective optimization. Therefore, in MO-MCDE$_{mv}$, the comparison criterion is based on the following rules:

(1) If the offspring dominate the parent or each other is a non-dominated solution, the offspring is considered to be better than the parent;

(2) If the parent dominates the offspring, it is considered that the parent is better than the offspring;

4.3 The Framework of MO-MCDE$_{mv}$

Based on the above description of NSGA-II, MCDE$_{mv}$ and MO-MCDE$_{mv}$, the overall illustration of the algorithm can be shown in the Fig. 1 below.

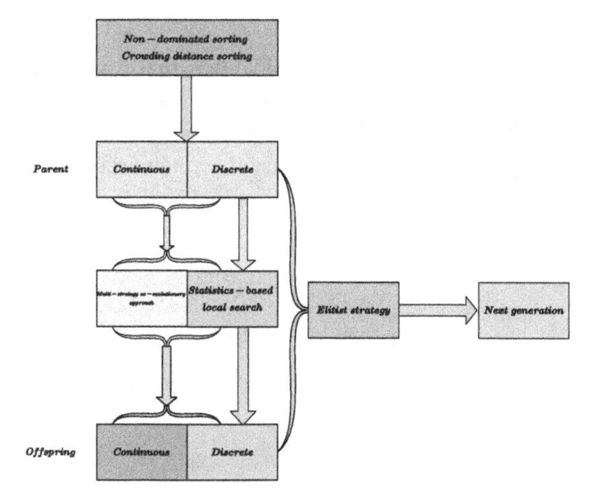

Fig. 1. Illustration of MO-MCDE$_{mv}$.

5 Numerical Experiment

5.1 Set Relevant Parameters

Set population scale NP = 100.

Set the number of function evaluations MaxFEs = 5000.

Set run times r = 30.

Set the parameters of local search *a, b, c* = 3, 3, 3

Win10 platform.

MATLAB R2016b.

PlatEMO v2.9 [15].

Pure diversity metric (PD) [16].

5.2 The Results of Two Practical MO-MVOPs

In this subsection, two practical MO-MVOPs are utilized to test the performance of the proposed MO-MCDE$_{mv}$. In the two practical problems, it is difficult for algorithm obtain a set of optimal trade-off designs or solutions due to the two objectives, many complex constraints and mixed types of decision variables that they contain. It is worth noting that the actual or analytical Pareto frontiers of these two actual optimization design problems are unknown.

Therefore, in order to study the performance of algorithm, the obtained Pareto optimal solution with the solution obtained using CMOEA/D [17] and NSGA-II (binary code) are compared with MO-MCDE$_{mv}$. In order to cope with the discrete variables in two practical MO-MVOPs, the mapping method in MCDE$_{mv}$ is used in these two algorithms. In addition, a diversity metric (PD) is used to test diversity of the obtained solutions. The larger the value of this index, the better the diversity of the solutions obtained by the algorithm. Moreover, this metric does not require any parameters, nor any reference.

Analytical MINLP Problem

The analytical MINLP problem (denoted as TP1) was first formulated by Dimkou [18] and should be obtained good distribution frontiers as much as possible. The analytical MINLP problem contains two objectives, nine constraints six decision variables: three continuous variables (x_1, x_2, x_3) and three binary discrete variables ($y_1, y_2, y_3 \in \{0, 1\}$). The definition of this problem is given in Formula 4 and 5.

Objective function:

$$\min \begin{array}{l} f_1 = x_1^2 - x_2 + x_3 + 3y_1 + 2y_2 + y_3 \\ f_2 = 2x_1^2 + x_2 - 3x_3 - 2y_1 + y_2 - 2y_3 \end{array} \tag{4}$$

Subject to:

$$g_1 = 3x_1 - x_2 + x_3 + 2y_1 \leq 0$$

$$g_2 = 4x_1^2 + 2x_1 + x_2 + x_3 + y_1 + 7y_2 - 40 \leq 0$$

$$g_3 = -x_1 - 2x_2 + 3x_3 + 7y_3 \leq 0$$

$$g_4 = -x_1 + 12y_1 - 10 \leq 0$$

$$g_5 = x_1 - 2y_1 - 5 \leq 0$$

$$g_6 = -x_2 + y_2 - 20 \leq 0$$

$$g_7 = x_2 - y_2 - 40 \leq 0$$

$$g_8 = -x_3 + y_3 - 17 \leq 0$$

$$g_9 = x_3 - y_3 - 25 \leq 0 \tag{5}$$

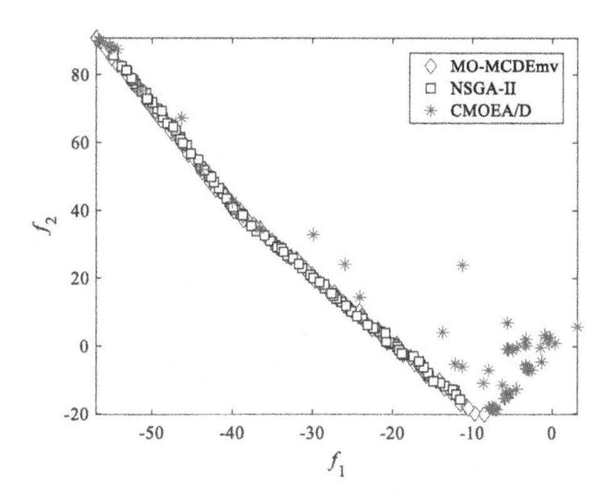

Fig. 2. The obtained Pareto optimal solutions by MO-MCDE$_{mv}$, NSGA-II and CMOEA/D for analytical MINLP problem.

Table 1. Diversity comparison of MO-MCDE$_{mv}$, NSGA-II and CMOEA/D for analytical MINLP problem.

Problem	NSGA-II	CMOEA/D	MO_MCDE$_{mv}$
TP1	1.2494e+5 (2.12e+4) =	1.1856e+5 (2.30e+4) =	1.2022e + 5 (1.90e+4)
±/ =	0/0/1	0/0/1	

As shown in Fig. 2, it shows the Pareto fronts obtained by MO-MCDE$_{mv}$, NSGA-II and CMOEA/D for analytical MINLP problem. Since the true Pareto frontier of this problem is unknown, it is not feasible to use quality indicators to test and analyze this performance of this algorithm. Therefore, we can observe the solutions graphically. By observing the overall distribution of the Pareto optimal solution obtained by MO-MCDE$_{mv}$ (rhombuses) and NSGA-II (squares), MO-MCDE$_{mv}$ gets better distribution in the tail. In addition, compared with the overall distribution of Pareto optimal solutions obtained by CMOEA/D (stars), this algorithm obtains a good distribution. As shown in Table 1, it presents the mean result of PD metric after run 30 times. The last row of the table is attached with the results of Wilcoxon's Rank sum test at the 0.05 significance level. From the results in the table, it can be concluded that the Pareto solutions obtained by the three algorithms have achieved similar performance in terms of diversity.

Disc Brake Design Problem

Disc brake design problem (denoted as TP2) was first formulated by Osyczka [19]. The objective of its optimized design is to minimize the stopping time and the mass of the brake. Disc brake design problem contains four decision variables: the inner radius of the discs (x_1), the outer radius of the discs (x_2), the engaging force (x_3) and the number of friction surfaces (x_4), respectively. Note that x_1, x_2 and x_3 are continuous variables, x_4

is a discrete variable. The values of x_4 must be an integer. The definition of this problem is given in Formula 6 and 7.

Objective function:

$$\min \begin{array}{l} f_1 = 4.9 \times 10^{-5}\left(x_2^2 - x_1^2\right)(x_4 - 1) \\ f_2 = \frac{9.82\times10^6\left(x_2^2 - x_1^2\right)}{x_3 x_3\left(x_2^3 - x_1^3\right)} \end{array} \quad (6)$$

Subject to

$$g_1 = 20 - (x_2 - x_1) \le 0$$

$$g_2 = 2.5(x_4 + 1) - 30 \le 0$$

$$g_3 = \frac{x_3}{\pi\left(x_2^2 - x_1^2\right)} - 0.4 \le 0$$

$$g_4 = \frac{2.22 \times 10^{-3} x_3\left(x_2^3 - x_1^3\right)}{\left(x_2^2 - x_1^2\right)^2} \le 0$$

$$g_5 = \frac{900 - 2.66 \times 10^{-2} x_3 x_4\left(x_2^3 - x_1^3\right)}{\left(x_2^2 - x_1^2\right)} \le 0$$

where

$$55 \le x_1 \le 80$$

$$75 \le x_2 \le 110$$

$$1000 \le x_3 \le 3000$$

$$2 \le x_4 \le 20 \quad (7)$$

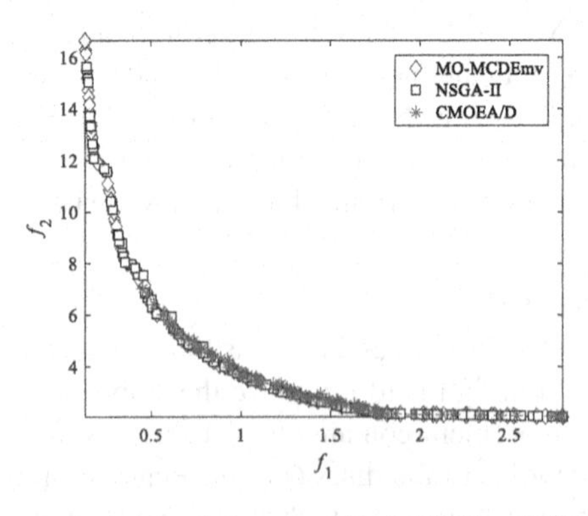

Fig. 3. The obtained Pareto optimal solutions by MO-MCDE$_{mv}$, NSGA-II and CMOEA/D for disc brake design problem.

Table 2. Diversity comparison of MO-MCDE$_{mv}$, NSGA-II and CMOEA/D for disc brake design problem.

Problem	NSGA-II	CMOEA/D	MO_MCDE$_{mv}$
TP2	8.6228e + 3 (1.37e + 3) =	4.1082e + 3 (1.05e + 3) -	9.4406e + 3 (1.97e + 3)
±/ =	0/0/1	0/1/0	

As shown in Fig. 3, it shows the Pareto fronts obtained by MO-MCDE$_{mv}$, NSGA-II and CMOEA/D for disc brake design problem. Since the true Pareto frontier of this problem is also unknown as analytical MINLP problem, it is not feasible to use quality indicators to test and analyze this performance of this algorithm. Therefore, as we do above, we can observe that the overall distribution of the Pareto optimal solution obtained by MO-MCDE$_{mv}$ (rhombuses) is better than the distribution obtained by NSGA-II (squares). By observing the Pareto optimal solution in the figure, it can be seen that NSGA-II has some discontinuous points. In the head part, the distribution of MO-MCDE$_{mv}$ is more uniform than NSGA-II. In addition, compared with the overall distribution of Pareto optimal solutions obtained by CMOEA/D (stars), this algorithm obtains more Pareto optimal solutions and good distribution. As shown in Table 2, MO-MCDE$_{mv}$ and NSGA-II obtain similar performance in the diversity of the obtained Pareto solutions, but compared with CMOEA/D, MO-MCDE$_{mv}$ obtains better diversity.

6 Conclusion

As a common and important type of problem in industrial optimization design, MO-MVOPs often include multiple complex functions, constraints and mixed types of decision variables. These features of MO-MVOPs make solving such problems face a big challenge. Hence, it is significative to design an efficient method to cope with this problem. In order to solve MO-MVOPs, a new algorithm, which can be named MO-MCDE$_{mv}$, are proposed. MO-MCDE$_{mv}$ is not a simple joint. In order to make MCDE$_{mv}$ better suitable for multi-objective technology and maintain the characteristics of handling with mixed-variable, we make important modifications in the optimal individual and statistics-based local search. In MO-MCDE$_{mv}$, the best solution is a random solution in the first rank after executing the fast non-dominated sorting approach in NSGA-II. In addition, the local search in MCDE$_{mv}$ is utilized to improve the parents. The quality of the newly generated individual depends on the dominance relationship between itself and its parent. The results of the experiment with several advanced multi-objective algorithm on two practical MO-MVOPs show that the MO-MCDE$_{mv}$ is competitive.

However, the multi-objective techniques used in this article come from NSGA-II, and these techniques may not be suitable for dealing with single-objective MVOPs. In the future, it is an interesting direction to design an effective multi-objective technology to effectively solving MO-MVOPs.

Acknowledgments. This work was supported by the National Natural Science Foundation of China (61763019, 61966018) and the Science and Technology Foundation of Jiangxi Province (20202BABL202019).

References

1. Tong, W., Chowdhury, S., Messac, A.: A multi-objective mixed-discrete particle swarm optimization with multi-domain diversity preservation. Struct. Multidiscip. Optim. **53**(3), 471–488 (2015). https://doi.org/10.1007/s00158-015-1319-8
2. Peng, H., Han, Y., Deng, C., Wang, J., Wu, Z.: Multi-strategy co-evolutionary differential evolution for mixed-variable optimization. Knowl.-Based Syst. **229**, 107366 (2021)
3. Altabeeb, A.M., Mohsen, A.M., Abualigah, L., Ghallab, A.: Solving capacitated vehicle routing problem using cooperative firefly algorithm. Appl. Soft Comput. **108**, 107403 (2021)
4. Cui, L., Deng, J., Wang, L., Xu, M., Zhang, Y.: A novel locust swarm algorithm for the joint replenishment problem considering multiple discounts simultaneously. Knowl-Based Syst. **111**, 51–62 (2016)
5. Fathi, M., Khakifirooz, M., Diabat, A., Chen, H.: An integrated queuing stochastic optimization hybrid Genetic Algorithm for a location-inventory supply chain network. Int. J. Prod. Econ. **237**, 108139 (2021)
6. Deb, K., Pratap, A., Agarwal, S., Meyarivan, T.: A fast and elitist multiobjective genetic algorithm: NSGA-II. IEEE Trans. Evol. Comput. **6**(2), 182–197 (2002)
7. Zhang, Q., Hui, L.: Moea/d: A multiobjective evolutionary algorithm based on decomposition. IEEE Trans. Evol. Comput. **11**(6), 712–731 (2008)
8. Zitzler, E., Künzli, S.: Indicator-based selection in multiobjective search. In: Yao, X., et al. (eds.) PPSN 2004. LNCS, vol. 3242, pp. 832–842. Springer, Heidelberg (2004). https://doi.org/10.1007/978-3-540-30217-9_84
9. Storn, R., Price, K.: Differential evolution–a simple and efficient heuristic for global optimization over continuous spaces. J. Global Optim. **11**(4), 341–359 (1997)
10. Li, H., Zhang, Q.: Multiobjective optimization problems with complicated Pareto sets, MOEA/D and NSGA-II. IEEE Trans. Evol. Comput. **13**(2), 284–302 (2008)
11. Peng, H., Guo, Z., Deng, C., Wu, Z.: Enhancing differential evolution with random neighbors based strategy. J. Comput. Sci. **26**, 501–511 (2018)
12. Qiu, X., Tan, K.C., Xu, J.: Multiple exponential recombination for differential evolution. IEEE Trans. Cybernet. **47**(4), 995–1006 (2016)
13. Liao, T., Socha, K., de Oca, M.A.M., Stützle, T., Dorigo, M.: Ant colony optimization for mixed-variable optimization problems. IEEE Trans. Evol. Comput. **18**(4), 503–518 (2013)
14. Deb, K.: An efficient constraint handling method for genetic algorithms. Comput. Meth. Appl. Mech. Eng. **186**(2–4), 311–338 (2000)
15. Tian, Y., Cheng, R., Zhang, X., Jin, Y.: PlatEMO: A MATLAB platform for evolutionary multi-objective optimization [educational forum]. IEEE Comput. Intell. Mag. **12**(4), 73–87 (2017)
16. Wang, H., Jin, Y., Yao, X.: Diversity assessment in many-objective optimization. IEEE Trans. Cybernet. **47**(6), 1510–1522 (2017)
17. Jain, H., Deb, K.: An evolutionary many-objective optimization algorithm using reference-point based non-dominated sorting approach, part II: Handling constraints and extending to an adaptive approach. IEEE Trans. Evol. Comput. **18**(4), 602–622 (2013)

18. Dimkou, T.I., Papalexandri, K.P.: A parametric optimization approach for multiobjective engineering problems involving discrete decisions. Comput. Chem. Eng. **22**, S951–S954 (1998)
19. Osyczka, A., Kundu, S.: A genetic algorithm-based multicriteria optimization method. In: Proceedings of the 1st World Congress of Structural Multidisciplinary Optimization, pp. 909–914 1995)
20. Khokhar, Z.O., et al.: "On the performance of the PSP method for mixed-variable multi-objective design optimization. J. Mech. Des. **132**(7), 071009 (2010)
21. Chen, Y., Zhou, A., Das, S.: utilizing dependence among variables in evolutionary algorithms for mixed-integer programming: a case study on multi-objective constrained portfolio optimization. Swarm Evol. Comput. **66**, 100928 (2021)

An Effective Data Balancing Strategy Based on Swarm Intelligence Algorithm for Malicious Code Detection and Classification

Dongzhi Cao, Zhenhu Ning(✉), Shiqiang Zhang, and Jianli Liu

Faculty of Information Technology, Beijing University of Technology, Beijing 100124, China
nzh41034@163.com

Abstract. Data balance has a great impact on the performance of neural network training model, and the model trained with high-quality data is often more robust. In the task of malicious code detection and classification, the number of samples is often very different among different malicious families. If the raw data is directly used for model training, it often brings over fitting problems. Aiming at this, this paper first analyzes the common data equalization methods in the classical machine learning model and neural network model, and then designs a hybrid dynamic sampling strategy based on swarm intelligence optimization algorithm to improve the performance of the model. Experimental results show that our malicious code data balancing strategy is effective.

Keywords: Malicious code detection · Unbalanced data · Swarm intelligence algorithms

1 Introduction

In classification applications, a common problem is that the number of samples in some categories is significantly higher than that in other categories. This phenomenon is called category imbalance. Data imbalance exists in various fields, such as medical diagnosis, fraud detection, image recognition, etc. [17,23]. The imbalance of data shows that there are great differences in the amount of data concentrated in different categories. For example, in the credit card fraud data set, the vast majority of transaction records are normal records, and only a few are fraudulent transaction data [19]. Previous studies have shown that data imbalance has a significant adverse impact on the training of traditional classifiers [14], which is reflected not only in the convergence of the model in the training stage, but also in the generalization of the model in the test stage. The classical machine learning model deeply studies the methods of dealing with data imbalance, including data level method and classifier level method.

The data level method samples the original data set and processes the training set before model training. The purpose is to make the training set tend

L. Pan et al. (Eds.): BIC-TA 2021, CCIS 1565, pp. 160–173, 2022.
https://doi.org/10.1007/978-981-19-1256-6_12

to balance or reduce imbalance by adding samples to the training set or deleting samples from the training set. It roughly includes both under-sampling and over-sampling. Classifier level equalization methods include threshold method and cost-sensitive learning. The former is also called threshold moving or post scaling method, which changes the output probability of categories by constantly adjusting the decision threshold of classifiers, while cost-sensitive learning uses the different costs generated after different categories of samples are incorrectly classified to solve the problem of data imbalance. In addition, there are one-class classification and mixing models, such as undersampling and oversampling.

Models trained with high-quality data tend to be more robust and converge quickly. However, in fact, most data sets are unbalanced. In this paper, malware usually belongs to multiple families, and the number of samples varies greatly between different malicious code families [11], as shown in Fig. 1. The model trained by using the unbalanced data often has low accuracy and poor robustness. To solve this problem, scholars have proposed many technologies, such as data augmentation and resampling, to improve data quality. In classification, resampling is a simple and effective method to deal with unbalanced training sets [3,9,15].

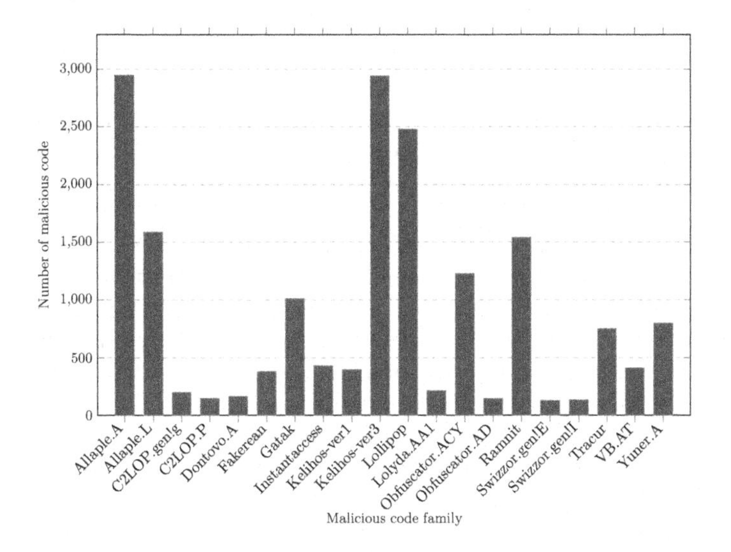

Fig. 1. Distribution of malicious code family samples

Aiming at the data imbalance of different malicious code families, a dynamic sampling method based on swarm intelligence algorithms is proposed. The rest of this article is organized as follows. Section 2 gives the swarm intelligence optimization model. Section 3 describes the dynamic sampling, including dynamic sampling model based on epoch, and dynamic sampling strategy based on swarm intelligence algorithms. Section 4 analyzes the performance of our dynamic sampling models. Finally, Sect. 5 concludes this paper.

2 Swarm Intelligence Optimization Model

Swarm intelligence algorithm belongs to the category of metaheuristic algorithm. There are three methods to solve problems in the optimization field: exact algorithm, heuristic algorithm, and metaheuristic algorithm. The accurate algorithm can solve the optimal solution of the problem [4,10]. Common models (such as linear programming, integer programming, etc.) can accurately obtain the problem's solution under constraints. However, as the problem size increases, the solution time of the exact algorithm grows exponentially, and it is difficult to obtain an optimal solution with a limited time constraint. A tradeoff between solution effectiveness and solution time is needed, i.e., obtaining suboptimal or acceptable solutions in finite time.

Heuristic algorithms and metaheuristic algorithms are imprecise algorithms [12]. The former is problem-oriented and designs solutions according to the specific problem structure, such as mountain climbing algorithm based on greedy strategy. Heuristic algorithms mainly solve local optimization problems. Compared with heuristic algorithms, metaheuristic algorithms are independent and are a general optimization process [1]. Metaheuristic algorithms usually include individual intelligence optimization algorithms and swarm intelligence optimization algorithms according to the size of the population [13,24]. Individual intelligent optimization algorithms include simulated annealing algorithm, tabu search algorithm, and so on. Swarm intelligence optimization algorithms include genetic algorithm, particle swarm optimization algorithm, bat algorithm, ant colony algorithm, cuckoo algorithm, etc. These algorithms have strategies or operators that jump out of local optimization, which are more suitable for solving global optimization problems [18].

The swarm intelligence algorithms simulate population behavior, such as reproduction, foraging, migration, and so on. The population contains multiple individuals, which depend on and influence each other, and have the same behavior patterns. According to the size of the search space, it can be simply divided into local search and global search. Local search refers to finding solutions in individual neighborhood spaces to obtain local optimal solutions. Global search refers to the individual jumping to a far region for solution, which is used to jump out of the local optimum. An algorithm includes global search and local search. Generally, in the early stage, it focuses on global search to better locate the dominant space, while in the later stage of the algorithm, it uses local search to find the optimal solution. The population of swarm intelligence algorithm is solved iteratively by continuously adjusting the 'global-local' operator until the termination condition is reached, such as finding a satisfactory solution or reaching the maximum number of iterations and finally finding an acceptable optimal solution.

Search strategy or operator is the key of swarm intelligence algorithm, which is used to adjust the local search and global search. In this paper, the sampling probability of different families is a complex combinatorial optimization problem. We employ three typical swarm intelligence algorithms for research, including particle swarm optimization algorithm (PSO) [10], bat algorithm (BA) [6], and cuckoo algorithm (CS) [5], in order to find the optimal sampling proportion.

3 Dynamic Sampling Strategy Based on Swarm Intelligence Algorithm

3.1 Dynamic Sampling Model

If the number of samples in one category in the dataset is several times or more than the number of samples in other malware categories, the dataset is an unbalanced dataset. This paper designs a dynamic oversampling model to adjust the training samples of each epoch dynamically. Epoch is the whole process of completing the forward propagation operation and backward propagation operation of all training samples in deep learning. The training data of an epoch is the sum of all training samples. The main idea of dynamic sampling is to adjust sampling weights in each epoch (reduce the weight of large samples and increase the weights of small samples), and then generate new training sets according to the resampling of new weights, as shown in Fig. 2.

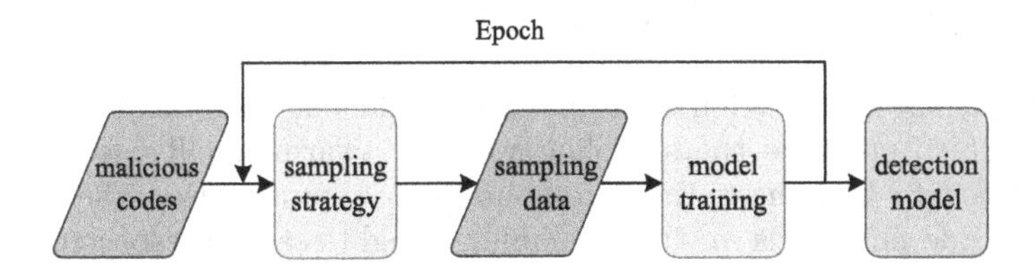

Fig. 2. Flowchart of dynamic oversampling sampling [8]

The sampling strategy is the core of the whole dynamic sampling, that is, how to adjust the sampling probability of samples in each epoch [21]. A simple method is the adaptive weights strategy (AWS). It is assumed that the data set used for classification has n samples belonging to m malware families. Firstly, the unbalanced data set is sampled according to the initial weight W^0 to obtain a relatively balanced sample set. Then, for the i-th malicious code family, update the sampling weight of its t-th epoch according to its initial sampling weight W_i^0 and preset final weight W_i^T, as follows:

$$W_i^t = r^{t-1} W_0^i + (1 - r^{t-1}) W_i^T \tag{1}$$

where, W_i^0 and W_i^T are respectively represented as initial sampling weight and preset final sampling weight. r is the factor used to adjust the weights, and t is the number of current epochs. Finally, the probability of each sample being selected as follows:

$$W_i = \frac{\sum_{m=1}^{M} g(c(i), m) W_m}{\sum_{n=1}^{N} \sum_{m=1}^{M} g(c(n), m) W_m} \tag{2}$$

where, W_m is the sampling weight of the m-th malicious code, and $c(i)$ is the family of the i-th malicious code. The function $g(x, y)$ is used to identify the family of malicious samples.

The resampled sample set is sent to the convolutional neural network for training. The sampling weight is updated according to the initial weight, the preset final weight, and the number of iterations. In short, we resample the training data of each epoch until the training model is stable. However, this method needs to reasonably set the initial value and final value of the weight. If the number of malicious code families is large, it is often difficult to determine the optimal initial weight. The sampling probability of each family can be regarded as a complex combinatorial optimization problem.

3.2 Dynamic Sampling Based on Swarm Intelligence Algorithm

Since the number of malicious code samples varies greatly among different malware families, classifiers trained directly with the original dataset can easily cause overfitting problems. Therefore, an appropriate sample ratio is very important for the training set. In the previous section, we proposed a dynamic oversampling framework and designed a dynamic sampling strategy with adaptive weights. This approach can effectively avoid overfitting of the model. However, the method is sensitive to parameter settings, and in practice, it is difficult to find the optimal initial weights for dozens of malicious code families.

In order to optimize the sampling weights, swarm intelligence algorithms are employed to find the optimal combination. Assuming that the number of malicious code families is m, the resampling weights can be respected as an m-dimensional one-dimensional vector. For this optimization, the combination of sampling weights can be considered as the position of individuals in the swarm intelligence algorithm, which can be given by:

$$position = W_1 W_2 \cdots W_n \tag{3}$$

Algorithm 1 gives a detailed resampling process [7]. The sampling weight of malicious samples can be calculated using the adaptive weight strategy or swarm intelligence algorithm. In the former (line 2–4), the function UpdateW is used to update the resampling weight. The latter (line 5-15) optimizes the sampling weight of malware families based on swarm intelligence algorithms.

4 Experimental Evaluation

This section evaluates the dynamic sampling model and conducts experiments on two real data set, MNIST [16] and malicious code gray image data sets (malware images, from vision research lab [20] and Microsoft malware classification challenge [22]). All network models are implemented by PyTorch.

Algorithm 1: Dynamic resampling

```
   Input: Raw data: imbdata
   Output: Optimized balanced data: bdata
 1 /*Calculate the sampling weight W using AWS or swarm intelligence
    algorithms*/;
 2 for i ∈ families do
 3 |  W_i^{t+1} ← UpdateW(W_i^t) as Eq.(1);
 4 end
 5 (p, p*) ← GetState(particles);
 6 while t < maxIterations do
 7 |  for p ⊂ particles do
 8 |  |  (p_i^{t+1}) ← UpdateState(p_i^t, p*);
 9 |  |  templeset ← ReSample(imbdata, p_i^{t+1});
10 |  |  accuracy_i^{t+1} ← TrainModel(templeset);
11 |  end
12 |  p* ← GetBest(particles, accuracy);
13 |  t = t + 1;
14 end
15 W ← p*;
16 for i ⊂ samples do
17 |  w_i ← Normalize(W) as Eq.(2);
18 end
19 bdata ← ReSample(imbdata, w);
20 Return(bdata);
```

4.1 Experimental Setup

In order to verify the dynamic oversampling strategy proposed in this paper, four data equalization methods during CNN training are set in this section:

1. Random Minority Oversampling (RMO)
2. Random Majority Undersampling (RMU)
3. Dynamic AWS-based Resampling (AWS-based DR)
4. Dynamic Resampling based Swarm intelligence optimization algorithm

To compare the advantages and disadvantages of each strategy and measure the generalization ability of the classification model trained by each sampling strategy, this section uses accuracy and ROC/AUC indicators.

When using CNNs for multi-class classification, the most extensive evaluation index is the total accuracy. According to its definition, it represents the percentage of samples correctly predicted by malicious code in the total samples. However, for unbalanced data, the accuracy often can not reasonably reflect the performance of the model. There are 6000 samples in total. As long as all large samples are predicted correctly, even if all small samples are predicted incorrectly, we can still get an accuracy of 83.33%. The total accuracy rate and the total recall rate can measure the performance of the multi-classification model from another perspective. The former measures the probability of predicting how many of the correct malicious code samples are actually positive samples. The latter measures the probability that they are actually positive samples and are correctly classified. In this paper, the malicious code family classification refers to the probability that the samples of a family are correctly classified into this family.

Another effective measure to solve the problem of unbalanced classification is ROC/AUC, which is the curve from false-positive rate to true-positive rate of all possible prediction thresholds. We used the ROC/AUC implementation provided in the Scikit-learn package. It calculates all threshold sensitivity and specificity defined by the classifier response in the test set and then uses the trapezoidal rule for AUC calculation. ROC/AUC is a widely studied and reliable discrimination method, which is widely used as the evaluation index of multiple classifiers.

4.2 Type of Unbalanced Data

Data imbalance has many forms, especially in using CNNs to deal with multi-classification problems. For example, in some cases, only one type of sample data is insufficient or too much, while in some cases, each category has a different number of samples. In this paper, we divide the data imbalance into two types: jump imbalance and linear imbalance [2]. For the jump imbalance type, there is a large difference in the data volume between the small sample category and the multi-sample category, while the internal data volume of the small sample category or the multi-sample category is relatively balanced. Its characteristics can be described by the following:

$$\mu = \frac{|C_{min}|}{N} \tag{4}$$

where, $|C_{min}|$ is the number of minority categories and N is the total number of categories. In addition, the difference of the data set can be defined by the maximum number of samples and the minimum number of samples, as shown in the following:

$$\rho = \frac{max\{|C_i|\}}{min\{|C_i|\}}, i \in \{1, 2, \cdots, n\} \tag{5}$$

Figure 3 shows an example of an unbalanced data set. In Fig. 3(a), there are 5 small sample categories and 5 large sample categories, a total of 6000 samples, belonging to the jump imbalance type. We can get $\mu = 5/10 = 0.5$ and $\rho = 1000/200 = 5$. Figure 3(b) has the same total number of samples as Fig. 3(a), but large samples and small samples show another distribution. There are more categories of small samples. And, $\mu = 0.8$, $\rho = 2$ can be calculated. It can be seen from Fig. 3(c) that the number of samples in different categories is linearly distributed. At this time, only the difference of data is measured, $\rho = 10$. Figure 3(d) shows a mixed distribution, with 5 small samples and 5 large samples, and the number of samples among small classes is different, but the difference is small. The large samples show a linear distribution. At this time, the data distribution can be roughly measured by the mean value, $\mu = 5/10 = 0.5$ and $\rho = 800/200 = 4$.

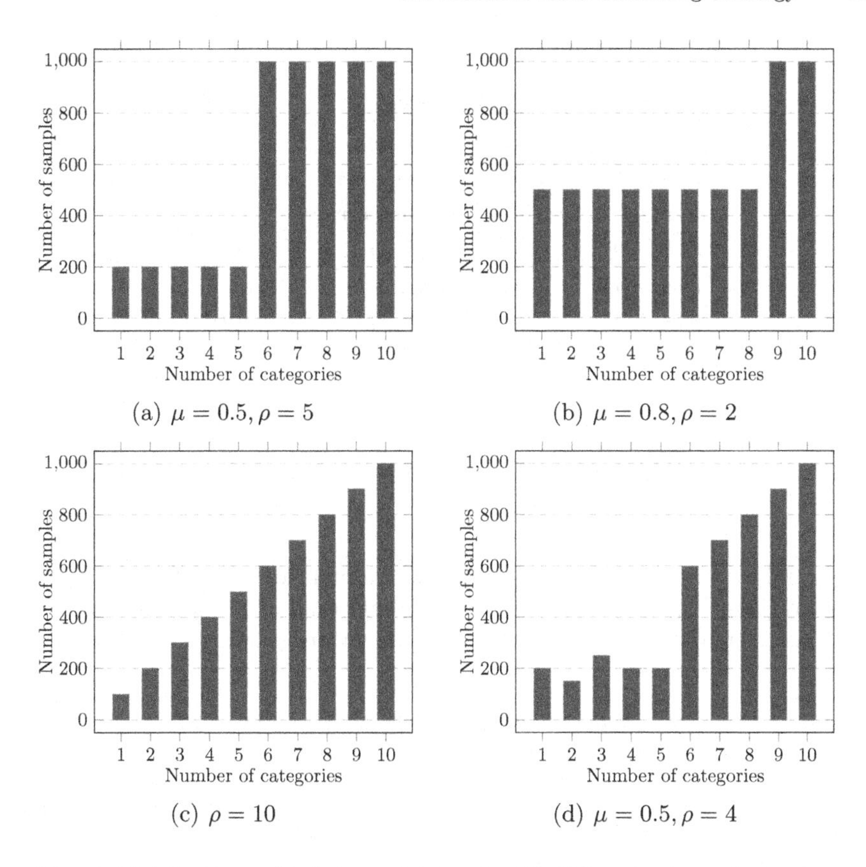

Fig. 3. Distributions of imbalanced dataset with different values of μ and ρ

4.3 Data Sets and Models

In this paper, we use two benchmark data sets: MNIST and malicious code gray image data set. MNIST is a simple and effective digital image classification data set, which consists of 28×28, there are ten categories, corresponding to numbers from 0 to 9. The malicious code gray image data set is converted from malicious code to gray image. We select the data of 20 families, as shown in Table 1.

Table 1. Malicious code dataset

Category	Number	Category	Number
Allaple.A	2949	Lollipop	2478
Allaple.L	1591	Lolyda.AA1	213
C2LOP.gen!g	200	Obfuscator.ACY	1228
C2LOP.P	146	Obfuscator.AD	142
Dontovo.A	162	Ramnit	1541
Fakerean	381	Swizzor.gen!E	128
Gatak	1013	Swizzor.gen!I	132
Instantaccess	431	Tracur	751
Kelihos-ver1	398	VB.AT	408
Kelihos-ver3	2942	Yuner.A	800

It can be seen that the sample data among malicious code families is unbalanced. We first resize it 28 × 28, and uniformly randomly select sub samples of each class. The data sets used in the experiment are shown in Table 2.

Table 2. Summary of the used datasets

Data set	Image feature			Number of categories	Data size	
	Length	Width	Height		Training set	Test set
MNIST	28	28	1	10	5000	1000
Malware images	28	28	1	20	10000	2000

The convolutional neural network used in the experiment is shown in Table 3. Since MNIST has 10 categories and malicious code image data has 20 categories, the former has 10 nodes in the full connection layer and the latter has 20 nodes. All networks use stochastic gradient descent (SGD), and the learning rate is $\eta_t = \eta_0(1 + \gamma t)^{-\alpha}$. η_0 is the initial learning rate, $\gamma = 0.0001$ and $\alpha = 0.75$ are the adjustment coefficients, and t is the current number of iterations.

4.4 Result Analysis

We conducted comparative experiments of jump imbalance and linear imbalance on MNIST and malware images. For the jump imbalance experiment, we compared the effectiveness of various data equalization methods under different combinations. For the swarm intelligence algorithm, we selected the PSO of strategy equalization. The experimental results are shown in Fig. 4. Figure 4(a) and Fig. 4(b) show the comparison results of MNIST. The former shows that the ROC-AUC scores of each method are similar and have good results. It is

Table 3. Architecture of the model used

Layer	Image feature			Kernel size	Stride
	Length	Width	Height		
Input	28	28	1	–	–
Convolution	24	24	20	5	1
Max Pooling	12	12	20	2	2
Convolution	8	8	50	5	1
Max Pooling	4	4	50	2	2
Fully Connected	1	1	500	–	–
ReLU	1	1	500	–	–
Fully Connected	1	1	10/20	–	–
Softmax	1	1	10/20	–	–

because the characteristics of handwritten digits in the MNIST dataset are obvious. When the imbalance rate increases to 1000, the equalization strategies are different, but the under-sampling basically does not change with the number of a few categories. This is because all classes are small sample equalization after under-sampling, and the classifier tends to small sample categories. Figure 4(c) and Fig. 4(d) are the experimental results of malware images dataset, which have more categories than MNIST. It can be seen from the experimental results that the overall score is worse than the former, because there is a problem that the characteristics of adjacent families are easy to be confused. In addition, it can be seen that oversampling is basically better than undersampling, and both adaptive dynamic sampling and PSO based dynamic sampling have achieved good results.

Figure 5 shows the results of linear imbalance on MNIST and malware images data sets. The maximum linear imbalance ratio of the MNIST dataset is 5000, which means that there is only one sample in the most underrepresented class. However, even in this case, the performance is not obvious according to the multi-class ROC-AUC score of the baseline model, as shown in Fig. 5(a). However, oversampling improves the scores of two data sets and all test values, while the subsampling score decreases approximately linearly with the imbalance ratio. For malware images data set, several over sampling schemes are better than under-sampling strategy, and with the increase of imbalance rate, adaptive strategy and intelligent optimization strategy achieve better results.

For the malicious code gray image data set, the number of samples of each family approximately presents a linear distribution, and the imbalance ratio is about 30. The experimental results of each equalization method are shown in Fig. 6(b). In order to further verify the impact of different swarm intelligence algorithms on the model, we compared the total accuracy of PSO, BA and CS algorithms, MNIST linear imbalance rate 1000 and malware images linear imbalance rate 30, as shown in Fig. 6.

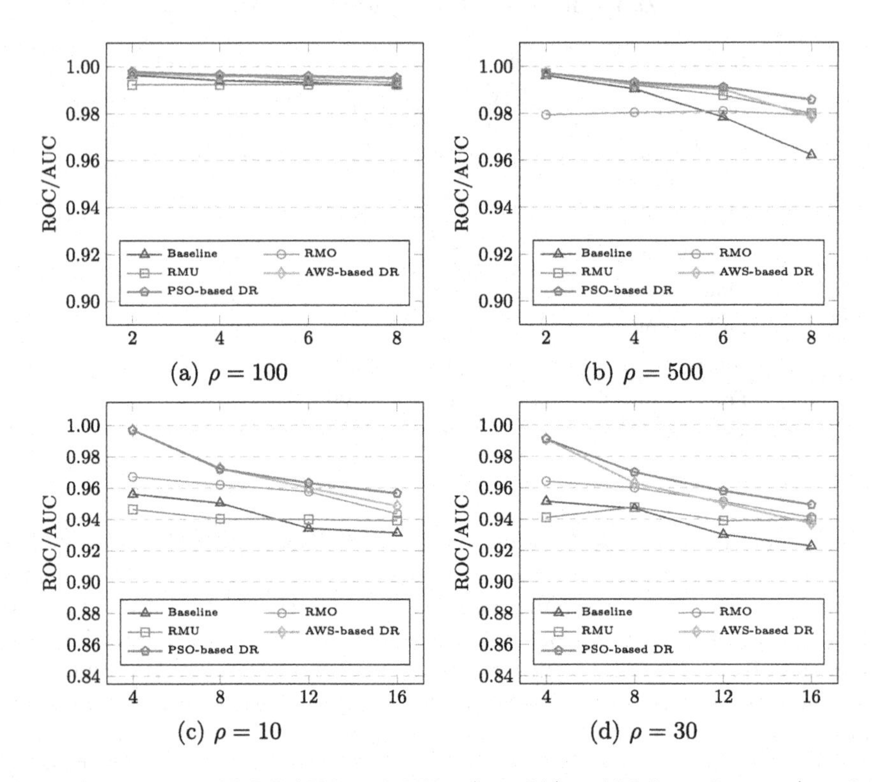

Fig. 4. Comparison of ROC AUC on MNIST (a and b) and Malware Images (c and d) for step imbalance with fixed ρ

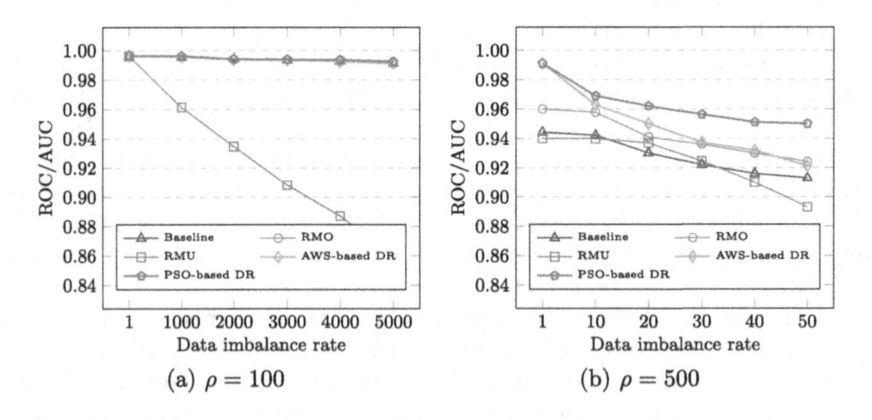

Fig. 5. Comparison of ROC AUC on MNIST and Malware Images for linear imbalance

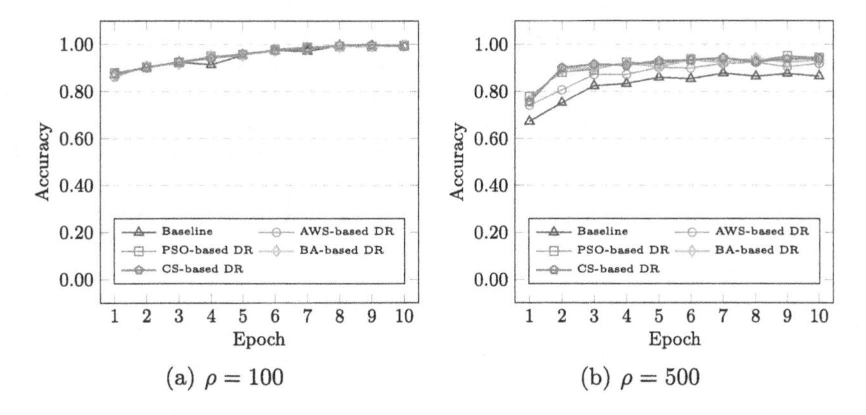

Fig. 6. Comparison of total accuracy on MNIST and Malware Images

Figure 6(a) shows the comparison of MNIST on accuracy. The effects of each algorithm are consistent. With the increase of training times, the accuracy continues to improve. Figure 6(b) shows the comparison of accuracy of malware images test. It can be seen that each equalization method has different effects compared with baseline (not applicable to equalization strategy). The intelligent optimization algorithm is slightly better than AWS, and there is little difference between different intelligent algorithms. In addition, although the intelligent optimization algorithm has achieved good results, the model training time is long due to the introduction of additional training times.

5 Conclusion

Aiming at the unbalanced number of malicious code family samples, this paper designs a dynamic sampling model based on epoch, and uses swarm intelligence algorithm to optimize the sampling weight of each family. We divide the types of data imbalance into jumping imbalance and linear imbalance, and then design comparative experiments on MNIST and malware images data sets to analyze the effects of each equalization strategy. The experimental results show that the unbalanced data set affects the performance of the training model, and our dynamic oversampling model can effectively solve the problem of data imbalance.

References

1. Beheshti, Z., Shamsuddin, S.M.H.: A review of population-based meta-heuristic algorithms. Int. J. Adv. Soft Comput. Appl **5**(1), 1–35 (2013)
2. Buda, M., Maki, A., Mazurowski, M.A.: A systematic study of the class imbalance problem in convolutional neural networks. Neural Netw. **106**, 249–259 (2018)
3. Burnaev, E., Erofeev, P., Papanov, A.: Influence of resampling on accuracy of imbalanced classification. In: Eighth international conference on machine vision (ICMV 2015), vol. 9875, p. 987521. International Society for Optics and Photonics (2015)

4. Cai, X., et al.: A sharding scheme based many-objective optimization algorithm for enhancing security in blockchain-enabled industrial internet of things. IEEE Trans. Ind. Inform. (2021)
5. Cai, X., et al.: An under-sampled software defect prediction method based on hybrid multi-objective cuckoo search. Concurr. Comput. Practice Exp. **32**(5), e5478 (2020)
6. Cai, X., Zhang, J., Liang, H., Wang, L., Wu, Q.: An ensemble bat algorithm for large-scale optimization. Int. J. Mach. Learn. Cybern. **10**(11), 3099–3113 (2019). https://doi.org/10.1007/s13042-019-01002-8
7. Cao, D., Zhang, X., Cao, Y., Wang, Y., Liu, W.: Detection and fine-grained classification of malicious code using convolutional neural networks and swarm intelligence algorithms. Int. J. Wireless Mobile Comput. **19**(1), 1–8 (2020)
8. Cao, D., Zhang, X., Ning, Z., Zhao, J., Xue, F., Yang, Y.: An efficient malicious code detection system based on convolutional neural networks. In: Proceedings of the 2018 2nd International Conference on Computer Science and Artificial Intelligence, pp. 86–89 (2018)
9. Cateni, S., Colla, V., Vannucci, M.: A method for resampling imbalanced datasets in binary classification tasks for real-world problems. Neurocomputing **135**, 32–41 (2014)
10. Cui, Z., Zhang, J., Wu, D., Cai, X., Wang, H., Zhang, W., Chen, J.: Hybrid many-objective particle swarm optimization algorithm for green coal production problem. Inf. Sci. **518**, 256–271 (2020)
11. Cui, Z., Zhao, Y., Cao, Y., Cai, X., Zhang, W., Chen, J.: Malicious code detection under 5g hetnets based on a multi-objective rbm model. IEEE Network **35**(2), 82–87 (2021)
12. Desale, S., Rasool, A., Andhale, S., Rane, P.: Heuristic and meta-heuristic algorithms and their relevance to the real world: a survey. Int. J. Comput. Eng. Res. Trends **351**(5), 2349–7084 (2015)
13. Dokeroglu, T., Sevinc, E., Kucukyilmaz, T., Cosar, A.: A survey on new generation metaheuristic algorithms. Comput. Ind. Eng. **137**, 106040 (2019)
14. He, H., Garcia, E.A.: Learning from imbalanced data. IEEE Trans. Knowl. Data Eng. **21**(9), 1263–1284 (2009)
15. Kubus, M.: Evaluation of resampling methods in the class unbalance problem. Ekonometria **24**(1), 39–50 (2020)
16. LeCun, Y., Bottou, L., Bengio, Y., Haffner, P.: Gradient-based learning applied to document recognition. Proc. IEEE **86**(11), 2278–2324 (1998)
17. Leevy, J.L., Khoshgoftaar, T.M., Bauder, R.A., Seliya, N.: A survey on addressing high-class imbalance in big data. J. Big Data **5**(1), 1–30 (2018). https://doi.org/10.1186/s40537-018-0151-6
18. Mavrovouniotis, M., Li, C., Yang, S.: A survey of swarm intelligence for dynamic optimization: algorithms and applications. Swarm Evol. Comput. **33**, 1–17 (2017)
19. Mrozek, P., Panneerselvam, J., Bagdasar, O.: Efficient resampling for fraud detection during anonymised credit card transactions with unbalanced datasets. In: 2020 IEEE/ACM 13th International Conference on Utility and Cloud Computing (UCC), pp. 426–433. IEEE (2020)
20. Nataraj, L., Karthikeyan, S., Jacob, G., Manjunath, B.S.: Malware images: visualization and automatic classification. In: Proceedings of the 8th International Symposium on Visualization for Cyber Security, pp. 1–7 (2011)
21. Pouyanfar, S., et al.: Dynamic sampling in convolutional neural networks for imbalanced data classification. In: 2018 IEEE Conference on Multimedia Information Processing and Retrieval (MIPR), pp. 112–117. IEEE (2018)

22. Ronen, R., Radu, M., Feuerstein, C., Yom-Tov, E., Ahmadi, M.: Microsoft malware classification challenge. arXiv preprint arXiv:1802.10135 (2018)
23. Thabtah, F., Hammoud, S., Kamalov, F., Gonsalves, A.: Data imbalance in classification: experimental evaluation. Inf. Sci. **513**, 429–441 (2020)
24. Yang, X.S.: Nature-inspired metaheuristic algorithms. Luniver press (2010)

Adaptive Multi-strategy Learning Particle Swarm Optimization with Evolutionary State Estimation

Jinhao Yu(✉) and Junhui Yang(✉)

School of Information Engineering, Jiangxi University of Science and Technology, Ganzhou 341000, China
6120200215@mail.jxust.edu.cn, jhyang@jxust.edu.cn

Abstract. Particle swarm optimizer (PSO) is a random search algorithm based on group cooperation applied to a wide range of practical problems. But its learning mode is relatively fixed, and it is easy to be trapped in the local optimum. Most variants of PSO use only one learning strategy in the entire evolution process, which may affect their performance when different learning strategies are required under different evolution states. In order to deal with search situations in various complex evolutionary environments, an adaptive multi-strategy learning PSO based on the evolutionary state evaluation called AMSLPSO is proposed. In AMSLPSO, a novel evolutionary state estimation (ESE) method is proposed. In order to meet various complex situations under different search stages, the algorithm judges the current evolution state according to the novel ESE method, and then adaptively selects the learning strategy, three state called exploration state, balance state and exploitation state are corresponding to three learning strategies respectively. The experimental results show that AMSLPSO performs better than or at least comparable with some other advanced PSO variants, on the 30 benchmark functions from CEC2017 real-parameter single-objective test suites.

Keywords: Evolutionary state estimation · Multi-strategy · Particle swarm optimization

1 Introduction

Optimization problems has always been an important research hotspot, attracting the attention of more and more scholars due to its obvious application potential in most practical problems in the real world, including project management system, resource scheduling, and path planning. In recent decades, scholars have proposed many valid evolutionary optimization strategies to deal with these problems. For example, In CMOABC, Ma *et al.* optimize multiple conflicting targets at the same time to obtain the optimal solution of MORNP [3]. Yang *et al.* proposed (ACSC), which separates the feature space with different density

L. Pan et al. (Eds.): BIC-TA 2021, CCIS 1565, pp. 174–186, 2022.
https://doi.org/10.1007/978-981-19-1256-6_13

regions to cluster data streams [4]. Ma *et al.* introduced reference vectors and adaptive strategies to analyze decision variables to solve two LSOPs [5].

In 1995, Kennedy and Eberhart proposed the particle swarm optimization (PSO) [1,2], which uses group collaboration to conduct random search, inspired by the survival behavior of social organisms. In Canonical PSO, all particles in the population are initialized randomly, and the learning objects of the particles are usually their own historical best experience and the best experience of all individuals. This single learning mode has greatly affected the performance of PSO. In order to overcome this shortcoming, a number of excellent multi-strategy PSO variants have emerged. For instance, Zhan *et al.* proposed the APSO, which estimates the evolution state in real time through the distribution of the population, and adaptively selects the learning strategy [6]. Zeng *et al.* proposed the SLEPSO. The learning mode is dynamically switched through calculation diversity and the best-so-far experience of the entire swarm [7]. Evolutionary state estimation (ESE) is a method of searching behavior and population distribution characteristics based on the PSO algorithm to analyze the current evolutionary state of the population. The most classic ESE method needs to calculate the Euclidean distance from each particle in the population to all other particles, and then calculate the evolutionary factor, and divide the evolution state according to the evolutionary factor [6]. In the classic ESE method, it takes a lot of time to calculate the Euclidean distance. In order to solve this problem, a novel ESE method is proposed, which only needs to calculate the Euclidean distance from each particle to the global worst particle ($gworst$) and the global best particle ($gbest$) each time. In summary, our purpose is to put forward an adaptive multi-strategy learning PSO (AMSLPSO) based on evolutionary state estimation to enrich the PSO community. The main contributions of this study are outlined as follows:

1) Different search stages will adaptively select appropriate learning strategies through a novel ESE method to take full advantage of beneficial information between individuals.

2) Different learning strategies correspond to different evolutionary states. The evolutionary state includes three situations, namely the exploration state, the balance state and the exploitation state. In the exploration state, we use the historical best experience of the particle and the neighbors' best-so-far position as learning exemplars to enhance diversity [8]. In the exploitation state, we introduce random elite exemplars and mainstream learning exemplars [9] to accelerate convergence, and in the balance state, comprehensively select the best historical experience of particles and mainstream learning examples to realize a balance between exploration and exploitation.

The other parts of this article are introduced as follows. Section 2 reviews the canonical PSO and some PSO variants. Section 3 describes the proposed AMSLPSO algorithm in detail. The experimental results from CEC2017 test suites between AMSLPSO and some advanced PSO variants are shown in Sect. 4. Finally, Sect. 5 draws the conclusions and future work.

2 Related Works

2.1 Canonical PSO

In canonical PSO, the individuals in the population are called particles, which are associated with two vectors. The vector $v_i = [v_i^1, v_i^2, ..., v_i^D]$ represents the velocity of the particle i, as the search direction and step length of the particle, the vector $x_i = [x_i^1, x_i^2, ..., x_i^D]$ represents the position of the particle i, as the candidate solution, and D represents the dimension of the problem. Each particle i has its personal historical best solution, called the personal best vector $pbest_i = [pb_i^1, pb_i^2, ..., pb_i^D]$, and the historical optimal position of the entire population, called the global best vector $gbest = [gb^1, gb^2, ..., gb^D]$. In each iteration, the acceleration direction of the particle depends on its own $pbest_i$ and the $gbest$ of the entire swarm. The velocity and position update formulas of particle i is as follows:

$$v_i(t+1) = \omega \cdot v_i(t) + c_1 \cdot r_1 \cdot (pbest_i(t) - x_i) + c_2 \cdot r_2 \cdot (gbest(t) - x_i(t)) \quad (1)$$

$$x_i(t+1) = x_i(t) + v_i(t+1) \quad (2)$$

where ω represents the weight of inertia, c_1 and c_2 represent the acceleration coefficients of particle self-cognition and social learning respectively, c_1 pulls the particle toward its own $pbest_i$ and c_2 pushes the particle toward the current $gbest$. r_1 and r_2 are two independent random numbers of [0, 1].

2.2 Variants of PSO

According to [2,10], PSO has two advantages. On the one hand, it is simple to implement and on the other hand it has fast convergence speed, making it more competitive than other evolutionary algorithms. Therefore, it has become a research hotspot, and many PSO variants have appeared in the past few decades. The main research branches of PSO are divided into three situations, including parameters tuning, learning model adjustment and hybrid strategy, but AMSLPSO mainly touch upon learning model part, so only this branch is discussed.

The learning model provides impetus for PSO to explore the problem space. Choosing appropriate learning exemplars can extremely enhance search performance. Zeng *et al.* proposed the SLEPSO. The learning mode is dynamically switched through calculation diversity and the best-so-far experience of the entire swarm [7]. Zhan *et al.* used orthogonal experimental design to discover more useful knowledge in the gbest and the particle neighbor's best [11]. In SLPSO, Cheng *et al.* integrated social learning mechanisms into the PSO community [12]. Liu *et al.* used random learning and mainstream learning exemplars to make up for the shortcomings of the canonical PSO learning objects $pbest$ and $gbest$ [9].

3 Proposed AMSLPSO Algorithm

3.1 Evolutionary State Estimation

The evolutionary state estimation (ESE) method has been used in previous studies [6,13]. However, these two methods need to calculate the Euclidean distance from each individual to all other individuals, which requires a lot of computing resources. Therefore, we propose a simple ESE method that only needs to calculate the Euclidean distance from each particle $pbest_i$ to the global worst particle ($gworst$) and the global best particle ($gbest$).

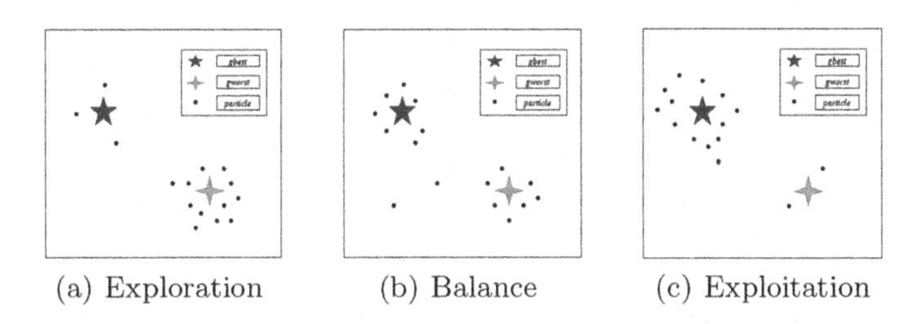

(a) Exploration (b) Balance (c) Exploitation

Fig. 1. Illustration of the evolutionary state based on the population distribution.

First, we calculate the Euclidean distances of all particles to $gbest$ and $gworst$, analyze the results of the comparison of the two distances to determine the distribution of particles, thereby determining the current evolutionary state. Consider three situations. When the number of particles around $gworst$ is far greater than the number of particles around $gbest$, as shown in Fig. 1(a), it represents the exploration state. At this time, it is necessary to maintain and improve the diversity of the population and enhance the global exploration ability of the particles. When the number of particles around $gworst$ is approximately equal to the number of particles around $gbest$, as shown in Fig. 1(b), it represents a balance state. At this time, it is necessary to better balance the global exploration and local exploitation capabilities of particles. When the number of particles around $gworst$ is much smaller than the number of particles around $gbest$, as shown in Fig. 1(c), it represents the exploitation state. At this time, it is necessary to accelerate the convergence speed of the population and enhance the local exploitation ability of the particles. The specific implementation steps are as follows:

Step 1: Calculate the Euclidean distance from $pbest_i$ to $gbest$ and $gworst$, for example:

$$d_{igb} = \sum_{k=1}^{D} (pbest_i^k - gbest^k)^2 \tag{3}$$

$$d_{igw} = \sum_{k=1}^{D} (pbest_i^k - gworst^k)^2 \tag{4}$$

Step 2: Let the number of particles around *gworst* be *ngworst* and the number of particles around *gbest* be *ngbest*. Comparing the values of d_{igb} and d_{igw}, if d_{igb} is greater than d_{igw}, we think that particle i is near *gworst*, and *ngworst* increases by 1. On the contrary, if d_{igb} is less than d_{igw}, then we think that particle i is near *gbest*, and *ngbest* increases by 1.

Step 3: After recording the neighborhood where all the particles belong, we will get the *ngworst* and *ngbest* after statistics, and then set $z = ngworst - ngbest$, set a threshold $\sigma = 0.3 * N$. When z is greater than σ, it is the exploration state, z is less than $-\sigma$, is the exploitation state, when the absolute value of z is less than σ, it is the balance state.

After evaluating the evolution state, the population will select a learning strategy suitable for the current state. The specific strategy will be described in detail in the follow-up summary. The particles will update their velocity and position according to the selected learning strategy. Note that the above evolutionary state estimation does not happen in every iteration, but depends on whether *gbest* is updated. If *gbest* does not improve within a few generations, the evolutionary state estimation will be carried out, otherwise the original state will be maintained.

Algorithm 1 gives the detailed process of evolutionary state estimation.

Algorithm 1. Evolutionary state estimation

Require: *pbest*, *gbest*, *gworst*, and threshold σ;
Ensure: Evolutionary state;
1: For i =1 to N Do
2: If $pbest_i$ not *gbest* and $pbest_i$ not *gworst*
3: Use formula 3, 4 to calculate Euclidean distance d_{igb} and d_{igw};
4: End If
5: If $d_{igb} > d_{igw}$
6: $ngworst = ngworst + 1$;
7: Else
8: $ngbest = ngbest + 1$;
9: End If
10: $z = ngworst - ngbest$;
11: If $z > \sigma$
12: State= Exploration;
13: Elseif $z < -\sigma$
14: State= Exploitation;
15: Elseif $-\sigma \leq z \leq \sigma$
16: State= Balance;
17: End If
18: End For

3.2 Random Elite and Mainstream Learning Exemplars

In many PSO variants, *pbest* and *gbest* are used as learning objects for particles. This update mechanism will cause the loss of population diversity and premature convergence. Therefore, designing effective learning exemplars to avoid these defects and improve convergence performance is an important aspect of current PSO research. In daily life, we often learn from excellent people to make up for our own shortcomings. In order to simulate this learning model, we introduce random elite exemplars. At each iteration, the entire population is sorted according to fitness values, and the top 20% of elite individuals are screened out. Two different *pbests* are randomly selected from the elite group. Compare the fitness values of the two and keep the better individual as the candidate elite best solution (*cepbest*). Then, compare the fitness value of the current $pbest_i$ and *cepbest*, and the better individual is used as the final random elite exemplars (*repbest*). Equations (5) and (6) describe the construction process of the exemplars.

$$cepbest(t) = argmin\{fit(epbest_c(t)), fit(epbest_d(t)), c \neq d \in \{1, 2, ..., EN\}\} \tag{5}$$

$$repbest_i(t) = \begin{cases} cepbest(t) & , fit(cepbest(t)) < fit(pbest_i(t)) \\ pbest_i(t) & , otherwise \end{cases} \tag{6}$$

where EN represents the size of the elite population, which is 20% of the population size N, *epbest* represents the personal best position of the elite particle, $epbest_c$ and $epbest_d$ are two randomly selected elite particles.

In addition, we are not only influenced by outstanding individuals, but also by some mainstream ideas in society. In order to simulate this social phenomenon, we introduced the mainstream learning exemplars [9], which is composed of the average value of each dimension of the personal best position of all particles, shown as:

$$msbest(t) = (\frac{1}{N}\sum_{i=1}^{N} pbest_i^1(t), \frac{1}{N}\sum_{i=1}^{N} pbest_i^2(t), ..., \frac{1}{N}\sum_{i=1}^{N} pbest_i^D(t)) \tag{7}$$

where $pbest_i^d(t), (d = 1, 2, ..., D))$ represents the value of the particle i on the d-th dimension of in the t-th iteration.

3.3 Choose Learning Strategy

After evaluating the evolutionary state of the population, we select the corresponding learning strategy for the particles according to the evaluation results. The three evolutionary states correspond to the following three learning strategies.

1) Exploration state: Ring topology can enhance the diversity of the population and the global exploration ability [8], so learn from *pbest* and *nbest* (neighbors' best-so-far position, neighborhood size is 2.). The velocity update formula in the exploration state is shown in (8):

$$v_i(t+1) = \omega(t) \cdot v_i(t) + c_1(t) \cdot r_1 \cdot (pbest_i(t) - x_i) + c_2(t) \cdot r_2 \cdot (nbest_i(t) - x_i(t)) \tag{8}$$

2) Balance state: The *pbest* of the particles maintains diversity, and the mainstream learning exemplars ensure the exploitation performance of the population, thereby achieving a balance between global exploration and local exploitation, so learn from *pbest* and *msbest* (mainstream learning exemplars). The velocity update formula in the balance state is shown in (9):

$$v_i(t+1) = \omega(t) \cdot v_i(t) + c_1(t) \cdot r_1 \cdot (pbest_i(t) - x_i) + c_2(t) \cdot r_2 \cdot (msbest(t) - x_i(t)) \quad (9)$$

3) Exploitation state: The elite strategy can effectively improve the convergence speed of the algorithm. The mainstream learning exemplars use the mean value of the dimensional information of each particle not only to help the particles effectively overcome the shortcomings of *gbest* premature convergence of the algorithm, but also to promote the communication of particles in the dimension, thereby enhancing the algorithm local exploitation capacity [9,14], so learn from *repbest* (random elite exemplars) and *msbest* (mainstream learning exemplars). The velocity update formula in the exploitation state is shown in (10):

$$v_i(t+1) = \omega(t) \cdot v_i(t) + c_1(t) \cdot r_1 \cdot (repbest_i(t) - x_i) + c_2(t) \cdot r_2 \cdot (msbest(t) - x_i(t)) \quad (10)$$

Algorithm 2 gives the selection process of learning exemplars.

Algorithm 2. Choose learning exemplars

Require: Evolutionary state;
Ensure: Learning exemplars;
1: If State == Exploration
2: exemplar 1= *pbest*; exemplar 2= *nbest*;
3: Elseif State == Exploitation
4: Use formulas (5), (6) and (7) to construct random elite and mainstream learning exemplars;
5: exemplar 1= *repbest*; exemplar 2= *msbest*;
6: Elseif State == Balance
7: Use formula (7) to construct mainstream learning exemplar;
8: exemplar 1= *pbest*; exemplar 2= *msbest*;
9: End If

3.4 Framework of AMSLPSO

In summary, Algorithm 3 gives the pseudo code of AMSLPSO.

In each generation, we evaluate the evolution state only when the initial and global optimal solutions have not been improved within a few generations, and then select learning exemplars according to Algorithm 2, update the entire population, and finally record whether *gbest* has been improved. So far, we have completed a loop sequence. Repeat these processes until the termination condition is reached.

Algorithm 3. AMSLPSO

Require: Population size N, elite population size EN, problem dimension D, maximum number of stagnant iterations $Stag_{max}$, $gbest$ stagnation iterations $Stag_g$;
Ensure: Optimal solution;
1: For i =1 to N Do
2: Initialize v_i and x_i;
3: Evaluate x_i; $pbest = x_i$
4: End For
5: Update $nbest$ and $gbest$;
6: Evolutionary state estimation according to Algorithm 1;
7: While stopping criterion not met Do
8: If $Stag_g > Stag_{max}$
9: Evolutionary state estimation according to Algorithm 1;
10: End If
11: Select learning exemplars according to Algorithm 2;
12: For i =1 to N Do
13: $v_i(t+1) = \omega(t) \cdot v_i(t) + c_1(t) \cdot r_1 \cdot (Exemplar1_i(t) - x_i) + c_2(t) \cdot r_2 \cdot (Exemplar2_i(t) - x_i(t))$;
14: $x_i(t+1) = x_i(t) + v_i(t+1)$;
15: Evaluate x_i; $pbest = x_i$
16: End For
17: Update $nbest$ and $gbest$;
18: If $gbest$ is improved
19: $Stag_g = 0$
20: Else
21: $Stag_g = Stag_g + 1$
22: End If
23: End While

4 Experimental Results and Analysis

4.1 Benchmark Functions and Comparison Algorithms

The real parameter optimization function derived from the CEC2017 competition is used to assess the performance of AMSLPSO. The benchmark consists of the following parts: unimodal function (*f1–f3*), simple multimodal function (*f4–f10*), hybrid function (*f11–f20*), composition function (*f21–f30*). Since the function *f2* will exhibit unstable behavior for higher dimensionality, and there will be significant performance changes when the same algorithm is implemented in Matlab, so we remove the evaluation of *f2* in the experiment [15]. Each function is run independently 51 times to reduce the error value and used to compare the average results. In order to make the result more intuitive, the mean result is $f(a) - f(a^*)$, where $f(a)$ is the optimal solution found in each iteration, and $f(a^*)$ is the actual optimal solution of each function. The fitness evaluation times of all algorithms for each test function are $10000 * D$, and the dimension D is 30. The state division threshold σ and the maximum number of stagnation iterations $Stag_{max}$ in AMSLPSO were $0.3 * N$, 5, respectively, where N represents the population size.

The effectiveness of AMSLPSO will be verified by comparing it with classic PSO and four advanced PSO variants, such as PSO [1], PSO-CF [16], HPSO-TVAC [17], LIPS [18] and XPSO [19]. The source codes of all comparison algorithms are obtained from the author, and the parameter settings of each algorithm are consistent with those given in the corresponding papers. The specific parameter settings are shown in Table 1.

Table 1. Some advanced variants of PSO

Algorithm	Year	Parameters settings	Reference
PSO	1995	$c_1 = c_2 = 2, \omega = 0.7298$	[1]
PSO-CF	2002	$c_1 = c_2 = 2, \omega = 0.9 \rightarrow 0.4$	[16]
HPSO-TVAC	2004	$\omega = 0.9 \rightarrow 0.4, c_1 = 2.5 \rightarrow 0.5, c_2 = 0.5 \rightarrow 2.5$	[17]
LIPS	2013	$\chi = 0.729, c = 2, Neighborhoodsize = 3$	[18]
XPSO	2019	$\mu_1 = \mu_2 = \mu_3 = 1, \sigma = 0.1, \eta = 0.2, p = 0.5$	[19]
AMSLPSO		$\sigma = 0.3 * N, Stag_{max} = 5$	**Proposed**

4.2 Experimental Results

In this experiment, using two indicators, mean value (Mean) and the standard deviation (Std), the performance test results of the solution accuracy are shown in Table 2. Boldface indicates the optimal result obtained from all algorithms, and Table 2 also gives the results of the Wilcoxon signed-rank test (WSRT) with a significance level of 0.05 [20]. The symbols ">", "≈", and "<" indicate that the performance of the AMSLPSO algorithm is significantly better than (>), similar to (≈), or significantly worse than (<) the comparative algorithm. Figure 2 shows the convergence characteristics of all algorithms.

Regarding accuracy, from the mean result, AMSLPSO's performance in functions *f5–f9*, *f11*, *f12*, *f14*, *f16–f18*, *f20*, *f21*, *f23*, *f24*, *f26*, *f29* stronger than the other five PSO variants. From the results of the standard deviation, the effect of AMSLPSO is stronger than five variants of PSO on functions *f4–f9*, *f11*, *f12*, *f14*, *f16–f18*, *f20*, *f21*, *f23–f27*, *f29*. From the results of the WSRT, AMSLPSO is far superior to other algorithms. In the worst case, it is also better than HPSO-TVAC in 21 functions, which shows the excellent performance of AMSLPSO.

In Fig. 2, 12 representative functions are given to show the convergence performance of AMSLPSO compared to the other five competitors. In some cases, such as in Fig. 2 g, j, and k, we can see that competitors converge faster than AMSLPSO, but they stop updating the optimal solution, and the accuracy of the solution is all worse than AMSLPSO. Therefore, AMSLPSO needs more evaluation times to converge to the best position and surpass these algorithms.

Table 2. Amslpso accuracy compared with other PSOS

Function	Criteria	AMSLPSO	PSO	PSO-CF	HPSO-TVAC	LIPS	XPSO
f1	Mean	2.50E+03	4.80E+06	4.63E+03	2.39E+03	**1.42E+03**	1.70E+07
	Std	3.01E+03	5.18E+06 (>)	5.49E+03 (>)	**2.41E+03 (>)**	3.17E+03 (<)	2.04E+07(>)
f3	Mean	2.69E+00	1.23E+04	3.97E+03	**2.68E-01**	2.20E+04	3.57E+03
	Std	4.07E+00	2.23E+03 (>)	1.19E+03 (>)	**4.98E-01 (<)**	6.31E+03 (>)	1.23E+03 (>)
f4	Mean	1.32E+02	2.10E+02	1.78E+02	**6.91E+01**	1.58E+02	1.71E+02
	Std	**1.07E+01**	3.13E+01 (>)	3.10E+01 (>)	1.79E+01 (<)	5.23E+01 (>)	1.44E+01 (>)
f5	Mean	**1.70E+01**	1.63E+02	5.80E+01	1.23E+02	6.44E+01	1.23E+02
	Std	**6.03E+00**	2.99E+01 (>)	9.02E+00 (>)	2.58E+01 (>)	1.85E+01 (>)	4.87E+01 (>)
f6	Mean	**8.36E-04**	3.00E+00	1.28E-02	5.19E+00	7.31E+00	2.79E+00
	Std	**1.57E-03**	1.55E+00 (>)	2.15E-02 (>)	5.33E+00 (>)	4.37E+00 (>)	1.70E+00 (>)
f7	Mean	**4.16E+01**	2.68E+02	1.03E+02	2.28E+02	9.12E+01	1.94E+02
	Std	**1.39E+01**	2.78E+01 (>)	2.64E+01 (>)	4.68E+01 (>)	1.74E+01 (>)	4.95E+01 (>)
f8	Mean	**2.12E+01**	1.60E+02	6.71E+01	8.99E+01	5.98E+01	1.47E+02
	Std	1.25E+01	3.95E+01 (>)	1.16E+01 (>)	2.56E+01 (>)	**1.01E+01 (>)**	5.49E+01 (>)
f9	Mean	**2.99E-01**	6.23E+02	2.30E+02	1.89E+03	4.19E+02	9.73E+01
	Std	**3.23E-01**	3.65E+02 (>)	2.18E+02 (>)	6.02E+02 (>)	2.71E+02 (>)	5.65E+01 (>)
f10	Mean	5.30E+03	5.50E+03	3.74E+03	3.19E+03	**2.83E+03**	4.99E+03
	Std	2.14E+03	6.10E+02 (>)	1.74E+03 (<)	6.79E+02 (<)	**3.89E+02 (<)**	1.59E+03 (<)
f11	Mean	**1.30E+01**	1.51E+02	6.31E+01	1.05E+02	1.49E+02	1.49E+02
	Std	**1.20E+01**	3.79E+01 (>)	4.65E+01 (>)	3.68E+01 (>)	6.24E+01 (>)	5.49E+01 (>)
f12	Mean	**4.47E+04**	3.83E+06	5.15E+05	2.16E+05	5.22E+05	2.31E+06
	Std	**6.27E+04**	8.71E+06 (>)	9.43E+05 (>)	1.38E+05 (>)	5.30E+05 (>)	2.27E+06 (>)
f13	Mean	9.74E+03	7.87E+03	1.80E+04	1.51E+04	**3.08E+03**	4.40E+04
	Std	1.33E+04	8.60E+03 (<)	1.55E+04 (>)	8.37E+03 (>)	**3.10E+03 (<)**	9.79E+04 (>)
f14	Mean	**5.99E+03**	1.91E+04	1.55E+04	1.03E+04	1.16E+04	8.57E+03
	Std	7.05E+03	1.89E+04 (>)	1.17E+04 (>)	1.16E+04 (>)	8.05E+03 (>)	**4.98E+03 (>)**
f15	Mean	2.57E+03	9.00E+03	2.35E+04	5.31E+03	**1.26E+03**	1.94E+03
	Std	4.33E+03	9.43E+03 (>)	6.94E+03 (>)	3.67E+03 (>)	**1.68E+03 (<)**	2.22E+03 (<)
f16	Mean	**1.59E+02**	7.26E+02	5.59E+02	9.94E+02	7.59E+02	6.66E+02
	Std	**1.44E+02**	2.64E+02 (>)	2.73E+02 (>)	1.91E+02 (>)	2.38E+02 (>)	3.87E+02 (>)
f17	Mean	**8.89E+01**	3.34E+02	2.53E+02	4.38E+02	2.28E+02	1.84E+02
	Std	**4.35E+01**	1.85E+02 (>)	1.19E+02 (>)	2.22E+02 (>)	6.89E+01 (>)	1.01E+02 (>)
f18	Mean	**1.27E+05**	7.43E+05	3.79E+05	1.73E+05	2.15E+05	5.61E+05
	Std	**7.95E+04**	6.04E+05 (>)	2.46E+05 (>)	1.39E+05 (>)	1.39E+05 (>)	6.12E+05 (>)
f19	Mean	7.12E+03	1.12E+04	1.95E+04	5.31E+03	**1.43E+03**	8.47E+03
	Std	7.30E+03	9.71E+03 (>)	1.34E+04 (>)	4.55E+03 (<)	**1.90E+03 (<)**	9.43E+03 (>)
f20	Mean	**9.70E+01**	2.76E+02	1.77E+02	2.53E+02	3.04E+02	2.77E+02
	Std	**9.55E+01**	1.29E+02 (>)	6.02E+01 (>)	5.55E+01 (>)	9.91E+01 (>)	1.18E+02 (>)
f21	Mean	**2.17E+02**	3.48E+02	2.66E+02	3.14E+02	2.63E+02	3.59E+02
	Std	**8.05E+00**	3.43E+01 (>)	1.38E+01 (>)	1.58E+01 (>)	1.21E+01 (>)	3.87E+01 (>)
f22	Mean	3.35E+02	4.13E+03	2.94E+03	1.08E+03	**1.00E+02**	1.10E+02
	Std	7.43E+02	2.80E+03 (>)	1.91E+03 (>)	1.58E+03 (>)	**0.00E+00 (<)**	6.19E+00 (<)
f23	Mean	**3.71E+02**	5.21E+02	3.98E+02	5.55E+02	4.50E+02	4.71E+02
	Std	**7.87E+00**	2.45E+01 (>)	1.60E+01 (>)	3.49E+01 (>)	2.19E+01 (>)	5.64E+01 (>)
f24	Mean	**4.38E+02**	5.88E+02	5.42E+02	7.15E+02	4.83E+02	5.88E+02
	Std	**6.40E+00**	2.21E+01 (>)	5.84E+01 (>)	8.05E+01 (>)	1.68E+01 (>)	3.15E+01(>)
f25	Mean	4.03E+02	4.25E+02	4.08E+02	**3.89E+02**	4.19E+02	4.13E+02
	Std	**1.09E+01**	2.54E+01 (>)	1.70E+01 (>)	1.89E+01 (≈)	1.49E+01 (>)	1.70E+01 (>)
f26	Mean	**1.13E+03**	2.80E+03	1.75E+03	2.90E+03	1.37E+03	1.52E+03
	Std	**1.08E+02**	2.91E+02 (>)	2.27E+02 (>)	1.93E+03 (>)	7.95E+02 (>)	9.14E+02 (>)
f27	Mean	5.25E+02	5.72E+02	5.53E+02	**4.75E+02**	6.14E+02	5.51E+02
	Std	**6.94E+00**	2.26E+01 (>)	2.25E+01 (>)	1.11E+01 (<)	2.48E+01 (>)	9.53E+00 (>)
f28	Mean	4.34E+02	5.03E+02	4.48E+02	**3.76E+02**	4.94E+02	4.57E+02
	Std	9.69E+01	4.46E+01 (>)	3.43E+01 (>)	5.09E+01 (<)	9.66E+01 (>)	**2.54E+01 (≈)**
f29	Mean	**5.05E+02**	7.37E+02	7.45E+02	6.71E+02	8.59E+02	7.21E+02
	Std	**8.44E+01**	1.74E+02 (>)	1.93E+02 (>)	1.94E+02 (>)	1.17E+02 (>)	9.72E+01 (>)
f30	Mean	7.81E+03	1.08E+05	1.05E+04	**2.08E+03**	8.98E+04	2.27E+04
	Std	3.38E+03	2.75E+05 (>)	3.33E+03 (>)	**2.34E+03 (<)**	1.31E+05 (>)	1.77E+04 (>)
> (AMSLPSO is significantly better)			28	28	21	23	24
< (AMSLPSO is significantly worse)			1	1	8	6	4
≈			0	0	1	0	1

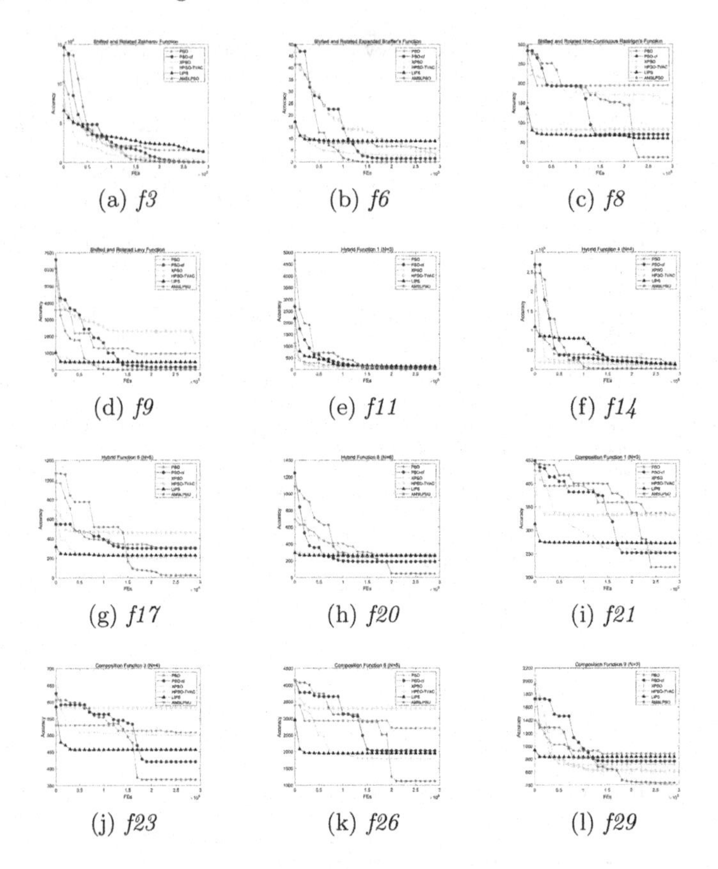

(a) *f3* (b) *f6* (c) *f8*

(d) *f9* (e) *f11* (f) *f14*

(g) *f17* (h) *f20* (i) *f21*

(j) *f23* (k) *f26* (l) *f29*

Fig. 2. Convergence curves of different PSO variants.

5 Conclusion

In this work, an adaptive multi-strategy learning PSO (AMSLPSO) with evolutionary state estimation is proposed. In AMSLPSO, A novel evolutionary state estimation (ESE) method is proposed. In order to meet various complex situations under different search stages, the algorithm judges the current evolution state according to the novel ESE method, and then adaptively selects the learning strategy, three state called exploration state, exploitation state and balance state are corresponding to three learning strategies respectively. In the exploration state, we use the historical best experience and the neighbors' best-so-far experience of the particles as learning exemplars to enhance the global exploration ability of the population. In the exploitation state, we introduce random elite exemplars and mainstream learning exemplars to strengthen the population's local exploitation capability. In the balance state, comprehensively select the historical best experience of the particle and the mainstream learning exemplars to help the particle achieve a balance between exploration and exploitation. In the CEC2017 benchmark test, AMSLPSO was compared with 5 well-known

PSO variants to verify its comprehensive performance on various optimization problems. The experimental results show that AMSLPSO is superior to these advanced PSO variants. However, the AMSLPSO algorithm also has certain shortcomings, such as slow convergence when solving complex multimodal and composition problems.

Regarding future work, we will use different metrics to judge the distribution of the population, and introduce more appropriate learning strategies to adapt to different evolutionary states. In addition, the division of evolutionary state is also worthy of further study.

References

1. Kennedy, J., Eberhart, R.C.: Particle swarm optimization. In: IEEE International Conference on Neural Networks, vol. 4, Perth, Australia, pp. 1942–1948 (1995)
2. Eberhart, R.C., Kennedy, J.: A new optimizer using particle swarm theory. In: Mhs95 Sixth International Symposium on Micro Machine & Human Science, pp. 39–43 (2002)
3. Ma, L., Hu, K., Zhu, Y.: Cooperative artificial bee colony algorithm for multi-objective RFID network planning. J. Network Comput. Appl. **42**, 143–162 (2014)
4. Fahy, C., Yang, S., Gongora, M.: Ant colony stream clustering: a fast density clustering algorithm for dynamic data streams. IEEE Trans. Cybern. **49**(6), 2215–2228 (2018)
5. Ma, L., Huang, M., Yang, S.: An adaptive localized decision variable analysis approach to large-scale multiobjective and many-objective optimization. IEEE Trans. Cybern. **PP**(99), 1–13 (2021)
6. Zhan, Z.H., Zhang, J., Yun, L.: Adaptive particle swarm optimization. IEEE Trans. Syst. Man Cybern. Part B: Cybern. **39**(6), 1362–1381 (2009)
7. Zeng, N., Hung, Y.S., Li, Y.: A novel switching local evolutionary PSO for quantitative analysis of lateral flow immunoassay. Expert Syst. Appl. **41**(4), 1708–1715 (2014)
8. Mendes, R., Kennedy, J., Neves, J.: The fully informed particle swarm: simpler, maybe better. IEEE Trans. Evol. Comput. **8**(3), 204–210 (2004)
9. Liu, H., Zhang, X.W., Tu, L.P.: A modified particle swarm optimization using adaptive strategy. Expert Syst. Appl. **152**, 113353 (2020)
10. Eberhart, R., Shi, Y.: Particle swarm optimization: developments, applications and resources. In: Proceedings of the 2001 Congress on Evolutionary Computation, vol. 1, pp. 81–86. IEEE (2001)
11. Zhan, Z.H., Zhang, J., Yun, L.: Orthogonal learning particle swarm optimization. IEEE Trans. Evol. Comput. **15**(6), 832–847 (2011)
12. Jin, Y.C., Cheng, R.: A social learning particle swarm optimization algorithm for scalable optimization. Inf. Sci. **291**, 43–60 (2015)
13. Bergh, F., Engelbrecht, A.P.: A study of particle optimization particle trajectories. Inf. Sci. **176**(8), 937–971 (2006)
14. Liu, H.R., Cui, J.C., Lu, Z.D.: A hierarchical simple particle swarm optimization with mean dimensional information. Appl. Soft Comput. **76**, 712–725 (2019)
15. Sallam, K.M., Elsayed, S.M., Sarker, R.A.: Multi-method based orthogonal experimental design algorithm for solving CEC2017 competition problems. In: 2017 IEEE Congress on Evolutionary Computation (CEC), pp. 1350–1357 (2017)

16. Eberhart, R.C.: Comparing inertia weights and constriction factors in particle swarm optimization. In: Proceedings of the 2000 IEEE Congress on Evolutionary Computation, La Jolla, vol. 1, pp. 84–88 (2002)
17. Ratnaweera, A., Halgamuge, S.K., Watson, H.C.: Self-Organizing hierarchical particle swarm optimizer with time-varying acceleration coefficients. IEEE Trans. Evol. Comput. **8**(3), 240–255 (2004)
18. Qu, B.Y., Suganthan, P.N., Das, S.: A distance-based locally informed particle swarm model for multimodal optimization. IEEE Trans. Evol. Comput. **17**(3), 387–402 (2013)
19. Xia, X.W., Gui, L., He, G.: An expanded particle swarm optimization based on multi-exemplar and forgetting ability. Inf. Sci. **508**, 105–120 (2020)
20. Gibbons, J.D., Chakraborti, S.: Nonparametric Statistical Inference, CRC Press (2020)

Water Wave Optimization with Distributed-Learning Refraction

Min-Hui Liao, Xin Chen, and Yu-Jun Zheng(✉)

School of Information Science and Technology, Hangzhou Normal University, Hangzhou, China
yujun.zheng@computer.org

Abstract. Water wave optimization (WWO) is a nature-inspired metaheuristic that simulates propagation, refraction and breaking of shallow water waves to search the global optimal solution in the search space. To suppress premature convergence and further accelerate the speed of the basic WWO, we propose an improved WWO with a new distributed-learning refraction operator (called DLWWO), which makes stationary waves learn from more better waves to increase the solution diversity. DLWWO also adopts a nonlinear dimension reduction strategy in the propagation operator to accelerate the search process. We test the DLWWO algorithm on 15 function optimization problems from the CEC2015 single-objective optimization test suite. Experimental results show that DLWWO exhibits very competitive performance compared to the original WWO and other comparative algorithms, which validates the effectiveness and efficiency of the two new proposed strategies.

Keywords: Water wave optimization (WWO) · Global optimization · Distributed learning · Nonlinear dimension reduction

1 Introduction

In recent years, inspired by nature, many researchers have proposed many new metaheuristic algorithms such as genetic algorithm (GA) [1], particle swarm optimization (PSO) [2], biogeography-based optimization (BBO) [3], ant colony optimization (ACO) [4], artificial bee colony (ABC) [5], etc., by imitating biological behaviors or natural phenomena.

Water wave optimization (WWO) algorithm [6] is a relatively new metaheuristics algorithm inspired by the shallow water wave theory. Due to its characteristics of simple algorithm framework and fewer control parameters, WWO has aroused great research interest. To improve the efficiency of the WWO algorithm, it has been extensively studied by some researchers. Zhang et al. [7] proposed a variation of WWO, named VC-WWO, which introduces a variable population size strategy and a comprehensive learning mechanism in the refraction process, thus providing a much better tradeoff between exploration and exploitation. Zheng et al. [8] developed a simplified version of WWO (SimWWO) by leaving

L. Pan et al. (Eds.): BIC-TA 2021, CCIS 1565, pp. 187–200, 2022.
https://doi.org/10.1007/978-981-19-1256-6_14

out the refraction operator and adding a linear population size reduction strategy to accelerate the convergence speed. Wu et al. [9] enhanced WWO with elite opposition-based (EOB) learning strategy, local neighborhood search strategy and the improved propagation operator. Zhang et al. [10] combined WWO with the sine cosine algorithm (SCA) in parallel to wave propagation and breaking, which has strong global search capability to improve WWO's exploitation and exploration capabilities, and also introduced the elite opposition-based learning strategy to increase the diversity of the population. Soltanian et al. [11] proposed a new kind of exploration parameter to increase the exploration ability of the algorithm, which helps the agent to escape the local optima more easily. Recently, Zheng et al. [12] proposed a systematic approach that consists of a set of basic steps and strategies for adapting WWO for different combinatorial optimization problems.

In this paper, we propose a new improved WWO algorithm, named DLWWO. The main contributions of this work are as follows:

1. We adopt a nonlinear dimension reduction strategy in the propagation operator to accelerate the search process.
2. We develop a new distributed-learning refraction operator, which generates new waves according to the distribution of the population, thus making stationary waves learn from more other better waves rather than only the best one.

Numerical experiments on the CEC 2015 benchmark suite [13] demonstrate that the improved algorithm DLWWO exhibits a better performance than the basic WWO and other comparative algorithms.

The rest of the paper is organized as follows. Section 2 briefly introduces the basic WWO algorithm, Sect. 3 describes the proposed improved WWO algorithm in detail, Sect. 4 presents the numerical experiments, and Sect. 5 concludes.

2 The Basic WWO Algorithm

Inspired by the shallow water wave theory, Zheng [6] proposes a relatively new evolutionary algorithm WWO, it through the simulation of the wave movement to solve the optimization problem. In the algorithm, each solution corresponds to a "water wave" with a wave height $h_{\boldsymbol{x}}$ (which is initialized to a constant integer $h_{\max}$) and a wavelength $\lambda_{\boldsymbol{x}}$ (which is initially set to 0.5). Problem solution space is analogous to the seabed area, as illustrated in Fig. 1, the fitness of a water wave is measured inversely by its seabed depth: the shorter the distance to the still water level, the higher the fitness $f(\boldsymbol{x})$ is. During the problem-solving process, WWO uses three types of operations on the waves: propagation, refraction, and breaking.

2.1 Propagation

At each generation, for each wave $\boldsymbol{x}$, the propagation operator once creates a new wave $\boldsymbol{x}'$ by adding a different displacement at each dimension d as:

$$\boldsymbol{x}'_d = \boldsymbol{x}_d + rand(-1,1) \cdot \lambda_{\boldsymbol{x}} L_d \tag{1}$$

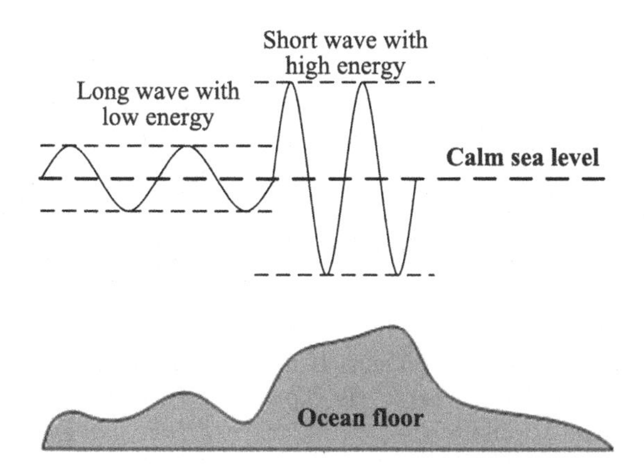

Fig. 1. Different wave shapes in deep and shallow water. [6]

where *rand* is a function generating a random number uniformly distributed in the specified range $[-1, 1]$, and L_d is the length of the dth dimension of the problem. Based on the evaluated fitness value, if $f(\boldsymbol{x}')$ is better than $f(\boldsymbol{x})$, the new wave $\boldsymbol{x}'$ will replace the old wave $\boldsymbol{x}$ in the population, and its wave height is reset to $h_{\max}$. Otherwise, $\boldsymbol{x}$ remains and the wave height is reduced by one to simulate the loss of energy.

Afterwards, the wave wavelength of each wave $\boldsymbol{x}$ is updated based on its fitness $f(\boldsymbol{x})$ at each generation as:

$$\lambda_{\boldsymbol{x}} = \lambda_{\boldsymbol{x}} \cdot \alpha^{-(f(\boldsymbol{x})-f_{\min}+\varepsilon)/(f_{\max}-f_{\min}+\varepsilon)} \tag{2}$$

where $f_{\max}$ and $f_{\min}$ denote the maximum and minimum fitness values respectively, α is the wavelength reduction coefficient, and ε is a very small value to avoid division by zero.

2.2 Refraction

In WWO, the refraction operator only performs on the wave whose height (initialized as $h_{\max}$) decreases to zero, which can avoid search stagnation. It makes the stagnant wave learn from the current best wave $\boldsymbol{x}^*$ at each dimension d by calculating a random number centered halfway between the original position and $\boldsymbol{x}^*$ as:

$$\boldsymbol{x}'_d = N(\frac{x^*_d + x_d}{2}, \frac{|x^*_d - x_d|}{2}) \tag{3}$$

where $N(\mu, \sigma)$ is a Gaussian random number with mean μ and standard deviation σ. After refraction, its height is reset to $h_{\max}$ and the wavelength is updated as:

$$\lambda_{\boldsymbol{x}'} = \lambda_{\boldsymbol{x}} \frac{f(\boldsymbol{x})}{f(\boldsymbol{x}')} \tag{4}$$

2.3 Breaking

WWO only performs breaking on a wave $\boldsymbol{x}$ that reaches a position better than the current best solution $\boldsymbol{x}^*$, where the breaking operator breaks it into a series of solitary waves. Each solitary wave $\boldsymbol{x}'$ of $\boldsymbol{x}$ is obtained by randomly selecting k dimensions (where k is a random number between 1 and a predefined number $k_{\max}$), and at each dimension d updating the component as:

$$\boldsymbol{x}'_d = \boldsymbol{x}^*_d + N(0,1) \cdot \beta L_d \tag{5}$$

where β is the breaking coefficient. If none of the solitary waves are better than $\boldsymbol{x}^*$, $\boldsymbol{x}^*$ remains; Otherwise $\boldsymbol{x}^*$ is replaced by the fittest one among the solitary waves in the population.

2.4 The Algorithmic Framework of WWO

Algorithm 1 presents the procedure of the basic WWO algorithm.

Algorithm 1. The basic WWO algorithm.

1: Randomly initialize a population P of n solutions (waves);
2: **while** the termination criterion is not satisfied **do**
3:     **for** each solution $\boldsymbol{x} \in P$ **do**
4:         Propagate $\boldsymbol{x}$ to a new $\boldsymbol{x}'$ according to Eq. (1);
5:         **if** $f(\boldsymbol{x}') < f(\boldsymbol{x})$ **then**
6:             **if** $f(\boldsymbol{x}) < f(\boldsymbol{x}^*)$ **then**
7:                 Break $\boldsymbol{x}'$ based on Eq. (5);
8:                 Update $\boldsymbol{x}^*$ with $\boldsymbol{x}'$;
9:             **end if**
10:         **else**
11:             $h_{\boldsymbol{x}} \leftarrow h_{\boldsymbol{x}} - 1$;
12:             **if** $h_{\boldsymbol{x}} = 0$ **then**
13:                 Refract $\boldsymbol{x}$ to a new $\boldsymbol{x}'$ according to Eq. (3);
14:                 Update $\lambda_{\boldsymbol{x}'}$ according to Eq. (4);
15:             **end if**
16:         **end if**
17:     **end for**
18:     Update the wavelengths of the waves according to Eq. (2);
19: **end while**
20: **return** $\boldsymbol{x}^*$.

3 The Proposed DLWWO Algorithm

In this section, we describe the improved DLWWO algorithm in detail.

3.1 Nonlinear Dimension Reduction

In the basic WWO, the number of dimensions D_P modified by propagation operation is fixed, which is lack of change. In general, it is desired for one algorithm to explore the whole solution space in early search stages, while exploit near a local space with the help of some strategies in later stages. Therefore, we design a dimension control mechanism, which decreases dimension nonlinearly with the population iteration. At each generation, the dimension D_P decreases from an upper limit $D_{\max}$ to a lower limit $D_{\min}$ as:

$$D_P = D_{\max} - (D_{\max} - D_{\min}) \cdot (\frac{t}{t_{\max}})^2 \tag{6}$$

where t is the number of fitness evaluations (NFEs) and $t_{\max}$ is the maximum allowable NFEs, $D_{\max}$ equals to the dimensions of optimization problem. When implementing the dimension reduction mechanism, we randomly select the dimension to be updated at the process of propagation. The random selection of some dimensions can avoid the population converging to a local optimum too fast.

3.2 Distributed-Learning Refraction

The refraction operation of the basic WWO is to make a wave that fails to find better positions for a certain period, learn from the current best wave. Such a learning mechanism may cause premature convergence, since the new wave may be strongly attracted by the current best wave which may be trapped in a local optimum.

Based on this observation, we hope to make stationary waves learn from more better waves to increase the solution diversity. Therefore, we propose a new distributed learning refraction operator, which generates new waves according to the distribution of the population. Before each refraction operation, we select the top M (M is a control parameter) best individuals in the current population to form an advanced set, and calculate their distribution (represented by the mean and standard deviation) at each dimension d:

$$\mu_d = \frac{\sum_{m=1}^{M} V_{m,d}}{M} \tag{7}$$

$$\sigma_d = \sqrt{\frac{\sum_{m=1}^{M} (V_{m,d} - \mu_d)^2}{M}} \tag{8}$$

Where $V_{m,d}$ is the value on the d dimension of the mth individual in the advanced set.

Based on the calculated mean and standard deviation, we use a simple way to generate new waves:

$$\boldsymbol{x}'_d = N(\mu_d, \sigma_d) \tag{9}$$

In this way, the new position of refracted solution is a random number according to the distribution of the advanced individuals in the population.

Besides, in order to solve the influence of excessive randomness on the effect of refraction operation, we use Eq. (9) to generate n (n is the population size) candidate solutions at the same time, and calculate the fitness of these candidate solutions respectively, and then select the best solution among them as the new solution.

Note that each time we perform a refraction operation, n candidate solutions are generated once, which requires additional computational resources. Therefore, we adopt a variable population size mechanism [14], which decreases the population size n from a maximum size $n_{\max}$ to a minimum size $n_{\min}$, to limit the number of candidate solutions. The mechanism of variable population size has been widely applied in many evolutionary algorithms and has demonstrated its effectiveness in balancing global search and local search on many problems.

3.3 The Framework of DLWWO

Algorithm 2 presents the framework of the DLWWO algorithm.

3.4 Runtime Complexity of DLWWO

In this section, the runtime complexity of the proposed algorithm is analyzed. The proposed algorithm, which is a meta-heuristic belonging to an iteration approach, has a requirement to take time to obtain a near optimal solution. The runtime complexity of the basic WWO algorithm is $O(n \cdot D \cdot t_{\max})$, where D is the dimension of optimization problem. Our method differs from the basic WWO mainly in the distributed-learning refraction operator.

For the distributed-learning refraction operator, the runtime complexity of finding the top M individuals depends on the sort method based on the individual fitness. By utilizing the heap sort algorithm, the sorting procedure can be completed in $O(n \cdot \log_2 n)$ time at each generation.

Considering the overall algorithm, the runtime complexity of DLWWO is $O(\max(n \cdot D \cdot t_{\max}, n \cdot \log_2 n \cdot t_{\max}))$.

4 Numerical Experiments

4.1 Experimental Parameter Setting

To evaluate the performance of the proposed DLWWO algorithm, we conduct numerical experiments on 15 benchmark functions, denoted by f_1–f_{15}, in the CEC 2015 learning-based benchmark suite [13]. The functions are high-dimensional problems, in which f_1 and f_2 are unimodal functions, f_3–f_5 are simple multimodal functions, f_6–f_8 are hybrid functions, and f_9–f_{15} are composition functions. Interested readers please refer to Liang et al. [13] for a more detailed survey of the test functions.

Algorithm 2. The DLWWO algorithm.

```
1: Randomly initialize a population P of n solutions (waves);
2: Let x* be the current best solution in P;
3: while the termination criterion is not satisfied do
4:     for each solution x ∈ P do
5:         Calculate D_P based on Eq. (6);
6:         for each dimension d in D_P dimensions do
7:             Propagate x_d to a new x'_d based on Eq. (1);
8:         end for
9:         if f(x') < f(x) then
10:            if f(x') < f(x*) then
11:                Break x' based on Eq. (5);
12:                Update x* with x';
13:            end if
14:            Replace x with x';
15:        else
16:            h_x ← h_x − 1;
17:            if h_x = 0 then
18:                Select the top M best solutions in P;
19:                for each dimension d in D_max dimensions do
20:                    Calculate μ_d and σ_d based on Eq. (7) and Eq. (8);
21:                end for
22:                Generate n candidate solutions based on Eq. (9);
23:                Select the best solution as x' from the candidate set;
24:            end if
25:        end if
26:    end for
27:    Update the wavelengths according to Eq. (2);
28: end while
29: return x*.
```

We compare the proposed DLWWO with WWO [6], PSO [2], CLPSO [15], BBO [3] and EBO [16]. To ensure a fair comparison, we set the maximum NFEs of all algorithms to 300,000 for each problem. The experimental environment is a computer with Intel Core i7-8700 processor (3.2 GHz) and 16 GB of memory. The control parameters of the algorithms are set as follows:

- PSO: Population size $n = 50$, inertia weight w ($w_{\max} = 0.9$, $w_{\min} = 0.4$), the acceleration coefficients $c_1 = 2$ and $c_2 = 2$.
- CLPSO: Population size $n = 50$, inertia weight w ($w_{\max} = 0.9$, $w_{\min} = 0.4$), the acceleration coefficients $c = 1.49445$.
- BBO: $n = 50$, maximum immigration rate $\lambda_{\max} = 1$, maximum emigration rate $\mu_{\max} = 1$, and mutation rate $m_p = 0.05$.
- EBO: $n = 50$, maximum immigration rate $\lambda_{\max} = 1$, maximum emigration rate $\mu_{\max} = 1$, immaturity index η ($\eta_{\max} = 0.7$, $\eta_{\min} = 0.3$).
- WWO: $h_{\max} = 12$, $\alpha = 1.0026$, $k_{\max} = 12$, β linearly decreases from 0.25 to 0.001, and $n = 10$.

- DLWWO: $\alpha = 1.0026$, $k_{\max} = 12$, β linearly decreases from 0.25 to 0.001, n nonlinearly decreases from 50 to 3, D_P nonlinearly decreases from 30 to 3.

Note that we simply use a fixed parameter setting for each algorithm on all the test problems instead of fine-tuning the parameter values on each problem.

4.2 Comparative Experiments

We run each algorithm 50 times (with different random seeds) on each test function, and record the maximum (max), minimum (min), median (mean), standard deviation (std) and t-test (h) of the results as presented in Table 1 and Table 2. On each test function, the best median values among the comparative algorithms are marked in bold (some values appear to be the same as the bold values because the decimals are omitted). We perform a t-test on the experimental results of DLWWO and other algorithms. We use 0 to denote that there is no significant difference between DLWWO and the corresponding algorithm, 1^+ to denote that the performance of DLWWO is significantly better than the corresponding algorithm, and 1^- vice versa (with a confidence level of 95%).

The experimental results show that, on the two unimodal group functions, DLWWO obtains the best median values on f_1 and f_2. On the three simple multimodal benchmark functions, DLWWO obtains the best median values on f_3, f_4 and f_5. On the three hybrid functions, DLWWO obtains the best median values on f_6, f_7 and f_8. On the seven composition functions, except that on f_{10} and f_{11}, EBO obtains the best median value, DLWWO obtains the best median values on all the remaining 5 functions.

According to the statistical test results, we could observe that:

- DLWWO performs significantly better than PSO on all 15 functions.
- DLWWO performs significantly better than CLPSO on 11 functions, while CLPSO performs significantly better than DLWWO only on two functions (f_{10} and f_{11}).
- DLWWO performs significantly better than BBO on 13 functions, while BBO performs significantly better than DLWWO only on one function (f_{10}).
- DLWWO performs significantly better than EBO on 10 functions, while BBO performs significantly better than DLWWO on four functions (f_2, f_7, f_{10} and f_{11}).
- DLWWO performs significantly better than the basic WWO on 11 functions, while there is no significant difference between DLWWO and the basic WWO on the remaining four functions.

Since two strategies are proposed to improve WWO in this paper, in order to better compare and analyze the improvement effect of the two strategies on WWO, we also test experiments with them, the improved WWO which only uses nonlinear dimension decreasing strategy is called WWO_D, while WWO_L is introduced the new distributed learning refraction operator, whose results as presented in Table 3 and Table 4.

Table 1. Experimental results on the unimodal and simple multimodal benchmark functions.

ID	Metric	PSO	CLPSO	BBO	EBO	WWO	DLWWO
f_1	max	3.72E+08	7.52E+07	1.75E+07	5.88E+06	5.54E+06	1.94E+06
	min	7.53E+06	3.58E+06	1.33E+06	1.13E+06	3.81E+05	3.19E+05
	mean	5.72E+07	1.36E+07	5.68E+06	3.02E+06	1.82E+06	**7.49E+05**
	std	6.06E+07	1.18E+07	4.29E+06	1.18E+06	9.46E+05	3.09E+05
	h	1^+	1^+	1^+	1^+	1^+	
f_2	max	2.07E+10	1.57E+04	3.84E+06	4.53E+03	1.41E+04	5.44E+03
	min	9.34E+07	2.16E+02	3.45E+05	2.05E+02	2.01E+02	2.00E+02
	mean	5.02E+09	6.42E+03	9.48E+05	6.45E+02	3.00E+03	**1.14E+03**
	std	4.73E+09	5.04E+03	5.80E+05	6.57E+02	3.24E+03	1.06E+03
	h	1^+	1^+	1^+	1^-	1^+	
f_3	max	3.21E+02	3.21E+02	3.20E+02	3.20E+02	3.20E+02	3.20E+02
	min	3.20E+02	3.21E+02	3.20E+02	3.20E+02	3.20E+02	3.20E+02
	mean	3.21E+02	3.21E+02	**3.20E+02**	**3.20E+02**	**3.20E+02**	**3.20E+02**
	std	1.34E−01	5.66E−02	2.98E−02	5.69E−02	2.84E−06	1.10E−04
	h	1^+	1^+	1^+	1^+	1^+	
f_4	max	6.03E+02	6.31E+02	5.02E+02	4.51E+02	6.28E+02	4.35E+02
	min	4.62E+02	5.75E+02	4.27E+02	4.22E+02	4.61E+02	4.09E+02
	mean	5.41E+02	6.04E+02	4.50E+02	4.36E+02	5.29E+02	**4.22E+02**
	std	3.28E+01	1.09E+01	1.38E+01	7.07E+00	3.57E+01	5.59E+00
	h	1^+	1^+	1^+	1^+	1^+	
f_5	max	5.19E+03	7.83E+03	3.39E+03	3.31E+03	6.22E+03	2.24E+03
	min	2.76E+03	6.15E+03	1.22E+03	1.29E+03	2.69E+03	6.32E+02
	mean	4.06E+03	7.31E+03	2.40E+03	2.59E+03	3.99E+03	**1.46E+03**
	std	5.96E+02	3.19E+02	4.62E+02	4.33E+02	6.83E+02	4.06E+02
	h	1^+	1^+	1^+	1^+	1^+	
f_6	max	3.14E+07	1.01E+07	7.74E+06	3.10E+06	3.63E+05	1.07E+05
	min	1.46E+05	1.66E+05	1.66E+05	3.13E+05	4.79E+03	1.16E+04
	mean	1.35E+06	2.77E+06	3.24E+06	1.36E+06	5.28E+04	**4.83E+04**
	std	8.00E+06	2.06E+06	1.67E+06	7.43E+05	5.41E+04	2.52E+04
	h	1^+	1^+	1^+	1^+	0	
f_7	max	9.41E+02	7.13E+02	7.78E+02	7.13E+02	7.23E+02	7.13E+02
	min	7.16E+02	7.08E+02	7.06E+02	7.05E+02	7.10E+02	7.06E+02
	mean	7.23E+02	7.10E+02	7.13E+02	**7.10E+02**	7.18E+02	**7.10E+02**
	std	3.86E+01	9.21E−01	9.51E+00	1.61E+00	3.04E+00	1.41E+00
	h	1^+	0	1^+	1^-	1^+	

Table 2. Experimental results on the unimodal and simple multimodal benchmark functions.

ID	Metric	PSO	CLPSO	BBO	EBO	WWO	DLWWO
f_8	max	2.79E+06	2.87E+06	2.28E+06	8.74E+05	1.35E+05	6.77E+04
	min	3.09E+04	4.15E+04	6.18E+04	6.77E+04	9.03E+03	9.34E+03
	mean	3.95E+05	6.82E+05	1.01E+06	2.77E+05	4.14E+04	**2.97E+04**
	std	5.90E+05	5.86E+05	5.94E+05	1.58E+05	2.88E+04	1.49E+04
	h	1^+	1^+	1^+	1^+	1^+	
f_9	max	1.28E+03	1.00E+03	1.01E+03	1.00E+03	1.00E+03	1.00E+03
	min	1.01E+03	1.00E+03	1.00E+03	1.00E+03	1.00E+03	1.00E+03
	mean	1.04E+03	**1.00E+03**	**1.00E+03**	**1.00E+03**	**1.00E+03**	**1.00E+03**
	std	5.29E+01	2.07E−01	4.39E−01	2.94E−01	4.35E−01	3.47E−01
	h	1^+	1^+	1^+	1^+	0	
f_{10}	max	2.41E+08	3.94E+03	5.62E+03	1.65E+03	2.51E+05	1.97E+05
	min	2.29E+03	1.90E+03	1.86E+03	1.22E+03	1.27E+04	1.07E+04
	mean	2.52E+07	2.23E+03	2.71E+03	**1.32E+03**	8.66E+04	7.01E+04
	std	3.78E+07	2.74E+02	5.80E+02	1.07E+02	6.24E+04	4.57E+04
	h	1^+	1^-	1^-	1^-	0	
f_{11}	max	2.31E+03	1.66E+03	2.02E+03	1.92E+03	2.32E+03	1.79E+03
	min	1.47E+03	1.53E+03	1.74E+03	1.41E+03	1.41E+03	1.51E+03
	mean	2.13E+03	1.58E+03	1.88E+03	**1.46E+03**	1.91E+03	1.65E+03
	std	1.73E+02	3.75E+01	5.01E+01	9.06E+01	3.19E+02	5.66E+01
	h	1^+	1^-	1^+	1^-	1^+	
f_{12}	max	1.40E+03	1.40E+03	1.31E+03	1.31E+03	1.32E+03	1.31E+03
	min	1.31E+03	1.31E+03	1.31E+03	1.30E+03	1.31E+03	1.30E+03
	mean	1.40E+03	1.36E+03	1.31E+03	1.31E+03	1.31E+03	**1.30E+03**
	std	4.06E+01	4.65E+01	1.26E+00	5.34E-01	2.07E+00	6.73E−01
	h	1^+	1^+	1^+	1^+	1^+	
f_{13}	max	1.30E+03	1.30E+03	1.30E+03	1.30E+03	1.30E+03	1.30E+03
	min	1.30E+03	1.30E+03	1.30E+03	1.30E+03	1.30E+03	1.30E+03
	mean	**1.30E+03**	**1.30E+03**	**1.30E+03**	**1.30E+03**	**1.30E+03**	**1.30E+03**
	std	1.01E−01	3.29E−04	3.40E−03	1.20E−03	2.41E−01	1.87E−02
	h	1^+	1^+	1^+	1^+	1^+	
f_{14}	max	6.98E+04	3.50E+04	3.64E+04	3.51E+04	4.06E+04	3.52E+04
	min	3.56E+04	3.45E+04	3.26E+04	3.27E+04	3.28E+04	2.61E+04
	mean	4.28E+04	3.47E+04	3.49E+04	3.36E+04	3.56E+04	**3.13E+04**
	std	6.45E+03	1.04E+02	8.70E+02	5.11E+02	2.11E+03	2.21E+03
	h	1^+	1^+	1^+	1^+	1^+	
f_{15}	max	3.95E+03	1.60E+03	1.60E+03	1.60E+03	1.60E+03	1.60E+03
	min	1.61E+03	1.60E+03	1.60E+03	1.60E+03	1.60E+03	1.60E+03
	mean	1.63E+03	**1.60E+03**	**1.60E+03**	**1.60E+03**	**1.60E+03**	**1.60E+03**
	std	4.18E+02	5.35E−12	1.35E−01	1.24E−12	0.00E+00	0.00E+00
	h	1^+	0	0	0	0	

Table 3. Experimental results on the unimodal and simple multimodal benchmark functions.

ID	Metric	WWO	WWO_D	WWO_L	DLWWO
f_1	max	5.54E+06	2.24E+06	2.58E+06	1.94E+06
	min	3.81E+05	1.27E+05	4.14E+05	3.19E+05
	mean	1.82E+06	9.39E+05	1.02E+06	**7.49E+05**
	std	9.46E+05	5.58E+05	4.27E+05	3.09E+05
	h	1^+	1^+	1^+	
f_2	max	1.41E+04	1.48E+04	5.20E+03	5.44E+03
	min	2.01E+02	2.01E+02	2.00E+02	2.00E+02
	mean	3.00E+03	3.14E+03	1.41E+03	**1.14E+03**
	std	3.24E+03	3.72E+03	1.21E+03	1.06E+03
	h	1^+	1^+	0	
f_3	max	3.20E+02	3.20E+02	3.20E+02	3.20E+02
	min	3.20E+02	3.20E+02	3.20E+02	3.20E+02
	mean	**3.20E+02**	**3.20E+02**	**3.20E+02**	**3.20E+02**
	std	2.84E−06	2.66E−06	8.27E−05	1.10E−04
	h	1^+	1^+	0	
f_4	max	6.28E+02	6.28E+02	4.44E+02	4.35E+02
	min	4.61E+02	4.61E+02	4.12E+02	4.09E+02
	mean	5.29E+02	5.35E+02	4.24E+02	**4.22E+02**
	std	3.57E+01	3.94E+01	6.98E+00	5.59E+00
	h	1^+	1^+	0	
f_5	max	6.22E+03	6.01E+03	2.42E+03	2.24E+03
	min	2.69E+03	2.45E+03	7.56E+02	6.32E+02
	mean	3.99E+03	3.53E+03	1.57E+03	**1.46E+03**
	std	6.83E+02	8.31E+02	3.94E+02	4.06E+02
	h	1^+	1^+	1^+	
f_6	max	3.63E+05	2.00E+05	2.45E+05	1.07E+05
	min	4.79E+03	6.64E+03	1.33E+04	1.16E+04
	mean	5.28E+04	**4.77E+04**	8.91E+04	4.83E+04
	std	5.41E+04	3.34E+04	4.70E+04	2.52E+04
	h	0	0	1^+	
f_7	max	7.23E+02	7.91E+02	7.13E+02	7.13E+02
	min	7.10E+02	7.08E+02	7.06E+02	7.06E+02
	mean	7.18E+02	7.19E+02	7.11E+02	**7.10E+02**
	std	3.04E+00	1.42E+01	1.29E+00	1.41E+00
	h	1^+	1^+	1^+	

Table 4. Experimental results on the unimodal and simple multimodal benchmark functions.

ID	Metric	WWO	WWO_D	WWO_L	DLWWO
f_8	max	1.35E+05	2.39E+05	5.82E+04	6.77E+04
	min	9.03E+03	4.10E+03	9.93E+03	9.34E+03
	mean	4.14E+04	5.01E+04	**2.53E+04**	2.97E+04
	std	2.88E+04	3.69E+04	1.18E+04	1.49E+04
	h	1^+	1^+	1^-	
f_9	max	1.00E+03	1.00E+03	1.00E+03	1.00E+03
	min	1.00E+03	1.00E+03	1.00E+03	1.00E+03
	mean	**1.00E+03**	**1.00E+03**	**1.00E+03**	**1.00E+03**
	std	4.35E−01	3.22E−01	4.68E−01	3.47E−01
	h	0	1^+	1^-	
f_{10}	max	2.51E+05	2.38E+05	3.05E+05	1.97E+05
	min	1.27E+04	4.29E+03	2.97E+04	1.07E+04
	mean	8.66E+04	**5.05E+04**	1.41E+05	7.01E+04
	std	6.24E+04	4.36E+04	6.24E+04	4.57E+04
	h	0	1^-	1^+	
f_{11}	max	2.32E+03	2.32E+03	1.78E+03	1.79E+03
	min	1.41E+03	1.40E+03	1.54E+03	1.51E+03
	mean	1.91E+03	1.78E+03	**1.65E+03**	**1.65E+03**
	std	3.19E+02	3.36E+02	5.64E+01	5.66E+01
	h	1^+	1^+	0	
f_{12}	max	1.32E+03	1.31E+03	1.31E+03	1.31E+03
	min	1.31E+03	1.31E+03	1.30E+03	1.30E+03
	mean	1.31E+03	1.31E+03	**1.30E+03**	**1.30E+03**
	std	2.07E+00	1.97E+00	6.03E−01	6.73E−01
	h	1^+	1^+	0	
f_{13}	max	1.30E+03	1.30E+03	1.30E+03	1.30E+03
	min	1.30E+03	1.30E+03	1.30E+03	1.30E+03
	mean	**1.30E+03**	**1.30E+03**	**1.30E+03**	**1.30E+03**
	std	2.41E−01	2.46E−01	1.74E−02	1.87E−02
	h	1^+	1^+	0	
f_{14}	max	4.06E+04	3.73E+04	3.52E+04	3.52E+04
	min	3.28E+04	3.26E+04	2.76E+04	2.61E+04
	mean	3.56E+04	3.45E+04	3.16E+04	**3.13E+04**
	std	2.11E+03	1.34E+03	1.87E+03	2.21E+03
	h	1^+	1^+	0	
f_{15}	max	1.60E+03	1.60E+03	1.60E+03	1.60E+03
	min	1.60E+03	1.60E+03	1.60E+03	1.60E+03
	mean	**1.60E+03**	**1.60E+03**	**1.60E+03**	**1.60E+03**
	std	0.00E+00	0.00E+00	0.00E+00	0.00E+00
	h	0	0	0	

By comparing the experimental results of DLWWO with WWO_D and WWO_L which use one of improved strategies separately, we can conclude that using the new distributed-learning refraction operator to improve the performance of WWO is more significant, and can obtain almost the best median value as DLWWO on 12 functions. Especially on f_8 and f_{11}, WWO_L obtains the best median value. As for the usage of the nonlinear dimension reduction strategy, WWO_D obtains good results in five functions, and obtains the best median value on f_6 and f_{10}. This makes DLWWO, which integrates the two improved strategies, achieve good results on these functions. Besides, on some functions, such as f_1, f_2, f_4, f_5, f_7 and f_{14}, the performance improvement is not significant when utilizing one of the improved strategies separately, but DLWWO can obtain the best value. This validates the effectiveness of the combination of the two strategies used in DLWWO.

In summary, the experimental results show that the overall performance of DLWWO is the best one among the comparative algorithms on the benchmark suite, which demonstrates the effectiveness and efficiency of the proposed strategies.

5 Conclusion

This paper presents an improved WWO algorithm, called DLWWO, which uses a nonlinear dimension reduction strategy in the propagation operator to accelerate the search process, and develops a new distributed-learning refraction operator to increase the solution diversity. Numerical experiments on the CEC 2015 benchmark suite show that the overall performance of DLWWO is better than the original WWO and other popular algorithms.

There is still room for further improving the performance of WWO. For the nonlinear dimension reduction strategy, we can adapt the dimension according to the search states of the algorithm. As for the new distributed-learning refraction operator, we can utilize the information of the original wave when calculating the new position after refraction. Further research also includes adapting WWO algorithm for high dimensional problems with equality and inequality constraints.

References

1. Holland, J.H.: Adaptation in Natural and Artificial Systems: An Introductory Analysis with Applications to Biology, Control and Artificial Intelligence. MIT Press, Cambridge (1975)
2. Kennedy, J., Eberhart, R.: Particle swarm optimization. In: Proceedings of the IEEE International Conference on Neural Networks, Perth, Australia, vol. 4, pp. 1942–1948, November–December 1995
3. Simon, D.: Biogeography-based optimization. IEEE Trans. Evol. Comput. **12**(6), 702–713 (2008). https://doi.org/10.1109/TEVC.2008.919004
4. Socha, K., Dorigo, M.: Ant colony optimization for continuous domains. Eur. J. Oper. Res. **185**(3), 1155–1173 (2008)

5. Karaboga, D., Basturk, B.: A powerful and efficient algorithm for numerical function optimization: artificial bee colony (ABC) algorithm. J. Global Optim. **39**(3), 459–471 (2007)
6. Zheng, Y.J.: Water wave optimization: a new nature-inspired metaheuristic. Comput. Oper. Res. **55**, 1–11 (2015)
7. Zhang, B., Zhang, M.-X., Zhang, J.-F., Zheng, Y.-J.: A water wave optimization algorithm with variable population size and comprehensive learning. In: Huang, D.-S., Bevilacqua, V., Prashan, P. (eds.) ICIC 2015. LNCS, vol. 9225, pp. 124–136. Springer, Cham (2015). https://doi.org/10.1007/978-3-319-22180-9_13
8. Zheng, Y.J., Zhang, B.: A simplified water wave optimization. In: Proceedings of the IEEE Congress on Evolutionary Computation (CEC 2015), Sendai, Japan, May 2015
9. Wu, X.L., Lu, Y.T.: Elite opposition-based water wave optimization algorithm for global optimization. Math. Probl. Eng. **2017**, 1–25 (2017)
10. Zhang, J.Z., Zhou, Y.Q., Luo, Q.F.: An improved sine cosine water wave optimization algorithm for global optimization. J. Intell. Fuzzy Syst. Appl. Eng. Technol. **34**(4), 2129–2141 (2018)
11. Soltanian, A., Derakhshan, F., Soleimanpour-Moghadam, M.: MWWO: modified water wave optimization. In: 2018 3rd Conference on Swarm Intelligence and Evolutionary Computation (CSIEC), pp. 1–5 (2018)
12. Zheng, Y.J., Lu, X.Q., Du, Y.C., Xue, Y., Sheng, W.G.: Water wave optimization for combinatorial optimization: design strategies and applications. Appl. Soft Comput. **83**, 105611 (2019)
13. Liang, J.J., Qu, B.Y., Suganthan, P.N., Chen, Q.: Problem definitions and evaluation criteria for the CEC 2015 competition on learning-based real-parameter single objective optimization
14. Wu, C., Xu, Y., Zheng, Y.: Water wave optimization with self-adaptive directed propagation. In: Pan, L., Liang, J., Qu, B. (eds.) BIC-TA 2019. CCIS, vol. 1159, pp. 493–505. Springer, Singapore (2020). https://doi.org/10.1007/978-981-15-3425-6_38
15. Liang, J.J., et al.: Comprehensive learning particle swarm optimizer for global optimization of multimodal functions. IEEE Trans. Evol. Comput. **10**(3), 281–295 (2006)
16. Zheng, Y.J., Ling, H.F., Xue, J.Y.: Ecogeography-based optimization: enhancing biogeography-based optimization with ecogeographic barriers and differentiations. Comput. Oper. Res. **50**(10), 115–127 (2014)

Adaptive Differential Privacy Budget Allocation Algorithm Based on Random Forest

Chong-yang Wang(✉), Si-yang Chen, and Xin-cheng Li

College of Computer Science and Engineering, Shandong University of Science and Technology, Qingdao, China
1337356463@qq.com

Abstract. In the context of big data, privacy protection in the process of data mining has become a hot issue in the field of security. The commonly used privacy protection method is to add differential privacy in the process of data mining, but the unreasonable privacy budget allocation leads to unsatisfactory classification results. To solve this problem, an adaptive differential privacy budget allocation algorithm RFDPP-weight based on random forest is proposed. Construct multiple decision trees with balanced features, calculate the Balance Error Rate (BER) of the data outside the bag in order to calculate the feature weight and then reconstruct the feature set. And the random forest with higher classification performance was constructed based on the new feature set. The decision tree weight was calculated according to the feature weight, and the privacy protection budget was adaptively allocated according to the decision tree weight. In the final prediction stage, the classification result with the largest weight of the corresponding decision tree is taken as the result of the random forest algorithm, so as to further improve the classification accuracy of the random forest algorithm. The experimental results show that the F1 score of our algorithm reaches 0.977 and 0.893 with a small number of decision trees in the Mushroom and Adult data sets, which proves that the algorithm can reasonably allocate the privacy budget and further improve the availability of data while protecting data privacy.

Keywords: Big data · Data mining · Privacy protection · Differential privacy · Random forest · Balance error rate

1 Introduction

With the development of computer and Internet technologies, all kinds of data are emerging in spurts and are collected and stored for analysis and research [1], and the era of big data has arrived. This has put forward higher requirements for data security and privacy protection. If the data cannot be effectively protected, the privacy information carried by the data will inevitably be exposed to the risk of leakage during data analysis and application [2, 3], such as the Facebook privacy leak in 2018 and the leak of the Swiss data management company Veeam, which have caused many adverse consequences. As people pay more attention to privacy protection issues, it has become an important issue to actively explore effective privacy protection methods.

L. Pan et al. (Eds.): BIC-TA 2021, CCIS 1565, pp. 201–216, 2022.
https://doi.org/10.1007/978-981-19-1256-6_15

Traditional privacy-preserving approaches such as the k-anonymity algorithm [4] and its variants are based on the assumption that the attacker model is not strictly defined and the attacker's background knowledge is not fully quantified [5]. When the attacker's background knowledge is sufficiently large, these methods become overwhelming. Dwork, a Microsoft Research scientist, proposes a differential privacy-preserving model [6, 7] that can quantify privacy by providing a strict mathematical definition, which addresses the shortcomings of traditional privacy-preserving models and can ensure data privacy even if the attacker has sufficient background knowledge [8]. Meanwhile, differential privacy protection can reduce the risk of privacy leakage while well guaranteeing the availability of data [9], which is applied in the field of data mining. Because of the advanced and practical nature of the technology, differential privacy technology was selected as one of the top 10 breakthrough technologies of 2020 by MIT Technology Review [10].

One privacy protection method is to combine differential privacy with random forest. However, unreasonable privacy budget allocation often leads to unsatisfactory classification results. This paper proposes an adaptive differential privacy budget allocation algorithm RFDPP-weight based on random forest. On the one hand, we use feature weights to reconstruct the dataset to improve the performance of decision trees to achieve better classification results with fewer target decision trees; on the other hand, we use decision tree weights to allocate privacy preserving budget, reduce the noise of high performance decision trees and improve the classification accuracy of decision trees; and finally we use decision tree weight integration to improve the classification accuracy of the whole random forest.

2 Related Work

Differential privacy plays a vital role in the process of data mining, it ensures that sensitive information is not leaked during the process of data being mined. Classification [11] is an important method of data mining, which can analyze the data well for prediction. Among them, decision tree is a typical representative of classification methods, which has many advantages such as high interpretability [12, 13], non-parametric design [14], low computational cost [11], etc., and it can be combined with differential privacy to better protect data privacy while classifying [15].

The first to propose an algorithm to fuse decision trees with differential privacy was Blum et al. [16], who developed the first interactive differential privacy decision tree algorithm SuLQ-based ID3, which is based on the decision tree algorithm ID3 and adds noise through Laplace mechanism to achieve data privacy protection, but it led to a significant decrease in the accuracy of the prediction results after adding noise. To address this problem McSherry et al. [17] proposed the PINQ-based ID3 algorithm, which uses the Partition operator to partition the dataset into disjoint subsets and improves the utilization of the privacy-preserving budget by exploiting its computationally parallel combinatorial nature. The algorithm directly evaluates the information gain using noise count values and then generates a decision tree using the ID3 algorithm. Since the count value of information gain needs to be computed for each feature, the privacy-preserving budget needs to be allocated to each query, which results in a small privacy-preserving budget per query and causes a lot of noises when faced with a problem with an overly large dataset.

To better solve the excessive noise problem Friedman et al. [18] introduced an exponential mechanism and proposed DiffP-ID3 algorithm, however, it can only handle discrete data. For the limitation of DiffP-ID3 algorithm on the dataset they further proposed DiffP-C4.5 differential privacy decision tree method. However, each iteration process invokes the exponential mechanism twice, which consumes excessive privacy-preserving budget.

The construction of decision trees is simple but not stable enough and generalizes poorly sometimes, and people started to consider replacing a single decision tree with a random forest. The first combination of random forest with differential privacy was proposed by Jagannathan et al. [19], and later, Patil and Singh [20] further proposed the DiffPRF algorithm, which guarantees multi-class classification under differential privacy, but it is based on ID3 decision tree and can only handle discrete features, and for continuous features it needs to pre-process the data first by discretization. To address this problem, Hailong Mu et al. [21] modified the random forest algorithm and proposed DiffPRFs, a differential privacy-preserving algorithm for random forests, which uses an exponential mechanism to select split points and split features during each decision tree construction and adds noise using the Laplace mechanism. Although DiffPRFs do not require discretization preprocessing of the data, similar to Diff-C4.5, the exponential mechanism is also invoked 2 times per iteration, which consumes more privacy-preserving budget and leads to a lower utilization of the privacy-preserving budget.

In order to improve the utilization of privacy budget and further enhance the classification effect, Yuanhang Li et al. [22] proposed a new algorithm RFDPP-Gini, which uses CART classification tree as a single decision tree in random forest, and uses the exponential mechanism and Laplace mechanism to handle continuous and discrete features respectively when selecting split features, which improves the classification effect but does not take into account the different classification performance of different decision trees. In addition, assigning the same privacy budget will reduce part of the classification effect and still does not make good use of the privacy preserving budget.

Therefore, in this paper, we propose a random forest-based adaptive differential privacy budget allocation algorithm to better ensure data availability while guaranteeing data privacy through a reasonable allocation of privacy-preserving budgets.

3 Algorithm-Related Definitions

Each decision tree requires randomly having put-back samples when generating, and when the samples are large enough, roughly 1/3 of the data will not be drawn, and these data are the out-of-bag data for that decision tree. The out-of-bag data can be used to verify the classification ability of this decision tree by using the mean value of the sum of the ratio of the number of misclassifications to the total number of positive and negative samples as the error balance rate of the out-of-bag data of the random forest [23]. The out-of-bag error balance rate provides a better measure of decision tree performance.

3.1 Solving for Feature Weights and Decision Tree Weights.

Since the out-of-bag dataset can be a good validation of the classification ability of decision trees, this paper calculates the out-of-bag error balance rate to solve for the weight of each feature by out-of-bag data.

Definition 1. Out-of-bag balance error rate. For a decision tree in a random forest, the out-of-bag balance error rate under its out-of-bag data *BER* is:

$$BER = \frac{1}{2}\left(\frac{P}{P_T} + \frac{N}{N_T}\right) \tag{1}$$

where P is the number of error prediction instances in the forward direction, N is the number of error prediction instances in the reverse direction. P_T is the total number of forward instances, and N_T is the total number of reverse instances. It is easy to see that the smaller the out-of-bag balance error rate of a decision tree represents the better the classification performance of that decision tree.

In order to measure the importance of a feature for the classification result of a decision tree, in this paper, the original out-of-bag data of the decision tree is randomly noise-added for a particular feature, and the out-of-bag balance error rate is calculated again for the decision tree using the noise-added out-of-bag data.

Definition 2. Out-of-bag balance error rate after feature noise addition. For a decision tree in a random forest, the out-of-bag balance error rate after noise addition for one of its features BER' is:

$$BER' = \frac{1}{2}\left(\frac{P'}{P'_T} + \frac{N'}{N'_T}\right) \tag{2}$$

where is the number of false prediction instances in the forward direction after noise addition P', N' is the number of false prediction instances in the reverse direction. $P^{'}_T$ is the total number of forward data, $N^{'}_T$ and is the total number of reverse instances. By comparing the change in out-of-bag balance error rate of the decision tree before and after feature noise addition, the importance of the feature to the classification result can be seen, but the classification performance of each decision tree is different, so calculating the mean value of the out-of-bag balance error rate of the feature in each decision tree can find the weight of the feature in the random forest. The specific definition of feature weights is as follows.

Definition 3. Feature weights. Features j The feature weights in a random forest are:

$$\overline{W_j} = \frac{\sum_{N} |BER' - BER|}{N} \tag{3}$$

where N is the number of decision trees, $|BER' - BER|$ is the change in out-of-bag balance error rate before and after feature noise addition.

Each decision tree is constructed with different selected features and has different classification ability. To measure the classification ability of the decision tree the decision tree t weighted as follows.

Definition 4. Decision tree weights. Decision tree t The weights in a random forest are:

$$W_t = \frac{\sum_{N} \overline{W_j}}{\sum_{N} W_t} \tag{4}$$

Where n represents the number of features selected in the decision tree, and N denotes the number of decision trees.

3.2 Filtering of Feature Sets

With the calculated feature weights, the feature set can be filtered to eliminate features of low importance and filter the new feature set. The new feature set will have a smaller number of features with higher weights than the previous algorithm feature set, so that each decision tree constructed will have a better performance and the number of decision trees can be reduced significantly. There are two methods of filtering:

Proportional Selection. The feature sets are ranked according to the feature weights from largest to smallest, and the new feature sets are selected proportionally for the construction of the decision tree.

Threshold Selection. A threshold value is set manually to filter out features with feature weights less than the threshold value to form a new feature set for decision tree construction.

3.3 Adaptive Allocation of Privacy Protection Budgets

The new set of features after filtering are the features with higher degree of importance, so the decision tree constructed from the new set of features (CART decision tree) has better classification performance. previous related algorithms such as DiffPRFs algorithm [21], PRDPP-Gini algorithm [22] in the process of noise addition is to distribute the privacy preserving budget to the decision trees equally, while the classification ability of each decision tree is different, the higher weighted decision tree proves to have better classification and assign higher and better privacy preserving budget to better ensure the availability of data.

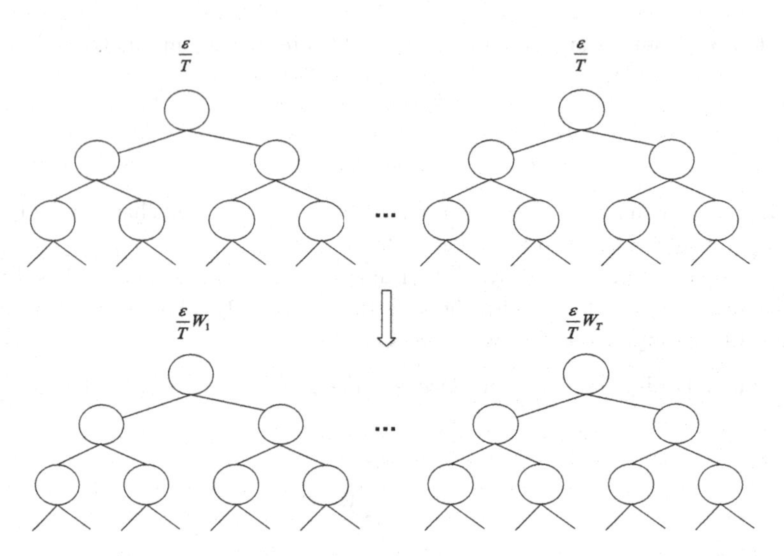

Fig. 1. The difference before and after the adaptive allocation of privacy protection budget in decision tree

As shown in Fig. 1, the upper part is the decision tree with average allocation of privacy budget, at this time, each decision tree is assigned the same privacy budget, which does not better take advantage of the decision tree with good classification ability. If the privacy preservation budget is adaptively assigned according to the weights of the decision trees, each decision tree is assigned the appropriate privacy budget, and the final decision tree is obtained as shown in the lower part of Fig. 1.

As shown in Fig. 1, we can see that after the allocation of adaptive privacy protection budget, the decision tree with higher weight can get more privacy protection budget, and the decision tree with low weight will get less privacy budget. So the decision tree with good classification ability is less affected by the noise and will have better prediction after adding noise; the decision tree with poor classification ability is more affected by the noise, but because of the small weight of the decision tree itself, it plays little role in the final decision process, and may even become better after adding noise due to the effect of excessive noise. In this way the whole random forest classification algorithm with the addition of differential privacy will have better classification results.

4 Algorithm Implementation and Analysis

4.1 Algorithm Flow

The algorithm is divided into two parts, the first part first builds a part of the forest and participates in the solution of the feature weights. The second part builds the rest of the forest, finds the weight of each tree in the forest based on the feature weights, performs the adaptive allocation of the privacy-preserving budget, and subsequently constructs the decision tree with weights in parallel to participate in the final decision.

The initial data set is first input and the initial data set is preprocessed. The data set is randomly divided into training set and test set in a ratio of 7:3. For the training set a random sampling with put-back is performed and the sampled data is used to construct a decision tree for out-of-bag error balance rate solving. For the unsampled data out-of-bag data is generated and the constructed decision tree is used to find the out-of-bag balance error rate for each feature attribute. The values of each feature attribute are noise added separately and the out-of-bag balance error rate is calculated again for each feature attribute. The change in out-of-bag balance error rate is calculated twice to find the feature weights and rank the feature importance, and a new set of features is selected based on a ratio or threshold.

For the new feature set the features of each decision tree are selected randomly, the weights of each decision tree are found, the privacy preserving budget is assigned according to the weights, and then the decision trees are constructed in parallel. The decision tree is a Cart classification tree, and the node splits are selected based on the size of the Gini index, combined with the privacy budget assigned to each decision tree for the index mechanism probability selection, and the split features and split values. When building to the leaf nodes to be combined with the privacy budget of the current leaf node using Laplace mechanism to add noise to the number of leaf nodes.

Thus the noise added random forest is constructed. The overall process is shown in Fig. 2. It utilizes a specific set of processed features and differential privacy noise addition for privacy budget allocation based on feature weights, and the weights of the corresponding decision trees are linearly summed at the time of classification prediction, and the one with the largest weight is taken as the classification result of the whole random forest algorithm. It is good to ensure both the security of the data and the high availability of the data.

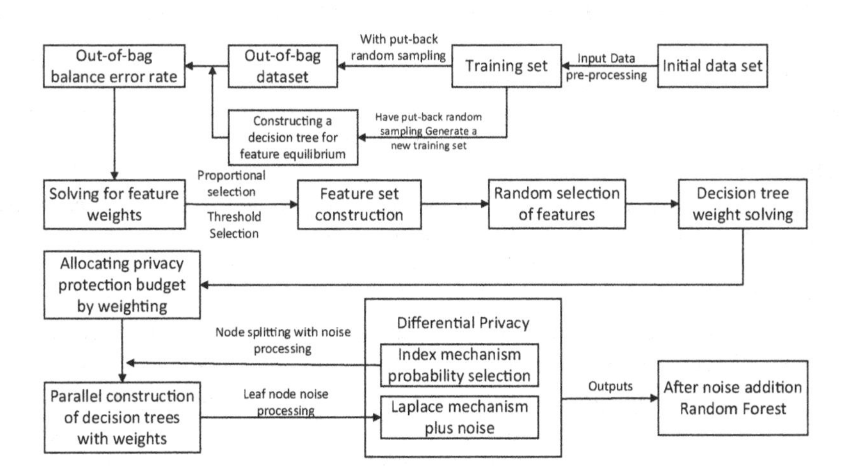

Fig. 2. RFDPP-weight algorithm flow

4.2 Algorithm Description

In this paper, we propose a random forest classification method with adaptive allocation of privacy-preserving budget under differential privacy that reconstructs the feature set, improves the performance of each decision tree, significantly reduces the number of decision trees required for random forest construction, and provides better classification accuracy while preserving the privacy of the data.

The construction of the first part of the forest provides an important support for the solution of the later weights and the final prediction classification. In order to make the decision trees in the first part of the forest efficient, we guaranteed the number of occurrences of each feature in this part of the forest is balanced (the difference in the number is 0 or 1), which ensure that each tree is constructed randomly and each feature of each tree is fully functional to achieve the goal of computing all feature weights with fewer decision trees. So it is necessary to give each feature a counter, the initial value of the counter is the same and the counter value is subtracted by 1 after each pick. When the counter value is 0, the feature is removed from the feature set and the feature set is updated. The initial value of the counter is calculated as follows.

Suppose the dataset has n features, and a decision tree $(m \leq n)$ is constructed by selecting m features at a time, and the total number of decision trees is T, the total number of features of T trees is mT, then the number of occurrences of each feature is guaranteed to be balanced under the condition that the number of occurrences of each feature is $\frac{m}{n}T$, and the initial value of the counter is $\lfloor \frac{m}{n}T \rfloor (\lfloor \frac{m}{n}T \rfloor \geq 1)$.

Because of the random nature of each feature extraction, the maximum difference of the feature counter after each feature extraction is specified to be less than 2 to avoid extracting the same feature several times in a row. The features with a large counter value are preferred for each feature extraction because a large counter value indicates that such features may not have been extracted before or may have been extracted only a relatively small number of times.

Algorithm 1: Multiple Decision Trees Construction Algorithm for Feature Equilibrium.

Input Training set D,feature set $M\{M_1, M_2,...,M_n\}$, set of feature countersC $= \{C_1, C_2,...,C_n\}$, number of decision trees$T$, maximum depth of decision trees$d$, number of randomly selected features at splitting m.

Output Feature balanced decision tree and out-of-bag dataset.

Stopping condition All samples on the decision tree nodes are classified consistently, or the maximum depth of the decision tree is reached d, or the number of feature set features is less than m.

Step 1 The training set D is sampled N samples as a training set D_N.

Step 2 For the current training set D_T, randomly select m a different feature from the feature M set M_T (priority is given to the feature with a large counter value), the corresponding feature counter $C_i = C_i - 1$, determine whether there is a feature with a counter value of 0, and update the feature set M.

Step 3 Calculate M_T the Gini index of each feature, select the best feature b as the splitting feature, divide the samples on the current node into different sub-nodes according to the different values of the features b, and generate the decision tree from the root node recursively.

Step 4 Repeat steps $1 \sim 3$ T times to obtain T training subsets $D_1, D_2, ..., D_T$, and generate T decision trees $rees\{tree_1, tree_2, ..., tree_T\}$. Out-of-bag dataset $D' = \{D - D_1, D - D_2, \ldots, D - D_T\}$.

After using Algorithm 1 to generate the decision tree for feature equilibrium and the set of corresponding out-of-bag datasets the feature weights can be calculated and the decision tree weights for constructing the random forest. The calculation process is shown in Algorithm 2 and Algorithm 3.

Algorithm 2: Algorithm for Solving Feature Weights.

Input out-of-bag dataset D', decision tree $Trees$, training set D,feature set $M\{M_1, M_2,...,M_n\}$.

Output The weights of all features.

Stop condition All features are calculated with weights.

Step 1 For the current feature of the feature set M_n, select t (initial value of counter) sets of decision trees $Trees'_n$ containing the features M_n and calculate their average out-of-bag balance error rate BER_n in the corresponding out-of-bag data set D'_n using the formula (1).

Step 2 For the out-of-bag dataset D'_n, add noise to the current features of all the datasets in M_n to form a new out-of-bag dataset D''_n.

Step 3 For the current feature M_n, use the decision tree set $Trees'_n$ to calculate its average out-of-bag balance error rateBER'_n in the new out-of-bag dataset D''_n using the formula (2).

Step 4 Calculate the weights $\overline{W_{M_n}}$ of the features M_n using the formula (3).

Step 5 Repeat steps 1 to 4 n times to obtain the set of weights for all features W.

After calculating the weights of all the features, we reconstruct the feature set by filtering them, and in this paper, we adopt the proportional selection method to reconstruct the feature set.

The feature weights required for each decision tree construction are computed in Algorithm 2, so that the decision tree weights can be further solved and the privacy-preserving budget can be adaptively assigned, and then the decision trees can be built in parallel and no additional time complexity is added due to the privacy budget assignment. The differential privacy-preserving oriented random forest algorithm for adaptive privacy-preserving budget allocation is shown below.

Algorithm 3: Adaptive Differential Privacy Budget Allocation Algorithm Based on Random Forest.

Input Training set, feature set M, feature set selection ratio a, privacy preserving budget A, number of decision trees$T$, maximum depth of decision trees d, number of randomly selected features at splitting m.

Output Random forests satisfying adaptive differential privacy-preserving budget allocation

Stop condition All samples on the node are classified consistently, or the maximum depth of the decision tree is reached d, or the privacy-preserving budget is exhausted

Step 1 Select the features M of the feature set in proportion a to their weights to form a new feature set M'

Step 2 Randomly select m features from the feature set M' to form the feature set M_1'

Step 3 Repeat step 2 T twice to calculate the weights W_T of each decision tree according to the formula (4).

Step 4 Assign the privacy-preserving budget to each decision tree by the weight of the decision tree $\varepsilon' = \frac{A}{T} W_i$.

Step 5 Distribute the privacy-preserving budget assigned to each tree equally to each level of the decision tree, and then divide the privacy-preserving budget of each node equally into 2 parts, $\varepsilon = \frac{\varepsilon'}{2(d+1)}$.

Step 6 Randomly select a training set of size $|D|$ from the training set D with put-back in D_i

Step 7 Recursively execute the decision tree in the random forest built in the following steps:

(1) If the node reaches the stop condition, set the current node as a leaf node and add noise using the Laplace mechanism to classify the current node.

(2) Calculate the number of samples in the training set of the current node D_j and add the noise $N_{D_j} = NoisyCount(|D_c|)$ using the Laplace mechanism, where D_j is a subset of D_i.

(3) The feature set for the current decision tree construction is updated to M_i'

(4) Determine whether the current feature set M'_i contains $n(n > 0)$ continuous attribute, and if so, perform step (5); otherwise, perform step (6).

(5) Allocate the privacy-preserving budget to each continuous-type feature and reserve a copy for the discrete-type features, $\varepsilon = \frac{\varepsilon}{n+1}$, select the best continuous-type feature and its split point with the following probabilities, and calculate the corresponding Gini index.

$$\frac{exp(\frac{\varepsilon}{2\Delta q} q(D_j, M'))|R_i|}{\sum_i exp(\frac{\varepsilon}{2\Delta q} q(D_j, M'))|R_i|} \tag{5}$$

where, $q(D_j, M')$ is the Gini index, Δq is the global sensitivity of the Gini index, R is the set of intervals consisting of the values that appear in the data set, and $|R_i|$ is the interval size.

(6) Calculate M_i' the Gini index corresponding to each discrete feature in the medium when it is split in different splitting ways, compare it with the Gini index corresponding to the best continuous-type feature, select the splitting feature and the splitting point that minimizes the Gini index in M_i', divide the current node into 2 sub-nodes according to the feature and its best splitting point, and perform steps (1) to (6) for each sub-node.

Step 8 Return the set of trees, the random forest satisfying the adaptive differential privacy-preserving budget allocation.

5 Experimental Results and Analysis

5.1 Experimental Design

This paper is written in Python to implement the random forest algorithm RFDPP-weight for differential privacy preservation based on attribute weights. The experimental operating system is Windows 8, CPU is i5-5200U@2.20GHz with 8G of memory. The data sets used in this experiment are real data sets in UCI database, Adult data sets and Mushroom data sets, which are commonly used in random forest algorithm.

The Adult dataset contains 48,842 sample data, and after removing the samples containing missing values, 45,222 sample data are obtained, 31655 data are used for training and 13567 data are used for testing. Each data contains 14 features, and the decision category is whether the salary is greater than 5K. 8124 data are contained in Mushroom dataset, 5686 data are used for training, and 2438 data are used for testing, and there are 22 features, and the decision category is whether the mushroom is edible.

This experiment uses F1 score as an index to evaluate the performance of random forest. The higher the F1, the more robust the model is, and the stronger the classifier's ability to correctly classify. Formula F1 score is as follows:

$$F_1 = \frac{2 * Pre * Rec}{Pre + Rec} \tag{6}$$

Where, *Pre* is precise and *Rec* is recall.

The RFDPP-weight algorithm aims to improve the classification accuracy of the random forest algorithm more efficiently while ensuring data privacy. In order to test the superiority of RFDPP-weight algorithm this paper sets up multiple sets of comparison experiments, each set of experiments is performed 100 times and the average of F1 score is taken as the final result: 1) comparison of F1 score before and after noise addition; 2) comparison between different decision tree depths; 3) comparison between different privacy preserving budgets. 4) Comparison between RFDPP-weight algorithm in this paper and DiffPRFs algorithm [21] and RFDPP-Gini algorithm [22].

In this paper, the experimental parameters for the part of the RFDPP-weight algorithm that calculates the feature weights are set as follows: the number of multiple decision trees for feature equilibrium is 1/3 of the total number of decision trees, the number of features selected for each decision tree construction is 1/2 of the total number of features, and the depth of the decision tree is set to 6.

For other experimental parameters set the total number of decision trees $T = 15$, the number of randomly selected features at node splitting $m = 5$, the proportion of new feature sets selected as 2/3, and the number of decision trees for calculating feature weights set the tree depths as 3, 4, 5, 6, 7, 8, 9 and 10 to compare the effect of different depths of decision trees on the classification accuracy. To compare the effect of different privacy preserving budgets ε on the classification accuracy, the privacy preserving budgets are set to 0.05, 0.10, 0.25, 0.5, 0.75 and 1.00, respectively, and the classification accuracy is calculated when different ε values are taken.

6 Experimental Results and Analysis

For the Adult dataset and the Mushroom dataset, the classification effect of the RFDPP-weight algorithm with different privacy-preserving budgets for the decision tree was compared, and the effect of different depths of the decision tree on the classification accuracy was compared, and the experimental results are shown in the line graphs Fig. 3 and Fig. 4.

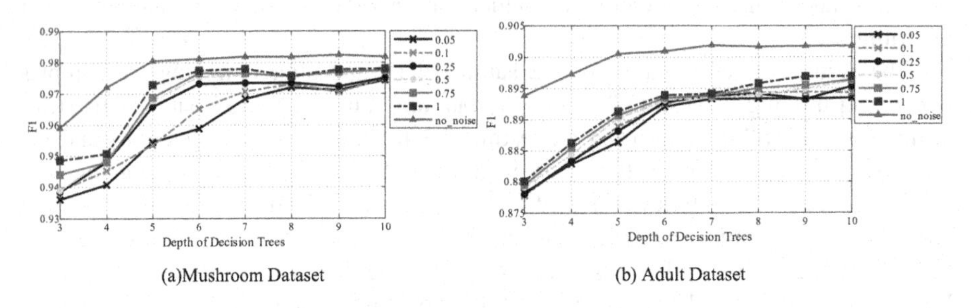

Fig. 3. F1 score changing with the depth of the tree

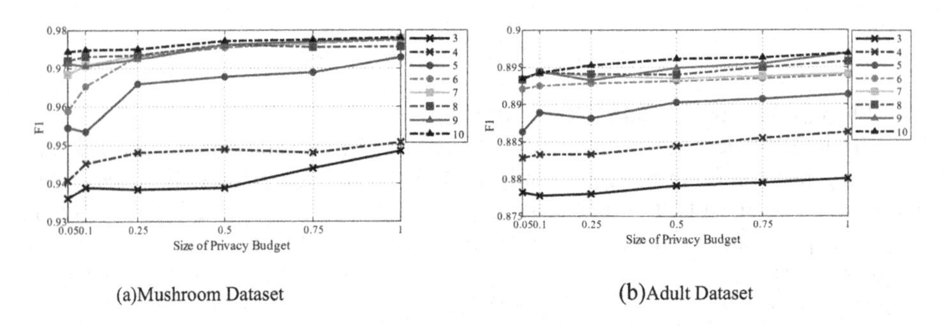

Fig. 4. F1 score changes with the privacy protection budget

Figures 3 and 4 represent the variation of F1 score at different tree depths and at tree depths of 3, 4, 5, 6, 7, 8, 9, and 10 for privacy-preserving budgets of 0.05, 0.1, 0.25, 0.50, and 1.00, respectively. The classification performance of RFDPP-weight algorithm after noise addition is shown through these two perspectives. It can be seen that the deeper the depth of the decision tree is, the higher the classification accuracy is when the privacy protection budget is certain. The classification accuracy increases with the increase of privacy preserving budget for a certain depth of decision tree. This is because the larger the privacy-preserving budget, the smaller the noise and thus the higher the classification accuracy of the decision tree, the higher the classification accuracy of the random forest algorithm.

To test whether the RFDPP-weight algorithm can achieve better classification results with a smaller number of decision trees compared with DiffPRFs algorithm [21] and RFDPP-Gini algorithm [22], the number of trees is set to T = 5, 10, 15, 20, 25, 30, 35, 40, 45, 50, the ratio of new feature set selection is 2/3, the node splitting when the number of randomly selected features m = 5, and the maximum depth of the tree is 6. The experimental results are shown in Fig. 5.

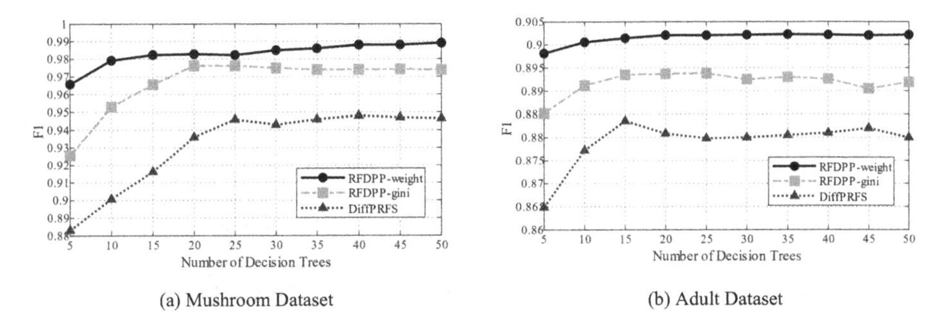

(a) Mushroom Dataset (b) Adult Dataset

Fig. 5. Classification performance under different number of decision trees

It can be seen from Fig. 5 that the RFDPP-weight algorithm has the highest values of 0.989 and 0.902 on F_1 Mushroom and Adult datasets, respectively, while the RFDPP-Gini algorithm [22] and DiffPRFs algorithm [21] F_1 have the highest values of 0.976, 0.893, 0.948 and 0.883, respectively. it is easy to see that the values of the RFDPP-weight algorithm stabilize quickly as the number of decision trees increases and are much higher than those of the RFDPP-Gini algorithm [21] under the same conditions. It is easy to see that the F_1 values of F_1 the RFDPP-weight algorithm stabilize quickly as the number of decision trees increases and are higher than the values F_1 of the RFDPP-Gini algorithm [22] and much higher than the values F_1 of the DiffPRFs algorithm [21] under the same conditions. DiffPRFs algorithm [21] condition of more number of decision trees for F1 score. This verifies that the decision trees constructed in the RFDPP-weight algorithm have good classification performance.

To better evaluate the performance of the algorithm in this paper, it is compared with DiffPRFs algorithm [21] and RFDPP-Gini algorithm [22] under the same conditions for the classification F1 score of the Mushroom and Adult datasets. The total number of decision trees T = 10, the number of randomly selected features at node splitting $m = 4$, the ratio of new feature set selection is 2/3, the maximum depth of the tree is set to 6, and the ε settings are 0.05, 0.10, 0.25, 0.5, 0.75, and 1.00. The experimental results are shown in Fig. 6.

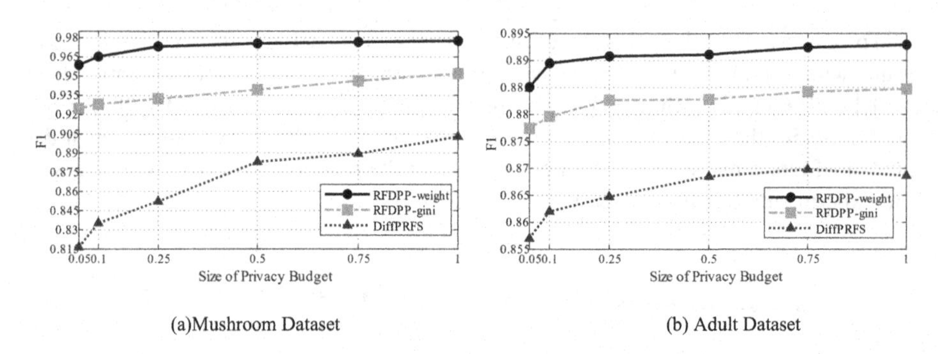

(a)Mushroom Dataset (b) Adult Dataset

Fig. 6. Comparison of classification performance of the three algorithms on Mushroom and Adult datasets

It can be seen from the figure that the RFDPP-weight algorithm has better classification performance than the other two algorithms when the privacy preserving budget is 0.05, 0.10, 0.25, 0.5, 0.75 and 1.00. The F1 scores are 0.958, 0.965, 0.973, 0.975, 0.976 and 0.977 on Mushroom dataset and are 0.885, 0.889, 0.890, 0.891, 0.892 and 0.893 on Adult dataset respectively. The algorithm in this paper achieves better results when the number of decision trees is small. This is due to 3 points. Firstly, the RFDPP-weight algorithm reconstructs the feature set, which makes the decision tree have better classification performance. Secondly, the RFDPP-weight algorithm gives weights to the decision trees, which gives more advantages to the decision trees with good decision performance. Finally, the RFDPP-weight algorithm adaptively assigns a privacy-preserving budget to the decision trees, which makes the decision trees with good decision performance less noisy and reduces the impact of noise on the high-weight decision trees. In summary, the algorithm in this paper has better classification effect and can ensure that the noise-added data has higher usability.

7 Conclusion and Discussion

In this paper, we propose an adaptive differential privacy budget allocation algorithm RFDPP-weight based on random forest, the algorithm selects better features to build random forest by solving for feature weights, which improves the decision tree performance, reduces the number of decision trees, avoids excessive computational overhead during model training, and avoids the low budget allocation for privacy protection of single decision tree due to the excessive number of decision trees, which leads to unsatisfactory prediction results. Assigning weights and adaptive privacy-preserving budgets to each decision tree, the experimental results demonstrate that the RFDPP-weight algorithm can further improve the accuracy of model classification after noise addition. However, in the process of solving the feature weights, the algorithm needs to add noise to the out-of-bag data and solve the out-of-bag balance error rate twice, which will consume some time. The next step will be to consider how to further optimize the time complexity of the algorithm, and further investigate the strategy of distributing the differential privacy-preserving budget among other integrated learning algorithms.

References

1. Kousika, N., Premalatha, K.: An improved privacy- preserving data mining technique using singular value decomposition with three-dimensional rotation data perturbation. J. Supercomput. **77**(6), 1–9 (2021)
2. Xie, X., Liang, Y., Wang, Z., Dong, X.: A quantitative evaluation method of social network uses' privacy leakage. Comput. Eng. Sci. **43**(08), 1376–1386 (2021)
3. Kairouz, P., Oh, S., Viswanath, P.: The composition theorem for differential privacy. IEEE Trans. Inf. Theory **63**(6), 4037–4049 (2017)
4. Sweeney, L.: k-anonymity: a model for protecting privacy. Int. J. Uncertain. Fuzziness Knowl.-Based Syst. **10**(05), 557–570 (2002)
5. Liu, X., Li, Q., Li, T., Chen, D.: Differentially private classification with decision tree ensemble. Appl. Soft Comput. **62**, 807–816 (2018)
6. Dwork, C.: Differential privacy. In: Bugliesi, M., Preneel, B., Sassone, V., Wegener, I. (eds.) ICALP 2006. LNCS, vol. 4052, pp. 1–12. Springer, Heidelberg (2006). https://doi.org/10.1007/11787006_1
7. Dwork, C.: A firm foundation for private data analysis. Commun. ACM **54**(1), 86–95 (2011)
8. Li, H., Xiaoping, W.: Network intrusion correlation method with differential privacy protection of alerts sequence. Comput. Eng. **44**(5), 128–132 (2018)
9. Hao, C., Peng, C., Zhang, P.: Selection method of differential privacy protection parameter ε under repeated attack. Comput. Eng. **44**(7), 145–149 (2018)
10. Breakthrough Technologies 2020. https://www.technologyreview.com/10-breakthrough-technologies, 2020–01–26/2020–06–06 (2020)
11. Murphy, K.P.: Machine Learning: A Probabilistic Perspective, 1st edn. MIT Press, Cambridge (2012)
12. Letham, B., Rudin, C., Mccormick, T.H., et al.: Interpretable classifiers using rules and Bayesian analysis: Building a better stroke prediction model. Ann. Appl. Stat. **9**(3), 1350–1371 (2015)
13. Han, J. Kamber, M., Pei, J.: Data Mining: Concepts and Techniques, pp. 585–631 (2006)
14. Huysmans, J., Dejaeger, K., Mues, C., et al.: An empirical evaluation of the comprehensibility of decision table, tree and rule based predictive models. Decis. Support Syst. **51**(1), 141–154 (2011)
15. Fletcher, S., Islam, M.Z.: Decision tree classification with differential privacy: a survey. ACM Comput. Surv. **52**(4), 1–33 (2016)
16. Blum, A., Dwork, C., Mcsherry, F., et al.: Practical privacy: the SuLQ framework. In: Proceedings of the 24th ACM SIGMOD-SIGACT-SIGART Symposium on Principles of Database Systems, pp. 128–138. ACM, Baltimore (2005)
17. Mcsherry, F.: Privacy integrated queries: an extensible platform for privacy-preserving data analysis. Commun. ACM **53**(9), 89–97 (2010)
18. Friedman, A., Schuster, A.: Data mining with differential privacy. In: Proceedings of the 16th ACM SIGKDD, pp. 493–502. ACM, New York (2010)
19. Jagannathan, G., Pillaipakkamnatt, K., Wright, R.N.: A practical differentially private random decision tree classifier. In: 2009 IEEE International Conference on Data Mining Workshops, pp. 114–121. Miami, FL, USA (2009)
20. Patil, A., Singh, S.:Differential private random forest. In: 2014 International Conference on Advances in Computing, Communications and Informatics (ICACCI), pp. 2623–2630. Delhi, India (2014)
21. Mu, H., Ding, L., Song, Y., et al.: DiffPRFs: random forest under differential privacy. J. Commun. **37**(9), 175–182 (2016)

22. Li, Y., Chen, X., Liu, L., et al.: Random forest algorithm for differential privacy protection. Comput. Eng. **46**(1), 93–101 (2020)
23. Chi, C., Liang, X.: Classification feature selection based on random forest and support vector machine. J. Univ. Sci Technol. Liaon. **39**(2), 146–151 (2016)

A Node Influence Based Memetic Algorithm for Community Detection in Complex Networks

Zhuo Liu[1], Yifei Sun[1(✉)], Shi Cheng[2], Xin Sun[1], Kun Bian[1], and Ruoxia Yao[2]

[1] School of Physics and Information Technology, Shaanxi Normal University, Xi'an 710119, China
{zhuoliu, yifeis, sunxin_, biankun}@snnu.edu.cn
[2] School of Computer Science, Shaanxi Normal University, Xi'an 710119, China
{cheng, rxyao}@snnu.edu.cn

Abstract. Community structure is a significant property when analyzing the features and functions of complex systems. Heuristic algorithm-based community detection treats finding the community structure as an optimization problem, which has received great attentions in a variety of fields these years. Several community detection methods have been proposed. To make an approach of detecting the community structure in a more efficient way, a node influence based memetic algorithm (NIMA), considering node influence, is proposed in this paper. The NIMA consists of three main parts. First of all, a transition probability matrix-based initialization is employed to accelerate the convergence speed and provide an initial population with great diversity. Secondly, a network-specific crossover and a node degree-based mutation are designed to enlarge the search space and keep effective information. Last, a multi-level greedy search is deployed to find the potential optimal solutions quickly and effectively. Extensive experiments on 28 synthetic and 6 real-world networks demonstrate that compared with 11 existing algorithms, the proposed NIMA has effective performance on detecting communities in complex networks.

Keywords: Memetic algorithm · Multi-level greedy search · Community detection · Complex networks

1 Introduction

Complex systems, which exist in domains like social relationships of sociology, animal activities of biology and power grid network of electricity, etc., can be represented as networks [1–4]. Based on the graph theory, entities and relationships of the systems are denoted as nodes and edges respectively [5]. Real-world networks are composed of several components which are called subnetworks or communities. As to such structure, nodes are tightly connected inside and sparsely linked between [6, 7]. To analyze such structures, community detection was proposed as an effective tool in the field of network science. Community detection aims to discover subnetworks in graph

L. Pan et al. (Eds.): BIC-TA 2021, CCIS 1565, pp. 217–231, 2022.
https://doi.org/10.1007/978-981-19-1256-6_16

structure. As its output, the nodes within the same subnetworks, or communities, are tightly connected, while the nodes in different communities are sparse connected [8, 9]. Therefore, the nodes within the same communities are regarded having similar attributes, and the properties of complex systems can be revealed by its community structure [10].

Many methods have been proposed in community detection. Modularity based optimization is one of the most popular approaches. The modularity is a quantity to evaluate the community structure in a network, considering the degree distribution of nodes [11–14]. It is demonstrated that by searching for a network partition with the maximal modularity, a relatively accurate community structure could be found.

It is demonstrated that the community detection of complex networks is a non-deterministic polynomial hard (NP-hard) problem, because the solution space grows with the scale of networks exponentially [15]. Heuristic algorithms (HAs) are thought as effective ways to solve NP-hard problems, including genetic algorithm (GA), simulated annealing (SA), ant colony optimization (ACO) and particle swarm optimization (PSO) [16], etc. Some HAs, called multi-objective evolutionary algorithms (MOEAs), are applied in solving Multi-objective optimization problems (MOPs), which aim to optimize several trade-off objective functions at the same time [17]. And community detection could also be designed as a MOP.

Many HA-based community detection methods have been proposed recent years. Ref. [18] proposed an algorithm using the framework of multi-objective particle swarm optimization (MOPSO) to solve community detection. Ref. [19] designed a density peak clustering and label propagation based algorithm with high performance in robustness. Ref. [20] proposed a multi-level learning memetic algorithm.

In this study, we propose a node influence based memetic algorithm (NIMA) as a tool to study the community structure of complex systems. The strategies employed in NIMA come from plenty of work in mathematics and network research. The NIMA consists of three main parts. First of all, a transition probability matrix-based initialization is employed to accelerate the convergence speed and create an initial population with high diversity. Secondly, a network-specific crossover and a node degree-based mutation are designed to enlarge the search space and keep effective information. Last, a multi-level greedy search is deployed to find the potential optimal solutions quickly and effectively. Extensive experiments on 28 synthetic and 6 real-world networks demonstrate that compared with 11 existing algorithms, the proposed NIMA has effective performance on detecting communities in complex networks.

The rest of this article is organized as follows. Section 2 illustrates the related fundamental notions about complex networks and community detection. Details of the proposed algorithm NIMA are introduced in Sect. 3. Experimental results and analysis are presented in Sect. 4. Section 5 concludes remarks and possible further studies.

2 Background Knowledge

In graph theory, a complex network can be described as a graph $G = (V, E)$, where $V = \{v_1, v_2, \ldots, v_N\}$ represents the set of N nodes (entities), and $E = \{e_{ij} = (v_i, v_j) | v_i, v_j \in V, i \neq j.\}$ denotes the set of edges (relationships) between the nodes. The connection of graph G can be further abstracted into an adjacent matrix $A = [a_{ij}]_{i,j \in V}$. And its elements a_{ij} equals 1 when there exists one link between node v_i and node v_j and 0 otherwise.

It is necessary to note that all complex networks mentioned in this paper are undirected.

2.1 Modularity

Community detection could be considered as a network clustering problem. In Ref. [11], M. E. J Newman and M. Girvan proposed modularity firstly to evaluate the quality of a community partition for a network. And it was devised to the complete version gradually in Ref. [12–14]. The modularity takes node degree distribution into consideration and can be described as follows:

$$Q = \sum\nolimits_{i=1}^{c} \left[\frac{l_i}{m} - \left(\frac{k_i}{2m} \right)^2 \right] \tag{1}$$

Where c and m are the total number of communities in the partition and edges in the network respectively; l_i is the sum of edges within the ith community; k_i is the sum of degree of nodes in ith community. The value of Q ranges from −0.5 to 1. And it is illustrated that when Q is between 0.3 and 0.7, the network partition is regarded as expected [13]. Generally, the larger Q is the better partition is.

2.2 Transition Probability Matrix

Markov chain (MC) is defined as the stochastic process existing in discrete index set and state space with Markov property, in the field of probability and mathematical statistics [21]. The structure of MC can be described by transition graph and transition matrix. The transition graph of a MC is illustrated as Fig. 1. For eight given states, a line with arrow is drawn if there is a probability for transition of two states. The arrows denote direction of the transition.

The probability between states can be represented by transition matrix or transition probability matrix (TPM) $P^{(k)} = [p_{ij}]_{i,j \in S}$, where k denotes the step of the TPM. The TPM has the following properties:

$$\sum\nolimits_j p_{ij} = 1 \tag{2}$$

$$P^{(k)} = \left(P^{(1)}\right)^k \tag{3}$$

The properties show that the sum of each row in TPM is 1; a k-step TPM can be obtained by calculating the kth power of 1-step TPM.

3 Proposed Algorithm

A node influence based memetic algorithm, named NIMA, is proposed in this paper. NIMA optimizes Eq. (1) for detecting communities. And its details will be illustrated in this section.

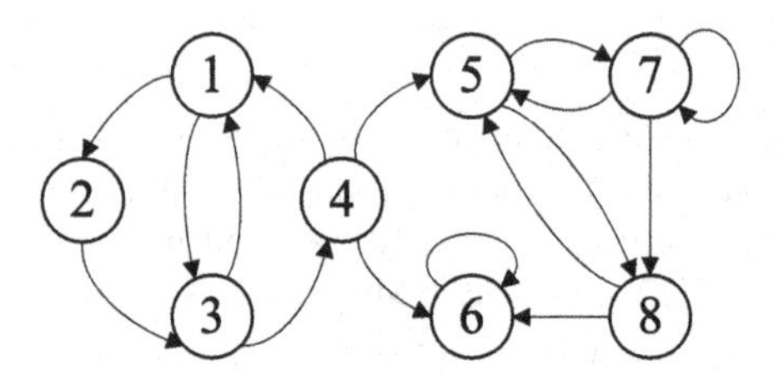

Fig. 1. Illustration of transition graph in Markov chain.

3.1 Representation

The solutions or chromosomes used in this study is denoted as $C_i = \left[x_i^1, x_i^2, \ldots, x_i^N\right]$, where $i \in \{1, 2, \ldots, N_p\}$ with N_p denotes the size of population, N represents the total number of nodes and $x_i^j \in \{1, 2, \ldots, N\}, j = 1, \ldots, N$ are the cluster label of node v_i. Each chromosome could be treated as a partition of the network, i.e., if $x_i^u = x_i^w$, the vertices v_u and v_w are divided into the same cluster and into different clusters otherwise. A visualized illustration of the representation can be seen in Fig. 2.

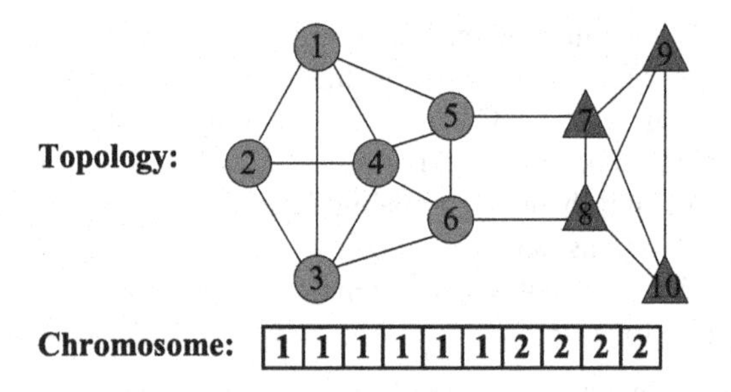

Fig. 2. Representation of chromosome.

3.2 Framework

The proposed algorithm NIMA consists of a TPM based initialization, the tournament selection, a network-specific crossover, a node degree-based mutation and a multi-level greedy search. The framework of NIMA is given in Algorithm 1.

Algorithm 1: The framework of NIMA

Input: The complex network G; the population size: N_p; the tournament size: N_o; the mating pool size: N_m; the crossover probability: p_c; the mutation probability: p_m; the maximum number of generations: g_{max}.
Output: The clustering divisions of the network with the largest modularity Q.
Step 1: Initialization:
1 Generate the population P consisting of N_p chromosomes, and set the iteration counter $t \leftarrow 0$.
Step 2: Selection:
2 $P_{parents} \leftarrow tournament\_selection(P, N_o, N_m)$;
Step 3: Genetic operation:
3 $P_{crossover} \leftarrow crossover(P_{parents}, p_c)$;
4 $P_{mutation} \leftarrow mutation(P_{crossover}, p_m)$;
Step 4: Greedy search:
5 $C_{greedy} \leftarrow greedy\_search(C_b, C_{globalbest})$; //$C_b$ and $C_{globalbest}$ are the chromosomes with the highest Q in $P_{mutation}$ and P respectively;
Step 5: Update:
6 Organize the new population by choosing the top N_p chromosomes with the highest value of modularity Q from P, $P_{mutation}$ and C_{greedy}.
Step 6: Judgement:
7 The algorithm will be stopped and output $C_{globalbest}$ and its converted clustering divisions if $t \geq g_{max}$. If not, $t \leftarrow t + 1$ and move to **Step 2.**

3.3 Initialization

Algorithm 2: Initialization

Input: The signed network G; the population size: N_p.
Output: The original population: P.
1 Generate the population $P = [C_1, C_2, \dots, C_{N_p}]$, and for each $C_i, 1 \leq i \leq N_p$, set its genes $x_i^j \leftarrow j, 1 \leq j \leq N$, where N represents the total number of the nodes;
2 Obtaining *TPM* by dividing the elements of A with the sum of the row they belonging, where A is the adjacent matrix of G;
3 $TPM \leftarrow TPM^3$;
4 Obtaining *cum* which is the cumulative sum of each row of *TPM*;
5 **for** each C_i in P **do**
6 Generate a random sequence $[r_1, r_2, \dots, r_N]$;
7 **for** each gene $x_i^{r_j}$ in chromosome C_i **do**
8 $R \leftarrow random(0,1)$;
9 $x_i^{r_j} \leftarrow x_i^{r_k}, \exists k \in \{k | cum(i, k-1) \leq R < cum(i, k)\}$;
10 **end**
11 **end**

To accelerate the process of convergence and take both quality and diversity of the population into consideration, a TPM based initialization is deployed. The pseudo code of initialization operation is given in Algorithm 2. In line 1, each node is distributed with a unique number to its community label. In line 2, a 1-step TPM is established with its elements of ith row is the reciprocal of the degree of node v_i. Then obtaining the 3-step TPM by calculating the third power of it. After that, the community label of each node is allocated from its 3-step neighbors by a roulette-like mechanism, as shown in line 4–11. The initialization takes the influence of 3-step neighbors into consideration. The neighbors with more influence are allocated with larger probability to be assigned into the same communities.

3.4 Selection

The classic tournament selection is used in NIMA to choose elite chromosomes with better performance on objective function. N_m tournaments are organized, and each one is composed of N_o competitors picked from population P randomly. N_m winners of the tournaments which have the higher value of Q than other competitors are sent to the mating pool as the parent population to execute the crossover operation.

3.5 Crossover

Algorithm 3: Crossover

Input: The parent population: $P_{parents}$; the crossover probability: p_c;
Output: The crossed population: $P_{crossover}$.

1 Generate a random sequence $[r_1, r_2, \ldots, r_{N_m}]$; // N_m is the size of mating pool;
2 **for** each pair of C_{r_i} and $C_{r_{i+1}}$ in $P_{parents}$ **do**
3   **if** $random(0,1) < p_c$ **then**
4   find the nodes within the same communities in C_{r_i} and $C_{r_{i+1}}$ respectively and set $g_{ki} = \{v_{ki_1}, v_{ki_2}, \ldots, v_{ki_{|g_{ki}|}}\}$, where $k \in \{1,2\}$ and $|g_{ki}|$ is the size of g_{ki};
5     $ctr \leftarrow 1$;
6   **for** each community g_{1i} of C_{r_i} **do**
7   **for** each community g_{2j} of $C_{r_{i+1}}$ **do**
8   find the consensus nodes of g_{1i} and g_{2j} and set as v_c;
9         $x^{v_c}_{crossover_i} \leftarrow ctr$;
10         $ctr \leftarrow ctr + 1$;
11   **end**
12   **end**
13   **else**
14     $C_{crossover_i} \leftarrow C_{r_i} or\ C_{r_{i+1}} randomly$;
15   **end**
16 **end**

Crossover is designed to prevent the solutions from being trapped in local optimal and keep the effective information of current solutions at the same time. The pseudo code of crossover is shown in Algorithm 3. After choosing a pair of chromosomes as the parents into the mating pool, the crossover is deployed. The main idea of it is to keep nodes divided into the same community if they have the same community label in both parent chromosomes, and into different communities otherwise.

3.6 Mutation

The mutation operation works on the population $P_{crossover}$ generated by crossover operation. For each node v_i in the chromosomes C_i, it starts with generating a random value ranging from 0 to 1. And if the value is smaller than the mutation probability p_m, generate another random value ranging from 0 to 1. Only when the second value is smaller than the node changing probability calculated by Eq. (4), mutate the cluster label of node v_i as that of its neighbors randomly. It should be noted that the neighbors with same label as v_i are not included.

$$p^i_{change} = \begin{cases} \frac{k_{max}-k_i}{2(k_{max}-\bar{k})}, & k_i \geq \bar{k} \\ \frac{\bar{k}-k_i}{2(\bar{k}-k_{min})} + 0.5 & k_i < \bar{k} \end{cases} \tag{4}$$

Where k_i is the degree of v_i, and k_{max}, k_{min} and $\bar{k}$ is the maximum, minimum and average degree of nodes in the network. In conclusion, the nodes with larger degree are more likely to stay unchanged. This mutation considers the influence of the nodes in network by holding the important nodes which has large degree to reduce unnecessary mutations and accelerate the convergence speed.

3.7 Multi-level Greedy Search

A multi-level greedy search, consisting of node, cluster and partition level, is adopted to find the potential optimum solution in a few iterations. It shares ideas from Louvain, proposed in Ref. [22]. It should be noted that there are similarities between these levels. Therefore, the node-level greedy search will be introduced first as an example frame of other two levels.

Algorithm 4: node-level greedy search

Input: The best chromosome of mutated population $P_{mutation}$: C_b;
Output: The node-level searched chromosome: C_{node}.
1 **repeat**
2 Generate a random sequence $[r_1, r_2, \ldots, r_N]$;
3 **for** each gene x_i in C_b **do**
4 $x_i \leftarrow \arg\max_{x_f} \left(\Delta Q\left(C_b|_{x_i \leftarrow x_f}\right)\right), \forall f \in \{\{f | a_{if} = 1\}\}, \Delta Q > 0;$
5 **end**
6 **until** each gene x_i in C_b is not changed.

As shown in Algorithm 4, the node-level greedy search is executed on the best chromosome of $P_{mutation}$. For every node in a random sequence, set its community label to that of its all neighbors. The transition with the largest gain in modularity Q will be chosen as the update. The procedure ends when the community labels of all nodes are not changed during the loop.

The cluster-level greedy search runs on the output of node-level C_{node}. It rebuilds the network by taking each community as a super node. The links within communities are set as the self-loops of the super nodes. Links between communities are set as weighted links between super nodes. Then, the node-level greedy search is executed on the new network.

To asist the first two level strategies jump out of the local optimum, the partition-level greedy search is employed. It starts with inputting the output of cluster-level greedy search $C_{cluster}$ and the best chromosome of the population $C_{globalbest}$ to cross-over operation. And its output will be sent to execute the first two level greedy search. The output of partition-level greedy search $C_{partition}$ and $C_{cluster}$ consist of the output of the greedy search C_{greedy}.

4 Experimental Study and Results Analysis

The proposed NIMA are tested on 28 synthetic and 6 real-world networks with comparison of 11 existing algorithms to show its performance. The details will be illustrated in this section.

4.1 Experimental Settings

All the experiments are simulated by MATLAB R2020a on a PC with AMD (R), Ryzen (TM), 7-4800U with Radeon Graphics CPU 1.80 GHZ, 16 GB DDR4 RAM, Microsoft (R) Windows 10 Home operating system with 30 independent runs.

To evaluate the degree of similarity between two clustering partitions, normalized mutual information (NMI) is adopted. The definition of NMI is given as follows:

$$I(S_1, S_2) = \frac{-2\sum_{i=1}^{N_1}\sum_{j=1}^{N_2} C_{ij}\log(C_{ij}N/C_{i\cdot}C_{\cdot j})}{\sum_{i=1}^{N_1} C_{i\cdot}\log(C_{i\cdot}/N) + \sum_{j=1}^{N_2} C_{\cdot j}\log(C_{\cdot j}/N)} \tag{5}$$

Where S_1 and S_2 represents two clustering partitions of the given network respectively; N_1 (N_2) is the number of clusters in S_1 (S_2); N denotes the sum of vertices in the network; C is a confusion matrix with its elements C_{ij} denoting the number of consensus vertices divided in cluster s_i of partition S_1 and cluster s_j of partition S_2; $C_{i\cdot}$ ($C_{\cdot j}$) represents the sum of elements in ith row (jth column) in C. The value of NMI ranges from 0 to 1. And if S_1 and S_2 are exactly the same partition, $I(S_1, S_2) = 1$, while if S_1 and S_2 are completely different, $I(S_1, S_2) = 0$.

In this paper, The NMIs between community partitions found by algorithms and the normal partitions of the experimental networks are chosen as one of the evaluation indexes.

Comparison Algorithms. 11 algorithms are chosen to compare with NIMA. They are MODPSO [18], MOPSO-CD-R [23], MOGA-net [24], MOEA/D-net [25], MOCD [26], Meme-net [27], GA-net [28], GN [29], CNM [30], Informap [31] and Louvain [22]. The parameter settings of these algorithm can be found in Table 1.

Table 1. Parameter settings of algorithms.

Algorithms	g_{max}	N_p	N_m	N_o	p_c	p_m
NIMA	200	300	15	2	0.9	0.15
Meme-net	200	300	15	2	0.9	0.15
MODPSO	200	300	15	2	–	0.15
MOGA-net	200	300	15	2	0.9	0.15
MOPSO-CD-r	200	300	15	2	0.9	0.15
MOEA/D-net	200	300	15	2	0.9	0.15
MOCD	200	300	15	2	0.9	0.15
GA-net	200	300	15	2	0.9	0.15
GN	–	–	–	–	–	–
CNM	–	–	–	–	–	–
Informap	–	300	–	–	–	–
Louvain	–	–	–	–	–	–

Experimental Networks. Proposed by Girvan and Newman, the GN benchmark networks [29] consist of 128 nodes with an average node degree of 16. The normal partition of GN networks are 4 communities composed of 32 nodes each. There is an index γ, called mixing parameter, reflecting the percentages of links connecting to nodes in other communities for each node. In brief, the larger γ is, the blurrier community structure is. 11 GN networks are generated with γ ranging from 0 to 0.5 with a step of 0.05.

Another type of synthetic networks is the LFR benchmark networks proposed by Lancichinetti and Fortunato in Ref. [32]. With more detailed index of nodes set, LFR networks can reflect more significant and practical features of the networks. The 17 LFR networks is set as follows: Each network contains 1000 nodes and the cluster size ranges from 10 to 50. Two exponents $\tau_1 = 2$ and $\tau_2 = 1$; the average degree for each node is 20 and the maximum is 50. There is also a mixing parameter μ, like GN networks, ranging from 0 to 0.8 with a step of 0.05.

Table 2. Topological features of the real-world networks.

Networks	V	E	$\overline{k}$	C	H
Karate	34	78	4.588	0.588	1.693
Dolphins	62	159	5.129	0.303	1.327
Football	115	613	10.66	0.403	1.007
Jazz	198	2742	27.70	0.633	1.395
Email	1133	5451	9.622	0.254	1.942
Power grid	4941	6594	2.669	0.107	1.450

Six real-world networks are chosen to show the application of NIMA and its comparison with other algorithms. The networks are Zachary's karate club network (Karate) [33], the Bottlenose Dolphins network (Dolphins) [34], the American College Football network (Football) [29], the Jazz musicians network (Jazz) [35], Emails within an university network (Email) [36] and Western U.S. Power Grid network (Power grid) [37]. Details of the networks are listed in Table 2. V, E, $\overline{k}$, C and H is number of nodes, number of edges, average node degree, clustering coefficient and degree heterogeneity respectively.

4.2 Experimental Results and Analysis

The experimental results of 11 GN benchmark networks are shown in Fig. 3. The results reflect that all algorithms are able to find the normal partition when $\gamma \leq 0.10$. Then, the NMIs start to fall with the increase of mixing parameter. The proposed NIMA finds the normal partition with $\gamma \leq 0.45$. And when the mixing parameter reaches 0.50, it performs an averaged NMI of 0.9875, which is the best among algorithms.

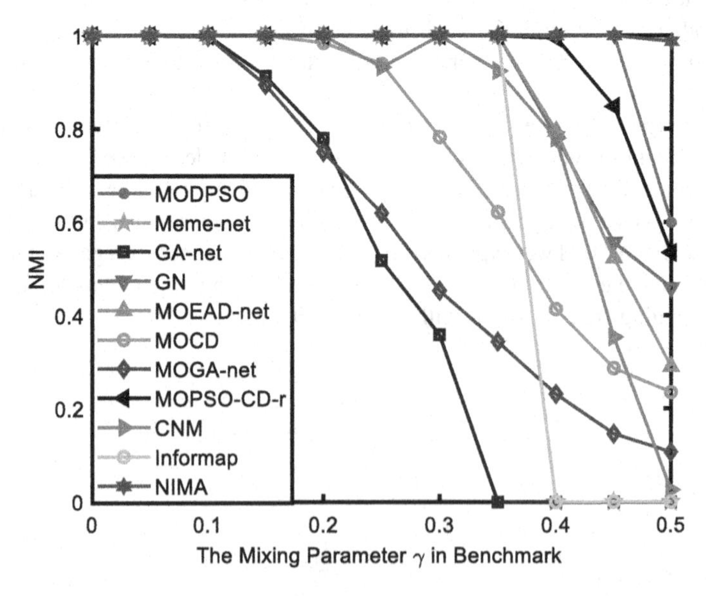

Fig. 3. Averaged NMI performance of algorithms in GN benchmark networks with different mixing parameters.

The experimental results of 17 LFR benchmark networks are shown in Fig. 4. Compared with GN networks, LFR networks are more complicated since many indexes are set in different configurations to simulate a variety of features. Therefore, a few algorithms cannot find the normal partition even when $\mu \leq 0.05$. Informap, MOPSO-CD-r, MODPSO and NIMA find the normal partition in every run when $\mu \leq 0.50$. Then, for Informap and MOPSO-CD-r, there are sharp declines when $\mu \geq 0.60$. The performance of NIMA deteriorates slightly when $0.55 \leq \mu \leq 0.70$. It performs the best when $\mu = 0.65\, and\, 0.70$ with the averaged NMI of 0.9641 and 0.9501, respectively. Last, it falls evidently when $\mu = 0.75\, and\, 0.80$, worse than MODPSO and MOEAD-net. As shown in the results, NIMA has strong performance in most of situations except those of $\mu = 0.75\, and\, 0.80$. The reason could be that most of the links of nodes are connected outside the communities in these two situations. Therefore, the node influence mechanism meets failures under the configurations that neighboring nodes are more likely to be divided into different communities.

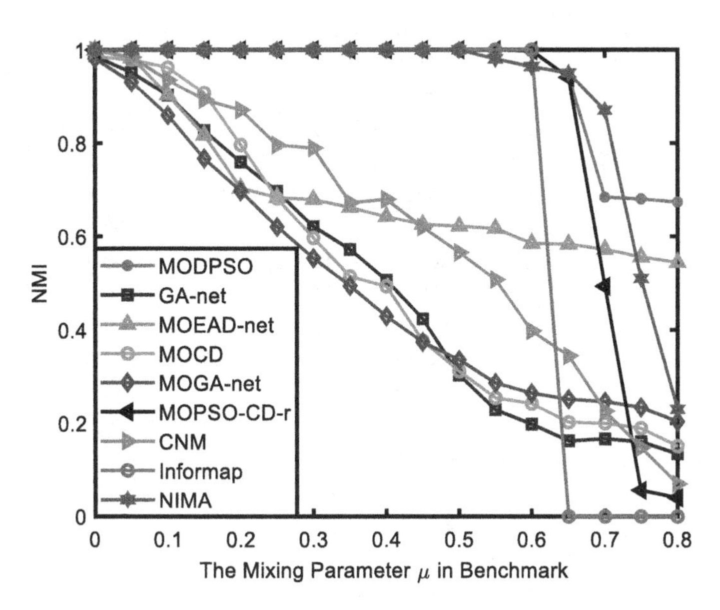

Fig. 4. Averaged NMI performance of algorithms in LFR benchmark networks with different mixing parameters.

The experimental results of six real-world networks are shown in Table 3. Overall, the proposed NIMA showed the best performance in all evaluation indexes including the maximum value, the average value and the standard deviation of modularity Q. Specifically, in small-scale networks like Karate, Dolphins, Football and Jazz, the proposed NIMA is able to find the clustering with the largest value of Q in every run. While MODPSO is able to find the same partition to NIMA in Karate and Football, but a unstable way, and so as Louvain performs in Football and Jazz. In middle-scale network Email and large-scale one Power grid, NIMA shows some fluctuations.

The experiments demonstrate that the proposed NIMA is efficient in modularity-maximization based community detection. And the modularity-based optimization methods are effective in finding the true community structure of complex networks.

Figure 5 is an illustration of topological community structure of the Power grid network detected by NIMA. 4941 nodes are divided into 41 communities, painted with 41 different colors. It can be summarized that, in Power grid network, there are few links between communities. Almost all the links are within the communities. Such phenomenon matches its high value of modularity which exceeds 0.9.

Table 3. Modularity Q optimization performance of algorithms in real-world networks.

Networks	Indexes	NIMA	MODPSO	MOGA-net	Louvain
Karate	Qmax	**0.4198**	**0.4198**	0.4159	0.4188
	Qave	**0.4198**	0.4182	0.3945	0.4165
	Qstd	**0**	0.0079	0.0089	0.0077
Dolphins	Qmax	**0.5222**	0.5216	0.5034	0.5168
	Qave	**0.5222**	0.5208	0.4584	0.5160
	Qstd	**0**	0.0062	0.0163	0.0029
Football	Qmax	**0.6046**	**0.6046**	0.4325	**0.6046**
	Qave	**0.6046**	0.6038	0.3906	0.6043
	Qstd	**0**	0.0011	0.0179	0.0009
Jazz	Qmax	**0.4451**	0.4421	0.2952	**0.4451**
	Qave	**0.4451**	0.4419	0.2929	0.4443
	Qstd	**0**	0.0001	0.0084	0.0023
Email	Qmax	**0.5828**	0.5193	0.3007	0.5412
	Qave	**0.5827**	0.3493	0.2865	0.5392
	Qstd	**0.0003**	0.0937	0.0075	0.0091
Power grid	Qmax	**0.9406**	0.8543	0.6914	0.7756
	Qave	**0.9397**	0.8510	0.6864	0.7688
	Qstd	**0.0003**	0.0056	0.0022	0.0102

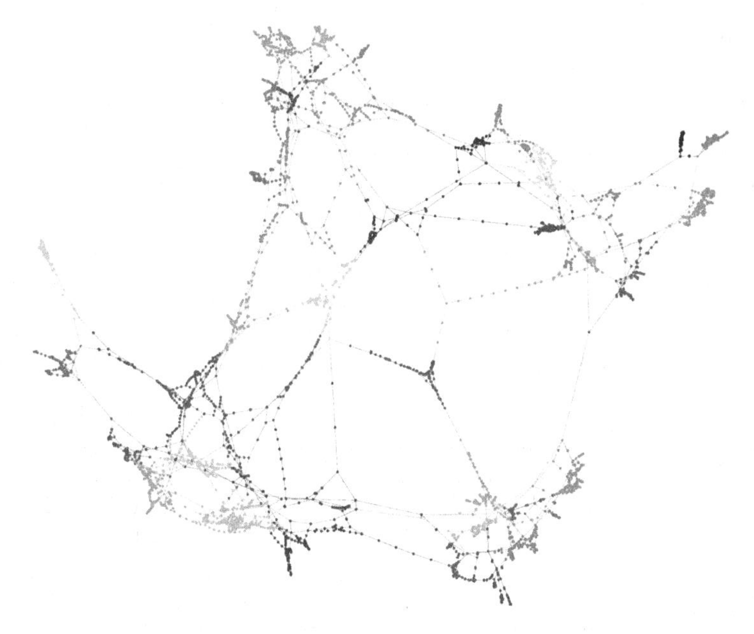

Fig. 5. The topological structure of the Power grid network after community detection.

5 Conclusions

Community structure is a significant property when analyzing the features and functions of complex systems. The community detection can be modeled as a modularity-maximization optimization problem. In this paper, a memetic algorithm, called NIMA, considering node influence is proposed for optimizing modularity as the community detection approach. The proposed NIMA consists of a transition probability matrix-based initialization, a network-specific crossover, a node degree-based mutation and a multi-level greedy search. Extensive experiments show that NIMA is an effective and efficient method in detecting communities of complex networks.

Despite of the promising performance of NIMA, there are still promotions need to be done. For instance, in the aspect of efficiency, more network-specific knowledge should be adopted as the strategies, while in the aspect of application, NIMA could be promoted into the study of dynamic and overlapped networks, which are more likely to reflect the real situations of complex systems.

Acknowledgements. This work was supported by the National Natural Science Foundation of China (Grant No. 61703256, 61806119), Natural Science Basic Research Plan in Shaanxi Province of China (Program No. 2017JQ6070), the Fundamental Research Funds for the Central Universities (Program No. GK201803020) and the Graduate Innovation Team Project of Shaanxi Normal University (Grant No. TD2020014Z).

References

1. Latora, V., Nicosia, V., Russo, G.: Complex Networks: Principles, Methods and Applications, 1st ed. Cambridge University Press, Cambridge (2017). https://doi.org/10.1017/9781316216002
2. Watts, D.J.: A twenty-first century science. Nature **445**(7127), 489 (2007)
3. Lazer, D., et al.: Life in the network: the coming age of computational social science. Science **323**(5915), 721–723 (2009)
4. Suweis, S., Simini, F., Banavar, J.R., Maritan, A.: Emergence of structural and dynamical properties of ecological mutualistic networks. Nature **500**(7463), 449–452 (2013)
5. Chakraborty, T., Ghosh, S., Park, N.: Ensemble-based overlapping community detection using disjoint community structures. Knowl.-Based. Syst **163**, 241–251 (2019). https://doi.org/10.1016/j.knosys.2018.08.033
6. Yang, L., Cao, X., He, D., Wang, C., Zhang, W.: Modularity based community detection with deep learning. In: Proceedings of International Joint Conference on Artificial Intelligence, pp. 2252–2258 (2016). https://doi.org/10.5555/3060832.3060936
7. You, X., Ma, Y., Liu, Z.: A three-stage algorithm on community detection in social networks. Knowl.-Based Syst. **187**, 104822 (2020). https://doi.org/10.1016/j.knosys.2019.06.030
8. Zhang, J., Ding, X., Yang, J.: Revealing the role of node similarity and community merging in community detection. Knowl.-Based Syst. **165**, 407–419 (2019). https://doi.org/10.1016/j.knosys.2018.12.009
9. Lu, M., Zhang, Z., Qu, Z., Kang, Y.: LPANNI: overlapping community detection using label propagation in large-scale complex networks. IEEE Trans. Knowl. Data Eng. **31**(9), 1736–1749 (2019). https://doi.org/10.1109/TKDE.2018.2866424
10. Fortunato, S.: Community detection in graphs. Phys. Rep. **486**(3–5), 75–174 (2010)
11. Newman, M.E.J., Girvan, M.: Finding and evaluating community structure in networks. Phys. Rev. E **69**(2), 026113 (2004). https://link.aps.org/doi/10.1103/PhysRevE.69.026113
12. Newman, M.E.J.: Fast algorithm for detecting community structure in networks. Phys. Rev. E **69**(6), 066133 (2004). https://doi.org/10.1103/PhysRevE.69.066133
13. Newman, M.E.J.: Modularity and community structure in networks. Proc. Natl. Acad. Sci. U.S.A. **103**(23), 8577–8582 (2006)
14. Newman, M.E.J.: Finding community structure in networks using the eigenvectors of matrices. Phys. Rev. E 74(3), 036104 (2006). https://link.aps.org/doi/10.1103/PhysRevE.74.036104
15. Lancichinetti, A., Fortunato, S.: Limits of modularity maximization in community detection. Phys. Rev. E **84**, 066122 (2011)
16. Zhan, Z., Shi, L., Tan, K., Zhang, J.: A survey on evolutionary computation for complex continuous optimization. Artificial Intell. Rev. **55**, 59–110 (2021). https://doi.org/10.1007/s10462-021-10042-y
17. Bian, K., Sun, Y., Cheng, S., Liu, Z., Sun, X.: Adaptive methods of differential evolution multi-objective optimization algorithm based on decomposition. In: Zhang, H., Yang, Z., Zhang, Z., Wu, Z., Hao, T. (eds.) NCAA 2021. CCIS, vol. 1449, pp. 458–472. Springer, Singapore (2021). https://doi.org/10.1007/978-981-16-5188-5_33
18. Gong, M., Cai, Q., Chen, X., Ma, L.: Complex network clustering by multiobjective discrete particle swarm optimization based on decomposition. IEEE Trans. Evol. Comput. **18**(1), 82–97 (2014). https://doi.org/10.1109/TEVC.2013.2260862

19. Li, C., Chen, H., Li, T., et al.: A stable community detection approach for complex network based on density peak clustering and label propagation. Appl Intell. **52**, 1188–1208 (2021). https://doi.org/10.1007/s10489-021-02287-5
20. Ma, L., Gong, M., Liu, J., Cai, Q., Jiao, L.: Multi-level learning based memetic algorithm for community detection. Appl. Soft Comput. **19**, 121–133 (2014). https://doi.org/10.1016/j.asoc.2014.02.003
21. Chen, D., Liu, C., Huang, X., Wang, D., Yan, J.: A probability transition matrix-based recommendation algorithm for bipartite networks. In: Liu, Y., Wang, L., Zhao, L., Yu, Z. (eds.) ICNC-FSKD 2019. AISC, vol. 1074, pp. 921–929. Springer, Cham (2020). https://doi.org/10.1007/978-3-030-32456-8_99
22. Blondel, V.D., Guillaume, J.L., Lambiotte, R., Lefebvre, E.: Fast unfolding of communities in large networks. J. Stat. Mech. Theory and Experiment (10), P10008 (2008)
23. Coello, C., Pulido, G., Lechuga, M.: Handling multiple objectives with particle swarm optimization. IEEE Trans. Evol. Comput. **8**(3), 256–279 (2004)
24. Pizzuti, C.: A multiobjective genetic algorithm to find communities in complex networks. IEEE Trans. Evol. Comput. **16**(3), 418–430 (2012)
25. Gong, M., Ma, L., Zhang, Q., Jiao, L.: Community detection in networks by using multiobjective evolutionary algorithm with decomposition. Phys. A **391**(15), 4050–4060 (2012)
26. Shi, C., Yan, Z., Cai, Y., Wu, B.: Multi-objective community detection in complex networks. Appl. Soft Comput. **12**(2), 850–859 (2012)
27. Gong, M., Fu, B., Jiao, L., Du, H.: Memetic algorithm for community detection in networks. Phys. Rev. E **84**(5), 056101 (2011)
28. Pizzuti, C.: GA-Net: A genetic algorithm for community detection in social networks. Proc. Parallel Problem Solving Nat. **5199**, 1081–1090 (2008)
29. Girvan, M., Newman, M.E.J.: Community structure in social and biological networks. Proc. Natl. Acad. Sci. U.S.A. **99**(12), 7821–7826 (2002)
30. Clauset, A., Newman, M.E.J., Moore, C.: Finding community structure in very large networks. Phys. Rev. E **70**(6), 066111 (2004)
31. Rosvall, M., Bergstrom, C.T.: Maps of random walks on complex networks reveal community structure. Proc. Natl. Acad. Sci. USA **105**(4), 1118–1123 (2008)
32. Fortunato, S., Lancichinetti, A.: Benchmarks for testing community detection algorithms on directed and weighted graphs with overlapping communities. Phys. Rev. E **80**(1), 016118 (2009)
33. Zachary, W.W.: An information-flow model for conflict and fission in small groups. J. Anthropol. Res. **33**(4), 452–473 (1997)
34. Lusseau, D., Schneider, K., Boisseau, O.J., Haase, P., Slooten, E., Dawson, S.M.: The bottlenose dolphin community of doubtful sound features a large proportion of long-lasting associations. Behav. Ecol. Sociobiol. **54**(4), 396–405 (2003)
35. Gleiser, P., Danon, L.: Community structure in jazz. Adv. Complex Syst. **6**(4), 565 (2003)
36. Guimerà, R., Danon, L., Díaz-Guilera, A.: Self-similar community structure in a network of human interactions. Phys. Rev. E **68**(6), 065103 (2003)
37. Watts, D.J., Strogatz, S.H.: Collective dynamics of 'small-world' networks. Nature **393** (6684), 440–442 (1998)

Adaptive Constraint Multi-objective Differential Evolution Algorithm Based on SARSA Method

Qingqing Liu, Caixia Cui, and Qinqin Fan(✉)

Logistics Research Center,
Shanghai Maritime University, Shanghai 201306, China
forever123@163.com

Abstract. The performance of constrained multi-objective differential evolution algorithm is mainly determined by constraint handling techniques (CHTs) and its generation strategies. Moreover, CHTs have different search capabilities and each generation strategy in a differential evolution is applicable to particular type of constrained multi-objective optimization problems (CMOPs). To automatically select appropriate CHT and generation strategy, an adaptive constrained multi-objective differential evolution algorithm based on state–action–reward–state–action (SARSA) approach (ACMODE) is introduced. In the ACMODE, the SARSA is used to select suitable CHT and generation strategy to solve particular types of CMOPs. The performance of the proposed algorithm is compared with other four famous constrained multi-objective evolutionary algorithms (CMOEAs) on 15 CMOPs. Experimental results show that the overall performance of the ACMODE is the best among all competitors.

Keywords: Constraint multi-objective optimization · Evolutionary computation · Reinforcement learning · SARSA method

1 Introduction

Constrained multi-objective optimization problems (CMOPs) are commonly found in the field of engineering optimization, such as robot's design optimization [1], compressed-air station scheduling problem [2] and scheduling optimization of micro-grid [3]. To effectively solve CMOPs, various improved CMOEAs have been proposed. For example, Wang et al. [4] proposed a cooperative multi-objective evolutionary algorithm with propulsive population (CMOEA-PP) to achieve a trade-off among the diversity, the convergence, and the feasibility in different evolutionary stages. Datta et al. [5] combined evolutionary multi-objective optimization method with penalty function method, and proposed a HyCon algorithm to deal with CMOPs. Yuan et al. [6] proposed an indicator-based evolutionary algorithm to prevent the population from falling into local areas. Cui et al. [7] proposed an adaptive constraint handling technique (CHT), which can adaptively select suitable CHT from three state-of-the-art CHTs via the Q-learning method.

L. Pan et al. (Eds.): BIC-TA 2021, CCIS 1565, pp. 232–246, 2022.
https://doi.org/10.1007/978-981-19-1256-6_17

Although various CMOEAs have been proposed to carry out adaptation selection of CHTs, their search strategies are generally constant during the entire evolutionary process. Therefore, it may not be effective when solving different types of CMOPs. To alleviate the above issues, an adaptive constraint multi-objective differential evolution algorithm based on SARSA method (named as ACMODE) is proposed. In the ACMODE, three effective and commonly used CHTs are selected and the SARSA [8] is utilized to select appropriate CHTs during different stages of the evolution. Moreover, three mutation strategies and two crossover strategies are chosen in differential evolution (DE) and the SARSA method is used to select appropriate generation strategies in the next iteration. Simulation experiments with other four CMOEAs are carried out on 15 constrained multi-objective test functions, and the results show that the ACMODE is more competitive than other four CMOEAs.

The rest of this paper is arranged as follows. Section 2 reviews the related work. Section 3 briefly introduces some definitions of constrained multi-objective optimization (CMO) and the basic concepts of DE. The details of the proposed algorithm are presented in Sect. 4. Subsequently, the experimental results are shown in Sect. 5. Section 6 draws some conclusions.

2 Literature Review

Recently, a large number of CMOEAs have been proposed to solve CMOPs. For example, Lin et al. [9] proposed a multi-objective differential evolution with dynamic hybrid constraint handling mechanism (MODE-DCH) to tackle CMOPs. In the MODE-DCH, the different search models combined with different CHTs are used. In [10], a coevolutionary framework was proposed to solve CMOPs via using two different populations. Fan et al. [11] introduced a novel framework, in which the entire search process is divided into push stage and pull stage. Constraints have not been considered in the push stage, while an improved epsilon-constraint method was used in the pull phase. A two-phase framework (ToP) was proposed in Ref. [12]. In the ToP, a CMOP was transformed into constrained single-objective optimization problems for locating promising regions in the first phase, and a specific and efficient CMOEA was employed to find feasible solutions in the second stage. Liu et al. [13] developed an indicator-based CMOEA framework, in which indicator-based MOEAs and CHTs are effectively combined to solve CMOPs. Based on the DE [14], a new DE variant named IMDE [15] was proposed, which used infeasible solutions to guide mutation operators and applied multiple combinations of mutation strategies and control parameters to enhance the search performance. Based on decomposition and the DE, Liu and Bi [16] presented an adaptive ε-constraint MOEA to make full use of the information of infeasible solutions.

3 Basic Concepts

3.1 Constrained Multi-objective Optimization Problems

CMOPs can be mathematically defined as follows:

$$\begin{array}{l} \min\ f(\boldsymbol{x}) = (f_1(\boldsymbol{x}), f_2(\boldsymbol{x}), \cdots, f_m(\boldsymbol{x}))^T \\ s.t.\ \ g_j(\boldsymbol{x}) \leq 0, j = 1, 2, \cdots, p \\ h_j(\boldsymbol{x}) = 0, j = p+1, \cdots, q \\ \boldsymbol{x} = (x_1, x_2, \cdots, x_D)^T \in O \end{array}, \tag{1}$$

where $\boldsymbol{x}$ is the D-dimensional vector of decision variables; $f(\boldsymbol{x})$ is an objective vector containing m objectives; $g_j(\boldsymbol{x})$ is the jth inequality constraint and p denotes the number of inequality constraints; $h_j(\boldsymbol{x})$ is the $(j-p)$th equality constraint and it has $(q-p)$ equality constraints; O is the decision space.

The constraint violation degree of the solution $\boldsymbol{x}$ can be computed as:

$$C_j(\boldsymbol{x}) = \begin{cases} \max(0, g_j(\boldsymbol{x}))\ ,\ 1 \leq j \leq p \\ \max(0, |h_j(\boldsymbol{x}) - \delta|)\ ,\ p+1 \leq j \leq q \end{cases}, \tag{2}$$

where δ is the tolerance value of equality constraints and is usually set as a small positive value. Therefore, the overall constraint violation degree of the solution $\boldsymbol{x}$ can be calculated as follows:

$$C(\boldsymbol{x}) = \sum_{j=1}^{q} C(\boldsymbol{x}). \tag{3}$$

For the Eq. (3), the solution $\boldsymbol{x}$ is a feasible solution when $C(\boldsymbol{x}) = 0$. In contrast, it is an infeasible solution.

3.2 Concepts in Multi-objective Optimization Problems

Basic concepts in multi-objective optimization are represented as follows [17]:

Definition 1 (*Dominance relation*): If there are two vectors $\boldsymbol{u}$ and $\boldsymbol{v}$ in the minimization optimization problem, $\forall n \in \{1, 2, \cdots, m\}$, $u_n \leq v_n$ and $\boldsymbol{u} \neq \boldsymbol{v}$, then $\boldsymbol{u}$ is said to dominate $\boldsymbol{v}$, denoted as $\boldsymbol{u} \succ \boldsymbol{v}$.

Definition 2 (*Pareto optimal set*): For a solution $\boldsymbol{x}^* \in R^D$, if and only if there is no other solution $\boldsymbol{x}$ such that $\boldsymbol{F}(\boldsymbol{x}) \succ \boldsymbol{F}(\boldsymbol{x}^*)$, it is called a Pareto optimal solution of a CMOP. All the Pareto optimal solutions form the Pareto set (PS), defined as $\boldsymbol{X}^*$.

Definition 3 (*Pareto front*): The Pareto front can be referred to $PF = \{F(\boldsymbol{x}^*) | \boldsymbol{x}^* \in \boldsymbol{X}^*\}$.

3.3 Performance Metrics

In the present work, two widely used performance metrics are employed: the inverted generational distance (*IGD*) and the hypervolume (*HV*).

IGD. IGD [18] is calculated as:

$$IGD(H, PF^*) = \frac{\sqrt{\sum_{z^* \in PF^*} d(z^*, H)^2}}{|PF^*|}, \quad (4)$$

where *H* represents the *PF* approximation; *PF** is a set of solutions obtained by evolutionary algorithms, which is uniformly distributed along the true *PF*; $d\ (z^*, H)$ is the minimum Euclidean distance between individual z^* in *PF** and *H*, $|PF^*|$ denotes the number of points in *PF**. Generally, *IGD* can simultaneously evaluate the convergence and diversity of *PF*.

HV. *HV* [19] Can Be Defined as Follows:

$$HV(H) = L\left(\bigcup_{z \in H} [z_1, z_1^r] \times \cdots \times [z_m, z_m^r]\right), \quad (5)$$

where L is the Lebesgue measure; $z = (z_1, \ldots, z_m)$ represents a solution in H; and $z^r = (z_1^r, \ldots, z_m^r)$ denotes a worst point dominated by all the Pareto optimal solutions.

3.4 Basics of DE

Differential evolution algorithm [14] is a simple and efficient meta-heuristic search algorithm. Its main operator steps are as follows:

Mutation Strategy. Three commonly seen mutation strategies are as follows:

$$\text{“DE/best/1”:} \quad \boldsymbol{v}_i^G = \boldsymbol{x}_{\text{best}}^G + F \cdot (\boldsymbol{x}_{r1}^G - \boldsymbol{x}_{r2}^G), \quad (6)$$

$$\text{“DE/current-to-rand/1”:} \quad \boldsymbol{v}_i^G = \boldsymbol{x}_i^G + F \cdot (\boldsymbol{x}_{r3}^G - \boldsymbol{x}_i^G) + F \cdot (\boldsymbol{x}_{r1}^G - \boldsymbol{x}_{r2}^G), \quad (7)$$

$$\text{“DE/rand-to-best/1”:} \quad \boldsymbol{v}_i^G = \boldsymbol{x}_{r1}^G + F \cdot (\boldsymbol{x}_{\text{best}}^G - \boldsymbol{x}_i^G) + F \cdot (\boldsymbol{x}_{r2}^G - \boldsymbol{x}_{r3}^G), \quad (8)$$

where $\boldsymbol{x}_i^G$ is the *i*th individual in the *G*th generation; indices r_1, r_2, and r_3, which are all different from *i* and are randomly generated from 1 to *NP* (population size). The scale factor is *F*, which is used to scale differential vectors. $\boldsymbol{x}_{\text{best}}^G$ represents the best individual in the *G*th generation.

Crossover Strategy. This operation generates the *j*th element of the *i*th new trial individual $\boldsymbol{u}_{ij}$:

$$\boldsymbol{u}_{ij}^{G} = \begin{cases} v_{ij}^{G}, R_j \leq CR \text{ or } j = j_{rand} \\ \boldsymbol{x}_{ij}^{G}, \text{otherwise} \end{cases}, j = 1, 2, \cdots D, \tag{9}$$

where R_j is a random number, which ranges from 0 to 1; CR is crossover probability; j_{rand} is a integer randomly generated within [1, D].

Selection. After the crossover strategy, the selection operation is performed to select the good solutions as the parents for the next generation, which can be defined as follows:

$$\boldsymbol{x}_{i}^{G+1} = \begin{cases} \boldsymbol{u}_{i}^{G}, f(\boldsymbol{u}_{i}^{G}) \leq f(\boldsymbol{x}_{i}^{G}) \\ \boldsymbol{x}_{i}^{G}, \text{otherwise} \end{cases}, \tag{10}$$

where $\boldsymbol{x}_{i}^{G+1}$ is the selected solution that can be used in next generation.

4 Proposed Algorithm

Different CHTs and generation strategies (crossover and mutation) have significant effects on the performance of CMOEAs. To further improve its performance, an adaptive constrained multi-objective differential evolution algorithm (ACMODE) is proposed in the present study. When solving different types of CMOPs in the ACMODE, suitable CHT and generation strategy can be adaptively selected during the whole evolutionary process.

The main operators in the ACMODE are as follows:

4.1 Adaptive Constraint Handling Technology

Different CHTs are suitable for solving different properties of CMOPs, thus an adaptive CHT is proposed in the current work. Three commonly used CHTs (SP [20], CDP [21] and ATM [22]) are selected and SARSA method is used to realize the adaptation of these three different CHTs. To evaluate the performance of each CHT, an improved *IGD* are given as follows:

$$mIGD(H, \overline{PF}) = \frac{\sqrt{\sum_{\overline{z} \in \overline{PF}} d(\overline{z}, H)^2}}{|\overline{PF}|}, \tag{11}$$

where $\overline{PF}$ is selected from all achieved *PF* approximations.

The pseudocode of the proposed adaptive CHT method is described in **Algorithm 1**. The action space can be defined as A = [SP, CDP, ATM], the state space can be expressed as S = [excellent, medium, poor], and the value of reward R is [1, 0, −1] [23]. In lines 1 to 2, according to the different CHTs selected by each individual, the population is divided into three subpopulations P_{SP}, P_{CDP}, and P_{ATM}. Therefore, *mIGD* value can be calculated by Eq. (11). In lines 3 to 7, the maximum *mIGD* value represents the individual chooses this CHT is in "poor" state and its reward is −1; the middle *mIGD* value indicates the state is "medium" and reward is 0; while the reward of "excellent" CHT is 1. Use the state *s*' to predict action *a*', and then update Q-table. Finally, action chain ***AC*** is updated.

Algorithm 1: Adaptive Constraint Handling Technique

Input: the state vector ***SV*** and the reward chain ***RC***
Output: action chain ***AC***
1 Divide the population into three subpopulations P_{SP}, P_{CDP}, and P_{ATM}.
2 Calculate $mIGD_{SP}$, $mIGD_{CDP}$ and $mIGD_{ATM}$ according to Eq. (11)
3 **for** i =1: NP **do**
4 Obtain the s' and r according to *mIGD* value, and update ***SV*** and ***RC***;
5 Use ε- greedy method to predict individual action a' according to s';
6 Update Q-table: $Q(s,a) = Q(s,a) + \alpha(r+\gamma Q(s',a')-Q(s,a))$;
7 **end**
8 Update action chain ***AC***;

4.2 Adaptive Mutation Strategy

Different mutation strategies play distinct roles in the search process. "DE/current-to-rand/1" has good exploration and search ability, while "DE/best/1" and "DE/rand-to-best/1" possess good local search capability and their convergence speed is faster than DE/rand/1. Consequently, these three mutation strategies are applied in the proposed algorithm. The process of adaptive mutation strategy is as follows:

Step 1: initialize the state s_0 for each individual and the corresponding Q-table. Set a as the action selected by the individual in the initial state using the ε-greedy method.

Step 2: perform the current action a in the current state s. According to the different mutation strategies selected by individuals, the population is divided into three subpopulations, which are denoted as $P_{current}$, P_{best} and P_{rand}. Their corresponding *mIGD* values are respectively denoted as $mIGD_{current}$, $mIGD_{best}$, and $mIGD_{rand}$.

Step 3: the minimum *mIGD* value represents that the individual choosing this mutation strategy is in an "excellent" state, and its reward is 1. A medium *mIGD* value indicates that the corresponding mutation strategy state is "medium" and the reward is 0. While individuals in a "poor" state are rewarded with −1. The new state and the corresponding reward can be obtained.

Step 4: according to *s*', a new action *a*' is selected.

Step 5: update the Q-table: $Q(s,a) = Q(s,a) + \alpha(r + \gamma Q(s',a') - Q(s,a))$.

Step 6: *s* = *s*'; *a* = *a*'.

4.3 Adaptive Crossover Strategy

The selection of crossover strategies significantly impacts the performance of CMODE. Simulated Binary Crossover (SBX) has strong local search ability, while Polynomial Crossover (PC) is good at global search. To balance the global search and local search, PC and SBX are selected and the SARSA method is used to predict the crossover strategy. The process of adaptive crossover strategy is as follows:

Step 1: initialize the state s_0 for each individual and the corresponding Q-table. Set *a* as the action selected by the individual in the initial state using the ε-greedy method.

Step 2: perform the current action *a* in the current state *s*. According to the different crossover strategies selected by individuals, the population is divided into two sub-populations, P_{SBX} and P_{PC}. Their corresponding *mIGD* values are denoted as $mIGD_{SBX}$ and $mIGD_{PC}$.

Step 3: the minimum *mIGD* value represents that the individual choosing this mutation strategy is in an "excellent" state, and its reward is 1. While individuals in a "poor" state are rewarded with −1. The new state and the reward can be obtained.

Step 4: according to *s*', a new action *a*' is selected.

Step 5: update the Q-table: $Q(s,a) = Q(s,a) + \alpha(r + \gamma Q(s',a') - Q(s,a))$.

Step 6: *s* = *s*'; *a* = *a*'.

4.4 Overall Implementation of the Proposed Algorithm

Our proposed algorithm ACMODE mainly includes three stages, namely initialization, population evolution and self-adaptation. The pseudocode of our proposed algorithm is described in **Algorithm 2**. Lines 1 to 6 are the initialization operator. Firstly, the population is divided into several sub-populations. The corresponding *mIGD* value can be calculated by Eq. (11). Therefore, individual state can be determined, and ***AC***, ***MC*** and ***CRC*** can be obtained. In lines 7 to 16, if $G > 0.2*G_{max}$, the mutation strategy and crossover strategy are used according to ***MC*** and ***CRC***. Otherwise, "DE/current-to-rand/1" is used. Lines 17–19 realize the adaption of CHTs, lines 20–21 realize the adaption of mutation strategies and crossover strategies. Finally, feasible solutions are chosen to enter in the external archive ***B***.

Algorithm 2: ACMODE

Input: *NP*: population size

G_{max}: the maximum number of iterations

x_i^G, v_i^G, u_i^G : *i*th target vector, mutant vector, trail vector under the *G*th generation

Output: final solution set *P* and corresponding objective vectors *H*

1 Initialize population $P^G = \{x_1^G, \ldots, x_i^G, \ldots x_{NP}^G\}$;

2 Initialize the external archive ***B***, Q-table, the state vector ***AV***, ***MV*** and ***CRV***, the action chain ***AC***, ***MC*** and ***CRC*** and the reward chain ***SRC***, ***MRC*** and ***RC***;

3 Use *k*-means clustering to divide P^G into two or three subpopulations;

4 Calculate the corresponding *mIGD* value of each subpopulation according to Eq. (11);

5 Use *mIGD* value to determine individual state and update ***CRV***, ***AV*** and ***MV***;

6 Use the ε-greedy method to obtain ***AC***、***MC***、***CRC***;

7 **for** G=1: G_{max} **do**

8   **for** i =1: *NP* **do**

9   Randomly select a F value from {0.6, 0.8, 1.0};

10   Randomly select a CR value from {0.1, 0.2, 1.0};

11   **If** $G< 0.2*G_{max}$ **then**

12 $$v_i^G = x_i^G + F*\left(x_{r_1}^G - x_i^G\right) + F*\left(x_{r_2}^G - x_{r_3}^G\right)$$

13   **else**

14   Mutate the individual according to ***MC***; Generate u^G by using ***CRC***;

15   **end**

16   **end for**

17   Divide population $P^G \cup u^G$ into three subpopulations P_{SP}^G, P_{CDP}^G and P_{ATM}^G;

18   Select half of the individuals from each subpopulation into P^{G+1} according to three CHTs;

19   Update ***AC*** by **algorithm 1**;

20   Implement the adaptive mutation strategy in section 4.2 to update ***MC***;

21   Implement the adaptive crossover strategy in section 4.3 to update ***CRC***;

22   Save the feasible solutions at the first level of non-dominated sorting in P^{G+1} to ***B***;

23 **end for**

24 if $|B|>NP$, use the non-dominated sorting method to select *NP* individuals from ***B*** as final solution set *P*; otherwise, *P* = ***B***; and calculate corresponding objective vectors *H*;

---

5 Experimental Studies

To evaluate the ACMODE algorithm, it was compared with other 4 CMOEAs on 15 CMOPs, which are ACHT-CMODE [7], AGS-CMODE, MOEA/D-CDP [24] and ANSGAIII [25]. In addition, two nonparametric statistical tests, the Wilcoxon rank sum test [26] and the Friedman test [27], are employed to analyze the search performance of all comparison algorithms. "+", "−" and "≈" respectively indicate that the performance of ACMODE is superior, inferior or similar to that of the comparison algorithm.

5.1 Test Instance

All experiments are performed on 15 widely used constrained multi-objective optimization test instances. Functions CTP1-CTP8 are taken from the famous test set CTPs [28], which has two objective functions. The remaining 7 test functions are actual optimization problems, including SRN [29], OSY [30], TNK [31], BNH [32], DBD [33], SRD [34] and CONSTR [35].

5.2 Parameter Settings

All algorithms run independently for 30 times, the maximum number of iterations of each test function is 500, and the population size is 100. Furthermore, the parameter settings of other comparison algorithms are consistent with the original literature.

5.3 Compared with Other Four CMOEAs

Two comprehensive performance indicators (*IGD* and *HV*) are used to evaluate the performances of all compared algorithms. The best results are in bold.

IGD: In this experiment, only feasible solutions can be utilized to calculate *IGD*. The *IGD* mean and standard deviation values as well as the Wilcoxon statistical results of all the comparison algorithms are shown in Table 1. As shown in Table 1, ACMODE is superior to ACHT-CMODE and AGS-CMODE in 10 and 11 test functions, and inferior to the ACMODE in 2 and 1 test functions, respectively. In addition, ACMODE has similar performance respectively to the above comparison algorithm in 3 test functions. Thus, it can be seen that the generation strategy adaptation can significantly improve the performance of CMODE. It is noted that the performance of the MOEA/D-CDP is worse than that of ACMODE on all test functions. However, the ANSGAIII algorithm is inferior to the proposed algorithm in 9 test functions and slightly better than the ACMODE in only 3 test functions. For the IGD values, the performance ranking of all comparison algorithms obtained by Friedman statistical analysis is shown in Fig. 1. The experimental results demonstrate that the proposed algorithm ranks the first among all the comparison algorithms, that is, the ACMODE has the best overall performance.

To sum up, the ACMODE outperforms other four comparison algorithms in most test functions. The reason is that our proposed ACMODE algorithm can automatically select the appropriate CHT and generation strategy in the iterative evolution process, to improve its ability to solve different types of CMOPs.

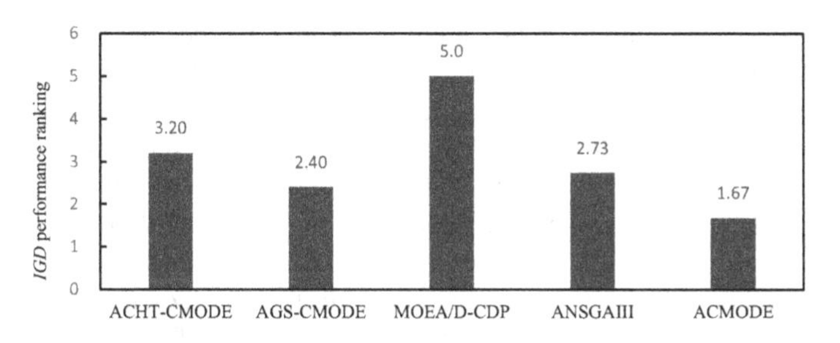

Fig. 1. Performance rankings of *IGD* of all comparison algorithms

Table 1. *IGD* results of all comparison algorithms

	ACHT-CMODE	AGS-CMODE	MOEA/D-CDP	ANSGAIII	ACMODE
CTP-1	4.56E−03 + (3.93E−03)	1.71E−03 + (2.48E−03)	1.16E−02 + (1.73E−03)	5.66E−03 + (3.27E−03)	**1.13E−03 (1.80E−03)**
CTP-2	2.36E−03 + (2.92E−03)	8.72E−04 + (9.01E−04)	1.49E−02 + (6.30E−04)	**1.86E−04− (4.32E−05)**	2.27E−04 (3.08E−05)
CTP-3	1.33E−02 + (2.13E−02)	**5.13E−03≈ (2.45E−03)**	2.12E−01 + (1.31E−02)	1.28E−02≈ (1.97E−02)	5.49E−03 (2.97E−03)
CTP-4	4.28E−02 + (4.50E−02)	3.00E−02 + (1.73E−02)	2.53E−01 + (3.65E−02)	7.30E−02 + (4.19E−02)	**2.56E−02 (4.31E−03)**
CTP-5	4.23E−03 + (5.82E−03)	**7.81E−04≈ (4.78E−04)**	1.09E−02 + (3.97E−04)	1.00E−03 + (1.12E−03)	8.71E−04 (4.96E−04)
CTP-6	2.02E−03− (1.21E−03)	1.42E−03− (8.07E−04)	4.82E−02 + (9.07E−03)	**8.78E−04− (4.93E−04)**	6.01E−03 (2.73E−02)
CTP-7	1.39E−02 + (1.34E−02)	3.95E−03 + (5.89E−03)	1.86E−02 + (3.88E−03)	4.07E−04 + (5.20E−04)	**1.34E−04 (1.40E−05)**
CTP-8	1.85E−02 + (3.23E−02)	6.41E−03 + (2.85E−02)	3.81E−02 + (8.97E−03)	7.27E−03≈ (8.44E−03)	**5.74E−04 (9.82E−05)**
SRN	3.64E−02 ≈ (5.13E−03)	4.56E−02 + (6.41E−03)	2.63E−01 + (2.09E−01)	**1.46E−02≈ (1.51E−04)**	3.65E−02 (4.38E−03)
OSY	2.08E−01 + (4.88E−01)	9.49E−02 + (3.52E−02)	2.94E + 00 + (2.27E−01)	2.50E−01 + (1.25E−01)	**7.35E−02 (2.77E−02)**
TNK	1.91E−03 + (4.00E−04)	1.12E−03 + (2.38E−04)	4.67E−02 + (1.00E−02)	**4.61E−04− (3.11E−05)**	8.56E−04 (8.67E−05)
BNH	**1.85E−02− (2.37E−03)**	2.27E−02 + (3.76E−03)	4.07E−01 + (1.52E−01)	2.28E−02 + (8.33E−03)	1.91E−02 (2.99E−03)
DBD	6.27E−03 ≈ (4.03E−03)	5.09E−03≈ (2.93E−03)	1.06E−01 + (2.69E−02)	2.00E−02 + (5.22E−03)	**3.80E−03 (1.91E−03)**
SRD	1.65E + 00≈ (3.39E + 00)	1.06E+00 + (2.10E−01)	2.43E+01 + (2.91E + 00)	1.56E + 01 + (6.54E + 00)	**1.04E+00 (2.23E−01)**
CONSTR	2.22E−03 + (3.13E−04)	2.50E−03 + (6.43E−04)	4.17E−02 + (1.75E−02)	1.83E−03 + (2.49E−04)	**1.54E−03 (4.72E−05)**
+	10	11	15	9	
−	2	1	0	3	
≈	3	3	0	3	

HV: In this experiment, the reference point is set to 1.1 times the upper limit of the Pareto Front, and only feasible solutions are selected to calculate *HV*. Table 2 provides mean and standard deviation values of *HV*. As can be seen from Table 2, ACHT-CMODE, AGS-CMODE, MOEA/D-CDP and ANSGAIII are superior to the proposed algorithm ACMODE in 1, 2, 1 and 5 test functions respectively, and inferior to ACMODE in 11, 7, 14 and 10 test functions respectively. In addition, the performance is similar to that of ACMODE algorithm in 3, 5, 0 and 0 test functions. Moreover, the performance ranking of *HV* of all comparison algorithms is shown in Fig. 2. Figure 2 demonstrates that in terms of *HV*, the overall performance of ACMODE is the best, followed by ANSGAIII, AGS-CMODE and other algorithms.

In conclusion, the convergence and diversity of the *PF* approximation obtained by ACMODE is the best. The main reason is that the ACMODE algorithm uses the SARSA method to predict the generation strategies respectively, and makes them adapt to the iteration of the population evolution.

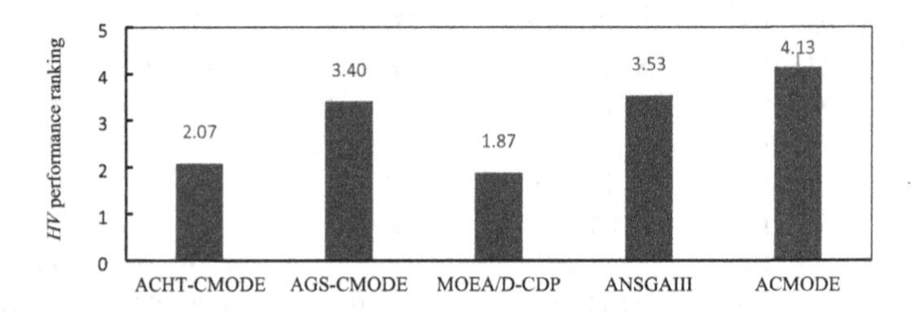

Fig. 2. Performance rankings of *HV* of all comparison algorithms

Table 2. *HV* results of all comparison algorithms

	ACHT-CMODE	AGS-CMODE	MOEA/D-CDP	ANSGAIII	ACMODE
CTP-1	3.14E−01 + (1.20E−01)	4.09E−01 + (8.65E−02)	3.31E−01 + (3.14E−02)	4.20E−01 + (2.81E−02)	**4.27E−01 (6.44E−02)**
CTP-2	9.15E−02 + (2.92E−02)	1.07E−01 + (9.81E−03)	4.62E−02 + (2.50E−03)	1.13E−01 + (3.89E−04)	**1.13E−01 (1.48E−04)**
CTP-3	4.44E−01 + (7.11E−02)	**4.69E−01≈ (5.12E−03)**	1.70E−02 + (3.01E−02)	4.55E−01 + (2.91E−02)	4.68E−01 (6.70E−03)
CTP-4	3.39E−01 + (9.72E−02)	3.67E−01≈ (5.46E−02)	1.23E−03 + (6.74E−03)	2.82E−01 + (7.26E−02)	**3.82E−01 (1.86E−02)**
CTP-5	4.30E−01 + (1.15E−01)	**4.82E−01≈ (3.95E−03)**	1.09E−01 + (9.20E−04)	4.57E−01 + (4.81E−02)	4.82E−01 (4.52E−03)
CTP-6	2.00E+00− (3.41E−02)	2.03E + 00− (2.41E−02)	8.00E−01− (5.26E−01)	**2.05E + 00− (1.60E−02)**	1.97E + 00 (3.72E−01)
CTP-7	3.22E−01 + (2.33E−01)	5.70E−01 + (1.80E−01)	2.75E−01 + (7.83E−02)	7.11E−01 + (7.19E−03)	**7.16E−01 (1.30E−04)**
CTP-8	1.10E+00 + (3.83E−01)	1.30E+00≈ (2.57E−01)	9.14E−01 + (3.87E−01)	1.28E+00 + (8.78E−02)	**1.36E+00 (2.39E−03)**

(continued)

Table 2. (*continued*)

	ACHT-CMODE	AGS-CMODE	MOEA/D-CDP	ANSGAIII	ACMODE
SRN	3.02E+04≈ (6.18E+01)	3.02E + 04 + (5.56E + 01)	2.70E+04 + (2.18E+03)	**3.06E+04− (4.44E+00)**	3.03E+04 (4.32E+01)
OSY	1.35E+04 + (1.50E+03)	1.39E + 04 + (8.94E+01)	6.07E+03 + (7.45E+02)	1.37E+04 + (2.69E+02)	**1.39E+04 (4.94E + 01)**
TNK	5.11E−01≈ (3.68E−03)	5.19E−01≈ (1.22E−03)	1.30E+00 + (4.84E−02)	**5.24E−01− (5.46E−04)**	5.20E−01 (7.95E−04)
BNH	2.82E+03 + (4.62E+00)	2.81E+03≈ (8.03E+00)	1.90E+03 + (3.67E+02)	**2.82E+03− (1.42E+01)**	2.82E+03 (5.45E+00)
DBD	4.28E+01 + (6.61E−02)	4.29E+01− (3.76E−02)	3.22E+01 + (7.03E+00)	4.28E+01 + (9.18E−02)	**4.28E+01 (7.23E−02)**
SRD	2.57E+06≈ (9.82E+03)	2.57E + 06 + (2.00E + 03)	2.63E+06 + (8.82E+04)	2.57E+06 + (7.59E+03)	**2.57E+06 (9.05E+ 02)**
CONSTR	5.18E+00 + (8.44E−03)	5.19E+00 + (1.06E−02)	5.54E+00+ (7.70E−01)	**5.21E+00− (3.31E−03)**	5.21E+00 (3.48E−03)
+	11	7	14	10	
−	1	2	1	5	
≈	3	6	0	0	

5.4 Experimental Analyses

The Effectiveness of the Adaptive Generation Strategy. To further validate the efficiency of the proposed adaptive generation strategy, the evolution curves of mutation and crossover strategies on two test functions (CTP-8 and OSY) of ACMODE are presented in this experiment (as shown in Fig. 3 and 4).

Figure 3(a) shows that in the adaptive stage of mutation strategy, "DE/current-to-rand/1" strategy is always in the dominant position, while other two mutation strategies play similar roles. Figure 3 (b) shows that "DE/rand-to-best/1" strategy plays an important role in the early stage of evolution, while" DE/current-to-rand/1" is mainly used in ACMODE. In the later period of evolution, "DE/best/1" occupies the main position, and the roles of other two mutation strategies are similar. Figure 4 shows that the number of the two crossover strategies changes in real time during evolution. Figure 4 (a) shows that "SBX" has good performance in the early stage. In contrast, "PC" is mainly used in the later stage. As shown in Fig. 4 (b), "PC" has good performance during the whole evolutionary process. It can be concluded that the ACMODE can adaptively select appropriate mutation strategy and crossover strategy according to different types of CMOPs or different evolutionary stages.

Compared with the ACHT-CMODE, the proposed algorithm ACMODE only has more generation strategy adaptive technology in the whole structure, and the experimental results in Sect. 5.3 demonstrate that the solution set obtained by ACMODE has better performance. Therefore, adaptive generation strategy can assist the algorithm to automatically select the appropriate mutation and crossover strategies in different search phases.

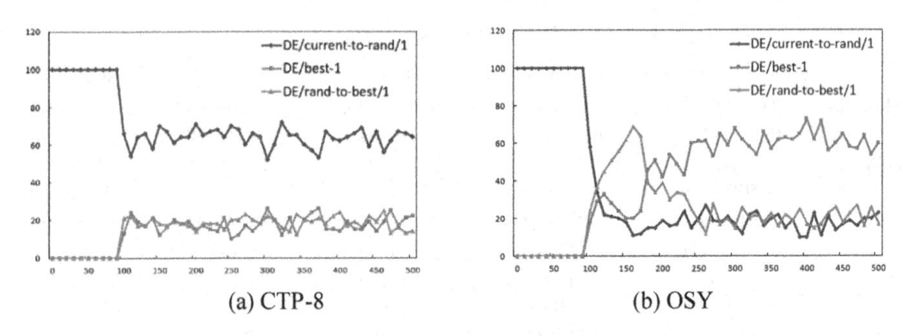

(a) CTP-8 (b) OSY

Fig. 3. The number changes of each mutation strategy of ACMODE

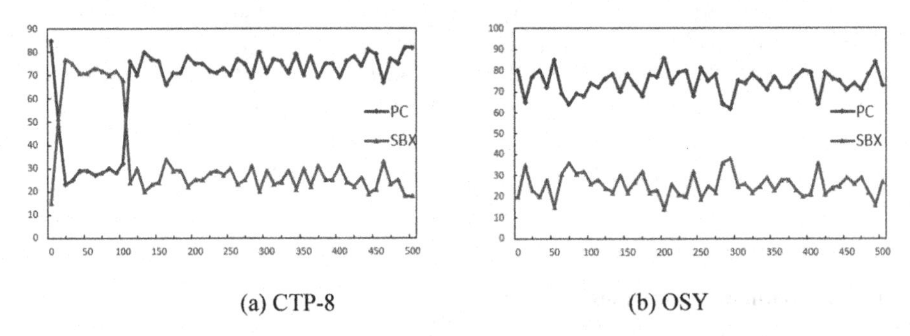

(a) CTP-8 (b) OSY

Fig. 4. The number changes of each crossover strategy of ACMODE

Parametric Analyses. In this experiment, the sensitivity of parameter ε in ε-greedy method is analyzed. 15 test functions are used to test the performance of the ACMODE under different ε values. Figure 5 provides the performance rankings of *HV* in terms of different ε values. as can be seen from Fig. 5, when ε = 0.5, the overall performance of the algorithm is the best, hence ε is set to 0.5 in this algorithm.

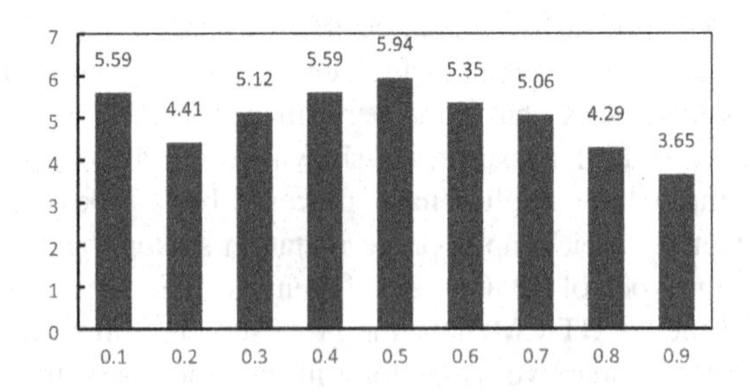

Fig. 5. Performance ranking of *HV* at different ε values

6 Conclusion

In the present work, an adaptive CMOEA based on SARSA approach (ACMODE) is proposed to realize the adaptive CHTs and generation strategies. In the proposed algorithm, three different CHTs: SP, CDP, ATM, three mutation strategies and two crossover strategies are selected and can be self-evolved via a SARSA method. Compared with the other four CMOEAs on 15 test functions, the ACMODE has better performance in solving CMOPs.

References

1. Maminov, A., Posypkin, M.: Constrained multi-objective robot's design optimization. In: 2020 IEEE Conference of Russian Young Researchers in Electrical and Electronic Engineering (EIConRus), pp. 1992-1995 (2020)
2. liu, J., Yang, Y., Tan, S., Wang, H.: Application of constrained multi-objective evolutionary algorithm in a compressed-air station scheduling problem. In: 2019 Chinese Control Conference (CCC), pp. 2023–2028 (2019)
3. Li,B., Wang, J., Xia, N.: Dynamic optimal scheduling of microgrid based on ε constraint multi-objective biogeography-based optimization algorithm. In: 2020 5th International Conference on Automation, Control and Robotics Engineering (CACRE), pp. 389–393 (2020)
4. Wang, J., Li, Y., Zhang, Q., Zhang, Z., Gao, S.: Cooperative multiobjective evolutionary algorithm with propulsive population for constrained multiobjective optimization. IEEE Trans. Syst. Man Cybernet. Syst. 1–16 (2021)
5. Datta, R., Deb, K., Segev, A.: A bi-objective hybrid constrained optimization (HyCon) method using a multi-objective and penalty function approach. In: 2017 IEEE Congress on Evolutionary Computation (CEC), pp. 317–324 (2017)
6. Yuan, J., Liu, H.L., Ong, Y.S., He, Z.: Indicator-based evolutionary algorithm for solving constrained multi-objective optimization problems. IEEE Trans. Evol. Comput. 1 (2021)
7. Cui, C.X., Fan, Q.Q.: Constrained multi-objective differential evolutionary algorithm with adaptive constraint handling technique. World Sci. Res. J. **7**, 322–339 (2021)
8. Richard, S.S., Andrew, G.B.: Temporal-difference learning. In: Reinforcement Learning: An Introduction, pp. 133–160, MIT Press (1998)
9. Lin, Y., Du, W., Du, W.: Multi-objective differential evolution with dynamic hybrid constraint handling mechanism. Soft. Comput. **23**(12), 4341–4355 (2018). https://doi.org/10.1007/s00500-018-3087-z
10. Tian, Y., Zhang, T., Xiao, J., Zhang, X., Jin, Y.: A Coevolutionary framework for constrained multiobjective optimization problems. IEEE Trans. Evol. Comput. **25**(1), 102–116 (2021)
11. Fan, Z., et al.: Push and pull search for solving constrained multi-objective optimization problems. Swarm Evol. Comput. **44**, 665–679 (2019)
12. Liu, Z.Z., Wang, Y.: Handling Constrained Multiobjective Optimization Problems With Constraints in Both the Decision and Objective Spaces. IEEE Trans. Evol. Comput. **23**(5), 870–884 (2019)
13. Liu, Z.Z., Wang, Y., Wang, B.C.: Indicator-based constrained multiobjective evolutionary algorithms. IEEE Trans. Syst. Man Cybernet. Syst. **51**(9), 5414–5426 (2021)

14. Storn, R., Price, K.: Differential evolution – a simple and efficient heuristic for global optimization over continuous spaces. J. Global Optim. **11**(4), 341–359 (1997)
15. Xu, B., Duan, W., Zhang, H., Li, Z.: Differential evolution with infeasible-guiding mutation operators for constrained multi-objective optimization. Appl. Intell. **50**(12), 4459–4481 (2020). https://doi.org/10.1007/s10489-020-01733-0
16. Liu, B.J., Bi, X.J.: Adaptive ε-constraint multi-objective evolutionary algorithm based on decomposition and differential evolution. IEEE Access **9**, 17596–17609 (2021)
17. Deb, K.: Multi-objective Optimization Using Evolutionary Algorithms. Wiley, Chichester (2001)
18. Bosman, P.A.N., Thierens, D.: The balance between proximity and diversity in multiobjective evolutionary algorithms. IEEE Trans. Evol. Comput. **7**(2), 174–188 (2003)
19. Zitzler, E., Thiele, L.: Multiobjective evolutionary algorithms: a comparative case study and the strength Pareto approach. IEEE Trans. Evol. Comput. **3**(4), 257–271 (1999)
20. Woldesenbet, Y.G., Yen, G.G., Tessema, B.G.: Constraint handling in multiobjective evolutionary optimization. IEEE Trans. Evol. Comput. **13**(3), 514–525 (2009)
21. Deb, K., Pratap, A., Agarwal, S., Meyarivan, T.: A fast and elitist multiobjective genetic algorithm: NSGA-II. IEEE Trans. Evol. Comput. **6**(2), 182–197 (2002)
22. Wang, Y., Cai, Z., Zhou, Y., Zeng, W.: An adaptive tradeoff model for constrained evolutionary optimization. IEEE Trans. Evol. Comput. **12**(1), 80–92 (2008)
23. Shahrabi, J., Adibi, M.A., Mahootchi, M.: A reinforcement learning approach to parameter estimation in dynamic job shop scheduling. Comput. Ind. Eng. **110**, 75–82 (2017)
24. Jan, M.A., Khanum, R.A.: A study of two penalty-parameterless constraint handling techniques in the framework of MOEA/D. Appl. Soft Comput. **13**(1), 128–148 (2013)
25. Jain, H., Deb, K.: An evolutionary many-objective optimization algorithm using reference-point based nondominated sorting approach, Part II: handling constraints and extending to an adaptive approach. IEEE Trans. Evol. Comput. **18**(4), 602–622 (2014)
26. Wilcoxon, F.: Individual comparisons by ranking methods. Biometrics Bull. **1**(6), 80–83 (1945)
27. Friedman, M.: The use of ranks to avoid the assumption of normality implicit in the analysis of variance. J. Am. Stat. Assoc. **32**(200), 675–701 (1937)
28. Deb, K., Pratap, A., Meyarivan, T.: Constrained test problems for multi-objective evolutionary optimization. In: Presented at the first international conference on evolutionary multi-criterion optimization (EMO), Zurich, Switzerland (2000)
29. Srinivas, N., Deb, K.: Multiobjective function optimization using nondominated sorting genetic algorithms. IEEE Trans. Evol. Comput. **2**(3), 1301–1308 (1994)
30. Osyczka, A., Kundu, S.: A new method to solve generalized multicriteria optimization problems using the simple genetic algorithm. Struct. Optim. **10**(2), 94–99 (1995)
31. Tanaka, M., Watanabe, H., Furukawa, Y., Tanino, T.: GA-based decision support system for multicriteria optimization. In: 1995 IEEE International Conference on Systems, Man and Cybernetics. Intelligent Systems for the 21st Century, vol. 2, pp. 1556–1561 (1995)
32. Binh, T.T., Korn, U.: MOBES: a multiobjective evolution strategy for constrained optimization problems. In: Presented at the third international conference on genetic algorithms, Mendel (1997)
33. Ray, T., Liew, K.M.: A swarm metaphor for multiobjective design optimization. Eng. Optim. **34**(2), 141–153 (2002)
34. Coello Coello, C.A., Pulido, G.T.: Multiobjective structural optimization using a microgenetic algorithm. Struct. Multidiscip. Optim. **30**(5), 388–403 (2005)
35. Justesen, P.D.: Multi-objective optimization using evolutionary algorithms. University of Aarhus, Department of Computer Science (2009)

A Hybrid Multi-objective Coevolutionary Approach for the Multi-user Agile Earth Observation Satellite Scheduling Problem

Luona Wei[1], Yanjie Song[1], Lining Xing[2(✉)], Ming Chen[1], and Yingwu Chen[1]

[1] College of Systems Engineering, National University of Defense Technology, Changsha 410073, China
xinglining@gmail.com
wlnelysion@163.com, songyj_2017@163.com, chenming_nudt@163.com, ywchen@nudt.edu.cn

[2] School of Electronic Engineering, Xidian University, Xi'an, China
xinglining@gmail.com

Abstract. Multi-user agile earth observation satellite scheduling problem (MU-AEOSSP) is an important combinatorial optimization problem for satellite daily management. In this study, a MU-AEOSSP is addressed to tackle the failure rate and the fairness of different users simultaneously. A hybrid multi-objective coevolutionary approach (HMOCA) is then proposed to handle the complicate constraints and to optimize the objectives. HMOCA evolves two populations to solve an original MU-AEOSSP considering all constraints and a helper problem without the transition time constraint. By the cooperation of the two population, both the convergence and the diversity performance can be significantly improved. To further enhance the performance of HMOCA, several specific variation operators and a local search operator considering the time-dependent transition time of the MU-AEOSSP are equipped. The HMOCA is extensively tested and compared with three classical multi-objective evolutionary algorithms (NSGAII, MOEA/D, IBEA) and two methods of the time-dependent multi-objective AEOSSP (D-MOMA-TD and I-MOMA-TD) on several instances which are generated based on real-word situation. Experiment results show that the proposed approach outperforms all the comparison methods on most of the instances in terms of convergence, solution quality and diversity.

Keywords: Coevolutionary algorithm · Agile satellite · Multi-objective

1 Introduction

Owing to the rapid development in space technology and the sharp increase in the number of orbiting satellites, the earth observation satellite (EOS) has become

Supported by the National Natural Science Foundation of China, Grant No. 71701203, 72001212 and 71901213.

L. Pan et al. (Eds.): BIC-TA 2021, CCIS 1565, pp. 247–261, 2022.
https://doi.org/10.1007/978-981-19-1256-6_18

one of the most significant platforms for collecting remote information, which is widely used in geographic information management, environmental monitoring, disaster rescue, military reconnaissance, etc. [3,18]. As the new generation of the EOS, the agile EOS (AEOS) has superior maneuverability and observing ability because of three rotational degrees of freedom in orbit. Thus, the AEOS is able to accomplish more complicated observing requests in prolonged visible time windows (VTWs) by maneuvering in three dimensions, rising a significantly superior scheduling efficiency for observation.

The multi-user agile earth observation satellite scheduling problem (MU-AEOSSP) in this paper is an extension of the conventional AEOS scheduling problem, which is proved to be a NP-hard problem [22]. The MU-AESSOP needs to consider the management process when several requests are emanated from different users, which is closer to the real-world situation [19]. Thus, two objective functions should be optimized, which are to minimize the failure rate of observation requests and ensure the fairness of resource sharing for all users simultaneously, leading the MU-AEOSSP to be modelled as a multi-objective optimization problem.

Previous studies always apply existing multi-objective evolutionary algorithms (MOEAs) to solve the AEOSSP with multiple optimization goals. However, there are several shortcomings which are neglected in these researches. First, most of the existing studies simplified the multi-objective AEOSSP (MO-AEOSSP) by neglecting its time-dependent characteristic, which is one of the most important features of the problem [19]. Without the time-dependent characteristic, the problem is severely distorted, greatly reducing the difficulty of solving the MO-AEOSSP. Second, the problem studied in this paper is not only a multi-objective optimization problem, but also a combinatorial optimization problem which has significant characteristics and complex constraints. Many existing MOEAs may be incapable of balancing constraints and objectives, especially when the feasible region is discrete [15]. This weakness may lead to bad performance in terms of convergence or diversity.

Due to the above analysis and inspired by the recent application of coevolutionary algorithms, a hybrid multi-objective coevolutionary approach (HMOCA) is proposed in this paper to tackle the MU-AEOSSP. The failure rate of the observation requests and the fairness between different users are taken as the objectives to be optimized. HMOCA coevolves two populations to solve an original problem with all the constraints and a helper problem with partial constraints simultaneously. Then by the cooperation of the two population, the convergence and the diversity of the obtained Pareto Front (PF) can be significantly improved. Besides, several operators and a local search procedure which takes the time-dependent feature of the MU-AEOSSP into account are designed to further improve the efficiency and the exploitation. Experimental results show that the HMOCA outperforms several comparison methods.

The reminder of the paper is presented as below. In Sect. 2, we first introduce the existing researches for MO-AEOSSP. Then the MU-AEOSSP is described and the mathematical model is established in the third section. In Sect. 4, the HMOCA is presented in detail. Computational studies are introduced in Sect. 5.

Finally, some discussion together with conclusion and future work are given in Sect. 6.

2 Related Work

A few researchers have studied multi-objective satellite scheduling problem. Wang et al. [17] applied a specific Strength Pareto Evolutionary Algorithm II (SPEA2) to addressing a imaging scheduling problem considering the mission planning together with the data transmission simultaneously. Tangpattanakul et al. [13] addressed a variant of the ROADEF 2003 challenge, proposed a biased random-key genetic approach to optimize the total profit of the scheduled tasks and the balance between users at the same time. Sun et al. [12] considered four criteria in their study and designed a modified SPEA2. Li et al. [6,7] combined the preference of decision makers and the multi-objective AEOSSP, then adopted several preference-based multi-objective optimization approaches to solve different instances.

Regarding the time-dependent characteristic, many studies have been targeted among the single-objective AEOSSP. Liu et al. [8] considered the time-dependent feature and the characteristic of the maneuvering time between two consective tasks for the first time, then proposed a specific Adaptive Large Neighborhood Search algorithm (ALNS). After that, He et al. [5] applied an improved ALNS with several new designed operators to address the multi-satellite AEOSSP.

As for constrained multi-objective optimization problems (CMOPs) which is similar as the MU-AEOSSP, various constraint handling techniques and coevolutionary approaches have been suggested. Woldesenbet et al. [21] modified the objective values of solutions by considering the constraint violation. Ning et al. [9] proposed a new nondominated rank by combining the standard nondominated rank sorting and the constraint rank. Fan et al. [4] developed the MOEAs with multiple stage to solve CMOPs, defining a push stage and a pull stage (PPS) to divide the search process. The push stage addresses the CMOP without considering any constraints while all the constraints and objectives are considered during the pull stage. Sorkhabi et al. [11] used more than one population to store the feasible and infeasible particles separately, and proposed a multi-objective particle swarm optimization algorithm with a migrating strategy. Tian et al. [15] proposed a coevolutionary approach for the CMOP, addressing a complex problem by a simple helper problem, utilizing a weak cooperation for two populations together with the offspring sharing mechanism to approximate the optimal PF.

3 Multi-user Agile Earth Observation Satellite Scheduling Problem (MU-AEOSSP)

In this section, the MU-AEOSSP is first described, then two objective functions and the time-dependent transition time are analyzed and introduced. Based on several concepts and assumptions, the mathematical model is established.

3.1 Problem Description

As discussed before, the AEOSSP aims to schedule a set of candidate requests given from different users under the condition of satisfying several constraints. All requests are provided by geographical latitude, longitude and altitude. A given AEOS may have more than one orbit (usually 15–16 orbits should be considered when scheduling per day), providing some requests with more than one visible window (VTW) during their working horizon. The MU-AEOSSP discussed in this section is a specific AEOSSP which need to maximize the profit of scheduled requests and balance the fairness of different users at the same time, considering the time-dependent characteristic and the task visibility.

3.2 Assumptions

Some reasonable assumptions and simplifications are given as below:

- Only spot requests or small area requests which can be observed in one pass are considered in this paper.
- On-board power is assumed to be sufficient in this study.
- Any executing request cannot be preempted and only one request can be observed at any moment.
- The downloading scheduling procedure and on-board memory management are not taken into account in this paper.

3.3 Notations and Variables

A set of candidate requests is given as the input of the MU-AEOSSP, denoted by $R = \{r_1, ..., r_N\}$, where N is the number of requests to be schedule. For each request $r_i \in R$, we define:

- p_i: the profit of request r_i;
- g_i: the index of the user that request r_i belongs to, $g_i \in G = \{1, ..., N_g\}$ where N_g is the number of users;
- du_i: the duration for observing r_i;
- $TW_i = [tw_i^1, ..., tw_i^q, ..., tw_i^{m_i}]$: the set of VTWs for r_i, where m_i is the number of VTWs.
  Each tw_i^q is denoted by $tw_i^q = [ts_i^q, te_i^q, o_i^q]$, where o_i^q is the index of orbit in which tw_i^q is located, ts_i^q and te_i^q represent the start and end moment of tw_i^q.

Orbits of the AEOS are defined as follows:

$O = \{o_1, ..., o_k, ..., o_{m_o}\}$: the set of orbits, which is calculated by the pre-processing procedure considering the satellite orbit trajectory.

Each orbit is denoted by $o_k = [o_k^s, o_k^e]$, where o_k^s and o_k^e represent the scheduling horizon of the given orbit.

The decision variables are defined below:

$$x_{il} = \begin{cases} 1, the \quad lth \quad VTW \quad of \quad task\ t_i\ is \quad chosen \\ 0, otherwise \end{cases} \tag{1}$$

h_i: the start-time of observing r_i.

3.4 Time-Dependent Transition Time

Similarly as discussed in [10], the transition time of the MU-AEOSSP is time-dependent, depending on the start moments of the two consecutive requests. More precisely, the transition time between two adjacent request r_i and r_j can be presented as follows:

$$Tr_{ij} = \begin{cases} 11.66, & \Delta\theta \leq 10 \\ 5 + \Delta\theta/v_1, & 10 \leq \Delta\theta \leq 30 \\ 10 + \Delta\theta/v_2, & 30 \leq \Delta\theta \leq 60 \\ 16 + \Delta\theta/v_3, & 60 \leq \Delta\theta \leq 90 \\ 22 + \Delta\theta/v_4, & \Delta\theta \geq 90 \end{cases} \tag{2}$$

where v_1, v_2, v_3, v_4 are angular transition velocities, $\Delta\theta$ indicates the transition angle between r_i and r_j. It can be easily found from the function that the larger $\Delta\theta$ is, the larger the transition time, revealing the characteristic of the maneuvering time between two consecutive requests.

$\Delta\theta = |\Delta\gamma| + |\Delta\pi| + |\Delta\phi|$, where the change of angle for each axis (roll, pitch and yaw) are represented by $\Delta\gamma$, $\Delta\pi$ and $\Delta\phi$. For each VTW, a given moment have been pre-calculated by second to link with the specific angles. Thus, once a start moment of a request is given, the corresponding transition time can be easily calculated.

3.5 Mathematical Modelling

The mathematical model of the MU-AEOSSP is formulated in this section.

The objective functions are first proposed:

$$\begin{aligned} \min F &= (f_1, f_2) \\ f_1 &= 1 - \sum_{i=1}^{N}\sum_{l=1}^{m_i} x_{il} \cdot p_i \Big/ \sum_{i=1}^{N} p_i \\ f_2 &= \frac{1}{\overline{L(G)}} \sqrt{\frac{\sum_{k=1}^{N_g} [L(g_k) - \overline{L(G)}]^2}{N_g - 1}} \end{aligned} \tag{3}$$

where f_1 denotes the failure rate of observation requests and f_2 represents fairness of resource for all users.

For f_2, $\overline{L(G)}$ is the mean occupation time of all users while $L(g_k)$ represents the occupation time for user g_k:

$$L(g_k) = \sum_{i=1}^{N}\sum_{l=1}^{m_i} x_{il} \cdot du_i, \ \ if\ g_k = g_i \tag{4}$$

$$\overline{L(G)} = \frac{1}{N_g}\sum_{k=1}^{N_g} L(g_k) \tag{5}$$

The constraints are as follows:

$$\sum_{l=1}^{m_i} x_{il} \leq 1 \tag{6}$$

$$o_k^s \leq ts_i^q \leq te_i^q \leq o_k^e, \forall o_i^q = k, k \in O \tag{7}$$

$$ts_i^q \leq h_i \leq te_i^q,\ \ if\ x_i^q = 1 \tag{8}$$

$$ts_i^q \leq h_i + du_i \leq te_i^q,\ \ if\ x_i^q = 1 \tag{9}$$

$$st_i + du_i + Tr_{ij} \leq st_j,\ \ if\ y_{ij} = 1 \tag{10}$$

Constraints (6) restricts that each request can be executed at most once.

Constraints (7) indicates that a VTW must be included in the horizon of its corresponding orbit, where the ts_i^q and te_i^q are the start and end moment of request r_i.

Constraint (8) and (9) state that a request should be completed in its corresponding VTW.

Constraints (10) shows that the observation time window (OTW) of any two requests must not overlap and there must have enough time between them for maneuvering.

4 A Hybrid Multi-objective Coevolutionary Approach (HMOCA)

4.1 Framework

Figure 1 shows the basic framework of the proposed HMOCA. Two populations are used for the coevolutionary process. One population is to solve the original MU-AEOSSP considering all constraints while the other population is evolved considering partial constrains to provide an assistance. The specific implementing process is as follows:

- **Initialization:** Two populations of the same size are generated through a uniform random distribution, which are denoted as *Population 1* and *Population 2.* In the following stages, the HMOCA evolves *Population 1* to solve the original MU-AEOSSP while *Population 2* is evolved to solve a helper problem which neglect the constraint of the transition time.
- **Original problem (*Population 1*):** *Population 1* is evolved to solve the original MU-AEOSSP considering all constraints. In each generation, a set of mating individuals are selected from the current *Population 1* through the mating selection strategy. Then, an offspring population is generated by the operators of HMOCA. After that, a specific local search procedure is

adopted to further enhance the performance and to improve the solutions. At the end of every iteration, both the individuals in *Population 1* and the offspring populations of the original and helper problems are combined and environmental selected to form the new *Population 1* if the terminal criterion is not satisfied.

- **Helper problem (*Population 2*):** *Population 2* is evolved to solve the MU-AEOSSP without the time-dependent transition time. The mating selection and offspring generation procedures are almost the same as them of *Population 1*. At the end of every generation, the individuals in *Population 2* and the offspring populations of the original and helper problems are combined and environmental selected to form the new *Population 2*. Note that the transition time constraint is not taken into account in all procedures of *Population 2*.
- **Output:** After the stopping criterion is satisfied, the final *Population 1* is returned as the final output.

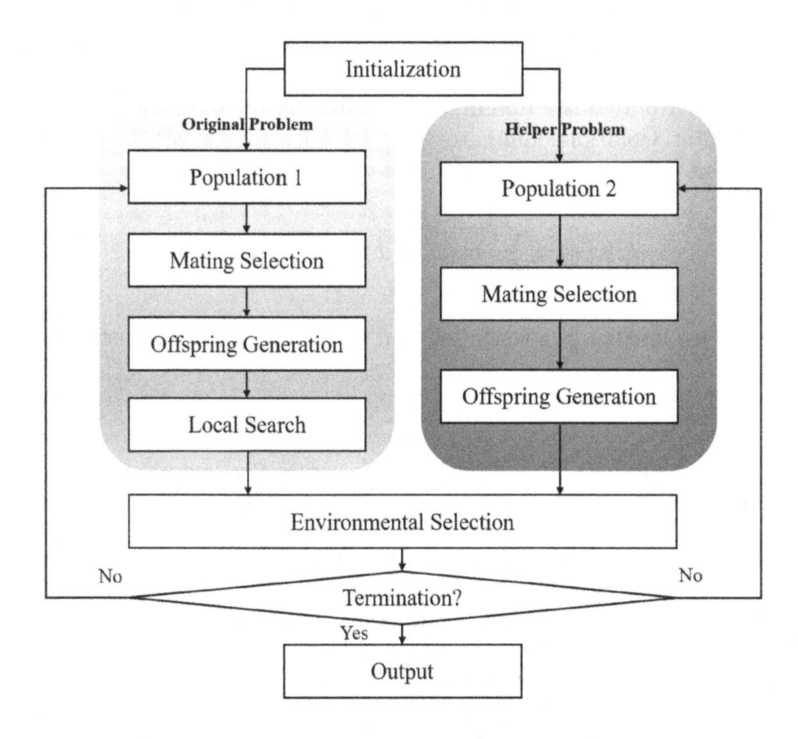

Fig. 1. Framework of the HMOCA for MU-AEOSSP

4.2 Individual Representation and Initialization

Based on [2,19,22], a specific permutation coding is applied:

1) Generating the chromosome $\boldsymbol{V} = \{V_1, ..., V_n\}$, which represents the assigned VTWs for candidate tasks.
2) The chromosome $\boldsymbol{V}$ is then decoded as a series of set for each orbit. Each set contains the candidate requests of their corresponding orbit. After assignment, each request is scheduled obeying a insertion order. The output schedule can be encoded as $\boldsymbol{Sch} = \{sch_1, ..., sch_k, ..., sch_{m_o}\}$, where sch_k contains the scheduled tasks in orbit k.

Both the $\boldsymbol{V}$ and the insertion order for each orbit are randomly generated at the initialization stage and generated through specific operators at other phases.

4.3 Selection Operation

Tournament selection is applied for the mating selection, forming the mating pool for offspring generation. In each generation, randomly selected individuals are evaluated and the better one is then put into the mating pool for the following stages. The non-dominated sorting method are adopted. Since the two populations consider different constraints and cooperated with each other, a Pareto dominance considering feasibility [1] is adopted when comparing individuals: 1) feasible individuals dominate infeasible individuals; 2) a solution with lower constraint violation dominates the one with higher constraint violation. The constraint violation (CV) of solution $\boldsymbol{x}$ can be calculated by

$$CV(\boldsymbol{x}) = \sum_{i=1}^{p} \max\{g_i(\boldsymbol{x}), 0\} + \sum_{j=1}^{q} |h_j(\boldsymbol{x})| \tag{11}$$

where $g_i(\boldsymbol{x})$ are inequality constraints and $h_j(\boldsymbol{x})$ are equality constraints.

4.4 Offspring Generation

Crossover. Inspired by [16,19], two specific crossover operators are designed, cooperating to balance the exploration and exploitation.

- Crossover operator 1 ($\boldsymbol{Cr1}$): Randomly select the VTW of each request from two parent VTW chromosomes to form a new VTW chromosome, then the new Sch chromosome is constructed as in initialization.
- Crossover operator 2 ($\boldsymbol{Cr2}$): Randomly select each sch_k from two parent sch chromosome then repair the whole schedule: 1) for a duplicated request, retain the VTW in the least-loaded orbit; 2) for an unselected request, select the VTW in the least-loaded orbit then put it at the end of the insertion order.

Mutation. The mutation operates by randomly choose a request and insert it to another position.

4.5 Local Search Operation

Local search operator is designed to further enhance the performance of HMOCA since the MU-AEOSSP is a complicated combinational optimization problem which is hard to approximate the true PF by genetic operators. As shown in Algorithm 1, a set of weight vectors are first generated to transfer the MU-AEOSSP into several single-objective optimization problem. For different weight vectors, three sorting methods are used for inserting the unscheduled requests. After the insertion, the individuals and their attributes are updated, updating the earliest and latest start time ($EALA^* \leftarrow UPDATE(EALA)$).

Algorithm 1. Local Search Operator

Input: population P^*, the depth of local search d
Output: population P
1: generate weight vectors $\boldsymbol{\delta}_1, ..., \boldsymbol{\delta}_i, ..., \boldsymbol{\delta}_u, \boldsymbol{\delta}_i = (\delta_{ix}, \delta_{iy})$
2: $P = \emptyset$
3: **while** $P^* \neq \emptyset$ **do**
4: randomly select one individual $\boldsymbol{x}$ from P^* and one weight vector $\boldsymbol{\delta}_i$
5: calculate the earliest and latest start-time of each request $EALA$
6: obtain the set of unscheduled requests U^* of $\boldsymbol{x}$
7: $P^* = P^* - \{\boldsymbol{x}\}$;
8: **if** $\delta_{ix} == 1$ **then**
9: sort the requests in U^* in descending order of their profit p
10: **for** each $\boldsymbol{u}$ in U^* **do**
11: $\boldsymbol{x}^* \leftarrow INSERT(\boldsymbol{x}, \boldsymbol{u})$ /*insert $\boldsymbol{u}$ in the sorted order
12: $EALA^* \leftarrow UPDATE(EALA)$
13: **end for**
14: **else if** $\delta_{iy} == 1$ **then**
15: first sort the requests in U^* in ascending order of the time occupied by each user, then sort requests issued by the same user in descending order of $(\frac{p}{du})$
16: **for** each $\boldsymbol{u}$ in U^* **do**
17: $\boldsymbol{x}^* \leftarrow INSERT(\boldsymbol{x})$
18: $EALA^* \leftarrow UPDATE(EALA)$
19: **end for**
20: **else**
21: randomly sort the requests in U^*
22: **for** depth = 1 to d **do**
23: $\boldsymbol{x}^* \leftarrow INSERT(\boldsymbol{x}, \boldsymbol{u})$
24: **if** $f(\boldsymbol{x}^*|\boldsymbol{\delta}_i) < f(\boldsymbol{x}|\boldsymbol{\delta}_i)$ **then**
25: $\boldsymbol{x} = \boldsymbol{x}^*$
26: **end if**
27: **end for**
28: **end if**
29: $P = P \cup \{\boldsymbol{x}^*\}$;
30: **end while**
31: return P

Because of the FIFO rule and triangle inequalities satisfied by the time-dependent MU-AEOSSP [10,19], the earliest and latest start time (t_{es} and t_{ls}) of any request can be calculated if the scheduling sequence of requests on its orbit is given. Then by judging whether t_{es} is greater than t_{ls} can determine whether the request can be inserted. The insertion and the calculation of t_{es} are illustrated by Algorithm 2 and Algorithm 3 (only the calculation of t_{es} is presented for simplicity).

Algorithm 2. The Insertion

Input: a VTW tw_λ^* of unscheduled request r_λ
an individual $\boldsymbol{x}$
Output: an improved individual $\boldsymbol{x}^*$
1: Target the orbit o_λ^* in which tw_λ^* is located
2: Find the scheduling sequence Seq on the orbit o_k
3: **for** each insect position in Seq **do**
4: Insertion attempting: assume r_λ is between task $Seq(j)$ and $Seq(j+1)$
5: $t_{es}^\lambda = EarliestStartTimeCalculation(Seq(j), r_\lambda)$;
6: $t_{ls}^\lambda = LatestStartTimeCalculation(Seq(j+1), r_\lambda)$;
7: **if** $t_{es}^\lambda \leq t_{ls}^\lambda$ **then**
8: $\boldsymbol{x}^* = \boldsymbol{x} \cup r_\lambda$;
9: **end if**
10: **end for**

In order to reduce the repeated calls of the dichotomy algorithm used in Algorithm 3, we pre-calculate t_{es} and t_{ls} of each request pair on the same orbit for each moment during their VTWs to form a table for the local search operator. During the local search operation, the scheduling interval can be obtained by checking the pre-calculated table.

5 Computational Experiment

5.1 Experimental Setting

The proposed HMOCA and comparison algorithms are programmed in the Evolutionary Multi-objective Optimization Platform (PlatEMO) [14] and all experiments are conducted through a i7-9700 CPU with 8 GB RAM. Instances in this paper are generated through a modified method used in [5,8]: Generate requests using a uniform random distribution through geographical area corresponding to 3°N-53°N and 74°E-133°E. MU-AEOSSPs with 100–300 requests are generated with an increment step of 50. Five users are adopted and each request is randomly assigned a user. "MU_NO" is used to denote instances of various size, where "MU" represents the MU-AEOSSP and "NO" represents the number of requests. The initial orbital parameters of the satellite in this study are shown in Table 1 and the time horizon of the scheduling is 24 h (15–16 orbits).

Algorithm 3. The t_{es} Calculation

Input: two consecutive requests r_i,r_j, the start-time h_i of r_i
Output: the earliest start-time t_{es}^j of r_j
1: **if** $h_i + du_i + Tr_{ij}(h_i, ts_j) \leq ts_j$ **then**
2: $t_{es}^j = ts_j$;
3: **else**
4: **if** $h_i + du_i + Tr_{ij}(h_i, te_j) \geq te_j$ **then**
5: There is no feasible observing chance for r_j
6: **else**
7: $ts_{jmin} \leftarrow ts_j; ts_{jmax} \leftarrow te_j$
8: **while** $ts_{jmax} - ts_{jmin} \geq 2$ **do**
9: $t^* \leftarrow \frac{ts_{jmin}+ts_{jmax}}{2}$;
10: **if** $ts_i + du_i + Trij(ts_i, t) \leq t^*$ **then**
11: $ts_{jmin} \leftarrow t^*$;
12: **else**
13: $ts_{jmax} \leftarrow t^*$;
14: **end if**
15: **end while**
16: $t_{es}^j = ts_{jmax}$;
17: **end if**
18: **end if**

We compare the optimization results of HMOCA with three classical MOEAs: NSGA-II [1], MOEA/D [23], IBEA [24], and two specific methods for MO-AEOSSP considering time-dependent feature: D-MOMA-TD and I-MOMA-TD [19]. All compared algorithms apply the same coding, decoding, variation operators as HMOCA. Other settings are as below: 1) the population size is set to 100; 2) the scaling factor for IBEA and I-MOMA-TD is set to 0.05; 3) the PBI approach is adopted for MOEA/D and T is set to 10; 4) the depth of local search is set to 10 and the max evaluation for all methods is set to 10000. The rest parameters and mechanisms of the compared methods are set according to their default settings.

Apart from the same setting as other methods, HMOCA is set as below: the number of weight vector is set to 20 for all instances and the vectors are uniformly generated; the original problem (*Population*1) is the MU-AEOSSP with all the constraints as described in Sect. 3; the helper problem (*Population*2) is set to the MU-AEOSSP without the transition time constraint (constraint 10 in Sect. 3).

Table 1. Orbital parameters of the satellite

Satellite	α	e	i	w	ω	m
AS-01	7141701.7	0.000627	98.5964	95.5069	342.307	125.2658

5.2 Experimental Results

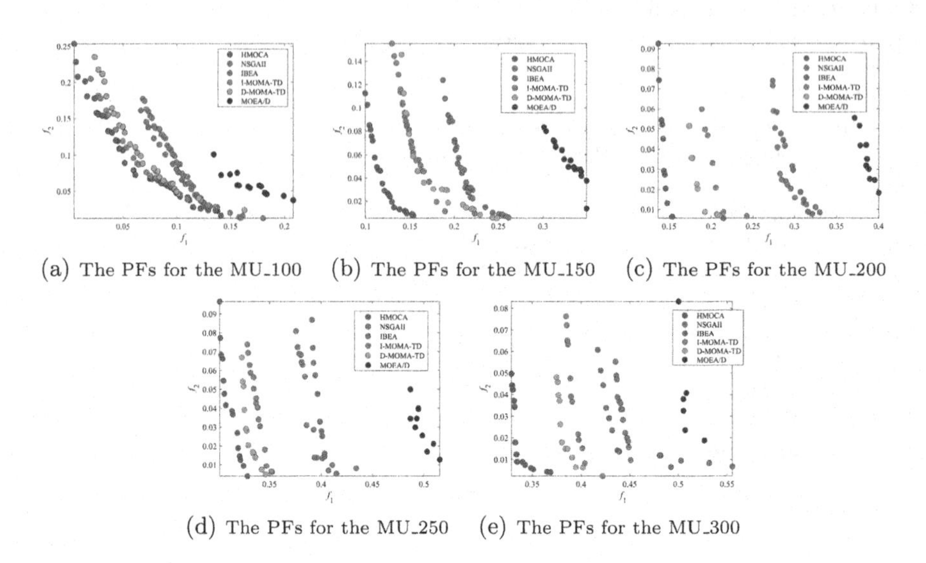

(a) The PFs for the MU_100 (b) The PFs for the MU_150 (c) The PFs for the MU_200

(d) The PFs for the MU_250 (e) The PFs for the MU_300

Fig. 2. Non-dominated solutions reserved in different scenarios of MU-AEOSSP using the proposed HMOCA and different comparison algorithms

Figure 2 shows the pareto fronts (PFs) which contain only the non-dominated solutions of all methods while preserving superior quality of spreading and diversity. For each method, a random run of each scenario is presented, where x-axis denotes the failure rate of observation requests and y-axis indicates the fairness of different users.

As Fig. 2 shows, the proposed HMOCA outperforms all other methods in terms of both convergence and diversity. Keeping the same level of failure rate for observation requests (f_1), HMOCA can output individuals which perform better in terms of the fairness between different users, and vice versa. For scenarios of small instance size, D-MOMA-TD and I-MOMA-TD seem to be competitive for obtaining high-quality solutions (e.g., in 100-request MU-AEOSSP). But when the scale of scenario increases, the gap between HMOCA and other methods become rather obvious. It is clear that in the 300-request MU-AEOSSP, HMOCA outputs more non-dominated solutions with lower values of both objectives, showing superior advantage than that in small-size scenarios. The level of the performance difference will be evaluated in the following.

To further demonstrate the performance of all the methods, the hypervolume (HV) [25] is adopted. The HV is the volume in the space enclosed by the non-dominated solutions and the reference point. In this study, the possible worst values of the objectives (1, 1) is applied as the reference point. Table 2 indicates the average HV values (of 30 runs) of all the methods. HMOCA succeeds in obtaining the highest HV values in all scenarios, implying the better overall performance than other comparison methods. As the scale of scenario increases, the HV value of HMOCA is obvious higher than other algorithms. Overall, HMOCA performs the best in all scenarios, followed by D-MOMA-TD and I-MOMA-TD, while NSGAII, IBEA, MOEA/D perform worst. When comparing MOMA-TDs with HMOCA, HMOCA is more superior in performance even all the three methods have been equipped with strategies to deal with the time-dependent characteristic of MU-AEOSSP.

Table 2. The average HVs of different algorithms.

Scenario	NSGA-II	IBEA	MOEA/D	I-MOMA-TD	D-MOMA-TD	HMOCA
MU_100	0.9259	0.9261	0.8250	0.9540	0.9532	**0.9612**
MU_150	0.7909	0.7875	0.7051	0.8505	0.8484	**0.9025**
MU_200	0.7274	0.7341	0.6588	0.7958	0.7965	**0.8592**
MU_250	0.6328	0.6268	0.5426	0.6779	0.6776	**0.7320**
MU_300	0.5833	0.5917	0.5148	0.6332	0.6339	**0.6896**

Statistics results summarizing the HV performance among all scenarios are illustrated by Table 3. From the results of the multi-problem Wilcoxon rank-sum test [20], there exist statistically significant differences between HMOCA and other methods, showing that the improvement of the non-dominated solutions is remarkable.

Table 3. Comparisons of the HMOCA against each comparative method with the $p-value$ of the Wilcoxon rank-sum test

$p-value$	NSGA-II	IBEA	MOEA/D	D-MOMA-TD	I-MOMA-TD
HMOCA	3.42E−04	4.60E-04	1.10E-11	0.00225	0.00400

In summary, this subsection indicates that the proposed HMOCA is of great overall performance when compared with several approaches which have already been applied before. The PFs and the HVs of different scenarios imply HMOCA is remarkable for solving the MU-AEOSSP.

6 Conclusion and Future Work

In this work, a multi-user agile earth observation satellite scheduling problem (MU-AEOSSP) is considered, where the failure rate of observation requests and the fairness between different users are targeted as the two objectives to be optimized simultaneously. To address this complicated constrained multi-objective optimization problem and to overcome the obstacle brought by its time-dependent characteristic, we proposed a hybrid multi-objective coevolutionary approach (HMOCA) equipped with several specifically designed operators. HMOCA solves an original problem and a helper problem to cooperate with each other, improving the convergence and the exploitation performance at the same time. HMOCA are implemented and compared with several classical MOEAs and two specific methods of the time-dependent multi-objective AEOSSP. Experimental results show that HOMCA outperforms other approaches in terms of overall performance.

For future work, we will analysis the characteristic of the MU-AEOSSP then enhance the problem with more real-world constraints. Besides, more mechanisms would be evaluated to further improve the efficiency of HMOCA.

References

1. Deb, K., Pratap, A., Agarwal, S., Meyarivan, T.: A fast and elitist multiobjective genetic algorithm: Nsga-ii. IEEE Trans. Evol. Comput. **6**(2), 182–197 (2002)
2. Dilkina, B., Havens, B.: Agile satellite scheduling via permutation search with constraint propagation. Actenum Corporation: Vancouver Canada, pp. 1–20 (2005)
3. Du, Y., Wang, T., Xin, B., Wang, L., Chen, Y., Xing, L.: A data-driven parallel scheduling approach for multiple agile earth observation satellites. IEEE Trans. Evol. Comput. **24**(4), 679–693 (2019)
4. Fan, Z., Li, W., Cai, X., Li, H., Wei, C., Zhang, Q., Deb, K., Goodman, E.: Push and pull search for solving constrained multi-objective optimization problems. Swarm Evol. Comput. **44**, 665–679 (2019)
5. He, L., Liu, X., Laporte, G., Chen, Y., Chen, Y.: An improved adaptive large neighborhood search algorithm for multiple agile satellites scheduling. Comput. Oper. Res. **100**, 12–25 (2018)
6. Li, L., Chen, H., Li, J., Jing, N., Emmerich, M.: Preference-based evolutionary many-objective optimization for agile satellite mission planning. IEEE Access **6**, 40963–40978 (2018)
7. Li, L., Yao, F., Jing, N., Emmerich, M.: Preference incorporation to solve multi-objective mission planning of agile earth observation satellites. In: 2017 IEEE Congress on Evolutionary Computation (CEC), pp. 1366–1373. IEEE (2017)
8. Liu, X., Laporte, G., Chen, Y., He, R.: An adaptive large neighborhood search metaheuristic for agile satellite scheduling with time-dependent transition time. Computers & Operations Research **86**, 41–53 (2017)
9. Ning, W., Guo, B., Yan, Y., Wu, X., Wu, J., Zhao, D.: Constrained multi-objective optimization using constrained non-dominated sorting combined with an improved hybrid multi-objective evolutionary algorithm. Eng. Optim. **49**(10), 1645–1664 (2017)
10. Peng, G., Dewil, R., Verbeeck, C., Gunawan, A., Xing, L., Vansteenwegen, P.: Agile earth observation satellite scheduling: An orienteering problem with time-dependent profits and travel times. Comput. Oper. Res. **111**, 84–98 (2019)
11. Ebrahim Sorkhabi, A., Deljavan Amiri, M., Khanteymoori, A.R.: Duality evolution: an efficient approach to constraint handling in multi-objective particle swarm optimization. Soft. Comput. **21**(24), 7251–7267 (2016). https://doi.org/10.1007/s00500-016-2422-5

12. Sun, K., Li, J., Chen, Y., He, R.: Multi-objective mission planning problem of agile earth observing satellites. In: Proceedings of the 12th International Conference on Space Operations, vol. 4, pp. 2802–2810. Citeseer (2012)
13. Tangpattanakul, P., Jozefowiez, N., Lopez, P.: Multi-objective optimization for selecting and scheduling observations by agile earth observing satellites. In: Coello, C.A.C., Cutello, V., Deb, K., Forrest, S., Nicosia, G., Pavone, M. (eds.) PPSN 2012. LNCS, vol. 7492, pp. 112–121. Springer, Heidelberg (2012). https://doi.org/10.1007/978-3-642-32964-7_12
14. Tian, Y., Cheng, R., Zhang, X., Jin, Y.: Platemo: a matlab platform for evolutionary multi-objective optimization [educational forum]. IEEE Comput. Intell. Mag. **12**(4), 73–87 (2017)
15. Tian, Y., Zhang, T., Xiao, J., Zhang, X., Jin, Y.: A coevolutionary framework for constrained multiobjective optimization problems. IEEE Trans. Evol. Comput. **25**(1), 102–116 (2020)
16. Wang, J., Ren, W., Zhang, Z., Huang, H., Zhou, Y.: A hybrid multiobjective memetic algorithm for multiobjective periodic vehicle routing problem with time windows. IEEE Trans. Syst. Man Cybern. Syst. **50**(11), 4732–4745 (2018)
17. Wang, J., Jing, N., Li, J., Chen, Z.H.: A multi-objective imaging scheduling approach for earth observing satellites. In: Proceedings of the 9th Annual Conference on Genetic and Evolutionary Computation, pp. 2211–2218 (2007)
18. Wei, L., Chen, Y., Chen, M., Chen, Y.: Deep reinforcement learning and parameter transfer based approach for the multi-objective agile earth observation satellite scheduling problem. Applied Soft Computing, p. 107607 (2021)
19. Wei, L., Xing, L., Wan, Q., Song, Y., Chen, Y.: A multi-objective memetic approach for time-dependent agile earth observation satellite scheduling problem. Comput. Ind. Eng. **159** (2021)
20. Wilcoxon, F.: Individual comparisons by ranking methods. In: Breakthroughs in Statistics, pp. 196–202. Springer (1992)
21. Woldesenbet, Y.G., Yen, G.G., Tessema, B.G.: Constraint handling in multiobjective evolutionary optimization. IEEE Trans. Evol. Comput. **13**(3), 514–525 (2009)
22. Wolfe, W.J., Sorensen, S.E.: Three scheduling algorithms applied to the earth observing systems domain. Manage. Sci. **46**(1), 148–166 (2000)
23. Zhang, Q., Li, H.: Moea/d: a multiobjective evolutionary algorithm based on decomposition. IEEE Trans. Evol. Comput. **11**(6), 712–731 (2007)
24. Zitzler, E., Künzli, S.: Indicator-based selection in multiobjective search. In: International Conference on Parallel Problem Solving from Nature, pp. 832–842. Springer (2004)
25. Zitzler, E., Thiele, L.: Multiobjective evolutionary algorithms: a comparative case study and the strength pareto approach. IEEE Trans. Evol. Comput. **3**(4), 257–271 (1999)

Surrogate-Assisted Artificial Bee Colony Algorithm

Tao Zeng[1], Hui Wang[1(✉)], Wenjun Wang[2], Tingyu Ye[1], and Luqi Zhang[1]

[1] School of Information Engineering, Nanchang Institute of Technology, Nanchang 330099, China
huiwang@whu.edu.cn

[2] School of Business Administration, Nanchang Institute of Technology, Nanchang 330099, China

Abstract. Search strategies play an essential role in the artificial bee colony (ABC) algorithm. Different optimization problems and search stages may need different search strategies. However, it is not easy to choose an appropriate search strategy efficiently. In order to select an appropriate search strategy with few evaluations, this paper proposes a surrogate-assisted ABC (called SAABC). Based on our previous work, we construct a strategy pool that contains three search strategies. Then, the radial basis function (RBF) network is applied to evaluate the offspring generated by the search strategies. The search strategy with the best evaluation value will be used to guide the population. A set of 22 classical benchmark problems with 30 dimensions are utilized to verify the performance of SAABC. Experimental results show that SAABC achieves better performance than five other ABC algorithms.

Keywords: Surrogate model · Artificial bee colony · Multiple search strategies · RBF network

1 Introduction

Swarm intelligence algorithms are inspired by biological behavior, and are broadly utilized to address assorted optimization problems. Swarm intelligence algorithms include particle swarm optimization [1,2], differential evolution [3,4], genetic algorithm [5,6], ant colony optimization [7,8], artificial bee colony [9,10] and so on.

ABC is simpler in structure and requires fewer control parameters than other swarm intelligence algorithms [11]. Therefore many scholars have been attracted to conduct research. Unfortunately, ABC also has some shortcomings, e.g., the convergence of ABC is slow. ABC is strong in exploration but weak in exploitation, which is the major contributor to this deficiency [12]. To solve this problem, some scholars have modified the search strategies. The best solution is applied to speed up convergence in [13]. Inspired by crossover operator

L. Pan et al. (Eds.): BIC-TA 2021, CCIS 1565, pp. 262–271, 2022.
https://doi.org/10.1007/978-981-19-1256-6_19

of GA [14], a novel search strategy is introduced to solve the "oscillation" phenomenon in GABC. Recently, multiple strategies have been applied to improved ABC. A strategy pool with several strategies is constructed in [15]. Each individual has distributed a search strategy randomly in the beginning. When the selected strategy fails to produce a better offspring, a new search strategy is randomly assigned to that solution. Kiran et al. [16] proposed five search strategies, and roulette was applied to choose the appropriate search strategies. The selection probability is calculated according to the number of successful updates. Ye et al. [17] adaptively select the appropriate search strategy from the strategy pool based on search status.

In swarm intelligence algorithms, surrogate models are used as a substitute for the time-consuming evaluation of exact function values. In general, it takes less time to build and use a surrogate model than it would for evaluating a computationally expensive problem. The surrogate model has been used in a variety of evolutionary algorithms [18–20]. The surrogate models often used are the support vector machine [21,22], Gaussian process [23,24], RBF network [25, 26], and artificial neural network [27,28].

The search strategy determines the search efficiency. Choosing an appropriate search strategy can be challenging. In order to save the computation cost and select a suitable search strategy, there is a surrogate-assisted artificial bee colony algorithm (called SAABC) proposed. To verify effectiveness of the strategy selection mechanism, five comparison algorithms are selected in this paper and compared with SAABC in 30-dimensional test functions.

The remaining sections are framed in the following way: the canonical ABC will be introduced briefly in the second section, detailed description of SAABC is given in section three, the experiments are presented in Sect. 4, and the conclusions will be given in the last section.

2 Artificial Bee Colony Algorithm(ABC)

ABC is proposed by Karaboga [29]. And it conducts the search by simulating behavior of bees. Furthermore, there are three kinds of bees in ABC. Each kind of bee performs its job to help the algorithm find the optimal solution. Based on different operation contents, ABC can be classified into the phases as follows:

(1) Initialization stage:
Assume that the $i-th$ solution in the population is $X_i = (x_{i,1}, x_{i,2}, \cdots, x_{i,D})$, where D is the dimension of the problem. An initial value is generated randomly within the search space.

(2) Employed bee stage:
The bees search around the current solution X_i, then an offspring V_i is obtained, the search strategy is as follows:

$$v_{i,j} = x_{i,j} + \phi_{i,j} \cdot (x_{i,j} - x_{k,j}) \tag{1}$$

where j is a randomly selected dimension, $j \in \{1, 2, \cdots, D\}$, $\phi_{i,j}$ is a random number within $[-1, 1]$, x_k is a solution randomly selected in current swarm,

$i \neq k$. If the offspring V_i is better than the parent individual X_i, then individual X_i is replaced by the offspring V_i, otherwise, the parent solution X_i remains unchanged.

(3) Onlooker bee stage:
The onlooker bees choose nectar sources to follow according to their quality(fitness). In ABC, the solution with better fitness is selected by a roulette selection. The selection probability can be described as follows:

$$p_i = \frac{fit(X_i)}{\sum_{i=1}^{SN} fit(X_i)} \tag{2}$$

where $fit(X_i)$ is the fitness value of $i-th$ individual. The fitness value is calculated below:

$$fit(X_i) = \begin{cases} \frac{1}{1+f(X_i)}, & \text{if } f(X_i) \geq 0 \\ 1 + |f(X_i)|, & \text{if } f(X_i) < 0 \end{cases} \tag{3}$$

(4) Scout bee stage:
Each individual X_i has a counter $trail_i$, when the individual X_i is not updated during the employed bee stage or onlooker bee stages, $trail_i$ will increase by 1, otherwise, $trail_i$ is set to 0. When $trail_i < limit$ (a pre-defined threshold), this means the solution may fall into a local optimum, then a new solution is initialized for X_i to help ABC skip out of the local optimum.

3 Proposed Approach

ABC is poor at exploitation because of the search strategy. It is helpful to apply multi-strategy to balance exploration and exploitation. The difficulty of adopting multi-strategy to improve ABC is to choose the appropriate search strategy accurately. Based on our previous work [15], we propose a surrogate-assisted artificial bee colony algorithm. The RBF model is trained by real function values. Then the surrogate model is utilized to select the appropriate search strategy for the current population.

3.1 Multi-strategy

In our previous work, the strategy pool contains three search strategies: ABC, GABC [13], and the modified MABC [30]. The strategy pool constructed in this paper replaces ABC with CABC [14], because the search strategy of CABC has better exploration capability. The search equations of CABC, GABC, and modified MABC are shown as follows:

$$v_{i,j} = x_{r1,j} + \phi_{i,j}(x_{r1,j} - x_{r2,j}) \tag{4}$$

$$v_{i,j} = x_{i,j} + \phi_{i,j}\left(x_{i,j} - x_{k,j}\right) + \psi_{i,j}\left(gbest_{\,j} - x_{i,j}\right) \tag{5}$$

$$v_{i,j} = x_{\text{best },j} + \phi_{i,j} \left(x_{\text{best },j} - x_{k,j} \right) \tag{6}$$

where x_{r1}, x_{r2} and x_k are three randomly chosen solutions, $r1, r2, k \in \{1, 2, \cdots, SN\}$, $r1 \neq r2 \neq i$, and $i \neq k$. $\phi_{i,j}$ and $\psi_{i,j}$ are random numbers within $[-1, 1]$ and $[0, C]$, respectively. C is a positive constant number. x_{best} is the best individual in current swarm.

Among the three strategies in the strategy pool, CABC has the slowest convergence speed (good exploration). In contrast the improved MABC shows the most rapid convergence speed (good exploitation), while GABC is between them.

3.2 Radial Basis Function Network

There are several kinds of surrogate models. In our study, radial basis function (RBF) network is applied.

The idea of RBF was first proposed by Hardy [31] to represent irregular surfaces. In 1985, Powell [32] proposed the RBF with multivariate interpolation. In 1988, Broomhead and Lowe [33] first applied RBF to the design of neural networks, and proposed a three-layer structured RBF network. Park and Sandberg [34] proved that the RBF network could approach any continuous function with any accuracy. In this paper, we use RBF network to assist algorithm in selecting the appropriate search strategy from the strategies pool. RBF network can be calculated as below:

$$\varphi(\boldsymbol{x}) = \sum_{i=1}^{CN} w_i \rho \left(\boldsymbol{x}, \boldsymbol{c}_i \right) \tag{7}$$

where CN is the number of hidden neurons, the center and weight of each hidden layer neuron are c_i and w_i, respectively, and x is the input vector that contains d dimensions. $\rho \left(\boldsymbol{x}, \boldsymbol{c}_i \right)$ is radial basis function. Gaussian, Multiquadric, and Reflected sigmoidal are a few widely utilized radial basis neural networks. The Gaussian radial basis function is given as:

$$\rho(\boldsymbol{x}, \boldsymbol{c}_i) = e^{\left(\frac{-||x - c_i||^2}{\sigma_i^2} \right)} \tag{8}$$

where σ_i is the width of the hidden layer neuron, and $\sigma_i > 0$

3.3 Surrogate-Assisted Strategy Selection Mechanism

In our previous work [15], a strategy is randomly assigned to each solution. If the currently selected strategy cannot produce better offspring for the solution, a new strategy is randomly selected for the solution. A randomly chosen strategy may not be suitable for the current search, but if we need to choose the optimal strategy precisely, this will waste evaluations. In order to balance these two contradictions, this paper uses RBF networks to approximate the function values around the current swarm, then uses the surrogate model to evaluate each strategy.

The training samples are stored in TS. We limit the size of TS to keep the complexity of building the RBF network from being too large. According to experience, the maximum capacity ($MaxTSSize$) of TS is $15 * D$. Then, we need to consider these two questions: firstly, what data should be included in the training set; secondly, what data should be deleted when the training set is full.

The purpose of using RBF network is to estimate the function values around current swarm as accurately as possible, so the data used to train the model needs to be as close to the current population as possible. The new solution is added to TS whenever the solution is updated successfully or reinitialized, because the new solution contains the information of the new swarm.

When TS has reached the storage limit, and a new sample needs to be added, the individual in the TS farthest from current swarm is selected for deletion, because we only need to model around the current population. The distance of an individual in TS from the current population is calculated as:

$$dist_i = \min \{d(TS_i, X_j) \mid j \in \{1, 2, \cdots, SN\}\} \tag{9}$$

where, $dist_i$ denotes the distance between $i - th$ solution in TS and current population, $d(TS_i, X_j)$ denotes the Euclidean distance between $i-th$ individual in TS and $j-th$ solution within the population, $i = \{1, 2, \cdots, TSSize\}$, $TSSize$ is the number of training samples within TS. The index of the element in TS that needs to be deleted is:

$$del\_index = \arg\max_i dist_i \tag{10}$$

According to the above description, update rules of training set TS can be described as:

$$TS = \begin{cases} TS \cup \{X_i\}, \text{if TSSize} < \text{MaxTSSize} \\ (TS - \{TS_{\text{del\_index}}\}) \cup \{X_i\}, \text{otherwise} \end{cases} \tag{11}$$

After the initialization and scout bee stage, RBF network is trained with the samples in TS. In the employed bee stage and onlooker bee stage, the search strategies are used to conduct neighborhood search for individual X_i, then the offsprings V_{1i}, V_{2i}, V_{3i} are obtained. The three offsprings are estimated by the RBF network, supposed that the best offspring selected by the RBF network is V_{best}, then the greedy selection is used to judge if the solution X_i needs to be updated by V_{best}.

4 Experimental Study

4.1 Test Functions and Parameter Settings

To verify the effectiveness of SAABC, we choose 22 classical test problems as the benchmark set. The dimensions of the test function are 30. The specific description of the test problem can be found in [35]. Where f_1-f_9 are unimodal problems and $f_{10}-f_{22}$ are unimodal problems. For the test problem f_{10}, it is a unimodal problem when the dimension is less than three and a unimodal problem when the dimension is greater than three.

For all test problems, the population size $SN = 40$, the $limit = 100$, and the maximum number of evaluations $MaxFEs = 10000$. In SAABC, the maximum capacity $MaxTSSize$ of the training set TS is $15 * D$, and the σ in Eq. 8 will adaptively changes according to the data within the TS, the detailed description can be found in [36].

4.2 Results on 30-Dimensional Problems

The algorithms involved in the comparison are ABC, CABC, GABC, MABC, and MEABC. The first three comparison algorithms are single-strategy algorithms, and there are two search strategies in MABC and three search strategies in MEABC. Table 1 presents test results from six ABC variants on 30-dimensional functions. The final line of Table 1 shows comparison results between each comparison algorithm and SAABC. Where w indicates SAABC outperforms the comparison algorithm on w problems, t represents SAABC gets the same results as the comparison algorithm for t test functions, and l implies SAABC are worse than the comparison algorithm on l problems.

As is shown in Table 1, for unimodal problems f_1-f_9, SAABC gets the best results than the comparison algorithm except for f_5. ABC achieved better results only on f_{10}, and was inferior to SAABC for the remaining 21 test problems. SAABC outperformed CABC on 14 functions, same results are obtained on f_8 and f_{20}, SAABC was slightly worse than CABC in the remaining 6 problems. Compared to GABC, SAABC obtained worse results only on f_5 and f_{10}, and better results were obtained for the remaining 20 test functions. MABC obtained the same results as SAABC on f_8 and f_{20}, but the results were worse for all the remaining problems. SAABC superior MEABC for all 22 test problems.

Table 2 lists the results from Friedman test, it is easy to find that SAABC has the best mean ranking against the five comparison algorithms. The results of Wilcoxon test are listed in Table 3, the table shows that SAABC is significantly better than comparison variants except for CABC. The convergence curve of each variant is displayed in Fig. 1. This figure illustrates that among the ABC variants, SAABC has the fastest convergence speed.

Table 1. Results on 30-dimensional problems.

Problem	ABC	CABC	GABC	MABC	MEABC	SAABC
f_1	7.96E+00	6.34E−01	1.18E−01	8.52E+01	5.83E+01	**1.79E−05**
f_2	1.74E+04	7.98E+04	8.10E+02	6.49E+05	4.50E+05	**3.17E+01**
f_3	1.95E+00	1.19E−01	3.20E−02	1.07E+01	8.73E+00	**1.11E−04**
f_4	4.75E−05	6.01E−08	1.92E−07	2.52E−07	5.49E−08	**3.31E−10**
f_5	3.74E−01	1.27E+00	**1.64E−01**	3.11E+00	2.40E+00	1.95E−01
f_6	6.61E+01	5.37E+01	5.91E+01	7.96E+01	6.26E+01	**4.52E+01**
f_7	7.10E+00	1.57E+00	9.00E−01	9.40E+01	7.39E+01	**0.00E+00**
f_8	4.11E−58	**7.18E−66**	7.22E−66	**7.18E−66**	7.39E−66	**7.18E−66**
f_9	6.17E−01	2.80E−01	3.68E−01	5.24E−01	4.82E−01	**2.09E−01**
f_{10}	9.62E+01	1.28E+02	**8.20E+01**	8.80E+02	9.57E+02	1.27E+02
f_{11}	6.60E+01	**1.55E+01**	4.43E+01	5.06E+01	4.44E+01	1.63E+01
f_{12}	4.25E+01	**1.56E+01**	2.74E+01	3.19E+01	2.71E+01	1.72E+01
f_{13}	9.24E−01	1.02E+00	9.61E−01	1.78E+00	1.60E+00	**1.45E−01**
f_{14}	3.43E+03	**1.01E+03**	2.86E+03	1.89E+03	2.30E+03	1.68E+03
f_{15}	1.85E+01	1.52E+01	1.52E+01	2.00E+01	1.63E+01	**9.15E+00**
f_{16}	1.93E−02	6.87E−02	3.07E−03	3.82E+00	2.15E+00	**2.81E−03**
f_{17}	9.29E−02	3.02E−01	1.31E−02	3.91E+02	1.29E+02	**4.42E−03**
f_{18}	3.46E+00	**1.12E−01**	1.70E+00	3.00E+00	1.67E+00	2.04E−E−01
f_{19}	1.49E+00	1.32E−01	1.65E−01	3.82E+00	3.11E+00	**6.40E−02**
f_{20}	3.62E+00	**0.00E+00**	3.78E−02	**0.00E+00**	1.44E−01	**0.00E+00**
f_{21}	8.34E+00	**1.55E−02**	2.76E+00	5.91E−01	1.01E+00	2.95E−02
f_{22}	8.94E+00	**5.63E+00**	8.00E+00	9.21E+00	7.74E+00	5.74E+00
w/t/l	21/0/1	14/2/6	20/0/2	20/2/0	22/0/0	–

Table 2. The results of Friedman test on 30-dimensional problems.

Algorithms	Mean ranking
ABC	4.59
CABC	2.48
GABC	2.98
MABC	5.09
MEABC	4.36
SAABC	**1.5**

Table 3. The results of Wilcoxon test on 30-dimensional problems.

SAABC vs.	p-value
ABC	**4.83E−04**
CABC	1.00E−01
GABC	**9.83E−04**
MABC	**8.90E−05**
MEABC	**4.00E−05**

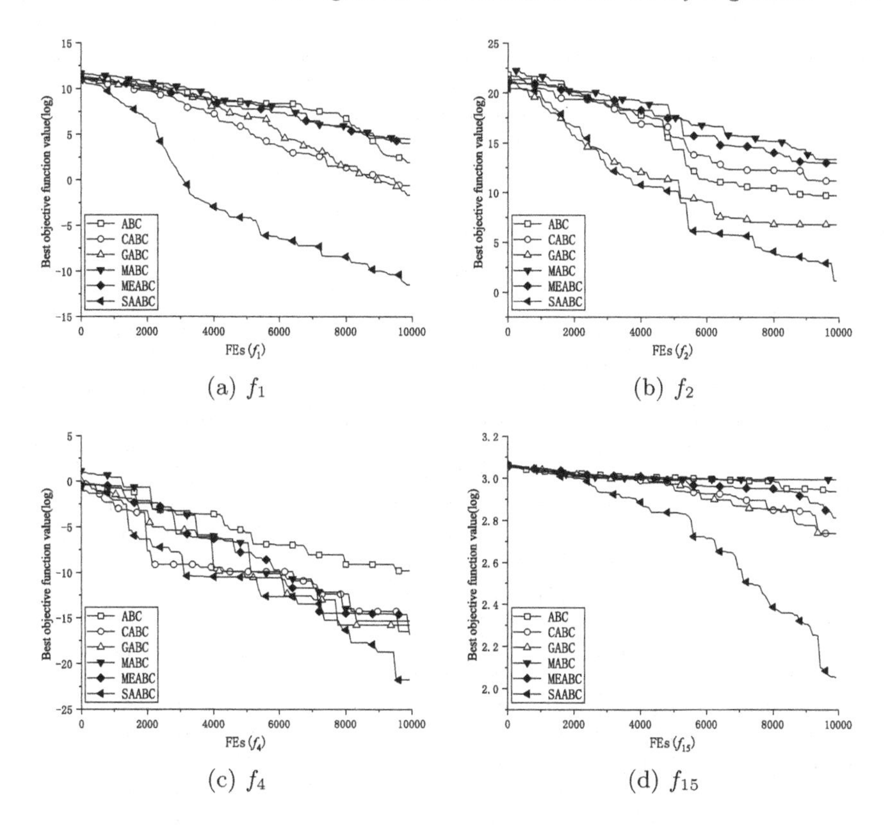

(a) f_1 (b) f_2

(c) f_4 (d) f_{15}

Fig. 1. Convergence curves of different ABC variants.

5 Conclusion

A surrogate-assisted artificial bee colony algorithm is proposed to select the appropriate search strategy as accurately as possible, and reduce evaluation times. Three search strategies with different characteristics are selected for the algorithm, and RBF network is applied to build surrogate model. Then the surrogate model is utilized to evaluate offsprings generated by three search strategies, so that each individual can have more chances to get updated.

To test effectiveness of SAABC, we chose 22 classical benchmark sets as test functions, selected five different ABC algorithms to participate in the comparison, and then tested the performance of these six algorithms in 30-dimensional test functions. Due to the limitation of article length, the test results of 100-dimensional test functions are not listed in this paper. The results demonstrate that SAABC can get better results than the comparison algorithm on most test problems in both 30 and 100 dimensions.

In this paper, the effectiveness of SAABC is verified in 22 classical test functions in 30 and 100 dimensions, and we will continue to improve SAABC and use it to solve some expensive computational problems in coming years.

Acknowledgment. This work was supported by the National Natural Science Foundation of China (No. 62166027), and Jiangxi Provincial Natural Science Foundation (Nos. 20212ACB212004 and 20212BAB202023).

References

1. Kennedy, J., Eberhart, R.: Particle swarm optimization. In: Proceedings of ICNN'95-International Conference on Neural Networks, vol. 4, pp. 1942–1948. IEEE (1995)
2. Tian, D., Shi, Z.: Mpso: modified particle swarm optimization and its applications. Swarm Evol. Comput. **41**, 49–68 (2018)
3. Price, K., Storn, R.M., Lampinen, J.A.: Differential evolution: a practical approach to global optimization. Springer Science & Business Media (2006)
4. Wu, G., Shen, X., Li, H., Chen, H., Lin, A., Suganthan, P.N.: Ensemble of differential evolution variants. Inf. Sci. **423**, 172–186 (2018)
5. Whitley, D.: A genetic algorithm tutorial. Stat. Comput. **4**(2), 65–85 (1994)
6. Metawa, N., Hassan, M.K., Elhoseny, M.: Genetic algorithm based model for optimizing bank lending decisions. Expert Syst. Appl. **80**, 75–82 (2017)
7. Dorigo, M., Birattari, M., Stutzle, T.: Ant colony optimization. IEEE Comput. Intell. Mag. **1**(4), 28–39 (2006)
8. Dorigo, M., Stützle, T.: Ant colony optimization: overview and recent advances. Handbook of metaheuristics, pp. 311–351 (2019)
9. Wang, H., Wang, W., Xiao, S., Cui, Z., Li, W., Zhu, H., Zhu, S.: Multi-strategy and dimension perturbation ensemble of artificial bee colony. In: 2019 IEEE Congress on Evolutionary Computation (CEC), pp. 697–704. IEEE (2019)
10. Zeng, T., Ye, T., Zhang, L., Xu, M., Wang, H., Hu, M.: Population diversity guided dimension perturbation for artificial bee colony algorithm. In: Zhang, H., Yang, Z., Zhang, Z., Wu, Z., Hao, T. (eds.) NCAA 2021. CCIS, vol. 1449, pp. 473–485. Springer, Singapore (2021). https://doi.org/10.1007/978-981-16-5188-5_34
11. Karaboga, D., Basturk, B.: A powerful and efficient algorithm for numerical function optimization: artificial bee colony (abc) algorithm. J. Global Optim. **39**(3), 459–471 (2007)
12. Cui, L., Li, G., Lin, Q., Du, Z., Gao, W., Chen, J., Lu, N.: A novel artificial bee colony algorithm with depth-first search framework and elite-guided search equation. Inf. Sci. **367**, 1012–1044 (2016)
13. Zhu, G., Kwong, S.: Gbest-guided artificial bee colony algorithm for numerical function optimization. Appl. Math. Comput. **217**(7), 3166–3173 (2010)
14. Gao, W.f., Liu, S.y., Huang, L.l.: A novel artificial bee colony algorithm based on modified search equation and orthogonal learning. IEEE Trans. Cybern. **43**(3), 1011–1024 (2013)
15. Wang, H., Wu, Z., Rahnamayan, S., Sun, H., Liu, Y., Pan, J.S.: Multi-strategy ensemble artificial bee colony algorithm. Inf. Sci. **279**, 587–603 (2014)
16. Kiran, M.S., Hakli, H., Gunduz, M., Uguz, H.: Artificial bee colony algorithm with variable search strategy for continuous optimization. Inf. Sci. **300**, 140–157 (2015)
17. Ye, T., Zeng, T., Zhang, L., Xu, M., Wang, H., Hu, M.: Artificial bee colony algorithm with an adaptive search manner. In: Zhang, H., Yang, Z., Zhang, Z., Wu, Z., Hao, T. (eds.) NCAA 2021. CCIS, vol. 1449, pp. 486–497. Springer, Singapore (2021). https://doi.org/10.1007/978-981-16-5188-5_35
18. Regis, R.G.: Particle swarm with radial basis function surrogates for expensive black-box optimization. J. Comput. Sci. **5**(1), 12–23 (2014)

19. Mallipeddi, R., Lee, M.: Surrogate model assisted ensemble differential evolution algorithm. In: 2012 IEEE Congress on Evolutionary Computation, pp. 1–8. IEEE (2012)
20. Sun, X.Y., Gong, D.W., Ma, X.P.: Directed fuzzy graph-based surrogate model-assisted interactive genetic algorithms with uncertain individual's fitness. In: 2009 IEEE Congress on Evolutionary Computation, pp. 2395–2402. IEEE (2009)
21. Loshchilov, I., Schoenauer, M., Sebag, M.: A mono surrogate for multiobjective optimization. In: Proceedings of the 12th Annual Conference on Genetic and Evolutionary Computation, pp. 471–478 (2010)
22. Herrera, M., Guglielmetti, A., Xiao, M., Filomeno Coelho, R.: Metamodel-assisted optimization based on multiple kernel regression for mixed variables. Struct. Multidiscip. Optim. **49**(6), 979–991 (2014). https://doi.org/10.1007/s00158-013-1029-z
23. Zhang, Q., Liu, W., Tsang, E., Virginas, B.: Expensive multiobjective optimization by moea/d with gaussian process model. IEEE Trans. Evol. Comput. **14**(3), 456–474 (2009)
24. Buche, D., Schraudolph, N.N., Koumoutsakos, P.: Accelerating evolutionary algorithms with gaussian process fitness function models. IEEE Trans. Syst. Man Cybern. Part C (Applications and Reviews) **35**(2), 183–194 (2005)
25. Zapotecas Martínez, S., Coello Coello, C.A.: Moea/d assisted by rbf networks for expensive multi-objective optimization problems. In: Proceedings of the 15th Annual Conference on Genetic and Evolutionary Computation, pp. 1405–1412 (2013)
26. Sun, C., Jin, Y., Zeng, J., Yu, Y.: A two-layer surrogate-assisted particle swarm optimization algorithm. Soft. Comput. **19**(6), 1461–1475 (2014). https://doi.org/10.1007/s00500-014-1283-z
27. Gaspar-Cunha, A., Vieira, A., et al.: A hybrid multi-objective evolutionary algorithm using an inverse neural network. In: Hybrid Metaheuristics, Citeseer, pp. 25–30 (2004)
28. Gaspar-Cunha, A., Vieira, A.: A multi-objective evolutionary algorithm using neural networks to approximate fitness evaluations. Int. J. Comput. Syst. Signals **6**(1), 18–36 (2005)
29. Karaboga, D.: An idea based on honey bee swarm for numerical optimization. Technical report (2005)
30. Gao, W.F., Liu, S.Y.: A modified artificial bee colony algorithm. Comput. Oper. Res. **39**(3), 687–697 (2012)
31. Hardy, R.L.: Multiquadric equations of topography and other irregular surfaces. J. Geophys. Res. **76**(8), 1905–1915 (1971)
32. Powell, M.J.D.: Radial Basis Functions for Multivariable Interpolation: A Review, pp. 143–167. Clarendon Press, USA (1987)
33. Broomhead, D.S., Lowe, D.: Multivariable functional interpolation and adaptive networks. Complex Syst. **2**(3), 321–355 (1988)
34. Park, J., Sandberg, I.W.: Universal approximation using radial-basis-function networks. Neural Comput. **3**(2), 246–257 (1991)
35. Cui, L., Li, G., Luo, Y., Chen, F., Ming, Z., Lu, N., Lu, J.: An enhanced artificial bee colony algorithm with dual-population framework. Swarm Evol. Comput. **43**, 184–206 (2018)
36. Sun, C., Jin, Y., Cheng, R., Ding, J., Zeng, J.: Surrogate-assisted cooperative swarm optimization of high-dimensional expensive problems. IEEE Trans. Evol. Comput. **21**(4), 644–660 (2017)

An Improved Bare-Bones Multi-objective Artificial Bee Colony Algorithm

Tingyu Ye[1], Hui Wang[1(✉)], Wenjun Wang[2], Tao Zeng[1], and Luqi Zhang[1]

[1] School of Information Engineering, Nanchang Institute of Technology, Nanchang 330099, China
huiwang@whu.edu.cn

[2] School of Business Administration, Nanchang Institute of Technology, Nanchang 330099, China

Abstract. Artificial bee colony (ABC) algorithm shows good performance on many optimization problems. However, most ABC variants focus on single objective optimization problems. In this paper, an improved bare-bones multi-objective artificial bee colony (called BMOABC) algorithm is proposed to solve multi-objective optimization problems (MOPs). Fast non-dominated sorting is used to select non-dominated solutions. The crowded-comparison operator is employed to maintain population diversity. To enhance the search ability, an improved bare-bones strategy is utilized. The fitness function is modified to handle multiple objective values. Then, a novel probability selection model is designed for the onlooker bees. To verify the effectiveness of BMOABC, five benchmark MOPs are employed in the experiment. Experimental results show that BMOABC is superior to three other multi-objective algorithms.

Keywords: Artificial bee colony · Multi-objective optimization · Multi-objective artificial bee colony · Selection probability

1 Introduction

Engineering optimization problems involve the multi-objective optimization problems (MOPs). For MOPs, it is hard to gain the optimal goal in the meantime. Therefore, how to solve effectively the MOPs has been a popular issue. Meanwhile, researchers are attracted by the genetic algorithm grew in the light of the theory of biological evolution. By integrating these two modes, it is helpful to avoid trapping the traditional MOP methods into the local optimal solution in the process of searching. Consequently, the genetic algorithm based multi-objective optimization method is used in distinct realms.

Artificial Bee Colony algorithm (ABC) [1] is a novel algorithm on the basic of swarm intelligence. Bees carry out distinct works in the light of their division of labor, and realize the sharing and exchange of bee colony information. It has received extensive attention, and many scholars have improved it [2–8]. For multi-objective problem, some scholars combine the standard ABC algorithm and a

L. Pan et al. (Eds.): BIC-TA 2021, CCIS 1565, pp. 272–280, 2022.
https://doi.org/10.1007/978-981-19-1256-6_20

multi-objective optimization algorithm: NSGA-II [9] to put forward different new multi-objective artificial Bee Colony algorithms [10–13].

This paper proposes an improved bare-bones multi-objective ABC algorithm called (BMOABC). In order to select a better solution set, the fast non-dominant sorting and crowded-comparison operator are used. And an improved bare-bones method were designed to strengthen the search capability of the algorithm. Moreover, the mode of calculating the selection probability is not applicable in the MOPs, a novel calculation method is proposed. In order to check the feasibility of BMOABC, five benchmark functions are tested and three different algorithms are compared. The final results demonstrate that BMOABC is superior to the compared algorithms.

The other parts of the paper is as below. The multi-objective optimization problem is depicted in Sect. 2, Sect. 3 introduces the standard artificial bee colony algorithm. Our approach BMOABC is described in detail in Sect. 4. The experimental part is in Sect. 5 and Sect. 6 is conclusion.

2 Multi-objective Optimization Problems

In the single objective optimization problem, it has only one objective. According to this objective, all solutions can be compared and an optimal solution can be gained. However, in the multi-objective optimization problem, it has changed from one objective to multiple objectives. The detailed description is as below.

$$\begin{array}{l} \min F(x) = [f_1(x), f_1(x), \ldots, f_u(x)] \\ \text{S.t. } x = (x_1, x_2, \ldots, x_d) \in S \end{array} \tag{1}$$

where u is number of objectives. S is the whole search range. x is the solution of search range. When one of the objectives is optimized, it will lead to the deterioration of the other objectives. So there is no best optimal solution for the MOPs. The solution of MOPs is compared by Pareto dominance. The specific description of Pareto dominance is as follows.

$$\{\forall i \in K : f_i(a) \leq f_i(b)\} \wedge \{\exists j \in K : f_j(a) < f_j(b)\} \tag{2}$$

where K is {1,2, ..., n}. a and b is decision variable. Set two decision variables a and b, a is said to dominate b ($v \prec w$) if and only if Eq. (2) establish. If there is no decision vector in the whole parameter space and Pareto dominates a decision vector, the decision vector is called Pareto Optimal Solution. Generally, there is more than one Pareto Optimal Solution of multi-objective optimization problem. The objective functions corresponding to all Pareto Optimal Solutions constitute the Pareto- Front (PF). The specific description of PF is as follows.

$$\text{PF} = [f_1(x), f_2(x), \ldots, f_u(x)] \tag{3}$$

where $f_u(x)$ is Pareto Optimal Solution.

3 Artificial Bee Colony Algorithm

As a novel swarm intelligence optimization algorithm, artificial bee colony (ABC) is extensively applied in many distinct problems. The types of bee include employed bee, onlooker bee and scout bee. And distinct bees assume diverse assignments to ensure the orderly implementation. The main step is introduced as below.

(1) The population is created in the given space.

$$x_{i,j} = Min_j + rand \cdot (Max_j - Min_j) \tag{4}$$

where Min_j and Max_j are the frontier of the search range, and $rand \in [0, 1]$ is a random number.

(2) Each employed bee is assigned to a distinct honey (food) source. They hunt for undeveloped honey source around the present honey source, and whether there is a superior one. The search formula is implemented as below.

$$v_{i,j} = x_{i,j} + \phi_{i,j} \cdot (x_{i,j} - x_{k,j}) \tag{5}$$

where X_i is the assigned honey source, X_k is selected in the population $(i \neq k)$ randomly, and $\phi_{i,j}$ is a random number between -1 and 1.

(3) Each onlooker bee is also assigned to a honey source. But they select the honey source on the basic of the probability p_i, which is defined as below.

$$p_i = \frac{fit(X_i)}{\sum_{i=1}^{SN} fit(X_i)} \tag{6}$$

where $fit(X_i)$ is the fitness value of the assigned honey source. Fitness value is utilized to evaluate the quality of a honey source. The greater the fitness value of honey source is, the better it is. The fitness value is calculated as below.

$$fit(X_i) = \begin{cases} \frac{1}{1+f(X_i)}, & \text{if } f(X_i) \geq 0 \\ 1 + |f(X_i)|, & \text{if } f(X_i) < 0 \end{cases} \tag{7}$$

where $f(X_i)$ is the objective function value. As a result of this selection method, a superior solution (honey source) has a greater probability to be chosen by the onlooker bees. For some better solutions, they are selected several times. When a solution is selected, it executes the same search by Eq. (5).

(4) For the scout bees, they are used to renewed the food sources that have been repeat searched multiple times but have not been renewed, it will be initialized by Eq. (4).

4 An Improved Bare-Bones Multi-objective ABC

For the single-objective problem, we only need to seek out a minimum value. In the multi-objective problem, global optimal value is unable to be found, rather a solution set that are closest to the real Pareto-Front (PF). In this section, an improved bare-bones multi-objective artificial bee colony algorithm is put forward. First, fast non-dominated sorting and crowded-comparison operator are referenced to effectively seek out a good PF, which are described in [9]. And based on the above the fast non-dominant sorting operation, a set of PF is obtained. Based on this PF, the original ABC can be modified. Detailed description is as follows

In the initialization phase of BMOABC, a given number of honey sources is created in a given search range by Eq. (4). Each bee is assigned a honey source to search. Then, objective function solution set is obtained and a set of Pareto Optimal Solutions is gained through the fast non-dominated sorting method. The obtained non-dominated solutions is stored in an external archive (called R).

In the employed bee phase, the search ability of bees determines the convergence speed. To boost the search capability of bees, the method of a bare-bones [14] is cited to improve the search equation as below.

$$v_{i,j} = \begin{cases} N\left(\frac{x_{p,j}-x_{i,j}}{2}, |x_{p,j} - x_{i,j}|\right), & \text{if rand } < 0.5 \\ x_{i,j}, & \text{otherwise} \end{cases} \tag{8}$$

Where $N()$ represents gaussian distribution and $rand \in [0, 1]$ is a random number. $x_{p,j}$ is a solution randomly chose from the external archive, $x_{i,j}$ is the current solution. By using the information of non dominated solutions, it is easier to find better non-dominated solutions.

In the onlooker bee stage, the way of calculating the selection probability is used by Eq. (6) in the original ABC. However, the objective function is not a single solution in the MOPs. This leads to the fitness value of honey source can not be obtained. The previous method is no longer applicable. Therefore, a new method of calculating the selection probability is proposed as follows.

$$fit\left(X_i\right) = \sum_{k=1}^{m} F_k\left(X_i\right) \tag{9}$$

$$p_i = 1 - \frac{fit\left(X_i\right)}{\text{Maxfit}\left(X_i\right)} \tag{10}$$

where $fit(X_i)$ is the fitness value of the current solution, $F_k(X_i)$ is the value corresponding to each objective function, p_i is the probability that each solution is chose and $Maxfit(X_i)$ is the maximum fitness value corresponding to all solutions. Unlike the original ABC, the smaller the fitness value, the greater the probability that the solution will be selected. After obtaining the probability of being chose for the solution, the roulette mode is used to choose the good individual for the search. The search equation is the same with the employed bee.

In the scout bee stage, *limit* is set. If a solution has not been updated more than *limit*, it will be discarded. The bee searching for the honey source will

Algorithm 1: BMOABC

```
1  Produce a population randomly, which the number of individuals is SN;
2  Compute apiece f(X_i), set trial_i = 0;
3  record the non-dominated solutions in R;
4  while Iter do
5      for apiece individual X_i in the colony do
6          Obtain V_i by Eq (8);
7          Calculate f(V_i) and fast non-dominated sorting;
8          record the non-dominated solutions in R
9          select new population by crowded-comparison operator;
10     end
11     for apiece individual X_i in the colony do
12         Obtain V_i by Eq. (8);
13         Calculate f(V_i) and fast non-dominated sorting;
14         record the solution with rank = 1 in R;
15         select new population by crowded-comparison operator;
16     end
17     if max{trial_i} > limit then
18         Obtain X_i by Eq. (4);
19         Set trial_i = 0;
20     end
21 end
```

become a scout bee. The discarded solutions will be regenerated by Eq. (4). It is equal with the origin ABC.

Algorithm 1 gives the key steps of BMOABC. *Iter* is number of iterations. *SN* is the number of population. *trial* is a monitor to judge whether the solution exceeds *limit*. At the beginning of the BMOABC, the objective function value of the solution is computed. Through fast non-dominated sorting, the solution with $rank = 1$ is recorded in external archiving (R). Then in the employed bee stage, a bare-bones strategy is improved to advance the convergence ability of the algorithm. *SN* new population is obtained by fast non-dominated sorting and crowded-comparison operator. In the onlooker bee stage, a new way of calculate fitness value and selection probability is designed. The scout bee is the same with the original ABC.

5 Experimental Study

To prove the expression of our algorithm BMOABC, six benchmarks for testing multi-objective problems are used, including ZDT1-ZDT4 and ZDT6. Their specific descriptions can be found in [15]. Based on these benchmarks, four different multi-objective algorithms are used to compare. In order to show the difference of comparison algorithm, two performance metrics are used.

The experimental machine is Intel Core i7-10875 2.30 GHz CPU 16.00 GB memory, and windows 10 operating system with python 3.8. Apiece algorithm is run 15 times independently in all problems. Population number (SN) and maximum iterations (MaxIter) are set to 50 and 150, respectively.

5.1 Performance Metrics

There are two main evaluation criteria: diversity and convergence. They are used to assess the performance of multi-objective optimization algorithm. Because a single performance index can not reflect the two evaluation criteria at the same time, two performance metrics to test the performance of multi-objective optimization algorithm is used in this section, including GD and SP. GD [16] is used to evaluate the degree to which the final solution set approaches the real frontier, SP [16] is applied to appraise the diversity of population distribution. The overall effect of the algorithm can be plainly reflected by testing the two metrics respectively.

Table 1. Comparison results for GD.

Problem		NSGA-II	MOABC	EMOABC	BMOABC
ZDT1	Min	8.71E−04	1.14E−03	1.83E−04	**1.51E−04**
	Mean	1.52E−03	8.39E−02	6.04E−04	**3.23E−04**
ZDT2	Min	2.18E−02	1.12E−01	8.83E-03	**2.53E−03**
	Mean	4.41E−02	7.32E−01	2.37E−02	**3.28E−03**
ZDT3	Min	6.01E−04	2.27E−02	2.06E-02	**2.39E−04**
	Mean	7.43E−04	4.56E−02	3.19E−02	**5.18E−04**
ZDT4	Min	8.19E−03	3.21E−01	3.21E-03	**2.27E−04**
	Mean	1.25E−02	6.52E−01	1.43E−03	**5.64E−04**
ZDT6	Min	4.86E−03	1.25E−03	8.93E−04	**1.54E−04**
	Mean	1.12E−02	5.28E−03	1.35E−03	**3.36E−04**

Table 2. Comparison results for SP.

Problem		NSGA-II	MOABC	EMOABC	BMOABC
ZDT1	Min	5.81E−03	9.17E−03	1.75E−03	**8.64E−04**
	Mean	6.05E−03	9.58E−03	4.56E−03	**9.75E−04**
ZDT2	Min	2.62E−02	1.24E−01	1.25E−02	**6.72E−03**
	Mean	5.45E−02	3.15E−01	3.35E−02	**9.18E−03**
ZDT3	Min	6.04E−03	5.98E−04	3.85E−03	**1.59E−04**
	Mean	7.23E−03	9.40E−04	6.02E−03	**4.18E−04**
ZDT4	Min	6.29E−03	1.30E−02	1.03E−02	**2.04E−03**
	Mean	1.15E−02	2.68E−02	1.25E−02	**3.91E−03**
ZDT6	Min	6.41E−03	6.32E−03	2.31E−03	**9.24E−04**
	Mean	1.45E−02	8.84E−03	5.78E−03	**1.31E−03**

5.2 Comparison of BMOABC with Other Algorithms

In order to analyze the performance of BMOABC, three different algorithms are compared, including NSGA-II, MOABC [10] and EMOABC [10]. Table 1

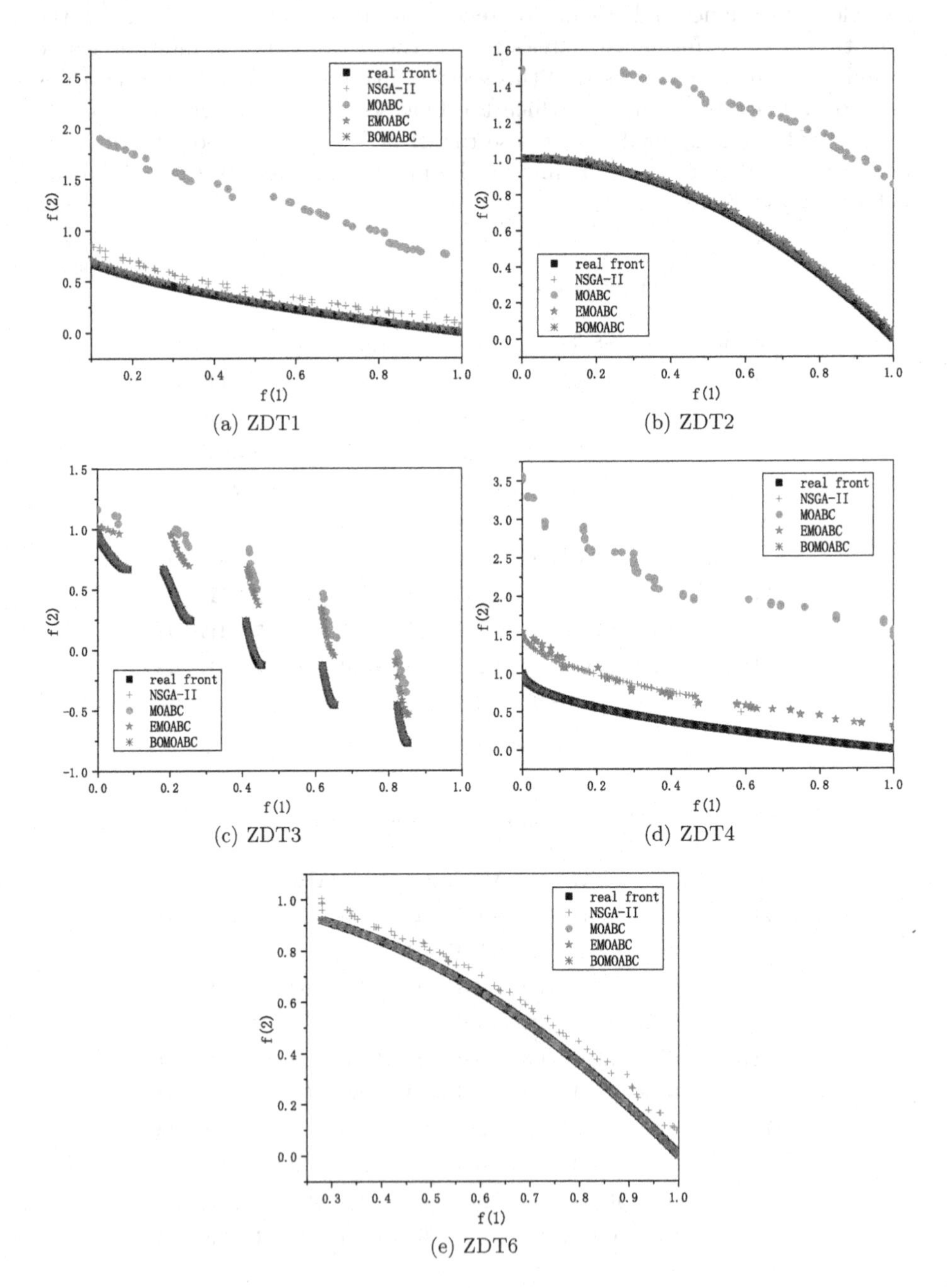

Fig. 1. Convergence curves of four multi-objective algorithms on ZDT1-ZDT4 and ZDT6.

and Table 2 shows the comparison results of NSGA-II, MOABC, EMOABC and BMOABC, where min (GD) and min (SP) is the minimum value obtained in the experiment, and mean (GD) and mean (SP) indicates the average value in multiple experiments. The best results are displayed in bold. In Table 1, obviously, it can be seen that the convergence ability of BMOABC is the best of the four algorithms in all problems. In Table 2, the distribution of BMOABC is also the best of the four algorithms in all problems.

In order to more intuitively show the differences of the four algorithms, Fig. 1 shows the real PF and PF of NSGA-II, MOABC, EMOABC and BMOABC in ZDT1-ZDT4 and ZDT6. NSGA-II can converge to the real PF in ZDT2 and ZDT3. MOABC can not converge to the real PF in any problem expect ZDT6. EMOABC performs well in ZDT1 and ZDT6, but it performs worse in other problems. BMOABC can converge to the real PF in all test problems. It proves that the performance of BMOABC is best in the three compared algorithms.

6 Conclusion

Though ABC has presented excellent effect on single objective optimization problems, it is rarely applied to multi-objective optimization problems (MOPs). This paper proposes an improved bare-bones multi-objective ABC algorithm (called BMOABC) to deal with MOPs. In BMOABC, four main modifications are made: 1) the fast non-dominated sorting is used to produce Pareto fonts; 2) the crowded-comparison operator is employed to maintain population diversity; 3) an improved bare-bones strategy is utilized to enhance the search ability; 4) a novel probability selection mode is designed. To validate the performance of BMOABC, five well-known benchmark MOPs are tested. The results present that BMOABC is superior to NSGA-II, MOABC, and EMOABC according to the convergence and distributions.

Experimental results show that the improved strategies can help ABC effectively solve MOPs. However, the test instances are old and simple. Some recent and complex benchmark MOPs will be investigated in the future work.

Acknowledgment. This work was supported by the National Natural Science Foundation of China (No. 62166027), and Jiangxi Provincial Natural Science Foundation (Nos. 20212ACB212004 and 20212BAB202023).

References

1. Karaboga, D.: An idea based on honey bee swarm for numerical optimization. Technical report-TR06, Erciyes University, Engineering Faculty, Computer Engineering Department (2005)
2. Wang, H., Wu, Z.J., Rahnamayan, S., Sun, H., Liu, Y., Pan, J.: Multi-strategy ensemble artificial bee colony algorithm. Inf. Sci. **27**, 587–603 (2014)
3. Wang, H., Wang, W.: A new multi-strategy ensemble artificial bee colony algorithm for water demand prediction. In: Peng, H., Deng, C., Wu, Z., Liu, Y. (eds.) ISICA 2018. CCIS, vol. 986, pp. 63–70. Springer, Singapore (2019). https://doi.org/10.1007/978-981-13-6473-0_6

4. Wang, H., et al.: Multi-strategy and dimension perturbation ensemble of artificial bee colony. In: IEEE Congress on Evolutionary Computation (CEC 2019), pp. 697–704 (2019)
5. Wang, H., Wang, W.J., Xiao, S.Y., Cui, Z.H., Xu, M.Y., Zhou, X.Y.: Improving artifificial Bee colony algorithm using a new neighborhood selection mechanism. Inf. Sci. **527**, 227–240 (2020)
6. Xiao, S., Wang, H., Wang, W., Huang, Z., Zhou, X., Xu, M.: Artificial bee colony algorithm based on adaptive neighborhood search and Gaussian perturbation. Appl. Soft Comput. **100**, 106955 (2021)
7. Ye, T., Zeng, T., Zhang, L., Xu, M., Wang, H., Hu, M.: Artificial bee colony algorithm with an adaptive search manner. In: Zhang, H., Yang, Z., Zhang, Z., Wu, Z., Hao, T. (eds.) NCAA 2021. CCIS, vol. 1449, pp. 486–497. Springer, Singapore (2021). https://doi.org/10.1007/978-981-16-5188-5_35
8. Zeng, T., Ye, T., Zhang, L., Xu, M., Wang, H., Hu, M.: Population diversity guided dimension perturbation for artificial bee colony algorithm. In: Zhang, H., Yang, Z., Zhang, Z., Wu, Z., Hao, T. (eds.) NCAA 2021. CCIS, vol. 1449, pp. 473–485. Springer, Singapore (2021). https://doi.org/10.1007/978-981-16-5188-5_34
9. Deb, K., Pratap, A., Agarwal, S., Meyarivan, T.: A fast and elitist multi-objective genetic algorithm: NSGA-II. IEEE Trans. Evol. Comput. **6**(2), 182–197 (2002)
10. Huo, Y., Zhuang, Y., Gu, J.J., Ni, S.R.: Elite-guided multi-objective artificial bee colony algorithm. Appl. Soft Comput. **32**, 199–210 (2015)
11. Xiang, Y., Zhou, Y.R.: A dynamic multi-colony artificial bee colony algorithm for multi-objective optimization. Appl. Soft Comput. **35**, 766–785 (2015)
12. Xiang, Y., Zhou, Y.R., Liu, H.L.: An elitism based multi-objective artificial bee colony algorithm. Eur. J. Oper. Res. **245**(1), 168–193 (2015)
13. Hu, Z.Y., Yang, J.M., Sun, H., Wei, L.X., Zhao, Z.W.: An improved multi-objective evolutionary algorithm based on environmental and history information. Neurocomputing **222**, 170–182 (2017)
14. Zhang, Y., Gong, D.W., Ding, Z.H.: A bare-bones multi-objective particle swarm optimization algorithm for environmental/economic dispatch. Inf. Sci. **192**, 213–227 (2012)
15. Zhang, M., Wang, H., Cui, Z., Chen, J.: Hybrid multi-objective cuckoo search with dynamical local search. Memetic Comput. **10**(2), 199–208 (2017). https://doi.org/10.1007/s12293-017-0237-2
16. Schott, J.: Fault tolerant design using single and multicriteria genetic algorithm optimization. Cell. Immunol. **37**(1), 1–13 (1995)

Fitness Landscape Analysis: From Problem Understanding to Design of Evolutionary Algorithms

Xinyu Zhou(✉), Junyan Song, Shuixiu Wu, Wenlong Ni, and Mingwen Wang

School of Computer and Information Engineering, Jiangxi Normal University, Nanchang 330022, China
xyzhou@jxnu.edu.cn

Abstract. As an effective kind of optimization technique, evolutionary algorithms (EAs) have been widely used in many fields. In the context of EAs, to determine which algorithm would be the most appropriate for the specific problem, the techniques of fitness landscape analysis (FLA) are often used to identify the landscape features of the optimization problem, and select the best algorithm for the problem according to the features. In fact, the landscape features can also be considered as useful feedback for guiding the search procedure of EAs. In recent years, there are some seminal works which use FLA techniques to design EAs, the key concept of them is to adaptively configure algorithms according to the landscape features. In this work, we provide a briefly survey on this emerging research topic, including two aspects: 1) Some representative FLA techniques and the applications to the design of EAs are introduced, with the aim of highlighting the possible ways of extending the FLA techniques from problem understanding to the design of EAs. 2) The constraint of the existing FLA techniques is discussed, and the possible solution, the online FLA techniques, is also presented. So, in this survey, the related FLA techniques and corresponding EAs variants are reviewed according to the answers to the two questions.

Keywords: Evolutionary algorithm · Fitness landscape analysis · Landscape features · Sampling technique

1 Introduction

Nowadays, the optimization techniques have been playing an increasingly important role in many fields of science and engineering, because many real-world problems can be treated as the optimization problems and then be solved by using the optimization techniques. However, many problems have some undesirable characteristics, like multimodal, strong constraints, or even non-differentiable, which are hard for the traditional optimization techniques, especially for the gradient-based techniques. As another alternative, evolutionary algorithms (EAs) are suitable for solving such problems, which are derived from the metaphors of biological evolution in nature. Compared with the traditional optimization techniques, EAs almost have no requirement on

L. Pan et al. (Eds.): BIC-TA 2021, CCIS 1565, pp. 281–293, 2022.
https://doi.org/10.1007/978-981-19-1256-6_21

the problem characteristics; instead, they have the advantages of simple structure yet good performance. As a result, EAs have attracted extensive interests from researchers in various fields, and have been successfully applied to solve kinds of complex optimization problems.

However, there exist many paradigms of EAs, such as genetic algorithm (GA) [1], particle swarm optimization (PSO) [2], and differential evolution (DE) [3]. Typically, different paradigms own different search behaviors, i.e., different paradigms are suitable for different types of optimization problems. In this scenario, to solve a specific optimization problem, how to choose the most appropriate paradigm is a challenging task, since there is no single algorithm that could always performs best on all types of problems, which has been theoretically verified by the well-known *No Free Lunch* (NFL) theorem [4]. In fact, the term "Algorithm Selection Problem (ASP)" has been defined to describe the task of selecting an algorithm from a given set to be expected to perform best [5].

In the context of EAs, as an effective way to solve the ASP, the techniques of *fitness landscape analysis* (FLA) are often used to understand problem before selecting the most appropriate algorithm [6]. Based on the fitness function, the FLA techniques aim at identifying landscape features (or structural properties) to characterize (or quantify) problem difficulty. The landscape features have significant impact on the algorithm performance, and the search behavior of different paradigms of EAs can be estimated through the FLA techniques in turn [7]. As new demands increase in the scenario of black-box optimization, highly effective EAs are required, and the landscape features can be considered as useful feedback for guiding the search procedure of EAs. However, the FLA techniques are rarely used in designing EAs, and the main reasons cover two aspects [8]: 1) The procedure of performing the FLA techniques is computational expensive, since a large number of samplings are consumed to identify landscape features; and 2) Specialized knowledge is required to use the FLA techniques, which is not suitable for the algorithm designer without any prior experience.

In recent years, it is encouraging that some seminal works have been emerged by integrating the FLA techniques into the design of EAs. The main purpose is to exploit the feedback provided by the landscape features for adaptively configuring algorithms, thus more efficient EAs can be obtained with the assistance of the FLA techniques. To some extent, the application of the FLA techniques is the most salient characteristic which discriminates the FLA-based EAs (FLA-EAs) from other types of EAs. Considering this, we are motivated to give a short survey to review some representative FLA-EAs, and highlight the possible ways of extending the FLA techniques from problem understanding to the design of EAs. Different from some previous surveys that only concern on the FLA techniques from the aspect of problem understanding, such as the work of Malan et al. [8], this work focuses more on the combination of the FLA techniques and EAs designing. To the best of our knowledge, we are the first attempt to survey the research topic of designing FLA-EAs. This short survey mainly includes two aspects. First, the basic definitions of some FLA techniques are given, and the related FLA-EAs are reviewed. The focus of this aspect is on how to combine the FLA techniques with EAs to obtain the FLA-EAs. Second, the constraint of the FLA techniques is discussed, which limits the application to the FLA-EAs. As the possible solution to this constraint, the online FLA techniques are presented. More importantly, the potential research direction is pointed out about this emerging research topic of FLA-EAs.

The rest of this short survey is organized as follows. The Sect. 2 gives the definition of the FLA. In the Sect. 3, some FLA techniques and the related FLA-EAs are introduced, which focuses on how the FLA techniques are applied to the EAs. Next, in the Sect. 4, we further discuss the constraint of the FLA techniques and the online versions, especially for the difficulties of how to modify the offline FLA techniques into online mode. Last, the Sect. 5 summarizes this short survey.

2 Brief Introduction to FLA

As for the FLA techniques, the concept of *fitness landscape* is the key, which was first proposed by Wright in 1932 to provide an intuitive picture for genetic evolution of biological genes [9]. Later, in the context of EAs, the concept is used to describe the topological structure of search space for a given optimization problem [10]. In the original concept proposed by Wright, there exists no formal definition or any other specific guidance; but in the context of EAs, a widely accepted formal definition is given by Stadler [11], which comprises three elements as follows:

- A set of solutions X in search space,
- A fitness function F: $X \rightarrow R$,
- A neighborhood N defined by a specific distance measure.

Based on the above three elements, the concept of fitness landscape can be expressed by a triple $FL = (X, F, N)$. In our opinion, the fitness landscape not only describes the corresponding relationship between a solution and its fitness value in the search space, but also the neighborhood relationships among different solutions. In fact, in the definition, the second element (i.e., the fitness function) and the third element (i.e., the neighborhood) correspond to the relationships, respectively. So, the two elements significantly influence the concept of fitness landscape. To be specific, the influence can be explained into two folds, which are listed as follows:

- On the one hand, the fitness function F typically has two kinds of models: continuous model and combinatorial model, and different models result in different types of fitness landscape. In general, the fitness landscape of the continuous model consists of an infinite number of solutions, while the combinatorial model has a finite number. Therefore, the FLA techniques of continuous model are often more complex than the ones of combinatorial model. To simplify the survey, in this work, we only focus on the FLA techniques of continuous model, though some of the techniques are also fit for both models.
- On the other hand, the neighborhood N relates to the accessibility among different solutions, which is determined by the solution encoding in combination with the employed distance measure [8]. For example, for the binary encoded solutions, the Hamming distance measure is usually employed, and any two solutions can be regarded as neighbors when their Hamming distance equals to one. However, if the solutions have real encoding, the Euclidean distance measure is often used, and the

neighborhood relationships among different solutions mainly rely on the evolutionary operators or strategies. In this case, any two solutions can be considered as neighbors when they can access one other after a single application of the evolutionary operators [8].

To characterize the topological structure of search space, a number of features have been introduced, and they are also known as the landscape features. Some common landscape features include modality, funnel, smoothness, ruggedness, neutrality, etc. [12–14]. In addition, for an optimization problem to be solved, the representations of some landscape features vary with the number of decision variables. For example, for the one-dimensional case, the landscape features can be illustrated by lines, while a 3-D figure can represent the landscape features for the two-dimensional case. However, for higher dimensional cases, no visualization method can be applied. The Fig. 1 illustrates the fitness landscapes of the well-known test function Schwefel 2.26 with two cases, including the one-dimensional case and two-dimensional case. As seen, the two subfigures show the fitness landscapes of the function Schwefel 2.26 with two cases, respectively.

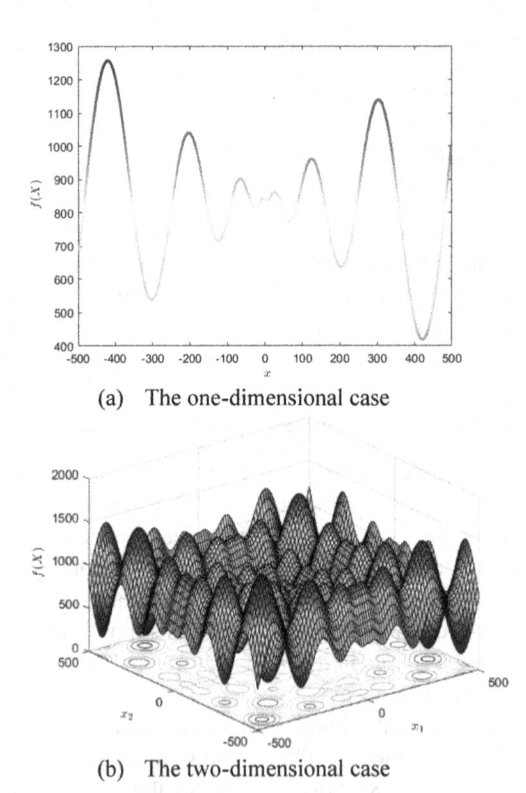

(a) The one-dimensional case

(b) The two-dimensional case

Fig. 1. The fitness landscapes of the function Schwefel 2.26 with two cases

3 Some FLA Techniques and the Applications to EAs

It can be inferred from the above example that landscape features could be visualized directly for low dimensional problems (the dimension size should be less than three), but it is not possible for higher dimensional problems. So, various FLA techniques have been designed in the past twenty years, which can be considered as an effective way to the ASP as well. The main purpose of the FLA techniques is to identify or quantify landscape features. However, it should be noted that different FLA techniques typically identify different landscape features, and it is not possible to identify all the kinds of landscape features by using a single FLA technique. In other words, it is infeasible to design a general FLA technique which could run in polynomial time, and He et al. have theoretically proven this [15]. Hence, a tentative conclusion can be drawn that multiple FLA techniques should be combined together, if we expect a deeper understanding for an optimization problem. Next, we briefly review some representative FLA techniques which prefer different landscape features.

3.1 Fitness-Distance Correlation (FDC)

(1) Basic definition

The FDC is a well-known FLA technique which is first proposed by Jones [16]. The responsibility of the FDC is to investigate the relationship between the problem difficulty and GA performance. It is used to predict GA performance on some well-studied (the global optimum is known) optimization problems [17, 18]. The difficulty of the problem can be quantified by the value of the correlation coefficient between f and the d, which the f denotes the fitness value of the solution and d denotes the hamming distance between the solution to the global best solution. To calculate the FDC, a set of solutions in the search space are randomly selected to be the sample. The value of FDC can be calculated as Eq. (1):

$$r = \frac{C_{FD}}{S_F S_D} \quad (1)$$

where $C_{FD} = \frac{1}{n}\sum_{i=1}^{n}\left(f_i - \overline{f}\right)\left(d_i - \overline{d}\right)$, n is the sample size, $\overline{f}$ and $\overline{d}$ are the mean value of f and d respectively, S_F and S_D are the standard deviations of f and d respectively. The FDC coefficient is range in $[-1, 1]$. It can be observed that the essence of FDC is a Pearson correlation coefficient which used to measure the linear relationship between the variables. The closer of r to -1 or 1, the stronger of the relationship between f and d. If r value close to 0, the relationship is weak. The difficulty of the maximization problems [16] is measured as follow: (1) $r < -0.15$ denotes the misleading landscape; (2) $-0.15 \leq r < 0.15$ are regarded as difficult landscape; (3) $r \geq 0.15$ is seemed as straightforward landscape. FDC is widespread used to predict the GA performance on the specific problem. However, the global best solution of the problem must be known which limits the application of the FDC.

To utilize the FDC for characterizing and illustrating the global topology of the continuous optimization problem, Muller et. al [23] made some modification on the basis of FDC and make verification on CEC 2005 Benchmark function suite. (1).

Employing other distance measure such as Euclidean distance to instead of original Hamming distance measure; (2). The global optimal is approximated by the best individual with lowest objective function value in sample. If $r > 0$, the fitness value decreases when the solution approaching to global minimum. On the contrast, the fitness value decreases with the distance to global minimum increasing when $r > 0$. The modified FDC can reflect the global structure of the continuous optimization problem to some extent.

(2) Application to EAs
Li et. al [26] proposed a self-feedback DE based on fitness landscape analysis (SFDE). The SFDE comprise a local FLA technique to measure the number of local optima where the population covered. The local FLA technique calculates the distance between best individual and other individuals in population. Then all the individuals are sorted as a sequence through the distance. The forepart of the sequence is the individuals close to global best. The most ideal situation is the individuals close to global best will have better fitness than the individuals far away to the global best. And the ruggedness of the problem can be quantified by the number of "non-ideal" situation. Counting the number of closer individuals have worse fitness value than further individual. The larger the count, the fitness landscape is more rugged. For the unimodal landscape, DE/current-to-best/1/bin is used as the mutation operator. Using DE/rand/2/bin as the mutation operator when landscape is multimodal.

Li et.al [29] presents a DE variant with an evolutionary state estimation method (DEET). In DEET, the evolutionary state is estimated by the FDC value. The FDC value r is calculated by the individuals of the population in current iteration. If $r > 0.85$, the evolutionary state is regarded as exploitation; If $r < 0.15$, the evolutionary state is seemed as exploration; otherwise, is balance. For the exploitation, algorithm randomly select mutation operator from DE/current-to-best/1 and DE/rand-to-best/1. Both mutation operator is global best guided. The mutation operator is random selected from two *p*best guided the DE/current-to-*p*best/1 and DE/rand-to-*p*best/1 when current evolutionary state is balance. As for exploration, the mutation pool consists of DE/current-to-rand/1 and DE/rand/1. The algorithm choosing the property mutation operator during optimization process can improve the search efficiency.

3.2 Information Landscape Measure (ILM)

(1) Basic Definition
The ILM is an improved landscape concept first presented by Borenstein [19]. The information landscape uses a pair comparison result function: $X \times X \rightarrow [0, 1]$ to instead the original fitness function F. Moreover, the ILM can quantify the problem hardness by the difference of the information landscape of current problem and an 'optimal' problem. The operation of ILM as follow steps: (1) Getting a set of candidate solution X in the search space and evaluate the fitness value. (2) Constructing the information matrix as follow Eq. (2):

$$m_{i,j} = \begin{cases} 1, & \textit{if } f(x_i) > f(x_j) \\ 0.5 & \textit{if } f(x_i) = f(x_j) \\ 0 & \textit{if } f(x_i) < f(x_j) \end{cases} \tag{2}$$

where each entry m in the information matrix is the comparison result of every pair of solution in X. The set X can be decomposed into $|X| \times |X|$ entries, where the $|X|$ denotes the sample size of the X. (3) Extracting the available information from the information matrix as a vector $V = (v_1, v_2, \ldots, v_m) = (m_{1,2}, m_{1,3}, \ldots, m_{|X-1|,|X|})$. The vector V is consist of the entry in the information matrix but it only includes the available elements. Due to the symmetry of information matrix, the entries above the diagonal are opposite to the entries below the diagonal. The entries on diagonal are all 0.5 and the row/column of the optimal solution is all 0 or 1. Hence the vector V only store the available entries in information matrix which not include the elements below the diagonal, elements on the diagonal and the elements in the row/column of the optimal solution. The size of V is $|V| = (|X| - 1)(|X| - 2)/2$. (4) Given two information vectors v_1 and v_2 for two problems respectively. The difficulty of the optimization problem can be quantified by the distance between v_1 and v_2 as follow Eq. (3):

$$D(v_1, v_2) = \frac{1}{m}\sum_{i=1}^{m} |v_{1i} - v_{2i}| \tag{3}$$

The problem corresponding to v_2 is set as an 'optimal' landscape. The 'optimal' landscape is the problem which is well-studied and the optimal value is easy to be found. The smaller the value of D, the problem easier to be solved. In [20, 21], the ILM is successfully used to measure the problem hardness.

(2) Application to EAs

Malan et. al [25] investigated the information landscape measure and the FC measure. And the modified information landscape measure and FC measure are used to characterize the searchability of PSO on continuous problem. The information landscape measure quantifies the landscape by difference between the landscape of problem and an 'optimal' landscape. The well-studied easy problem always used to be 'optimal' landscape. To utilize the information landscape measure on continuous problems. The modified information landscape measure uses the shifted Spherical function ($f(x) = \sum_{i=1}^{D} (x_i - x_i^*)^2$) as the 'optimal' landscape. The reasons of select the sphere function as the reference landscape are as follow: (1) The sphere function have an 'ideal' landscape, which can be expressed in if any point has a lower fitness value than other solutions, then the point must be closer to optimum than other solutions. (2) The sphere function can be defined up to any dimensions that as fit as a reference for any real-value problem. (3) The sphere function can be shifted by moving the position of optimum so it can measure the problems that the optimum position is not zero.

Sallam et. al [27] developed a DE variant with landscape-based adaptive operator selection (LAOS-DE). In LAOS-DE, the Latin Hypercube design is used to generate the initial population. Five different search operators are randomly applied to population in every generation. After CS iterations, the information landscape measure and the history performance of each operator is calculated and quantified. Based on the two measures, the search operator with highest success rate and lowest information landscape value is selected and used to evolve entire population in next CS generations. Sallam et. al [28] proposed another landscape-based DE (FDC-DE) to solve the constrained optimization problem, the framework of FDC-DE is similar with LAOS-DE. The biggest difference between two DE variants is the FDC-DE utilize the FDC value as the landscape measure instead of information landscape measure in LAOS-DE.

3.3 Information Characteristics Analysis (ICA)

(1) Basic Definition

The ICA is proposed by Vassilev to analyze the ruggedness of the fitness landscape [22]. The ICA consists of three indicators: information content; partial information content; information stability. The three indicators are defined based on a sequence of $\{1, 0, \overline{1}\}$, and the sequence is constructed as follow steps: (1) Executing the random-walk procedure in search space. Recording the fitness values of the solution in each step. For example, the fitness value of i th step are recorded as $f(x_i)$. (2) Generating a $[-1, 0, 1]$ sequence $S(\epsilon)$. The $S(\epsilon)$ is calculated by Eq. (4):

$$S_i(\epsilon) = \begin{cases} -1 & \textit{if}\, f(x_i) - f(x_{i-1}) < -\epsilon \\ 0 & \textit{if}\ |f(x_i) - f(x_{i-1})| \leq \epsilon \\ 1 & \textit{if}\ \textit{if}\, f(x_i) - f(x_{i-1}) > \epsilon \end{cases} \tag{4}$$

where the ϵ is a parameter to control the fluctuation intensity of $S(\epsilon)$. If the ϵ is set as a large number, so that the values in $S(\epsilon)$ are more likely to be zero. (3) Calculating the three ensembled indicators based on $S(\epsilon)$. The information content is an entropic measure calculated by follow Eq. (5):

$$H(\epsilon) = -\sum_{p \neq q} P_{[pq]} log_6 P_{[pq]} \tag{5}$$

The information content is inspired by the information entropy in information theory, which denotes the frequency of the sub-block $[pq]$ that the p is not equal to q in $S(\epsilon)$. The information content can capture the ruggedness by the frequency. If the sub-block $[pq](p \neq q)$ occurred frequent in the $S(\epsilon)$, it denotes the landscape of the problem is rugged. The partial information content is calculated on the basis of the sequence obtained in step (2). Different with information content, the partial information content removes all zero value in the $S(\epsilon)$. The zero value denotes the fitness value equal to the preceding value in the sequence. A new sequence $S'(\epsilon)$ is obtained after removed the zero value. The partial information content is defined as Eq. (6):

$$M(\epsilon) = \frac{\mu}{n} \tag{6}$$

where the μ is the length of new generated $S'(\epsilon)$ and the n denotes the size of original sequence $S(\epsilon)$. The partial information content indicates the count of change slope in random walk. It can reflect the modality of the problem fitness landscape. The information stability is defined as follow:

$$\epsilon^* = \min\{\epsilon | S_i(\epsilon) = 0\}$$

The ϵ can control the fluctuation intensity of $S(\epsilon)$, so the fluctuation intensity of the fitness value sequence can reflect the ruggedness of the fitness landscape. The minimal value of ϵ is the information stability degree. The rugged fitness landscape tends to have relatively large ϵ. On the contrast, the smooth landscape always has relatively small ϵ.

Malan et al. [24] developed a modified ICA to characterize the ruggedness of continuous problem. The original information stability $S(\epsilon)$ is considered with periodic boundary conditions. So that the n-step walks result in n symbols in $S(\epsilon)$, because the comparison result of the last fitness value in time series with first fitness value can be as the last value in $S(\epsilon)$. For continuous problems landscape without the statistically isotropic, the n-step walk result in $n-1$ elements in $S(\epsilon)$. The visual analysis of $H(\epsilon)$ is also provided based on nine values of ϵ: 0, $\frac{\epsilon^*}{128}$, $\frac{\epsilon^*}{64}$, $\frac{\epsilon^*}{32}$, $\frac{\epsilon^*}{16}$, $\frac{\epsilon^*}{8}$, $\frac{\epsilon^*}{4}$, $\frac{\epsilon^*}{2}$, ϵ^*. The $\epsilon^* = \min\{\epsilon | S_i(\epsilon) = 0\}$ which is same as the information stability in the original ICA.

(2) Application to EAs

Huang et al. [30] developed a multi-objective DE based on FLA and reinforcement learning strategy (LRMODE). The landscape of optimization problem can be classified into unimodal and multimodal according to the quantify result of ICA measure. Five different mutation operators are employed in the LRMODE. The reinforcement strategy can control the selection probability of each mutation operator. If a strategy has fast convergence on specific landscape, then this strategy will have greater possibility to be selected on the similar landscape. On the contrast, if a strategy has poor performance on specific landscape, the selection probability of this search strategy will decrease on the similar landscape.

Tan et.al [31] proposed an adaptive mutation strategy DE, called FLDE. In FLDE, the relationship between three mutation operator and nine landscape features is established by a random forest classification model. Three different mutation operators marked 1, 2, 3 are applied to training set. The training set is classified into three classed according to operation performance. For example, if the first operation has better performance on the specific problem than other two operation, then the problem is classified into first class. Nine features of landscape, FDC value r and the ICA entropy value $H(\epsilon)$ with eight different ϵ, are selected to train the random forest model. 45 test functions in CEC2013 and CEC2015 Benchmark are used as training set. The trained random forest can predict the best mutation operation according to the landscape features. The historical parameter adaption and the linear population reduction are introduced into FLDE to enhance the performance.

4 Further Discussion

There are various FLA techniques are proposed in the past decades. Most of FLA techniques can extract and analyze the characterize of the landscape features. However, the main responsibility of FLA is algorithm selection which means predict the algorithm performance on specific optimization problem and select the best algorithm to execute the search process. Unfortunately, rare work uses the fitness landscape information to enhance the performance of the algorithm. In fact, the performance of the search strategy can be influenced by the landscape of the optimization problems. For example, using the global best individual to guide the search direction can effectively improve the convergence speed on unimodal landscape. However, global best guided search strategy always leads the entire population fall into the local optimal on multimodal landscape.

However, it is not easy to combine the landscape information into the search process of the algorithms. Most of FLA techniques only concentrate on characterize the optimization problem itself but pay no attention to fit the properties of specific algorithm. The existing FLA techniques have the ability which can accurately extract the features of the optimization problems but they are incompatible with the search process. The main reason of the situation is the work pattern of the most FLA techniques are offline mode. The offline mode FLA techniques extract the features of the landscape by a mass of sample in search space. This work pattern always results in large computation cost. The computation cost of FLA technique even greater than the search cost of algorithm, so that the execution of the traditional FLA techniques is independent of the evolutionary algorithms. The FLA techniques works on offline mode cannot embedded into the algorithm and it is necessary to transfer the offline work pattern to the online work pattern. The online work pattern means the FLA techniques extract the landscape information during the search process of the EAs. In fact, the ultimate goal of the combination of FLA techniques and algorithm design is to provide the available landscape information to control the search behavior of the algorithm.

For the above problems, the online sampling [7, 32] is a promising way to transfer the FLA techniques from offline mode. Online sampling denotes using the known solutions generated by the algorithm instead of the original mass random sampling. There are various solutions are evaluated during the search process of the algorithm, these solutions can be used to analyze the landscape feature of the optimization problem. The online sampling can effectively save the extra function evaluation and the computation cost. The online sampling has become one of the most popular random sample techniques in recent years. Online sample may cannot extract the global structure of optimization problems accurately, but it can analyze the local fitness landscape of the area which covered by population. The local fitness landscape can reflect the landscape characterizes of the current population location. Which can provide the valuable information to help the population estimate the evolutionary state of current population. The biggest advantage of local fitness landscape with online sampling is the sample size is small so that the computation cost is smaller than the traditional FLA techniques. It is worth mention that, the online sampling can also measure the global structure through some specific method, such as weighed resample [32]. In a word, the online FLA techniques with online sampling is a promising way to combine the FLA techniques with algorithm.

5 Conclusion

In this paper, we have made a concise survey on the FLA techniques and the combination of the landscape information and algorithm design. First, six classic FLA techniques are summarized detailed. All the six typical FLA techniques can extract landscape features through different analysis method. Second, a survey of related work on combination of FLA and specific algorithm is presented. Various FLA techniques are designed for combinatorial optimization problems and assume the problem has discrete representation. To analyze the continuous optimization problem, several previous works made modification on the classic FLA techniques to extract features of continuous problems. A brief introduction of the typical works is given in this paper. Most of these works consider the correlation of the performance of the search operators with the landscape of the optimization problems. Finally, we discuss the feasible of the FLA techniques work in online mode. The online sampling, which is one of the random sample methods is mentioned is a promising way to transfer the offline FLA techniques to online mode. The online sampling utilizes the solution generated by specific algorithm as a sample to analyze the features of landscape. It can extract the local landscape structure where the population covered. It can effectively reduce the unnecessary evaluation and computation cost of the FLA techniques.

Acknowledgment. This work is supported by the National Natural Science Foundation of China (Nos. 61966019, 61866017 and 61876074), the Science and Technology Foundation of Jiangxi Province (No. 20192BAB207030).

References

1. Zhang, X.Y., et al.: Kuhn-Munkres parallel genetic algorithm for the set cover problem and its application to large-scale wireless sensor networks. IEEE Trans. Evol. Comput. **20**(5), 695–710 (2016)
2. Blackwell, T., Kennedy, J.: Impact of communication topology in particle swarm optimization. IEEE Trans. Evol. Comput. **23**, 689–702 (2018)
3. Colutto, S., et al.: The CMA-ES on riemannian manifolds to reconstruct shapes in 3-D voxel images. IEEE Trans. Evol. Comput. **14**(2), 227–245 (2010)
4. Wolpert, D.H., Macready, W.G.: No free lunch theorems for optimization. IEEE Trans. Evol. Comput. **1**(1), 67–82 (1997)
5. Rice, J.R.: The algorithm selection problem. Adv. Comput. **15**, 65–118 (1976)
6. Kerschke, P., et al.: Automated algorithm selection: survey and perspectives. Evol. Comput. **27**, 3–45 (2019)
7. Janković, A., Doerr, C.: Adaptive landscape analysis. In: Genetic and Evolutionary Computation Conference Companion 2019, pp. 2032–2035. ACM, New York (2019)

8. Malan, K.M., Engelbrecht, A.P.: A survey of techniques for characterising fitness landscapes and some possible ways forward. Inf. Sci. **241**, 148–163 (2013)
9. Wright, S.: The roles of mutation, inbreeding, crossbreeding, and selection in evolution. In: The Sixth International Congress on Genetics (1932)
10. Li, W., Meng, X., Huang, Y.: Fitness distance correlation and mixed search strategy for differential evolution. Neurocomputing **458**, 514–525 (2020)
11. Stadler, P.F.: Fitness landscapes. In: Biological Evolution and Statistical Physics, pp. 187–207. Springer, Heidelberg (2002). https://doi.org/10.1007/978-3-662-04726-2_2
12. Sutton, A.M., Whitley, D., Lunacek, M.: PSO and multi-funnel landscapes: how cooperation might limit exploration. In: The 8th Annual Conference on Genetic and Evolutionary Computation, pp. 75–82. ACM, New York (2006)
13. Garnier, J., Kallel, L.: How to detect all maxima of a function. In: Kallel, L., Naudts, B., Rogers, A. (eds.) Theoretical Aspects of Evolutionary Computing. Natural Computing Series, pp. 343–370. Springer, Heidelberg (2001). https://doi.org/10.1007/978-3-662-04448-3_17
14. Beaudoin, W., Verel, S., Collard, P.: Deceptiveness and neutrality the ND family of fitness landscapes. In: The 8th Annual Conference on Genetic and Evolutionary Computation, pp. 507–514. ACM, New York (2006)
15. He, J., Reeves, C., Witt, C., Yao, X.: Note on problem difficulty measures in black-box optimization: classification, realizations and predictability. Evol. Comput. **15**(4), 435–443 (2007)
16. Jones, T.C., Forrest, S.: Fitness distance correlation as a measure of problem difficulty for genetic algorithms. In: The Sixth International Conference on Genetic Algorithms, pp. 184–192. Pittsburgh (1995)
17. Altenberg, L.: Fitness distance correlation analysis: an instructive counterexample. In: The Seventh International Conference on Genetic Algorithms, pp. 57–64 (1997)
18. Tomassini, M., et al.: A study of fitness distance correlation as a difficulty measure in genetic programming. Evol. Comput. **13**(2), 213–239 (2005)
19. Yossi, B., Poli, R.: Information landscapes. In: the 7th Annual Conference on Genetic and Evolutionary Computation, pp. 1515–1522. ACM, New York (2005)
20. Yossi, B., Poli, R.: Information landscapes and the analysis of search algorithms. In: The 7th Annual Conference on Genetic and Evolutionary Computation, pp. 1287–1294. ACM, New York (2005)
21. Yossi, B., Poli, R.: Information landscapes and problem hardness. In: The 7th annual Conference on Genetic and Evolutionary Computation, pp. 1425–1431. ACM, New York (2005)
22. Vassilev, V.K., Fogarty, T.C., Miller, J.F.: Information characteristics and the structure of landscapes. Evol. Comput. **8**(1), 31–60 (2000)
23. Müller, C.L., Sbalzarini, I.F.: Global characterization of the CEC 2005 fitness landscapes using fitness-distance analysis. In: European Conference on the Applications of Evolutionary Computation, pp. 294–303 (2016)
24. Malan, K.M., Engelbrecht, A.P.: Quantifying ruggedness of continuous landscapes using entropy. In: IEEE Congress on Evolutionary Computation 2009, pp. 1440–1447. IEEE, Scandinavia (2009)
25. Malan, K.M., Engelbrecht, A.P.: Characterising the searchability of continuous optimisation problems for PSO. Swarm Intell. **8**(4), 275–302 (2014). https://doi.org/10.1007/s11721-014-0099-x
26. Li, W., Li, S., Chen, Z., Zhong, L., Ouyang, C.: Self-feedback differential evolution adapting to fitness landscape characteristics. Soft. Comput. **23**(4), 1151–1163 (2017). https://doi.org/10.1007/s00500-017-2833-y

27. Sallam, K.M., Elsayed, S.M., Sarker, R.A.: Landscape-based adaptive operator selection mechanism for differential evolution. Inf. Sci. **18**, 383–404 (2017)
28. Sallam, K., Elsayed, S., Sarker, R.: Landscape-based differential evolution for constrained optimization problems. In: IEEE Congress on Evolutionary Computation 2018, pp. 1–8. IEEE, Rio de Janeiro (2018)
29. Li, Y., Li, G.: Differential evolutionary algorithm with an evolutionary state estimation method and a two-level selection mechanism. Soft. Comput. **24**(15), 11561–11581 (2019). https://doi.org/10.1007/s00500-019-04621-z
30. Huang, Y., Li, W., Tian, F.: A fitness landscape ruggedness multiobjective differential evolution algorithm with a reinforcement learning strategy. Appl. Soft Comput. **96**(1), 66–79 (2020)
31. Tan, Z., Li, K., Wang, Y.: Differential evolution with adaptive mutation strategy based on fitness landscape analysis. Inf. Sci. **5**(4), 142–163 (2021)
32. Munoz, M.A., Kirley, M., Halgamuge, S.K.: Landscape characterization of numerical optimization problems using biased scattered data. In: 2012 IEEE Congress on Evolutionary Computation 2012, pp. 1–8. IEEE, Brisbane (2012)

Optimal Overbooking Appointment Scheduling in Hospitals Using Evolutionary Markov Decision Process

Wenlong Ni[1(✉)], Jue Wang[2], Ziyang Liu[1], Huaixiang Song[1], Xu Guo[1], Hua Chen[1], Xinyu Zhou[1], and Mingwen Wang[1]

[1] School of Computer Information Engineering, JiangXi Normal University, Nanchang, China
{wni,dawnn,songhuaixiang,guoxu,chenhua,xyzhou,mwwang}@jxnu.edu.cn

[2] Department of Rehabilitation, The Second Affiliated Hospital of Nanchang University, Nanchang 330006, China
cny9707@dingtalk.com

Abstract. This research proposes an algorithm to solve overbooking scheduling problem for outpatient hospitals with multiple providers and high patient demand. Assuming there is reward and system cost for serving each patient, and the decision maker in the hospital can decide the amount of resources to assign to each patient. Regardless of the random patient arrivals and departures, a novel model of the Continuous-Time Markov Decision Process (CTMDP) is explored, our objective in this paper is to find an optimal policy to achieve maximum total discounted expected reward starting from any initial states. Further more, to solve the computation complexity of CTMDP models when the action and system state space is large, genetic algorithm (GA) is proposed to search the optimal solution, which can be calculated in a parallel way thus reducing the computation time for the optimal policy.

Keywords: Overbooking appointment scheduling · Outpatient services · Genetic algorithms · Optimal control policy · Continuous-Time Markov Decision Process (CTMDP)

1 Introduction

Hospitals with outpatient services have grown significantly in the past decade [2]. Nowadays there are a number of short-term appointments for outpatients provided at the hospital. Due to the increasing demands for service providers, many patients may have to wait for extended periods of time, and thus increase the burden for both patients and hospitals, generating different kinds of successive problems [8]. It is well known that an efficient appointment scheduling can

---

W. Ni and J. Wang—These authors contributed equally to this work. This work was supported in part by JiangXi Education Department under Grant No. GJJ191688.

L. Pan et al. (Eds.): BIC-TA 2021, CCIS 1565, pp. 294–301, 2022.
https://doi.org/10.1007/978-981-19-1256-6_22

greatly improve patient satisfaction and hospital revenue. Therefore, scheduling in hospital is very important and hard to solve due to random patient arrivals and cancellations.

Scheduling model helps to schedule each patient to an appointment slot. Due to the possibility of unexpected events like appointment cancellation, the patient may not be served on his/her scheduled slot, which makes the situation in a chaos state and hard to predict. To solve this problem, many studies have been proposed on scheduling models within a variety of healthcare practices like patient rules, appointment time, and patient/provider preference, etc. [1,3,5, 7,13]. Overall these models can be classified to 3 types: unit-level, FCFS, and batch. Here unit-level splits the time period to many units and checks the system status at each time unit. FCFS follows the strategy of first-come-first-serve for incoming patients, while batch collects a batch of arriving patients and then serve them after a random period.

To find the optimal strategy of overbooking appointment scheduling in the hospital, in this paper we propose a novel algorithm using CTMDP model, with the goal of maximizing accumulated discounted rewards in infinite horizons. The CTMDP models consists of five elements like state space, action space, transition probability, reward function and decision epoch. It is well know that when the state space and action space is large, it can take a long time to reach the optimal solution. To save the computation time of a CTMDP model when the parameters, our contribution in this paper is the introduction of solving CTMDP with GA search algorithm. Based on the GA, the proposed calculation process for optimal policy can easily be searched in a parallel way starting from different initiap policies.

The remainder of this article is organized as follows: Sect. 2 lists the details of research background. Section 3 shows the proposed algorithm. Section 4 places the experimental results and analyses. Finally, Sect. 5 describes conclusions and future work.

2 Background

In this research, a hospital providing outpatient services is modelled as CTMDP model. There are random arrivals and departure of patients. Due to the fact that the resources for providing outpatient services are limited, it is an common approach to use admission control [6,9] to avoid congestion. for a better quality of service and maximum revenue, the decision maker of the hospital will decide what actions to take when there is a patient coming for service.

2.1 CTMDP Model

As explained, each CTMDP model consists of five elements:(1) State Space; (2) Action Space; (3) Reward function; (4) Transition function and (5) Decision epochs. For a hospital with multiple classes of patients, the decision epochs are those time points of patients arriving/leaving the hospital. The other assumptions are described below.

1. System States: The system state S in the model for a hospital can be described by the number of patients, the resources being used for services. Assume there are M types of resources in the hospital, each type of resource has the capacity $C_i, i = 1, \ldots, M$, then the state space can be defined as $(n_1, \ldots, n_M)$, with $n_i \leq C_i, i = 1, \ldots, M$.
2. Actions: Upon receiving a patient, the decision maker in the hospital can decide whether to accept or transfer to another clinic based on current system state, potential waiting time, possible revenue, etc. When there is an patient arrival, the decision maker in the hospital will decide whether he should accept or reject the patient, so the action space can be defined as $A = \{a_A, a_R\}$, with a_A be the action to accept and a_R be the action to reject.
3. Reward function: With a certain state and the selected action, we can denote the time duration to the next state and then find out the total reward and cost during this epoch. In this research we focus on the discounted reward model defined in [10,11]. The reward function can be determined by the decision maker in the hospital.
4. Transition function: For a certain system state, after an action is taken, the probability of the next incoming state it may arrive can be determined, which is the transition probability.

After determination of the five elements in a CTMDP model, we can search through the state space and action space to find out the optimal policy which defines the actions for each state. Our goal is that for any initial state in the system, find out the policy (the actions for each state) that can bring maximum reward.

2.2 Policy Iteration (PI)

To solve a CTMDP problem, generally we can use either Policy Iteration (PI) or Value Iteration (VI) methods. In this paper we focus on the PI method. Here during the PI iteration it computes the optimal Policy, it the Policy does not change any more we conclude that the optimal policy is found. When the state space is large, it becomes impractical to do the PI or VI directly, and the heuristic algorithm like GA is incorporated.

2.3 Genetic Algorithm (GA)

GA has been used widely to solve search and optimization problems. It is based on the ideas of natural selection and genetics. The solutions are coded as genes, in each generation there are a number of genes (solutions), the best are saved to next generation, others are muted and exchanged to find the better one.

As shown in the Fig. 1, general GA is summarized by the flow chart. Each gene is calculated with a fitness value, during the selection process the genes with higher fitness values has a higher probability to reproduce and more chances to survive to the next generation. As the generations goes on, genes with best fitness values may be found out and used as the solution to the problem.

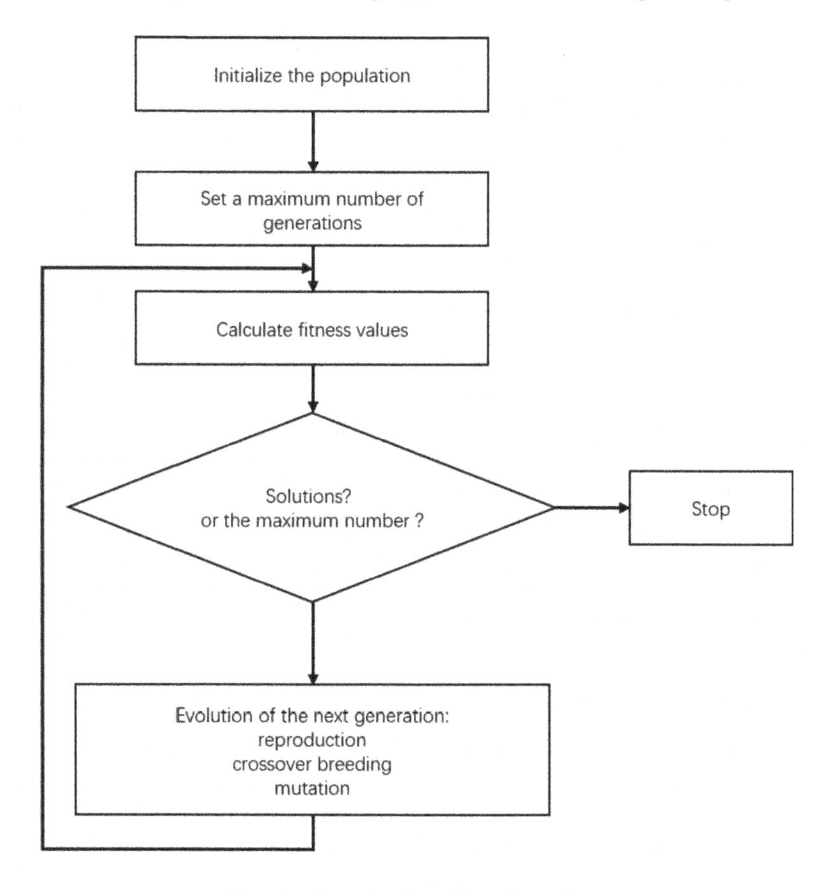

Fig. 1. Genetic algorithm flow chart

3 Evolutionary CTMDP

With the concepts introduced in the previous sections, we can now define the proposed algorithm of evolutionary CTMDP for the overbooking appointment scheduling problem in a hospital. Similar as a GA algorithm, it mainly consists of three steps: (1) initialization; (2) policy switching and (3) policy mutation.

3.1 Initialization

During initialization, a population of policies are randomly generated. Each policy is initialized with random actions for each state, and will be updated during the PI process. A number of such polices are created, each works as a member in the population.

3.2 Policy Switching

There is an "crossover" operation in the GA, in which members in the population are selected to exchanges part of their genes, thus generating new members, making the whole population diversified. Here we can create a similar operation of CTMDP polices, some of the polices are selected to exchange its actions on some randomly selected states, generating new policies.

3.3 Policy Mutation

Mutation is an very important step in GA. Through mutation step, some genes in the member are randomly selected and changed, making the whole population diversified, avoid being trapped in the local optima criteria. Here we can also incorporate such action into the evolution PI process. Here mutation means select new actions for some randomly selected states, thus generating totally new polices which helps search the optimal policy through a nearby search. A well designed mutation rate can help reach the global optima in a faster way, while it may causes large permutation if the rate is too large and hard to find new members if the rate is too small.

3.4 Proposed Algorithm

For the proposed algorithm, its procedure imitates that of general GA [4,12,14], as listed in the previous sections, polices are chosen as member in population, they are updated through PI processes, and then take the GA operations like switch, mutation etc. By doing this, other than waiting one PI iteration to finish, we can start from different random polices and search for the optimal policy parallelly thus saving the computation time. The details of the proposed algorithm is listed below:

- Initialization:
  Generating a number of random polices
  Define the rate for switch
  Define the rate for mutation
  Define the steps to update policy in PI
- Repeat in each generation:

  • Policy Evaluation
    Calculate fitness values
  • Policy Improvement
    Update the policy through policy iteration
  • Policy Switch
    Crossover randomly selected policies to generate new policies
  • Policy Mutation
    Change the actions of some randomly selected states in a policy

- Optimal Policy
  - Find the optimal policy with best fitness value
  - Stop the search if optimal policy stays unchanged

4 Numerical Analysis

To show the effectiveness of the proposed algorithm, we consider a simple case study, like a small clinic providing dental treatments with multiple providers serving two types of patients. Each type the patient has its own arrival/departure rates.

4.1 Case Study

Assuming the arrival processes of patients are poisson process [11], which is a general random process. Their departure process follow negative exponential distribution, there is reward for serving the patients and cost for making them waiting in the hospital, we can build the corresponding CTMDP model and solve it with the proposed algorithm and compare it with the regular PI method.

For better explanation, we list all the parameters needed for numerical analysis in Table 1.

Table 1. Parameter values

Parameter	Value	Memo
λ	2,4	Arrival rates
μ	6,8	Service rate
α	0.5	Discount factor
C	6	Provider capacity
R	3,2	Reward
P	5	Population
m	0.2	Mutation probability

And, the cost function is

$$f(n_1, n_2) = n_1 + n_2$$

To save some time we choose a simple case with two types of arrivals, the system capacity is limited as 6, thus the corresponding state space and action space is not that large. And the population of evolutionary PI is set to 5, and only mutation operation is taken during generations.

4.2 Optimal Policy

After the calculation of proposed algorithm, to save the space, only the actions for Type-1 arrival of optimal policy is listed in Table 2.

Table 2. Actions of optimal policy for Type-1 arrivals

	1			$n_2 \rightarrow$			6
1	1	1	1	1	1	1	0
	1	1	1	1	1	0	−1
	1	1	1	1	0	−1	−1
$n_1 \downarrow$	1	1	1	0	−1	−1	−1
	1	1	0	−1	−1	−1	−1
	1	0	−1	−1	−1	−1	−1
6	0	−1	−1	−1	−1	−1	−1

In Table 2 and Table 3, the actions for two types of arrivals are listed. '−1' means there is no such state in the system, '0' means the system will take 'reject' action for the arrival, '1' means the system will take the 'accept' action to accommodate the service for such arrival. It is seen that since the more reward for Type-1 arrivals than Type-2 arrivals, the system accepts more Type-1 arrivals into the system.

Table 3. Actions of optimal policy for Type-2 arrivals

	1			$n_2 \rightarrow$			6
1	1	1	1	1	1	0	0
	1	1	1	1	0	0	−1
	1	1	1	0	0	−1	−1
$n_1 \downarrow$	1	1	0	0	−1	−1	−1
	1	0	0	−1	−1	−1	−1
	0	0	−1	−1	−1	−1	−1
6	0	−1	−1	−1	−1	−1	−1

By checking the values from PI method and GA algorithm, both lead to the same optimal policy, which ascertains the correctness of proposed method.

5 Conclusion

Due to the random arrivals and possible cancellations of patients in a hospital, it is very complex problem to find an optimal overbooking appointment schedule to maximize its revenue. In this paper a novel model of the Continuous-Time Markov Decision Process (CTMDP) is explored with an objective to find an optimal policy to achieve maximum total discounted expected reward. Further more, to solve the computation complexity of MDP process when the action and system state space is large, the search algorithm of Genetic Algorithm is proposed, which can be processed in a parallel way thus reduces the computation time for the optimal policy. The numerical analysis of a simple case proved the correctness of our proposed algorithm and we plan to carry it to further complex situation of hospital overbooking appointment schedule issues.

References

1. Bosch, P.M.V., Dietz, D.C.: Minimizing expected waiting in a medical appointment system. IIE Trans. **32**(9), 841–848 (2000)
2. De Lathouwer, C., Poullier, J.: How much ambulatory surgery in the world in 1996–1997 and trends? Ambul. Surg. **8**(4), 191–210 (2000)
3. Erdogan, S.A., Denton, B.: Dynamic appointment scheduling of a stochastic server with uncertain demand. INFORMS J. Comput. **25**(1), 116–132 (2013)
4. Goldberg, D.E., Holland, J.H.: Genetic algorithms and machine learning. Mach. Learn. **3**, 95–99 (1988). https://doi.org/10.1023/A:1022602019183
5. Gupta, D., Denton, B.: Appointment scheduling in health care: challenges and opportunities. IIE Trans. **40**(9), 800–819 (2008)
6. Li, W., Chao, X.: Call admission control for an adaptive heterogeneous multimedia mobile network. IEEE Trans. Wireless Commun. **6**(2), 515–525 (2007). https://doi.org/10.1109/TWC.2006.05192
7. Liao, C.J., Pegden, C.D., Rosenshine, M.: Planning timely arrivals to a stochastic production or service system. IIE Trans. **25**(5), 63–73 (1993)
8. Muthuraman, K., Lawley, M.: A stochastic overbooking model for outpatient clinical scheduling with no-shows. IIE Trans. **40**(9), 820–837 (2008)
9. Ni, W., Li, W., Alam, M.: Determination of optimal call admission control policy in wireless networks. IEEE Trans. Wireless Commun. **8**, 1038–1044 (2009). https://doi.org/10.1109/TWC.2009.080349
10. Ni, W., Li, W.W.: Optimal resource allocation for brokers in media cloud. In: Chen, X., Sen, A., Li, W.W., Thai, M.T. (eds.) CSoNet 2018. LNCS, vol. 11280, pp. 103–115. Springer, Cham (2018). https://doi.org/10.1007/978-3-030-04648-4_9
11. Puterman, M.: Markov Decision Processes: Discrete Stochastic Dynamic Programming (2005). https://doi.org/10.1002/9780470316887
12. Reeves, C.R.: Genetic algorithms for the operations researcher. INFORMS J. Comput. **9**(3), 231–250 (1997)
13. Rohleder, T.R., Klassen, K.J.: Rolling horizon appointment scheduling: a simulation study. Health Care Manag. Sci. **5**(3), 201–209 (2002)
14. Srinivas, M., Patnaik, L.M.: Genetic algorithms: a survey. Computer **27**(6), 17–26 (1994)

A Multi-direction Prediction Multi-objective Hybrid Chemical Reaction Optimization Algorithm for Dynamic Multi-objective Optimization

Hongye Li[1,2,3](✉), Xiaoying Pan[1,2,3], Wei Gan[4], and Lei Wang[5,6]

[1] School of Computer Science and Technology, Xi'an University of Posts and Telecommunications, Xi'an 710121, Shaanxi, China
lihongye@xupt.edu.cn

[2] Shaanxi Key Laboratory of Network Data Analysis and Intelligent Processing, Xi'an 710121, Shaanxi, China

[3] Xi'an Key Laboratory of Big Data and Intelligent Computing, Xi'an 710121, Shaanxi, China

[4] Xi'an Shiyou University, Xi'an 710065, Shaanxi, China

[5] School of Computer Science and Engineering, Xi'an University of Technology, Xi'an 710048, China

[6] Shaanxi University of Technology, Han Zhong 723001, Shaanxi, China

Abstract. The challenge of solving dynamic multi-objective optimization problems is to trace the varying Pareto optimal front and/or Pareto optimal set quickly and efficiently. This paper proposes a multi-direction prediction strategy using a hybrid chemical reaction optimization, aimed at finding the dynamic Pareto optimal front and/or Pareto optimal set as quickly and accurately as possible before the next environmental change occurs. The proposed method, multi-direction prediction multi-objective hybrid chemical reaction optimization algorithm which mainly includes a hybrid chemical reaction optimization algorithm is proposed for solving dynamic multi-objective problems, which can guide population trace the optimum. When the environment has changed, the population is divided into several subpopulations. In subpopulation, a center point was found to construct multi-direction prediction model. As a result, this approach enhances the diversity of algorithm. While the environment has not changed, the hybrid chemical reaction algorithm with particle swarm optimization algorithm can efficiency find the optimal solution, it can achieve good diversity as well as guarantee the avoidance of local optimal solutions. The proposed algorithm is measured on several benchmark test suites with various dynamic characteristics and different difficulties. Experimental results show that this algorithm is very competitive in dealing with dynamic multi-objective optimization problems when compared with four state-of-the-art approaches.

Keywords: Dynamic multi-objective optimization · Hybrid chemical reaction optimization algorithm · Multi-direction prediction

L. Pan et al. (Eds.): BIC-TA 2021, CCIS 1565, pp. 302–316, 2022.
https://doi.org/10.1007/978-981-19-1256-6_23

1 Introduction

In fact, dynamic multi-objective optimization problems require a fast convergence to tracking the PF before the next change appears. Many real-world multi-objective optimization problems (MOPs) are dynamic in nature, whose objective functions, constraints, and/or parameters may change over time. Due to the dynamisms character of dynamisms MOPs (DMOPs), multi-objective evolutionary algorithms (MOEAs) pose big challenges to handle with dynamic MOPs. Since any environmental change may affect the objective function, constraints, and/or parameters, the Pareto-optimal set (POS), which is a set optimal solution of decision space, the Pareto-optimal front (POF) is the POS imaging to the objective space. Dynamic MOEAs (DMOEAs)'goal is to track the moving POF and/or POS and obtain a sequence of approximations over time. Evolutionary algorithms can efficiently solve DMOPs. Although many excellent multi-objective evolutionary algorithms have been proposed, when solving DMOPs, the MOEAs lack a rapid change response mechanism in a dynamic environment, which shows that it cannot track and predict the real Pareto front in time. Therefore, when the dynamic multi-objective optimization problem is changed, how to design prediction strategies of DMOEAs to quickly obtain the POS and/POF is important work. Recently, many DMOEAs have proposed for solving DMOPs, according to the dynamic characteristics, those approaches have been proposed which include diversity approaches [1, 2], prediction approaches [3–6], multiple population [7–9] and some matching learning approaches [10–12].

Although some research have been done in the dynamic optimization field, there are still many works have to be improve the diversity or/and convergence of DMOEAs. Some prediction approaches may guide the algorithm search optimal solution set, but for complex Pareto front or Pareto sets, some prediction approaches cannot search the whole complex Pareto front or Pareto sets. Considering the characteristics of DMOPs and the shortcoming of present methods, this paper proposes a new DMOEA, which include a multi-direction prediction strategy (MDPS) and hybrid chemical reaction optimization. MDPS can n search the variable space from multiple directions to update outdated solutions. When the environment changes, MDPS is utilized to explore in multiple directions and discover new optimal solutions, so as to enhance the diversity of the population and update the population using solutions which have been found. The main contributions of this research are as follows.

1. In order to enhance the diversity and multi-direction prediction model is proposed. The population is divided into several subpopulation according there characters in decision space, each subpopulation find an optimal direction avoid misguide the algorithm escape local optimal solution.
2. A hybrid chemical reaction optimization is proposed to guide the algorithm find the POS or POF.

2 Background

2.1 Dynamic Multi-objective Optimization

In a dynamic multi-objective optimization problem (DMOP), there exist two or more conflicting objectives, where the objectives, constraints or parameters of the problem change over time. Whenever a change happens and is detected in the environment of a MOP, usually at least the Pareto optimal set (POS) or the Pareto optimal front POF) may change. The dynamic POF is the set of non-dominated solutions with respect to the objective space, while the dynamic POS is the set of non-dominated solutions with respect to the decision space, at the given time. Since a change in the DMOP affects the existing solutions of the DMOP, the goal is to track the dynamic POF or POS. Therefore the new POF and the new POS must be found as fast as possible before the next change happening.

The DMOP can be described as follows

$$\min_{x\in\Omega} F(x,t) = (f_1(x,t), f_2(x,t), \ldots, f_m(x,t))^T$$
$$s.t. \begin{cases} g_i(x,t) \leq 0, i = 1,2,\ldots,p \\ h_j(x,t) = 0, j = 1,2,\ldots,q \\ x \in \Omega_x,\ t \in \Omega_t \end{cases} \tag{1}$$

where t is the time variable, the mathematical definition of time parameter t is $t = \frac{1}{n_t}\left\lfloor \frac{\tau}{\tau_t} \right\rfloor$, τ is the s the iteration counter, n_t is the number of distinct steps in t, and τ_t is the number of iterations for which t remains the same. n_t and τ_t determine the severity level and frequency value of t, respectively. $x = (x_1, x_2, \ldots, x_n)$ is the n-dimensional decision variables bounded by the decision space Ω_x, $F = (f_1, f_2, \ldots, f_m)$ presents the set of m objectives to be minimized the functions of $g_i \leq 0, i = 1,2,\ldots,p$ and $h_j = 0, j = 1,2,\ldots,q$ presents the set of inequality and equality constraints, $X(t) = \{x|g(x,t) \leq 0, h(x,t) = 0\}$ present the feasible set at time t.

Definition 1 (Dynamic Pareto Dominance): Given two individuals p and q in the population, p is said to dominate q, written as $f(p) \prec f(q)$ if $f_i(p) \leq f_i(q)\ \forall i \in 1, 2,\ldots, m$ and $f_j(p) < f_j(q)\ \exists\, j \in 1, 2,\ldots, m$.

Definition 2 (Dynamic Pareto-Optimal Set): A dynamic Pareto-optimal set at time t, denoted as PS(t), includes all solutions that are not dominated by any other solutions at time t, and that can be defined mathematically as follows:

$$\text{PS}(t) = \{x|\neg\exists y \in \Omega : x(t) \prec y(t)\}.$$

Definition 3 (Dynamic Pareto-Optimal Front): A dynamic Pareto-optimal front at time t, denoted as PF(t), which is the set of all non-dominated solutions in the objective space, and which is the mapping of the solutions in PS(t) in the objective space, and that can be defined mathematically as follows:

$$\mathrm{PF}(t) = \{F(x,t)|x \in \mathrm{PS}(t)\}.$$

Due to the dynamic Pareto-optimal set and dynamic Pareto-optimal front, a DMOP can be divided into four types [13, 14] according to the dynamic characteristics of PF(t) and PS(t) as follows:

Type I, the optimal PS(t) changes while the optimal PF(t) remains fixed.

Type II, both PS(t) and PF(t) change over time.

Type III, PF(t) changes with time, while PS(t) remains invariant.

Type IV, Both PF(t) and PS(t) remain fixed, but the problem changes over time.

2.2 Dynamic Multi-objective Evolutionary Algorithms

Recently, more and more DMOEAs have been proposed to solve the DMOPs [14], since DMOEAs are a kind of efficient tool for solving DMOPs. Consequently, a great deal of progress has been made about the design of DMOEAs in recent decades [15]. These DMOEAs in most existing DMO literatures are classified two groups, that is, diversity introduction and prediction based approaches.

Diversity introduction approaches mainly enhance the diversity, which take into account the potential diversity loss of the population in a dynamic environment. Those approaches mainly introduce randomized or mutated individuals detected an environmental changed, which can help an algorithm jump out of current optimum. Thus, a effectively diversity introduction method can accelerate the DMOEAs track the varying POF and/or POS. For example, Deb et al. [1] proposed two DMOEAs (NSGAII-A and DNSGA-II-B) which based on the Non-dominated Sorting Genetic Algorithm-II (NSGA-II) [2]. The NSGAII-A enhance diversity by randomly reinitializes 20% of the current individuals, and the DNSGA-II-B enhance the diversity by randomly mutating 20% of the current individuals, when a change is detected.

To accelerate the adaptation of the population to the dynamic environments, prediction-based approaches have been proposed to generate a promising population in a new environment. For example, Zhou et al. [13] proposed PPS to divide the population into a center point and a manifold. Muruganantham et al. [16] applied a Kalman filter [17] in the decision space to predict the new Pareto-optimal set. Ruan et al. [4] applied a gradual search to predict the ideal position of the individuals in the new environment. Jiang and Yang [18] introduced an SGEA, which guides the search of the solutions by a moving direction from the centroid of the nondominated solution set to the centroid of the entire population. Zhao Q et al. [19] proposed Evolutionary Dynamic Multiobjective Optimization via Learning From Historical Search Process.

In this article, we proposed multi-direction prediction to solve most complex DMOPs. The proposed approach can divide the population into several subpopulations according to their characteristics, which can efficiency enhance the diversity. It is worth highlighting that the method of division is used the matching method self-organization mapping (SOM), which is the first attempt to using a matching method-SOM solving DMOPs. In this way, we can construct direction in each subpopulation, thus the method can search whole complex POS and/or POF.

2.3 Hybrid Chemical Reaction Optimization Algorithm

CRO is inspired by the chemical reaction process where atoms of substances change to form new atoms. To solve an optimization problem using CRO, a population of molecules is created, where each molecule holds an answer to the problem. Then, some elementary chemical reactions are applied iteratively to alter the population, so the size of population is change during the reaction. The CRO contains four operators, On-wall Ineffective Collision, Decomposition operator, Inter-molecular Ineffective Collision operator, Synthesis operator. The operators of On-wall Ineffective Collision and Inter-molecular Ineffective Collision carry on local search. The operators of Decomposition and Synthesis carry on global search. Except that, there has an energy system which keeps the chemical reaction system balancing. However, CRO has poor global searching energy in complex problems. PSO has a good performance on global searching. Thus, this paper a hybrid chemical reaction optimization algorithm (HCRO) is proposed in this paper.

The framework of HCRO is shown in Fig. 1 and Algorithm 1. In Fig. 1, the number of big circles represents the number of subpopulations. This hybrid algorithm is combination advantages of CRO and PSO. The CRO has a good ability of global search, and PSO has a good ability of local exploitation. This hybrid algorithm only uses the On-wall Ineffective Collision operator and Synthesis operator of CRO, therefore, the size of population does not changes and parameters of α has not used. In this algorithm the On-wall Ineffective Collision operator use the polynomial mutation (PM) and the Synthesis operator use the simulated binary crossover (SBX).

In HCRO, the velocity of particle i ($i = 1, 2, \ldots, N$) is update as defined in Eqs. (2).

$$\begin{cases} v_i(t+1) = wvi(t) + c_1 r_1 (x_{pbest_i} - x_i(t)) \text{ if } r_3 < \delta \\ v_i(t+1) = wvi(t) + c_2 r_2 (x_{gbest_i} - x_i(t)) \; else \end{cases} \tag{2}$$

Where r_3 is a uniformly distributed random number in [0,1]. The appropriate setting of δ can keep the balance between exploitation and exploration. δ is generally set in [0.5, 0.9] to put more attention on the exploitation of the current search region.

Algorithm 1: Hybrid chemical reaction algorithm with particle swarm optimization algorithm

Input P (a set of solutions (particles/ molecules))
Output *minimum point*

1	**if** particles or molecules meet the condition of PSO update
2	| Carry on the PSO ( ) in Eqs (2)
3	**else**
4	| Carry on the CRO ( )
5	**endif**

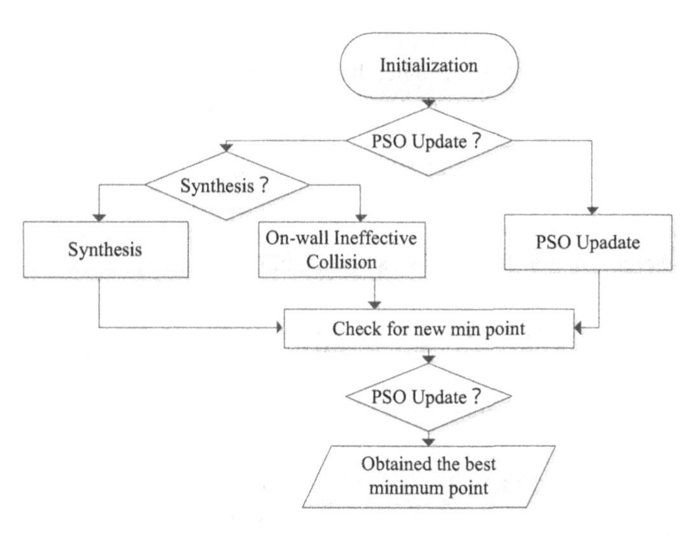

Fig. 1. The flow chart of the hybrid chemical reaction algorithm with particle swarm optimization algorithm

3 Multi-direction Prediction Based Multi-objective Hybrid Chemical Reaction Algorithm

This paper proposes a model based on exponential smoothing and multi-directional prediction to solve dynamic multi-objective optimization problems. For the dynamic multi-objective optimization problem with irregular changes in the front of Pareto, the method in this paper can quickly and accurately track the dynamic Pareto optimal solution set and can effectively cope with this change.

3.1 Exponential Smoothing Forecasting Model

Exponential smoothing is a method of pre-processing data [21]. It was originally mainly used for economic forecasting. It has now been extended to other engineering application neighborhoods and has become a commonly used forecasting method. The exponential smoothing method is divided into primary, secondary and tertiary exponential smoothing methods, which are respectively applicable to three types of time series with no obvious change, linear change and non-linear change. Different exponential smoothing methods are used according to the changes of the actual situation. This article uses three exponential smoothing methods.

The definition of the three-time exponential smoothing method, suppose the time series value at different moments is $x_i(i = 1, 2,\ldots)$, and the three-time exponential smoothing value at time t is: s_t^1, s_t^2, s_t^3, the calculation formula is:

$$\begin{cases} s_t^1 = \sigma x_t + (1 - \sigma) s_{t-1}^1 \\ s_t^2 = \sigma s_t^1 + (1 - \sigma) s_{t-1}^2 \\ s_t^3 = \sigma s_t^2 + (1 - \sigma) s_{t-1}^3 \end{cases} \tag{3}$$

Among them, x_t is the actual data value at time t; σ is the smoothing coefficient, and its value range is (0, 1). On the other hand, the three-time exponential smoothing method predicts the data at time q after time t as follows:

$$x_{t+q} = a_t + b_t q + c_t q^2 \tag{4}$$

In the formula (4), a_t, b_t, c_t, represent the prediction parameters at time t, respectively, and the calculation formula is as shown in (5):

$$\begin{cases} a_t = 3s_t^1 - 3s_t^3 + s_t^3 \\ b_t = \frac{\sigma}{2(1-\sigma)^2}\left[(6-5\sigma)s_t^1 - 2(5-4\sigma)s_t^2 + (4-3\sigma)s_t^3\right] \\ c_t = \frac{\sigma^2\left(s_t^1 - 2s_t^2 + s_t^3\right)}{2(1-\sigma)^2} \end{cases} \tag{5}$$

The determination of the initial value is also important for the entire forecasting process. The initial parameters of the cubic exponential smoothing method are s_0^1, s_0^2 and s_0^3. In the traditional exponential smoothing method, there are two ways to select the initial value. One is to use the actual value at the initial moment as the initial parameter when there are many time series. It is as formula 9. When the data is small, the average of the first three periods is selected as the initial value, it is shown in formula 7.

$$s_0^1 = s_0^2 = s_0^3 = x_0 \tag{6}$$

$$s_0^1 = s_0^2 = s_0^3 = \frac{x_1 + x_2 + x_3}{3} \tag{7}$$

3.2 Multi-directional Prediction Strategy

When the PS center point prediction method is used to solve the equidistant change of PS over time, the method shows better performance as shown in Fig. 1. However, the actual dynamic optimization problem will also show unequal-distance non-linear changes as shown in Fig. 2. At this time, the method of center point prediction will not be able to track the complete evolution direction of PS on this type of problem.

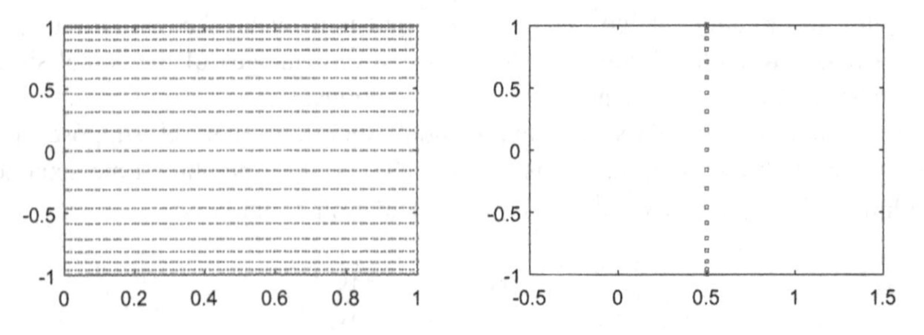

Fig. 2. The changes in the first two dimensions of PS and center point over time in the FDA1 test function

Therefore, this article uses a multi-directional prediction model, and the algorithm can still search for the problems in Fig. 3. When the prediction direction adopts one direction to predict the local optimal solution that will lead to the rapid convergence of the algorithm, this paper uses the method of multi-directional prediction to make the algorithm track the optimal solution more accurately.

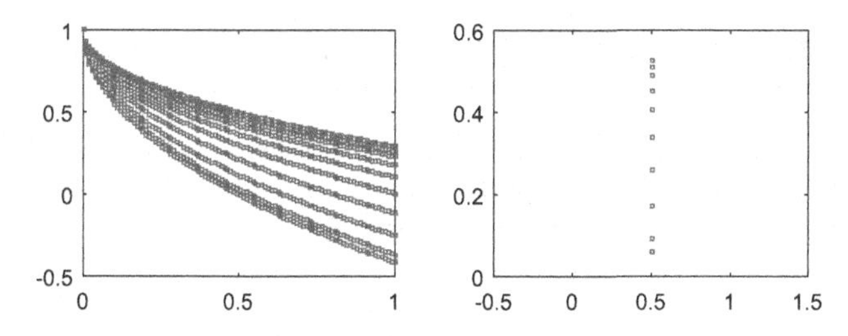

Fig. 3. The changes in the first two dimensions of PS and center point over time in a test function

This paper first divides the evolutionary population into H classes through a self-organizing mapping method [20], and then calculates the center point of each class as the evolution direction. The calculation method of the center point of the i-th category in PS at time t is calculated by formula (8).

$$c_i^t = \frac{1}{|P_i^t|} \sum\nolimits_{x_i^t \in P_i^t} x_i^t (i = 1, \ldots, H) \tag{8}$$

Among them, $P_i^t = \{x_i^t\}$ is the output at time t is approximate PS_i^t, $|P_i^t|$ is the basis of P_i^t of the i-th type in the entire PS at time t.

The center point of the i-th cluster at time t is c_i^t, the center point of the i-th cluster at time t-1 is c_i^{t-1}, then the calculation formula for the evolution direction of the i-th cluster is as follows (9).

$$\Delta c_i^t = c_i^t - c_i^{t-1} \tag{9}$$

3.3 Proposed Framework and Implementation

The basic description of the dynamic multi-objective evolutionary algorithm framework based on exponential smoothing and multi-directional prediction model (ESMDPS) is shown in Algorithm 2. ESMDPS first initializes the evolution group *POP*, the external elite population Archive, the number of clusters H, the weight vector, the training solution set S, the parameters and the parameters of exponential smoothing. At the beginning of each generation of evolution, ESMDPS judges the environment has changed or not firstly. If it changes, the ESMDPS is used the multi-direction model predict the evolution population at the next moment, otherwise using the static A self-organizing multi-objective evolutionary algorithm (SMEA) [20] to generate offspring.

Algorithm 2 ESMDPS()

Input: The scale of the population POP is N_p; External elite archive (Archive) scale N; the maximum number of iterations g_{max}; parameter t, τ, n_t, n_τ
Output: Approximate Pareto optimal solution set
Initialize the population of N_p, set up t=0, g=0
if $g<g_{max}$
Use the SOM method to divide the evolutionary group into H classes, and calculate the center position of each class to monitor environmental changes
If Environmental change
Calculate the search direction Δc_i^t of each cluster center at current t and t-1
Use exponential smoothing to predict Pareto optimal solution set POP_{esp} at t+1
New evolutionary population $POP_{t+1}=POP_{esp}+\sum_i^K \Delta c_i^t + POP_{rand}$
else
for u=1: N_p
A choice of matching pool Q for one solution x^u $Q = \begin{cases} \bigcup_{k=1}^{H}\{x^{u^k}\} & \text{if rand} < \beta \\ POP & \text{otherwise} \end{cases}$
Generate a new solution y through differential evolution algorithm y=Generate(Q,x^u)
Update evolution population P and external files *Archive*
endfor
Update the training data set
endif
endif

4 Experimental Study

4.1 Test Instances and Performance Metrics

The algorithm proposed in this paper is mainly to track such dynamic multi-objective optimization problems. FDA1 means that the PF does not change when the environment changes, while the Pareto optimal solution set changes with the environment. FDA2 means that when the environment changes, the Pareto frontier changes and the Pareto solution set remains unchanged. As FDA3 changes over time, both PF and PS have changed. The Pareto front surface of FDA1 and FDA3 is a concave curved surface. The Pareto front surface of FDA4 is a convex surface. The Pareto front surface changes with the change of the environment, and the Pareto solution set remains unchanged. This article also uses the DMOP1-DMOP3 [22] and F5-F10 [13] test sets.

(1) Modified Inverse Generation Distance, MIGD

Inverse Generation Distance, IGD is defined as measuring the diversity and convergence of the algorithm, when IGD is small, P must be very close to PF and no part of the entire PF can be lost.

The MIGD indicator [4] defines the average IGD value of the algorithm running at some time point t, which is defined as follows:

$$MIGD = \frac{1}{|T|}\sum\nolimits_{t \in T} IGD(PF^{t*}, P^{t}) \tag{10}$$

Among them, T is the set of discrete time points set in one run, $|T|$ is the base of T.

(2) Modified Hyper-volume Difference, MHVD

Hyper-volume Difference, HVD [4] is the super-volume difference between the solution set obtained by the measurement algorithm and the ideal PF. The ideal PF solution set uniformly distributed at time t is denoted as PF_t, P_t represents the real Pareto solution set searched by the algorithm.

$$HVD(PF_t, P_t) = |HV(PF_t) - HV(P_t)| \tag{11}$$

Among them, $HV(S)$ is the hyper-volume of the solution set S, $|\cdot|$ represents the absolute value of the resulting super volume.

MHVD is a revised version of HVD, the same as MIGD is a revised version of IGD. MHVD is the average value of HVD over a period of time in one run.

$$\text{MHVD} = \frac{1}{|T|}\sum\nolimits_{t \in T} HVD(PF_t, P_t) \tag{12}$$

Among them, T is a series of discrete time points in a run, $|T|$ is the basis of T.

When calculating the hyper-volume, the reference point is set to $(Z_1^t + 0.2, Z_2^t + 0.2, \ldots, Z_m^t + 0.2)$, among them Z_j^t is the maximum value of the jth target of the ideal PF at time t, m is the objective number. MHVD is a comprehensive index that evaluates the DMOEAs of convergence and diversity performance. The smaller the value of MHVD, the performance of DMOEAs is better.

4.2 Parameter Settings

All experimental tests are run in a desktop computer room configured with Inter(R) Core(TM) i7–3770 CPU @ 3.40 GHz and 4.00 GB. The test software is Matlab 2016 under the Windows 10 operating system. The maximum function evaluation times are Max_FES as the termination condition of all comparison algorithms. In order to reduce errors, each running algorithm is independently run 30 times on each test function. The decision variable of all comparison algorithms and the proposed algorithm EMDPS is 20. The scale of the evolution group is 100 for the two-objective test set, 200 for the three-objective test set, the number of function evaluations is 300000, the amplitude of environmental change is 10, and the frequency of environmental change is 25. The parameters of the comparison algorithm are the same as those set in the paper.

4.3 Experimental Results

This section compares the performance of ESMDPS algorithm with RIS, FPS, PPS and CKPS. The experiment in this article includes the comparison of two groups of experiments, the comparison of MIGD index and MHVD index. The method of rank sum test is used to compare the experimental results of the proposed algorithm and the comparison algorithm. The " + ", "-", "≈" in the table indicate that the proposed algorithm ESMDPS has better results compared with the corresponding comparison algorithm. Compared with the corresponding comparison algorithm, ESMDPS has a bad result, and compared with the corresponding comparison algorithm, the result of ESMDPS is similar. The table is bolded and the gray background is used to indicate that the algorithm has the best data result on the test function. In the comparison, the entire environment is divided into 120 environments, the frequency of change is 25, the number of function evaluations is 30000, and the evolution algebra is 3000. If the environment changes once in the 25th generation, there will be a total of 120 environments. In this experiment, the 120 environments are divided into three stages. The first stage StageI, the second stage StageII and the third stage StageIII are the environments of [0 20], [21 70] and [71 120] respectively. Table 1 show the mean and variance of MIGD and MHVD obtained after 30 runs of the three stages.

(1) MIGD matric

In this set of experiments, the maximum number of iterations is used as the condition for the termination of the algorithm. Table 1 shows the MIGD statistical mean and variance of the solutions obtained by the four dynamic comparison algorithms in 13 dynamic benchmark test functions for 30 independent runs, and the significance of the ESMDPS algorithm and the comparison algorithm in performance indicators. This article uses the method of rank sum test to test the significance of the data obtained by the algorithm. In the text † indicates that the ESMDPS algorithm performance is better than the comparison algorithm, ≈ indicates that the ESMDPS algorithm performance index is similar to the comparison algorithm, and-indicates that the ESMDPS algorithm performance index is inferior to the comparison algorithm.

From the results in Table 1, it is concluded that the IGD performance indicators of ESMDPS in the test functions FDA1, FDA4, F6 and F10 are more effective than the other four comparison algorithms. When time has changes, the ESMDPS can effectively response environmental changes. It can also be seen from Fig. 4 that the algorithm shows better rapid convergence than the other four comparison algorithms when solving FDA4 and F8.

(2) MHVD matric

The MHVD index is a measure of the difference between the actual super-volume and the super-volume composed of the ideal Pareto front. If the MHVD is smaller, the algorithm performance is better, that is, the super-volume composed of the frontier obtained by the algorithm is closer to that of the ideal Pareto front. The super-volume difference.

Table 1. Comparison of MIGD mean and variance of ESMDPS and comparative Algorithm

		RIS	FPS	PPS	CKPS	ESMDPS
FDA1	Total	1.2989E+00(3.3412E-	7.2265E-02(1.9322E-02)†	7.6392E-02(1.1218E-02)†	1.6025E-02(1.2989E-02)†	**2.7694E-03(8.2351E-03)**
	Stage I	1.2472E+00(1.4056E-	1.2502E-01(2.5444E-02)†	4.2213E-01(2.0411E-02)†	1.4985E-01(3.5602E-02)†	**1.2965E-02(2.4642E-03)**
	Stage II	1.3545E+00(2.8483E-	1.3228E-02(2.6942E-02)†	8.3096E-03(2.0414E-02)†	8.4491E-03(2.6096E-03)†	**7.5813E-04(3.9436E-04)**
	Stage	1.2639E+00(1.1602E-	1.2999E-02(3.6074E-02)†	6.1798E-03(7.0971E-02)†	8.1724E-03(2.5630E-02)†	**7.0255E-04(1.3811E-03)**
FDA3	Total	2.2407E+00(6.6954E-	3.2955E-01(7.1610E-02)†	3.7155E-01(1.9142E-02)†	3.6631E-01(1.4411E-02)†	**6.0231E-02(2.0864E-02)**
	Stage I	1.4239E+00(4.3601E-	3.7759E-01(1.7039E+00)†	2.6325E-01(4.1928E-02)†	5.6919E-01(2.1803E-02)†	**3.2015E-02(4.2678E-03)**
	Stage II	2.5191E+00(1.0821E-	3.18834E-01(8.1811E-	3.0206E-01(6.2930E-02)†	**3.2075E-02(1.7873E-02)-**	6.5779E-02(4.3338E-02)
	Stage	2.2237E+00(1.8422E-	3.2394E-01(7.1759E-02)†	3.5355E-01(2.8486E-03)†	3.3071E-01(3.1933E-02)†	**6.5969E-02(4.6712E-03)**
FDA4	Total	4.6006E-01(2.0365E-	1.4047E-01(3.0335E-01)†	1.3943E-01(1.6531E-03)†	1.1266E-01(1.8570E-02)†	**1.7126E-02(3.0593E-02)**
	Stage I	4.6726E-01(2.7698E-	1.42243788(4.7075E-02)†	2.2025E-01(3.2493E-02)†	1.4336E-01(3.9974E-02)†	**2.5621E-02(5.2923E-02)**
	Stage II	4.5707E-01(2.2674E-	1.3005E-01(3.3831E-02)†	1.2455E-01(3.7233E-02)†	1.0373E-01(2.1821E-03)†	**1.5498E-02(1.0956E-02)**
	Stage	4.6001E-01(1.5278E-	1.3005E-01(4.0236E-03)†	1.2198E-01(2.2575E-02)†	1.0992E-01(1.1420E-02)†	**1.5202E-02(2.5601E-03)**
DMOP1	Total	6.6605E-01(5.1978E-	1.1046E-01(5.5462E-01)†	3.9764E-01(3.6252E-02)†	1.0944E-01(4.9113E-03)†	**2.6483E-02(1.7265E-02)**
	Stage I	6.1045E-01(8.4110E-	1.0230E-01(9.3990E-02)†	1.3686E+00(2.1365E-03)†	4.5327E-02(2.1339E-02)-	**4.5442E-02(1.9719E-02)**
	Stage II	7.0339E-01(3.1761E-	1.1842E-01(2.9895E-02)†	3.0129E-01(5.2557E-02)†	1.1856E-01(2.5287E-03)†	**2.1441E-02(3.4619E-02)**
	Stage	6.5095E-01(1.9531E-	1.0575E-01(1.4715E-03)†	1.0559E-01(4.2007E-02)†	1.0585E-01(1.8677E-02)†	**2.3943E-02(2.3847E-03)**
DMOP2	Total	1.7994E+00(5.1152E-	1.3576E-01(2.1905E-02)†	2.2964E-01(9.6534E-02)†	1.4901E-01(2.6409E-02)†	**2.1191E-02(3.1213E-02)**
	Stage I	1.6476E+00(2.0904E-	2.5001E-01(5.1113E-02)†	6.0694E-01(1.0247E-02)†	1.2681E-01(6.0044E-02)†	**1.8983E-02(4.3282E-04)**
	Stage II	1.9703E+00(1.1651E-	1.2045E-01(6.0369E-02)†	1.6841E-01(1.5682E-02)†	1.1899E-01(2.0378E-02)†	**2.2068E-02(3.4393E-02)**
	Stage	1.6893E+00(3.5256E-	1.0538E-01(2.7707E-02)†	1.3995E-01(3.7573E-03)†	1.0521E-01(1.5136E-03)†	**2.1196E-02(3.4546E-02)**
DMOP3	Total	1.2221E+00(2.8332E-	3.2824E-02(3.7697E-02)≈	5.3253E-02(7.2724E-02)≈	1.0130E-02(5.2189E-02)†	**2.8951E-02(2.1568E-02)**
	Stage I	1.2267E+00(3.8281E-	1.3499E-01(1.6148E-02)†	2.8084E-02(2.2294E-02)†	**2.1119E-02(1.0152E-02)-**	1.3732E-02(1.9223E-02)
	Stage II	1.2569E+00(5.1166E-	1.5593E-02(4.5856E-03)†	9.2311E-03(3.2570E-02)≈	8.1068E-03(4.7662E-02)†	**7.7424E-03(4.3769E-02)**
	Stage	1.1987E+00(2.4973E-	1.3274E-02(4.9094E-02)†	6.2387E-03(5.0991E-02)≈	8.1992E-03(3.1795E-03)†	**6.8103E-04(3.6939E-03)**
F5	Total	1.1754E+00(1.2989E-	4.0351E-01(7.5347E-01)†	4.1191E-01(5.0175E-02)†	2.8946E-01(5.9005E-03)†	**7.8224E-02(1.2001E-03)**
	Stage I	9.9052E-01(2.2331E-	9.3122E-01(2.6806E-02)≈	1.0578E+00(1.1607E-02)†	2.3288E-01(7.9007E-03)-	**2.3616E-01(1.3001E-03)**
	Stage II	1.3370E+00(7.8472E-	3.2670E-01(2.9482E-02)†	2.9640E-01(3.2491E-02)†	2.7033E-01(9.3159E-03)†	**6.6778E-02(2.5145E-02)**
	Stage	1.0878E+00(3.0648E-	2.6924E-01(4.5994E-03)†	2.6903E-01(2.9479E-03)†	2.6723E-01(4.8706E-02)†	**2.6495E-02(1.1691E-02)**
F6	Total	5.8651E-01(1.9409E-	3.1786E-01(4.2490E-02)†	2.7560E-01(4.5621E-02)†	2.7955E-01(3.8884E-02)†	**5.7018E-02(2.4833E-02)**
	Stage I	7.6513E-01(2.2449E-	4.9975E-01(1.2246E-02)†	3.0184E-01(2.9085E-02)≈	3.3581E-01(2.6523E-02)†	**1.5874E-01(1.6964E-03)**
	Stage II	5.8682E-01(4.1493E-	2.9851E-01(1.0993E-03)†	2.7190E-01(2.8743E-02)†	2.6903E-01(7.5021E-02)†	**4.4755E-02(1.1062E-02)**
	Stage	5.1475E-01(1.2300E-	2.6446E-01(4.1972E-02)†	2.6881E-01(1.4343E-02)†	2.6757E-01(2.0542E-03)†	**2.8591E-02(1.5940E-04)**
F7	Total	6.1521E-01(2.3407E-	3.2134E-01(4.3692E+00)†	3.2392E-01(6.6402E-02)†	2.7397E-01(7.6965E-02)†	**3.6454E-02(2.0699E-02)**
	Stage I	6.7352E-01(4.0020E-	4.5182E-01(1.9452E-02)†	5.9379E-01(1.8629E-02)†	1.0379E-01(2.5998E-02)-	**1.6522E-01(1.2246E-02)**
	Stage II	6.1234E-01(3.5281E-	2.9396E-01(4.8523E-02)†	2.7162E-01(4.3770E-02)†	2.6869E-01(3.9134E-02)†	**3.0227E-02(1.0993E-02)**
	Stage	5.9475E-01(1.6435E-	2.9652E-01(7.2787E-01)†	2.6828E-01(5.5435E-03)†	2.6732E-01(2.7481E-03)†	**2.6853E-02(1.9452E-04)**
F8	Total	9.6057E-01(3.3775E-	1.4192E-01(2.9218E-02)†	1.6682E-01(3.3317E-02)†	1.5472E-01(1.6767E-02)†	**1.9639E-02(1.8701E-02)**
	Stage I	9.3703E-01(3.2409E-	2.3878E-01(1.7396E-02)†	3.5685E-01(3.5179E-02)†	3.1945E-01(4.4614E-02)†	**5.0215E-02(2.3662E-04)**
	Stage II	9.3321E-01(2.5145E-	1.1815E-01(6.2003E-04)†	1.3076E-01(4.6498E-02)†	1.2368E-01(1.2413E-04)†	**1.2641E-02(1.0711E-02)**
	Stage	9.9735E-01(1.1691E-	1.2694E-01(1.6529E-02)†	1.2688E-01(8.3526E-04)†	1.1987E-01(6.0886E-03)†	**1.3951E-02(3.8269E-02)**
F9	Total	1.1332E+00(2.4834E-	6.4266E-01(3.2842E-02)†	8.6422E-01(4.4608E-02)†	3.0650E-01(3.8272E-02)†	**5.7153E-02(2.5561E-02)**
	Stage I	1.1056E+00(2.4075E-	1.3432E+00(2.2050E-02)†	3.6487E+00(2.0109E-02)†	1.2485E-01(1.5874E-02)-	**1.8676E-01(2.3098E-03)**
	Stage II	1.1508E+00(1.6964E-	5.4205E-01(7.5919E-02)†	3.6868E-01(4.0634E-02)†	2.6751E-01(4.6440E-02)†	**3.7192E-02(4.2490E-02)**
	Stage	1.1267E+00(1.1062E-	4.6302E-01(1.5425E-02)†	2.4596E-01(2.3912E-02)†	2.5814E-01(2.6313E-02)†	**2.5268E-02(1.9725E-04)**
F10	Total	1.0739E+00(1.5940E-	3.9603E-01(3.0673E-03)†	4.5825E-01(4.7898E-02)†	2.9229E-01(8.8097E-02)†	**7.6819E-02(2.9931E-02)**
	Stage I	1.1095E+00(3.5579E-	5.2817E-01(9.8763E-02)†	1.0465E+00(2.285E-02)†	1.1201E-01(9.2388E-02)-	**1.1375E-01(1.8714E-02)**
	Stage II	1.0996E+00(3.4795E-	4.996E-01(1.8714E-02)†	3.5662E-01(5.1314E-03)†	2.7986E-01(8.4018E-02)†	**3.2303E-02(4.1562E-03)**
	Stage	1.0340E+00(2.6998E-	3.2801E-01(2.3991E-03)†	3.2454E-01(4.1841E-02)†	2.6804E-01(3.2523E-03)†	**2.5104E-02(8.1953E-03)**

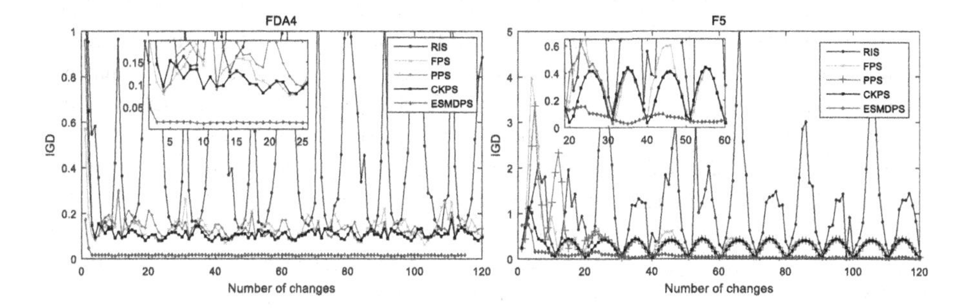

Fig. 4. Evolution curves of average IGD values on FDA4 and F5

In this set of experiments, the conditions for the termination of all algorithms are the same as the conditions for the termination of the MIGD indicator. Similarly, the mean and variance obtained after each algorithm is independently run 30 times are listed in Table 1. The comparison algorithm has the best effect in bold. Compared with the comparison algorithm, the symbols †, – and ≈ respectively indicate that the SEMDPS algorithm has significant performance, insignificant performance and similar performance compared with the comparison algorithm.

It can be seen from Table 1 that for FDA1, FDA4, it has better performance than other comparison algorithms in terms of convergence. ESMDPS algorithm in general has better MHVD than other comparison algorithms for the four optimization problems at each stage, and other algorithms are better than the comparison algorithm MHVD in a certain stage on the FDA3 and F5-F10 test problems. As shown in Fig. 5, the ESMDPS algorithm solves the FDA1 problem when the algorithm converges to the optimal solution when t changes to around 10. When solving the FDA4 problem, the algorithm did not maintain a stable convergence curve when solving FDA1, but it also showed better performance in dealing with environmental changes than the other four comparison algorithms. When solving the DMOP3 problem, the ESMDPS algorithm can also maintain a better stable ability to respond to environmental changes.

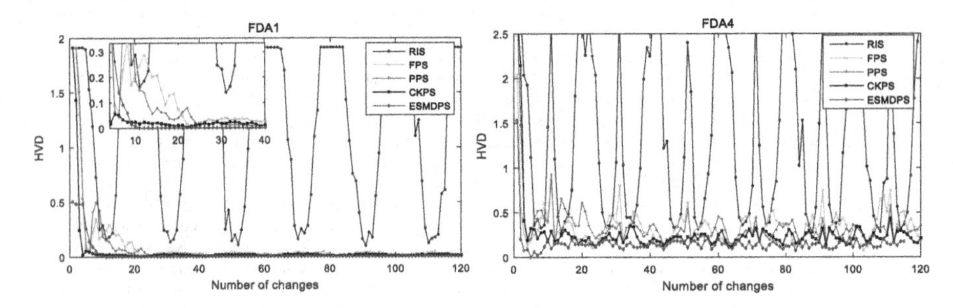

Fig. 5. Evolution curves of average *HVD* values on FDA1 and FDA4

5 Experimental Study

The key point of the dynamic multi-objective optimization problem is that when the environment changes over time, the algorithm should maintain effective diversity and rapid convergence balance ability to deal with this change. Aiming at the problem that the time series does not change with time and the phenomenon of unequal distance is often encountered in practical problems, this paper proposes a multi-directional prediction model for this type of problem to predict the evolutionary population after environmental changes, so as to quickly and accurately predict the evolution direction of the evolution group. Secondly, in order to effectively predict the next moment, the optimal solution that changes with time, this article uses exponential smoothing as a prediction model to predict the evolution of the next moment. In addition, in order to ensure the diversity of the evolutionary group, this paper also uses a re-initialization of a set of solutions to increase the diversity of the evolutionary group.

The multi-directional prediction model plays a role in accelerating the accurate convergence of the search algorithm in the algorithm proposed in this paper. The model first uses a self-organizing mapping machine learning method to cluster the solution space, and cluster the sub-categories of each sub-category. Directional guidance can speed up the convergence speed of the algorithm in the new environment. In addition, individuals predicted by the exponential smoothing method and the individuals predicted by the multi-directional prediction model Random initialization to maintain the diversity of the evolutionary group. The algorithm is tested with 12 standard test functions and compared with other 5 compared algorithms. The results show that the proposed algorithm can maintain good performance when dealing with dynamic multi-objective optimization problems.

Acknowledgements. This work was supported by the research projects: Natural Science Foundation of Shaanxi Province nos. 2021JQ-711, Shaanxi Provincial Department of Education Fund nos. 20JK0910. The thesis work was funded by the special fund construction project for key disciplines of ordinary colleges and universities in Shaanxi Province.

References

1. Deb, K., Pratap, A., Agarwal, S., Meyarivan, T.: A fast and elitist multi-objective genetic algorithm: NSGA-II. IEEE Trans. Evol. Comput. **6**(2), 182–197 (2002)
2. Deb, K., Rao, N.U.B., Karthik, S.: Dynamic multi-objective optimization and decision-making using modified NSGA-II: a case study on hydro-thermal power scheduling. In: Obayashi, S., Deb, K., Poloni, C., Hiroyasu, T., Murata, T. (eds.) Evolutionary Multi-Criterion Optimization. EMO 2007. LNCS, vol. 4403, , pp. 803–817. Springer, Berlin, Heidelberg. https://doi.org/10.1007/978-3-540-70928-2_60
3. Jin, Y., Yang, C., Ding, J., Chai, T.: Reference point based prediction for evolutionary dynamic multiobjective optimization. In: 2016 IEEE Congress on Evolutionary Computation (CEC), pp. 3769–3776. IEEE (2016)
4. Ruan, G., Yu, G., Zheng, J., Zou, J., Yang, S.: The effect of diversity maintenance on prediction in dynamic multi-objective optimization. Appl. Soft Comput. **58**, 631–647 (2017)
5. Guo, Y., Yang, H., Chen, M., Cheng, J., Gong, D.: Ensemble prediction-based dynamic robust multi-objective optimization methods. Swarm Evol. Comput. **48**, 156–171 (2019)
6. Rong, M., Gong, D., Zhang, Y., Jin, Y., Pedrycz, W.: Multidirectional prediction approach for dynamic multiobjective optimization problems. IEEE Trans. Cybern. **49**(9), 3362–3374 (2018)
7. Liu, R., Li, J., Mu, C., Jiao, L.: A coevolutionary technique based on multi-swarm particle swarm optimization for dynamic multi-objective optimization. Eur. J. Oper. Res. **261**(3), 1028–1051 (2017)
8. Azzouz, R., Bechikh, S., Ben Said, L.: Dynamic multi-objective optimization using evolutionary algorithms: a survey. In: Bechikh, S., Datta, R., Gupta, A. (eds.) Recent Advances in Evolutionary Multi-objective Optimization. Adaptation, Learning, and Optimization, vol. 20, pp. 31–70. Springer, Cham (2017). https://doi.org/10.1007/978-3-319-42978-6_2
9. Liu, X.F., Zhou, Y.R., Yu, X.: Cooperative particle swarm optimization with reference-point-based prediction strategy for dynamic multiobjective optimization. Appl. Soft Comput. **87**, 105988 (2020)

10. Wang, F., Li, Y., Liao, F., Yan, H.: An ensemble learning based prediction strategy for dynamic multi-objective optimization. Appl. Soft Comput. **96**, 106592 (2020)
11. Jiang, M., Huang, Z., Qiu, L., Huang, W., Yen, G.G.: Transfer learning-based dynamic multiobjective optimization algorithms. IEEE Trans. Evol. Comput. **22**(4), 501–514 (2017)
12. Cao, L., Xu, L., Goodman, E.D., Bao, C., Zhu, S.: Evolutionary dynamic multiobjective optimization assisted by a support vector regression predictor. IEEE Trans. Evol. Comput. **24**(2), 305–319 (2019)
13. Zhou, A., Jin, Y., Zhang, Q.: A population prediction strategy for evolutionary dynamic multiobjective optimization. IEEE Trans. Cybern. **44**(1), 40–53 (2013)
14. Azzouz, R., Bechikh, S., Ben Said, L.: Dynamic multi-objective optimization using evolutionary algorithms: a survey. In: Bechikh, S., Datta, R., Gupta, A. (eds.) Recent Advances in Evolutionary Multi-objective Optimization. Adaptation, Learning, and Optimization, vol. 20, pp. 31–70. Springer, Cham. https://doi.org/10.1007/978-3-319-42978-6_2
15. Liu, R., Yang, P., Liu, J.: A dynamic multi-objective optimization evolutionary algorithm for complex environmental changes. Knowl.-Based Syst. **216**, 106612 (2021)
16. Muruganantham, A., Tan, K.C., Vadakkepat, P.: Evolutionary dynamic multiobjective optimization via Kalman filter prediction. IEEE Trans. Cybern. **46**(12), 2862–2873 (2015)
17. Goh, C.K., Tan, K.C., Liu, D.S., Chiam, S.C.: A competitive and cooperative co-evolutionary approach to multi-objective particle swarm optimization algorithm design. Eur. J. Oper. Res. **202**(1), 42–54 (2010)
18. Jiang, S., Yang, S.: A steady-state and generational evolutionary algorithm for dynamic multiobjective optimization. IEEE Trans. Evol. Comput. **21**(1), 65–82 (2016)
19. Zhao, Q., Yan, B., Shi, Y., Middendorf, M.: Evolutionary dynamic multiobjective optimization via learning from historical search process. IEEE Trans. Cybern. (2021)
20. Zhang, H., Zhou, A., Song, S., Zhang, Q., Gao, X.Z., Zhang, J.: A self-organizing multiobjective evolutionary algorithm. IEEE Trans. Evol. Comput. **20**(5), 792–806 (2016)
21. Gardner, E.S., Jr.: Exponential smoothing: the state of the art. J. Forecast. **4**(1), 1–28 (1985)
22. Deb, K., Rao, N.U.B., Karthik, S.: Dynamic multi-objective optimization and decision-making using modified NSGA-II: a case study on hydro-thermal power scheduling. In: Obayashi, S., Deb, K., Poloni, C., Hiroyasu, T., Murata, T. (eds.) Evolutionary Multi-Criterion Optimization. EMO 2007. LNCS, vol. 4403, pp. 803–817. Springer, Berlin, Heidelberg (2007). https://doi.org/10.1007/978-3-540-70928-2_60

Automatic Particle Swarm Optimizer Based on Reinforcement Learning

Rui Dai, Hui Zheng(✉), Jing Jie(✉), and Xiaoli Wu

Zhejiang University of Science and Technology, Hangzhou, China
{huizheng, jingjie}@zust.edu.cn

Abstract. As an efficient search technique based on population, particle swarm optimizer (PSO) has been widely used to deal with practical optimization problems in different fields. To improve the generalization ability and accuracy of PSO, this paper proposes an automatic PSO based on reinforcement learning (RLAPSO). In RLAPSO, reinforcement learning is introduced to conduct the global search. By designing state, action, 3-dimensional Q table, and reward function to determine the generation strategy that is more suitable for the current process characteristics. Meanwhile, the parameters of optimizers are adjusted linearly in the process of optimization. To avoid prematurity, the global search in the later stage of the search is transformed into a local one to find a fine solution. Finally, the performance of RLAPSO is tested on five notable benchmark functions. The experimental results show that RLAPSO is competitive with the state-of-the-art PSO variants.

Keywords: Reinforcement learning · Automatic PSO · Global search

1 Introduction

As an efficient and powerful population-based random search technique, swarm intelligence optimization algorithms (SIs) are widely applied in many scientific and engineering fields, such as scheduling design [1], robot path planning [2], and so on. Particle swarm optimizer (PSO) is one of the most popular optimization techniques in SIs. It has gradually become a research hotspot in intelligent computing due to its self-organization, self-learning, and self-adaptability. However, PSO cannot perform well on all optimization problems. Especially when optimization problems have numerous locally optimal solutions or high dimensions, the performance of PSO is poor [3]. This is mainly because PSO tends to fall into local optimality when dealing with multimodal and high-dimensional problems, which leads to premature convergence [4].

As we all know, convergence speed and global search ability are two important indicators to measure PSO performance [5]. To balance the trade-off between them to alleviate premature convergence, scholars have proposed many PSO variants, which are roughly divided into four categories:

(1) **Parameter control.** In the optimization process, the parameters ω and c_i ($i = 1, 2$) have a significant impact on PSO. To find a real-time parameter adjustment mechanism and maintain the best performance of PSO in solving optimization

L. Pan et al. (Eds.): BIC-TA 2021, CCIS 1565, pp. 317–331, 2022.
https://doi.org/10.1007/978-981-19-1256-6_24

problems, more and more scholars are committed to adaptive parameter control. For example, Liu et al. [6] proposed a PSO based on reinforcement learning (QLPSO), which uses Q-learning to online control the parameters of PSO. Subsequently, Lu et al. [7] proposed an adaptive online closed-loop parameter control (CLPC) strategy. The PID controller is used to adjust the parameter values adaptively. The mean shift clustering method obtains the convergence entropy and extended entropy as feedback to monitor and reflect the evolution state in real-time.

(2) **Topology modification.** Some scholars introduce the topology modification strategy to improve the diversity during the optimization process. For example, Cheng et al. [5] introduced a pairwise competition mechanism to balance PSO exploration and development capabilities (CSO). Then, Jin et al. [8] introduced the social learning mechanism into PSO and proposed a social learning PSO (SLPSO). Each particle learns from any better particle in the current population instead of based on historical information. Zheng et al. [9] proposed a reinforcement-learning-based PSO (RLPSO), which adds the learning dimension of predicting experience by elite networks and introduces reinforcement learning strategies to guide particle behavior effectively.

(3) **Multi-swarm mechanism.** The idea of the multi-swarm is mainly to enhance diversity by exchanging information between different swarms. For example, in [10], a dynamic multi-swarm PSO (DMS-PSO) is proposed to obtain better population diversity by dynamically changing the neighborhood structure. Inspired by multi-swarm information sharing and elite disturbance guidance, Zhao et al. [11] proposed a multi-swarm cooperative multistage perturbation guiding PSO (MCpPSO). The exchange and sharing of information among different subpopulations with different evolutionary mechanisms effectively improve evolutionary efficiency. Wang et al. [12] proposed a heterogeneous comprehensive learning dynamic multi-swarm PSO with two mutation operators (HCLDMS-PSO), enhancing the performance through dynamic multi-swarm search with decreasing nonlinear adaptive inertia weight and global search based on Gaussian mutation.

(4) **Ensemble learning.** As known from the "No Free Lunch Theorem" [13], a single algorithm cannot perform well on all optimization problems, and different search techniques have different advantages. Thus, Lynn et al. [14] proposed an ensemble-based PSO (EPSO), encouraging elite management in competitive algorithms using an adaptive evaluation strategy. What's more, Olorunda and Engelbrecht [15] proposed a heterogeneous collaborative algorithm based on the cooperative model [16], which combines differential evolutionary (DE), genetic algorithm (GA), and PSO to find effective solutions for different problems. Meanwhile, Liang et al. [17] also proposed a hybrid PSO with crisscross learning strategy (PSO-CL), which uses a crossover-based comprehensive learning strategy (CCL) and a stochastic example learning Strategy (SEL) to avoid early convergence.

Anyway, many PSO variants are developed to improve performance. The development of adaptive PSO that can solve a variety of complex problems has become the primary goal of the current research. Therefore, this paper proposes an automatic PSO based on reinforcement learning (RLAPSO). Firstly, four PSO variants are adopted to construct the optimizer pool. Then, based on the state of the decision space (DS) and the objective space (OS) to match different search stages, introduce the Q-learning strategy to adaptively determine the generation strategy which is more suitable for the current process characteristics from the optimizer pool. Then the optimizer parameters are adjusted linearly in optimization to avoid the expensive calculation cost caused by repeated trial and error. Finally, the global search in the later search stage is transformed into a local landscape to find a better solution. Additionally, RLAPSO is tested on five popular benchmark functions. The experimental results show that RLAPSO has better accuracy and generalization ability than other state-of-the-art algorithms.

This paper is organized as follows: Sect. 2 provides a brief overview of the principles of four PSO variants utilized in this paper. Section 3 introduces the proposed RLAPSO in detail. Section 4 verifies the effectiveness of the proposed algorithm through five benchmark functions. Finally, Sect. 5 gives the conclusion and future work.

2 Preliminary

2.1 PSO Variant Principle

Particle swarm optimizer (PSO), first proposed by Kennedy and Eberhart in 1995 [18], is a computational model to simulate the predation behavior of birds or fish. Each particle represents a possible solution in the search space, and the population is a set of potential solutions. Let there be N randomly placed particles in the search space. Each particle has its position and velocity, expressed as $x_i = [x_i^1, x_i^2, ..., x_i^d]$ and $v_i = [v_i^1, v_i^2, ..., v_i^d]$ (i = 1, …, N), respectively. When the t_{th} iteration, the position and velocity of particle i are updated as follows:

$$v_{i,t+1}^d = v_{i,t}^d + c_1 r_1 (pbest_{i,t}^d - x_{i,t}^d) + c_2 r_2 (gbest_t^d - x_{i,t}^d) \tag{1}$$

$$x_{i,k+1}^d = x_{i,k}^d + v_{i,k+1}^d \tag{2}$$

Where d is the dimension; c_1 is the coefficient of cognitive acceleration; c_2 is the coefficient of social acceleration; r_1 and r_2 are random numbers uniformly distributed in the range of [0,1].

Later, to improve the search performance of PSO, a variety of PSO variants are proposed. This paper selects four PSO variants, i.e., standard PSO (SPSO) [19], fitness-distance-ratio based PSO (FDR-PSO) [20], self-organizing hierarchical PSO with time-varying acceleration coefficients (HPSO-TVAC) [21] and distance-based locally informed PSO (LIPSO) [22]. Their velocity update formulas are shown in Table 1. The velocity update formulas of four PSO variants:

Table 1. The velocity update formulas of four PSO variants

Optimizer	Velocity update formula	
SPSO	$v_{i,k+1}^{d} = \omega v_{i,k}^{d} + c_1 r_1 (pbest_{i,k}^{d} - x_{i,k}^{d}) + c_2 r_2 (gbest_k^{d} - x_{i,k}^{d})$	(3)
FDR-PSO	$v_{i,k+1}^{d} = \omega v_{i,k}^{d} + c_1 r_1 (pbest_{i,k}^{d} - x_{i,k}^{d}) + c_2 r_2 (gbest_k^{d} - x_{i,k}^{d}) + c_3 (nbest_{i,k}^{j} - x_{i,k}^{d})$	(4)
HPSO-TVAC	$\begin{cases} v_{i,k+1}^{d} = c_1 r_1 (pbest_{i,k}^{d} - x_{i,k}^{d}) + c_2 r_2 (gbest_k^{d} - x_{i,k}^{d}) & v_i^d \neq 0 \\ v_{i,k+1}^{d} = rand_i^d v_{i,k}^{d} & v_i^d = 0 \end{cases}$	(5)
LIPSO	$v_{i,k+1}^{d} = \chi (v_{i,k}^{d} + \varphi (p_{i,k}^{d} - x_{i,k}^{d}))$	(6)

Where the linearly decreasing inertia weight ω is introduced in (3) to control the weight of historical velocity adaptively better to realize the effective balance between exploitation and exploration. Equation (4) adds a social learning dimension that learns from the experience of neighborhood particle $nbest_i = [nbest_i^1, nbest_i^2, \ldots, nbest_i^d]$, and the neighborhood particles are determined by the fitness distance ratio (FDR), $FDR_i = (f(pbest_i) - f(x_i)) / |pbest_i^d - x_i^d|$, $i = 1, \ldots, N$. In HPSO-TVAC, the particle velocity will be reinitialized whenever the particle stagnates in the search space ($v_i^d = 0$), as shown in (5). Meanwhile, c_1 and c_2 in (5) are linear and time-varying. LIPSO completely dependents on the information from neighboring particles $nbest_j$ to guide the search, so $p_i = \left(\sum_{j=1}^{nsize} (\varphi_j nbest_j) / nsize\right) / \sum_{j=1}^{nsize} \varphi_j$ and $nsize$ is the neighborhood size. As shown in (6), shrinkage coefficient $\chi = 0.7298$ and φ_j is a random number uniformly distributed in the range of [0, 4.1/*nsize*].

2.2 Reinforcement Learning

Reinforcement learning (RL) [23] is a popular unsupervised learning in machine learning, which mainly interacts with the environment through trial and error and uses maximizing cumulative reward to learn the optimal strategy. The RL model comprises five essential elements, i.e., agent, environment, state (s), action (a), and reward (r). Its operation can be described as a cycle of state, action, and reward. As shown in Fig. 1, Q-learning is a value-based and model-free reinforcement learning strategy in which agents are rewarded according to the quality of each given action.

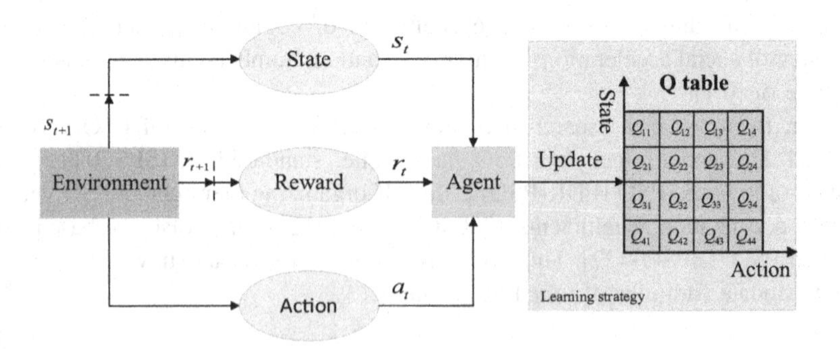

Fig. 1. Illustration of reinforcement learning and Q-learning

Set $S = \{s_1, s_2,\ldots, s_m\}$ is a state set of the environment, $A = \{a_1, a_2, \ldots,a_n\}$ is an executable action set. The agent selects action with the maximum Q value according to the information accumulated in the Q table. After the action occurs, the environment gives the reward and the new state, and the agent learns the knowledge to evaluate the expected value of each given action. The update formula is as follows [24]:

$$Q(s_{t+1}, a_{t+1}) = (1-\alpha)Q(s_t, a_t) + \alpha\left[R(s_t, a_t) + \gamma \max_a Q(s_{t+1}, a_t)\right] \tag{7}$$

Here, the learning rate α and the discount factor γ are usually selected within the range of [0,1], and $R(s_t, a_t)$is the real-time reward for performing action a_t in state s_t.

3 Automatic PSO Based on Reinforcement Learning

The proposed RLAPSO can realize the online adaptive selection of optimizers based on Q-learning strategy to improve the generalization ability and accuracy effectively. The Q-learning strategy and RLAPSO are described in detail below.

3.1 Q-Learning Strategy

The role of the Q-learning strategy is to adaptively select an optimizer with the maximum Q value from the optimizer pool (SPSO, FDR-PSO, HPSO-TVAC, and LIPSO) at each iteration based on the states of individuals in the DS and OS, updated with the reward (or punishment) received at the end of each iteration. Therefore, the design of state, action, reward, and Q table is the key of RLAPSO.

Table 2. The definition of the DS state and OS state.

Relative distance (Δd)	DS state	Relative fitness value(Δf)	OS state
$0 \leq \Delta d < 0.25\Delta D$	Nearest	$0 \leq \Delta f < 0.25\Delta F$	Smallest
$0.25\Delta D \leq \Delta d < 0.5\Delta D$	Nearer	$0.25\Delta F \leq \Delta f < 0.5\Delta F$	Smaller
$0.5\Delta D \leq \Delta d < 0.75\Delta D$	Farther	$0.5\Delta F \leq \Delta f < 0.75\Delta F$	Larger
$0.75\Delta D \leq \Delta d \leq \Delta D$	Farthest	$0.75\Delta F \leq \Delta f \leq \Delta F$	Largest

State Design. The Euclidean distance $d = \|x_i - gbest\|$ between particle i and the global optimal position $gbest$ is determined in the DS. Then four states are defined according to the relative distance $\Delta d = |d - \Delta D|$ of d relative to the scope of the search space ΔD, i.e., nearest, nearer, farther, and farthest. Similarly, the fitness difference $f = \|f(x_i) - f(gbest)\|$ between particle i and the global optimal value $f(gbest)$ is first determined in the OS. Then four states are defined according to the relative fitness value $\Delta f = |f - \Delta F|$ of f relative to the objective space range ΔF, i.e., smallest, smaller, larger, largest, as shown in Table 2.

Action Design. The purpose of introducing the Q-learning strategy in this paper is to realize the online adaptive selection of optimizers, so this paper uses four PSO variants, i.e., SPSO, FDR-PSO, HPSO-TVAC, and LIPSO, to build an optimizer library. Among them, SPSO and HPSO-TVAC are suitable for solving unimodal problems, while CLPSO, LIPS, and FDR-PSO are suitable for solving multimodal problems.

Therefore, this paper designs the four behaviors of particles, i.e., using SPSO for coarse-grained search, FDR-PSO for fine-grained search, HPSO-TVAC for fast convergence, and LIPSO for slow convergence.

Q Table Design. The traditional Q table is 2 dimensions, but this paper uses a 3-dimensional Q table to consider both the DS and OS. According to each particle's state to determine their behavior. Among them, DS state, OS state, and action are three dimensions of the Q table.

First, a specific position in the state plane, such as s = {nearer, larger}, is determined according to the state s of individuals in the DS and OS. Then, the associated Q column is extracted according to this specific position. Finally, the optimizer with the maximum value in the extracted Q column is selected to update the population information, as shown in Fig. 2.

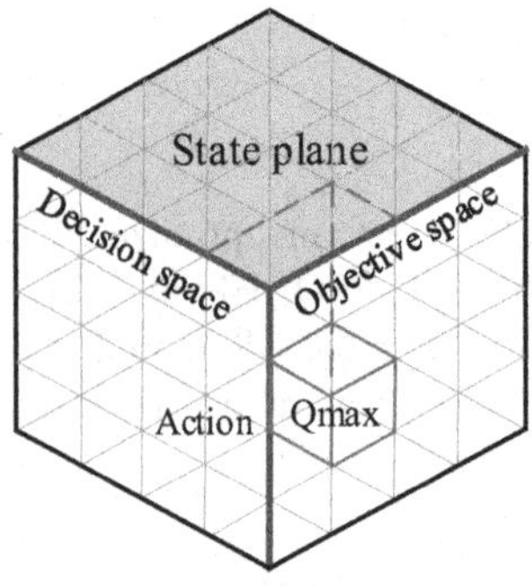

Fig. 2. Illustration of 3-dimensional Q table

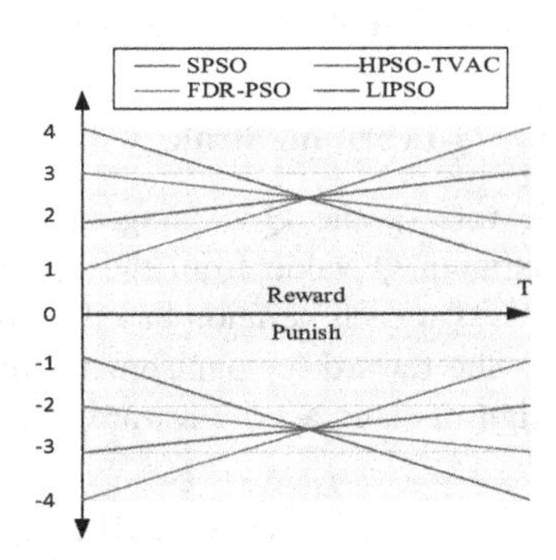

Fig. 3. Illustration of the reward function

Reward Function Design. All particles share a 3-dimensional Q table, and individuals can contribute to updating the Q table, but it is known from (7) that the reward function (RF) is also critical to update it. Design of the RF requires to consider not only the fitness of particles but also the influence on the search behavior of particles. To maintain the consistency of the reward system throughout the optimization process, the RF, utilized in this paper, is designed as shown in Fig. 3 [6]. After acting, if the fitness improves, the behavior should be rewarded. That is, continue to use the optimizer to update the population information. On the contrary, if the fitness worsens or remains, the action should be punished. That is, the optimizer is re-selected for population update.

3.2 RLAPSO

Abundant experiments verify that local exploration is helpful to find a more accurate solution. Therefore, in the last 10% generations, RLAPSO mainly conducts the local search to sustain the robustness of the algorithm. Moreover, the parameters of each optimizer are dynamically linearly adjusted in the optimization process to reduce the impact on the proposed algorithm. The workflow of the RLAPSO is shown in Fig. 4.

First, initialize the swarm information, 3-dimensional Q table, and determine the state s of the individual. Later, based on the Q table, choose the most suitable optimizer for the current search phase and update population information. Finally, employ the RF to update the Q table and repeat in each generation. If the number of iterations reaches the 90% maximum generation, stop the global exploitation and start the local exploration using SPSO only with $c_2 = 3$ until the end, as shown in Algorithm. Note that SPSO is chosen for local search instead of FDR-PSO because a large number of experiments show that the local search accuracy of SPSO and FDR-PSO is similar and the convergence speed of SPSO with $c_2 = 3$ is faster.

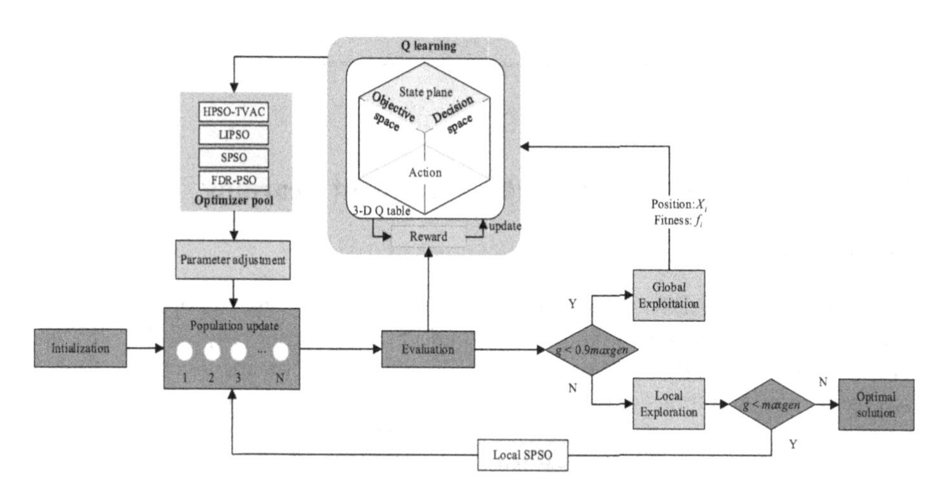

Fig. 4. Illustration of RLAPSO workflow.

Algorithm: RLAPSO	
1	Initialize swarm information, 3-dimensional Q table, $pbest_i$ and $gbest$, i=1, 2, ..., N;
2	while $g \leq maxgen$
3	if $g \leq 0.9maxgen$
4	for i = 1: N //Global exploitation
5	Determine the current state s_i according to the position x_i of particles in the DS and OS;
6	Find the $max(Q_i)$ in the relevant column of the Q table and determine the action a according to the state s. That is the optimizer used to update the particle information;
7	Update the position x_i and velocity v_i of particle i;
8	Evaluate the fitness value by the objective function;
9	Update individual optimal and global optimal information;
10	Update Q table using the RF;
11	end for
12	else // Local exploration
13	Update particles information using local SPSO;
14	Update individual optimal and global optimal information;
15	end if
	t++;
16	end while

4 Automatic PSO Based on Reinforcement Learning

4.1 Experimental Conditions

To further study RLAPSO, five benchmark functions widely used in optimization are selected, and the results are compared with QLPSO [6], EPSO [14], SLPSO [8], and HPSO-TVAC [21]. Table 3 shows the introduction of benchmark functions. Table 4 lists the parameter settings of four PSO variants. Suppose that the population size of all algorithms is N = 20, and the maximum generation is $maxgen$ = 100, i.e., the maximum number of fitness evaluations is $maxFEs$ = 5000. In addition, the other parameter settings of comparative algorithms are the same as in the original paper. The experimental results are obtained in MATLAB ®R2019b under Windows10 system and run 20 times independently on an Intel (R) Core (TM) i5–7200 U CPU@2.50 GHz laptop.

Table 3. Characteristic summary of benchmark functions

Function	Dimension	Characteristic	Global optimum
Sphere	10, 30, 50	Unimodal	0
Rosenbrock	10, 30, 50	Multimodal with narrow valley	0
Ackley	10, 30, 50	Multimodal	0
Griewank	10, 30, 50	Multimodal	0
Rastrigin	10, 30, 50	Multimodal	0

Table 4. Parameter settings of PSO variants

Algorithm	Inertia weight ω	Constriction coefficients χ	Acceleration coefficients c, c_1, c_2, c_3	Neighborhood size
SPSO	0.9–0.2	–	$c_1 = 2$, $c_2 = 2$	–
FDR-PSO	0.9–0.2	0.729	$c_1 = 1$, $c_2 = 1$, $c_3 = 2$	–
HPSO-TAVC	–	–	$c_1 = 2.5–0.5$, $c_2 = 0.5–2.5$	–
LIPS	–	0.729	$c = 2$	3
Local SPSO	0	–	$c_1 = 0$, $c_2 = 3$	–

5 Experimental Results

Table 5, 6 and Table 7 records the optimization results of the five algorithms in different functions. Through the statistical data such as the best value (*best*), the worst value (*worst*), the mean value (*mean*), and the standard deviation value (*std.*), the proposed RLAPSO is compared with EPSO, QLPSO, SLPSO, and HPSO-TVAC. In all cases, the best average results are highlighted in bold. Furthermore, Fig. 5 describes the dynamic curves of five algorithms running once in different function optimization.

Table 5. Optimization results of three algorithms on 10-dimensional test functions.

Function	Algorithm	*best*	*mean*	*std*	*worst*
Sphere	RLAPSO	0.00	**0.03**	0.06	0.18
	QLPSO	0.00	0.68	1.44	5.93
	EPSO	0.10	0.35	0.20	0.75
	SLPSO	0.00	0.05	0.05	0.02
	HPSO-TVAC	0.51	4.88	3.66	14.60
Rosenbrock	RLAPSO	6.14	**7.84**	1.09	9.60
	QLPSO	6.12	8.42	0.85	9.61
	EPSO	23.40	15.70	3.99	10.10
	SLPSO	4.13	29.20	62.40	238.00
	HPSO-TVAC	6.85	8.64	0.69	10.30
Ackley	RLAPSO	0.03	2.42	1.29	5.29
	QLPSO	0.03	2.52	1.00	4.17
	EPSO	7.14	6.01	0.83	4.05
	SLPSO	0.08	**0.32**	0.21	0.78
	HPSO-TVAC	0.63	2.79	1.22	4.79

(*continued*)

Table 5. (*continued*)

Function	Algorithm	*best*	*mean*	*std*	*worst*
Griewank	RLAPSO	0.00	**0.02**	0.02	0.06
	QLPSO	0.03	0.27	0.19	0.63
	EPSO	3.85	2.33	0.70	1.44
	SLPSO	0.00	0.08	0.11	0.48
	HPSO-TVAC	0.17	0.65	0.39	1.78
Rastrigin	RLAPSO	1.99	**8.59**	4.43	18.90
	QLPSO	2.99	11.20	4.66	21.90
	EPSO	53.40	42.90	6.19	28.30
	SLPSO	4.27	9.36	4.18	19.20
	HPSO-TVAC	3.71	11.20	4.88	21.90

For 10-dimensional test functions, RLAPSO effectively optimizes Sphere, Griewank, and Rastrigin functions, which is an order of magnitude less than the average results of QLPSO, EPSO, and HPSO-TVAC. In other functions, the average performance of RLAPSO is similar to QLPSO and HPSO-TVAC, and slightly better than EPSO. In addition, it can be observed that the average result of SLPSO optimizing Ackley is better than RLAPSO, and the results on other functions (except Rosenbrock) are also comparable to RLAPSO.

Table 6. Optimization results of three algorithms on 30-dimensional test functions

Function	Algorithm	*best*	*mean*	*std*	*worst*
Sphere	RLAPSO	**0.04**	**0.21**	0.13	0.67
	QLPSO	0.56	2.94	2.25	8.05
	EPSO	6.55	4.92	1.00	2.82
	SLPSO	0.48	1.15	0.57	2.98
	HPSO-TVAC	364.00	850.00	265.00	1250.00
Rosenbrock	RLAPSO	**30.40**	**37.30**	9.56	74.00
	QLPSO	33.20	79.50	30.40	147.00
	EPSO	156.00	115.00	20.40	82.00
	SLPSO	93.90	567.00	678.00	2520.00
	HPSO-TVAC	50.20	84.00	23.90	135.00
Ackley	RLAPSO	4.73	**6.07**	0.90	7.75
	QLPSO	6.75	9.22	1.63	12.50
	EPSO	10.40	9.16	0.64	8.23
	SLPSO	**4.24**	7.24	0.60	6.26
	HPSO-TVAC	6.69	8.82	1.13	10.70

(*continued*)

Table 6. (*continued*)

Function	Algorithm	*best*	*mean*	*std*	*worst*
Griewank	RLAPSO	**0.02**	**0.08**	0.06	0.28
	QLPSO	2.08	8.59	4.87	22.60
	EPSO	22.10	16.60	3.71	8.86
	SLPSO	0.68	0.88	0.08	1.02
	HPSO-TVAC	4.74	8.47	1.66	11.10
Rastrigin	RLAPSO	**17.40**	**39.20**	15.70	70.20
	QLPSO	44.00	67.30	15.00	91.80
	EPSO	233.00	209.00	13.30	185.00
	SLPSO	116.00	147.00	20.60	205.00
	HPSO-TVAC	57.00	90.10	29.30	95.50

For 30-dimensional test functions, when RLAPSO optimizes Sphere and Griewank functions, the average result is at least one order of magnitude lower than the remaining algorithms. In terms of Rosenbrock and Rastrigin, the average performance of RLAPSO is similar to QLPSO and HPSO-TVAC but slightly better than EPSO and SLPSO. The average results of comparative algorithms are similar when optimizing Ackley. Because the complexity of the Ackley landscape increases significantly with the increase of dimension, and it is easy to fall into local optimization. However, the optimization results of comparative algorithms are still ideal.

Table 7. Optimization results of three algorithms on 50-dimensional test functions

Function	Algorithm	*best*	*mean*	*std*	*worst*
Sphere	RLAPSO	**2.94**	11.70	5.24	20.90
	QLPSO	5.62	**11.30**	3.77	19.80
	EPSO	15.90	12.40	1.59	9.67
	SLPSO	24.20	66.70	31.00	142.00
	HPSO-TVAC	2010.00	3730.00	907.00	5330.00
Rosenbrock	RLAPSO	**14.60**	**27.40**	69.40	421.00
	QLPSO	63.60	84.80	16.10	131.00
	EPSO	393.00	282.00	41.90	224.00
	SLPSO	2210.00	14400.00	22200.00	83800.00
	HPSO-TVAC	167.00	229.00	35.60	292.00
Ackley	RLAPSO	**5.56**	**7.24**	0.90	9.07
	QLPSO	9.61	12.30	1.25	15.30
	EPSO	11.70	10.70	0.67	9.06
	SLPSO	69.70	89.60	0.82	10.70
	HPSO-TVAC	10.10	11.10	0.65	12.10

(*continued*)

Table 7. (*continued*)

Function	Algorithm	*best*	*mean*	*std*	*worst*
Griewank	RLAPSO	21.50	47.20	16.90	83.90
	QLPSO	21.90	**41.90**	13.10	69.30
	EPSO	58.70	44.50	5.70	36.70
	SLPSO	76.10	88.80	84.50	11.30
	HPSO-TVAC	**17.20**	45.50	7.48	45.40
Rastrigin	RLAPSO	**22.50**	**46.40**	19.90	117.00
	QLPSO	113.00	173.00	27.60	225.00
	EPSO	431.00	398.00	19.30	363.00
	SLPSO	266.00	338.00	31.40	408.00
	HPSO-TVAC	176.00	225.00	31.10	281.00

For 50-dimensional test functions, when RLAPSO optimizes Ackley and Rastrigin functions, the average result is an order of magnitude smaller than the other algorithms. On the Rosenbrock function, the average performance of RLAPSO is similar to QLPSO but outperforms EPSO, SLPSO, and HPSO-TVAC. The average results of comparative algorithms in optimizing Sphere and Griewank functions are similar. Although the average value of QLPSO is slightly better than RLAPSO, the best value of RLAPSO is better than QLPSO. It is worth mentioning that the complexity of the Griewank function decreases with the increase of dimension, which shows that the performance advantages of these algorithms are equal when dealing with high-dimensional simple functions.

In a word, the proposed RLAPSO has advantages in three different dimensions of five functions, and its optimization average result is better than or comparable to the other algorithms. Meanwhile, Fig. 5 shows the dynamic performance of a complete optimization process performed by five algorithms, proving the above conclusion. From the curve, although the convergence speed of RLAPSO is slow at the initial search stage, with the progress of the search, RLAPSO can quickly converge to a more accurate solution. Since the previous global search is mainly based on reinforcement learning, it is necessary to select the optimizer suitable for the current process characteristics and dynamically linearly adjust the optimizer parameters in each iteration to slow down the convergence speed. Nevertheless, the automatic optimizer selection with Q-learning can effectively improve the global convergence performance of the proposed algorithm, so in the later local search process, RLAPSO can converge quickly and improve the quality of the search solution.

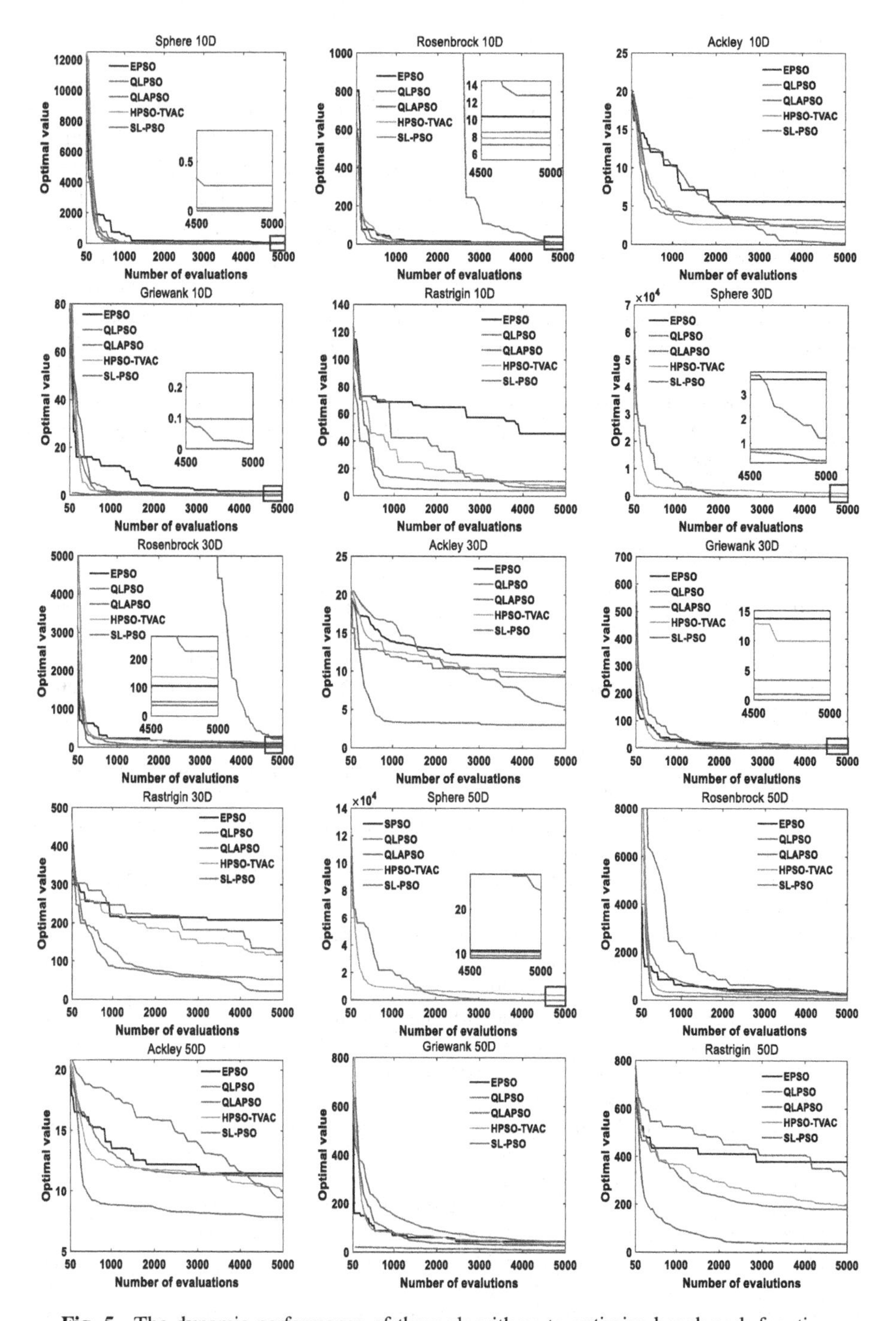

Fig. 5. The dynamic performance of three algorithms to optimize benchmark function.

6 Conclusion

This paper proposes an automatic PSO based on reinforcement learning (RLAPSO). Q-learning realizes the automatic online selection of PSO variants suitable for the current process features through the design of state, action, 3-dimensional Q table, and reward function. Thus, the generalization ability and accuracy are effectively improved. The performance of RLAPSO is tested on five popular benchmark functions with different fitness landscapes. The experimental results show that the proposed algorithm has a better performance compared with the four state-of-the-art algorithms. The future work also includes applying the proposed algorithm in real-world applications and researching multi-objective PSO based on reinforcement learning.

Acknowledgment. This work was supported in part by Pinghu Science and Technology Plan Project (No. GY202112) and Zhejiang Public Welfare Technology Application Research Project (No. LGG19F030005).

References

1. Park, H.J., Cho, S.W., Lee, C.: Particle swarm optimization algorithm with time buffer insertion for robust berth scheduling. Comput. Ind. Eng. **160**, 107585 (2021)
2. Song, B., Wang, Z., Zou, L.: An improved PSO algorithm for smooth path planning of mobile robots using continuous high-degree Bezier curve. Appl. Soft Comput. **100**(1), 106960 (2021)
3. Yang, Y., Pedersen, J.: A comparative study on feature selection in text categorization. In: Proceedings of International Conference on Machine Learning, pp. 412–420. Morgan Kaufmann Publishers (1997)
4. Chen, W., Zhang, J., Lin, Y., Chen, E.: Particle swarm optimization with an aging leader and challengers. IEEE Trans. Evol. Comput. **17**(2), 241–258 (2013)
5. Cheng, R., Jin, Y.: A competitive swarm optimizer for large scale optimization. IEEE Trans. Cybern. **45**(2), 191–204 (2015)
6. Liu, Y., Lu, H., Cheng, S., et al.: An adaptive online parameter control algorithm for particle swarm optimization based on reinforcement learning. In: IEEE Congress on Evolutionary Computation (CEC), pp. 815–822 (2019)
7. Lu, H., Liu, Y., Cheng, S., et al.: Adaptive online data-driven closed-loop parameter control strategy for swarm intelligence algorithm. Inf. Sci. **536**, 25–52 (2020)
8. Cheng, R., Jin, Y.: A social learning particle swarm optimization algorithm for scalable optimization. Inf. Sci. **291**, 43–60 (2015)
9. Lu, L., Zheng, H., Jie, J., Zhang, M., Dai, R.: Reinforcement learning-based particle swarm optimization for sewage treatment control. Complex Intell. Syst. **7**(5), 2199–2210 (2021). https://doi.org/10.1007/s40747-021-00395-w
10. Liang, J., Suganthan, P.: Dynamic multi-swarm particle swarm optimizer. In: Proceedings of IEEE Swarm Intelligence Symposium, pp. 124–129. IEEE (2005)
11. Zhao, X., Liu, Z., Yang, X.: A multi-swarm cooperative multistage perturbation guiding particle swarm optimizer. Appl. Soft Comput. **22**, 77–93 (2014)
12. Wang, S., Liu, G., et al.: Heterogeneous comprehensive learning and dynamic multi-swarm particle swarm optimizer with two mutation operators. Inf. Sci. **540**, 175–201 (2020)

13. Wolpert, D., Macready, W.: No free lunch theorems for optimization. IEEE Trans. Evol. Comput. **1**(1), 67–82 (1997)
14. Lynn, N., Suganthan, P.: Ensemble particle swarm optimizer. Appl. Soft Comput. **55**, 533–548 (2017)
15. Olorunda, O., Engelbrecht, A.: An analysis of heterogeneous cooperative algorithms. In: IEEE Congress on Evolutionary Computation, pp. 1562–1569 (2009)
16. Potter, M.: The Design and Analysis of a Computational Model of Cooperative Coevolution. Citeseer (1997)
17. Liang, B., Zhao, Y., Li, Y.: A hybrid particle swarm optimization with crisscross learning strategy. Eng. Appl. Artif. Intell. **105**, 104418 (2021)
18. Eberhart, R., Kennedy, J.: A new optimizer using particle swarm theory. In: Proceedings of the Sixth International Symposium on Micro Machine and Human Science, vol. 1, pp. 39–43 (1995)
19. Shi, Y., Eberhart, R.: A modified particle swarm optimizer. In: Proceedings of the IEEE Congress on Evolutionary Computation, pp. 69–73 (1999)
20. Peram, T., Veeramachaneni, K., Mohan, C.: Fitness-distance-ratio based particle swarm optimization. In: Proceedings of the 2003 IEEE Swarm Intelligence Symposium, pp. 174–181 (2003)
21. Ratnaweera, A., Halgamuge, S., Watson, H.: Self-organizing hierarchical particle swarm optimizer with time-varying acceleration coefficients. IEEE Trans. Evol. Comput. **8**, 240–255 (2004)
22. Qu, B., Suganthan, P., Das, S.: A distance-based locally informed particle swarm model for multimodal optimization. IEEE Trans. Evol. Comput. **17**, 387–402 (2013)
23. Shahrabi, J., Adibi, M., Mahootchi, M.: A reinforcement learning approach to parameter estimation in dynamic job shop scheduling. Comput. Ind. Eng. **110**, 75–82 (2017)
24. Huynh, T., Do, D., Lee, J.: Q-Learning-based parameter control in differential evolution for structural optimization. Appl. Soft Comput. **107**(11), 107464 (2021)

A Multi-UUV Formation Control and Reorganization Method Based on Path Tracking Controller and Improved Ant Colony Algorithm

Bin Yang[1], Shuo Zhang[2], Guangyu Luo[3(✉)], and Dongming Zhao[3]

[1] The Second Military Representative Office of Seafarer in Wuhan Area, Wuhan 430064, China
[2] China Ship Development and Design Center, Wuhan 430064, China
[3] School of Automation, Wuhan University of Technology, Wuhan 430070, China
{luoguangyu,dmzhao}@whut.edu.cn

Abstract. Aiming at the path selection problem in the reorganization of multi-UUV formations, a path tracking controller and an improved ant colony algorithm are proposed. First, a path tracking controller based on the UUV kinematics model is designed to realize the formation maintenance of multiple UUVs. Secondly, a new local pheromone update method is introduced to improve the convergence of traditional ant colony algorithm. The simulation proves the effectiveness of the proposed control algorithm.

Keywords: Unmanned underwater vehicle · Path tracking · Formation control · Ant colony algorithm · Formation reorganization

1 Introduction

Unmanned underwater vehicle (UUV) is a submersible that relies on its own energy, self-propelled, and autonomously controlled. Compared with a single UUV, a multi-UUV system has advantages in terms of distribution, applicability, coordination, security, and intelligence. Teamwork to complete tasks can effectively improve efficiency, expand capabilities, save costs, and improve accuracy [1–3]. The existing control strategies for multi-UUV cooperative formation mainly include leader-following method [4–6], artificial potential field method [6,7], and virtual structure method [8]. However, the formation will be disrupted when multi-UUVs perform autonomous obstacle avoidance. In the process of reorganization, choosing an optimal path is worth exploring. There exist different approaches to solve the optimization problems [9,10], including ant colony algorithm [1,11], particle swarm algorithm [12], bee colony algorithm [13], simulated annealing algorithm [14]. Ant colony algorithm has been widely used in path planning due to its strong robustness. Therefore, this paper applies the ant colony algorithm to the reorganization and formation of multiple UUVs. At

L. Pan et al. (Eds.): BIC-TA 2021, CCIS 1565, pp. 332–341, 2022.
https://doi.org/10.1007/978-981-19-1256-6_25

the same time, the traditional ant colony algorithm needs to be improved to overcome the local minimum problem.

In this paper, a path tracking controller is designed to realize the formation maintenance of multi-UUVs based on the kinematic model. The local pheromone update method is introduced to improve the convergence of the standard ant colony algorithm, which can solve the local minimum problem. The remainder of this paper is organized as follows: Sect. 2 designs a path tracking controller. In Sect. 3, the formation reorganization method is introduced and the improved ant colony algorithm is proposed. In Sect. 4, the proposed algorithms are verified by many simulation experiments. Section 5 concludes the paper.

2 Multi-UUV Formation Control Method

2.1 UUV Kinematics and Dynamics Model

When multi-UUVs maintain a fixed formation, the kinematic and dynamical model can be expressed as [15]:

$$\begin{cases} \dot{x} = u\cos\psi - v\sin\psi \\ \dot{y} = u\sin\psi - v\cos\psi \\ \dot{\psi} = r \\ \dot{u} = \dfrac{m_2 vr - X_u u + \tau_u}{m_1} \\ \dot{v} = \dfrac{-m_1 ur - Y_v v}{m_2} \\ \dot{r} = \dfrac{(m_1 - m_2)uv - N_r r + \tau_r}{m_3} \end{cases} \tag{1}$$

where $[x, y, \psi]^T$ is the position and heading vector, $[u, v, r]^T$ is the velocity vector, X_u, Y_v, and N_r denote the hydrodynamic coefficients, m_1, m_2, and m_3 denote the mass inertia coefficients, τ_u and τ_r denote the thrust and steering torque.

2.2 Multi-UUV Formation Controller Design

In multi-UUVs system, one UUV is selected as the master and the remaining UUVs are slaves. A master-slave path tracking controller is designed so that each slave UUV tracks the master UUV to keep a fixed formation. Without considering external interference and communication constraints, the controller is designed based on Lyapunov and backstepping theory [16]. The Lyapunov function is designed as formula (2):

$$V_{xy} = \frac{1}{2}x_e^2 + \frac{1}{2}y_e^2 \tag{2}$$

where x_e and y_e are the path tracking errors, which is expressed as formula (3):

$$\begin{cases} \dot{x}_e = ry_e + u - u_F \cos\psi_e \\ \dot{y}_e = -rx_e + u_F \sin\psi_e + v \\ \dot{\psi}_e = r - r_F \end{cases} \tag{3}$$

where u_F and r_F denote the velocity and angular velocity of the desired target point, ψ_e is the bow angle tracking error. Derivation and simplification of formula (2) is shown as formula (4):

$$\dot{V}_{xy} = x_e\left(u - u_F \cos\psi_e\right) + y_e\left(v + u_F \sin\psi_e\right) \tag{4}$$

The control law u_c and ψ_c can be designed as formula (5):

$$\begin{cases} u_c = -k_1 x_e + u_F \cos\psi_e \\ \psi_c = -\arcsin\left(\dfrac{k_2 x_e}{\sqrt{1+\left(k_2 x_e\right)^2}}\right) \end{cases} \tag{5}$$

where k_1, k_2 denote constant parameters. The control law r_c is shown as formula (6):

$$r_c = r_F + \psi_c \tag{6}$$

The Lyapunov equation is shown as formula (7):

$$\dot{V}_{\mathrm{xy}} = -k_1 x_{\mathrm{e}}^2 - k_2 u_F \frac{1}{\sqrt{1+\left(k_2 x_e\right)^2}} x_{\mathrm{e}}^2 + x_e y \tag{7}$$

In summary, the controller can be designed as formula (8):

$$\begin{cases} \tau_u = m_1 \dot{u}_c - m_2 v r + X_u u_c \\ \tau_r = m_3 \dot{r}_c - (m_1 - m_2) u r + N_r r \end{cases} \tag{8}$$

where τ_u and τ_r denote the thrust force and steering torque.

3 Formation Reorganization Method Based on Ant Colony Algorithm

There are obstacles, communication interruptions, information transmission packet loss, and other unexpected situations in the course of travel. Consequently, the original formation will be disrupted. After passing obstacles or restoring communication, a new formation will be reassembled according to the formation controller. In the process of reorganizing the formation, it will often travel a long distance, which may increase energy consumption. Therefore, in order to reduce the distance traveled in the reorganization process, an improved ant colony algorithm (IACA) is used to re-plan the formation and optimize the path.

3.1 Problem Description

As shown in Fig. 1, the multi-UUVs keep a parallel formation before obstacle avoidance. The original formation is disrupted when fixed obstacles appear. After completing the obstacle avoidance process, multi-UUVs need to reform a parallel formation according to the designed controller. If UUVs adopt the previous control law, they will follow the path of the solid line in Fig. 1. In this paper, IACA is introduced to calculate and optimize the path, so that the multi-UUVs can replan the path of the dashed line in Fig. 1.

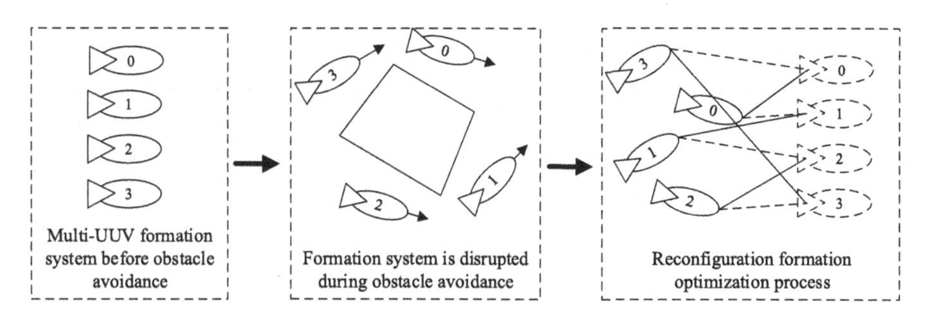

Fig. 1. Multi-UUV formation reorganization process

Suppose the set of positions coordinates of multi-UUVs before reorganization is $\Gamma = \{\eta_1, \eta_2, \ldots, \eta_i, \ldots, \eta_n\}$, the set of position coordinates after reorganization is $\Gamma^* = \{\eta_1{}^*, \eta_2{}^*, \ldots, \eta_i{}^*, \ldots, \eta_n{}^*\}$, where $\eta_i = (x_i, y_i)$, $\eta_i{}^* = (x_i{}^*, y_i{}^*)$. Then the set of distance is shown as formula (9):

$$D = \{d_{ij} : d_{ij} = \left\|\eta_j^* - \eta_i\right\|, i, j \in \{1, 2, \ldots, n\}\} \tag{9}$$

where d_{ij} denotes the distance between coordinates within the set of different path types. Then the problem can be described as the minimum driving path of all UUVs.

$$S^* = \min \sum_{i=1}^{n} d_{ij_i} \tag{10}$$

3.2 Standard Ant Colony Algorithm

The ant colony algorithm is proposed by [17] in 1996, which simulates the foraging behavior of ant colonies in nature. The ants with shorter walking paths have higher concentrations of released pheromone. More ants will choose shorter paths so that all ants will concentrate in the shortest path to arrive at the optimal solution.

Suppose the number of ants is m, the number of ant colony is n, the distance between path point i and path point j is $d_{ij}(i, j = 1, 2, ..., n)$, and the pheromone concentration on the path connecting ant i and ant j at time t is $\tau_{ij}(t)$. At the initial moment, the pheromone concentration is set to $\tau_{ij}(0) = \tau_0$. Ant k decides the next point to visit based on the pheromone concentration on the connected paths of each path point. Suppose $P_{ij}^k(t)$ denote the probability of ant k reaching path point j from path point i at time t:

$$P_{ij}^k(t) = \begin{cases} \frac{[\tau_{ij}(t)]^\alpha \times [\gamma_{ij}(t)]^\beta}{\sum\limits_{s \in allow_k} [\tau_{is}(t)]^\alpha \times [\gamma_{is}(t)]^\beta}, s \in allow_k \\ 0, s \notin allow_k \end{cases} \tag{11}$$

where $\gamma_{ij}(t) = \frac{1}{d_{ij}}$ denotes the heuristic function, α denotes the pheromone importance factor, β is the heuristic function importance factor, $\rho(0 < \rho < 1)$ denotes the pheromone volatility degree, $allow_k(k = 1, 2, ..., m)$ denotes the set of path points to be visited by ant k. As time increases, the elements in $allow_k(k = 1, 2, ..., m)$ keep decreasing. When ant k releases pheromone, the pheromone on the connected paths of each path point gradually decreases. When the ant completes one iteration, the pheromone concentration needs to be updated according to formula (12):

$$\begin{cases} \tau_{ij}(t+1) = (1-\rho)\tau_{ij}(t) + \rho\Delta\tau_{ij} \\ \Delta\tau_{ij} = \sum\limits_{k=1}^{n} \Delta\tau_{ij}{}^k \end{cases} \tag{12}$$

where $\Delta\tau_{ij}{}^k$ denotes the pheromone concentration released by ant k at path point i and j, $\Delta\tau_{ij}$ denotes the sum of the pheromone concentrations released by all ants at path point i and j. In general, the ant cycle system model is applied to express $\Delta\tau_{ij}$ as formula (13):

$$\Delta\tau_{ij}{}^k = \begin{cases} Q/L_k, \text{ant } k \text{ from } i \text{ to } j \\ 0, others \end{cases} \tag{13}$$

where Q is a constant parameter, L_k denotes the length of the path passed by ant k.

3.3 Improved Ant Colony Algorithm

In order to solve the problem of local minima, an improved method is proposed based on the standard ant colony algorithm. The main improvement is the pheromone update method. In this paper, a new local updates method is introduced to accelerate the convergence speed to avoid falling into the local optimum. The new method is designed as formula (14):

$$\tau_{ij}(t+1) = (1-\theta)\tau_{ij}(t) + \theta\tau_0 \tag{14}$$

where τ_0 denotes the initial value of pheromone and θ denotes the local pheromone volatility factor. The value of the pheromone on the path (i, j) is reduced by the volatility factor when the UUV selects the next location j from the current location i. Then, the probability of the remaining UUVs selecting the path (i, j) is reduced to avoid duplicate search paths, which can increase the probability of searching the remaining paths and improve the search speed of the algorithm. The global update is performed by $\tau_{ij}(t+1) = (1-\rho)\tau_{ij}(t) + \Delta\tau_{ij}$.

The steps of the IACA to solve UUV reorganization problem are described as follows:

- Step 1: Initialization parameters. The relevant parameters are initialized, including the number of UUVs n, the pheromone importance factor α, the heuristic function importance factor β, the local pheromone volatility factor θ, the global pheromone volatility factor ρ, the total number of pheromone releases Q, the maximum number of iterations $iter\_max$, the initial value of the number of iterations $iter$, the initial position coordinates of each UUV and the target point location coordinates.
- Step 2: Calculate the distance between each UUV. The mutual distances are calculated based on the coordinates of the initial positions of each UUV. The elements on the diagonal are corrected to 10^{-4} to ensure that the denominator is not zero.

$$d_{ij} = \begin{cases} \sqrt{(x_i - x_j)^2 + (y_i - y_j)^2}, i \neq j \\ 10^{-4}, i = j \end{cases} \tag{15}$$

- Step 3: Constructing solution spaces. Each UUV is randomly placed at a departure point. The next target position is calculated according to formula (16):

$$P_{ij}^{k}(t) = \begin{cases} \frac{[\tau_{ij}(t)]^{\alpha} \times [\eta_{ij}(t)]^{\beta}}{\sum\limits_{s \in allow_k} [\tau_{is}(t)]^{\alpha} \times [\eta_{is}(t)]^{\beta}}, s \in allow_k \\ 0, s \notin allow_k \end{cases} \tag{16}$$

- Step 4: Path update. Calculate the path length $L_k (k = 1, 2, ..., n)$ and record the shortest formation path length $L_{\min}$.
- Step 5: Partial update of pheromones. When each UUV selects the target location once, the pheromone of the last traveled path (i, j) is updated according to the formula $\tau_{ij}(t+1) = (1-\theta)\tau_{ij}(t) + \varepsilon\tau_0$ until the path construction is completed. Otherwise, return to Step 4 to continue updating the path.
- Step 6: Global update pheromone. The global pheromone concentration is updated according to formula (17) and (18):

$$\begin{cases} \tau_{ij}(t+1) = (1-\rho)\tau_{ij}(t) + \rho\Delta\tau_{ij} \\ \Delta\tau_{ij} = \sum\limits_{k=1}^{n} \Delta\tau_{ij}{}^{k} \end{cases} \tag{17}$$

$$\Delta\tau_{ij}{}^{k} = \begin{cases} Q/L_k, \text{ant } k \text{ from } i \text{ to } j \\ 0, others \end{cases} \tag{18}$$

- Step 7: Algorithm termination. Let $iter = iter + 1$ before the number of iteration reaches $iter\_max$. Otherwise, the algorithm terminates and outputs the result.

The main steps of IACA is shown in Fig. 2.

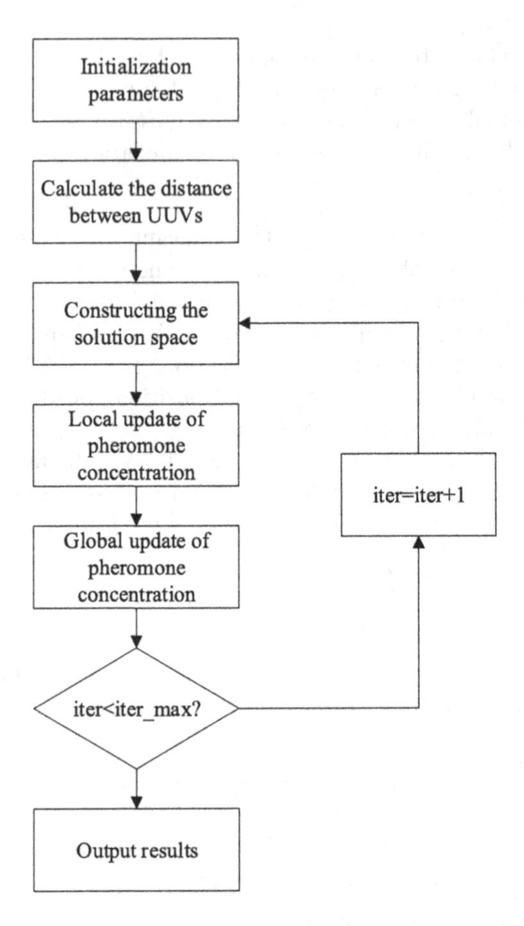

Fig. 2. Improved ant colony algorithm

4 Simulation Experiments

In this section, simulation experiments are held to verify the effectiveness of the algorithm. Ten UUVs are set up with random initial positions within a given range. After the master-slave path tracking controller is applied, a parallel formation at a fixed depth is formed. The formation process is shown in Fig. 3. Then, the initial formation is disrupted after the UUV formation passes through the fixed obstacles. At the same time, the initial coordinates of UUVs are obtained. The coordinates of the target point are set as Table 1 and Fig. 4.

In Fig. 4, blue points are initial points and red points are target points. The IACA is applied to optimize the re-formation path decision. The experimental parameters are set as follows.

Table 1. UUV initial and target point coordinates

UUV numbers	UUV formation initial coordinates	UUV formation target point coordinates
1	(0.5, 0.82)	(0.9, 0.1)
2	(0.64, 0.8)	(0.9, 0.2)
3	(0.82, 0.6)	(0.9, 0.3)
4	(0.6, 0.7)	(0.9, 0.4)
5	(0.43, 0.74)	(0.9, 0.5)
6	(0.3, 0.65)	(0.9, 0.6)
7	(0.4, 0.3)	(0.9, 0.7)
8	(0.22, 0.1)	(0.9, 0.8)
9	(0.7, 0.21)	(0.9, 0.9)
10	(0.33, 0.43)	(0.9, 0.1)

$$\begin{cases} n = 10 \\ \alpha = 1 \\ \beta = 5 \\ \theta = 0.1 \\ \rho = 0.5 \\ Q = 1 \\ iter\_max = 200 \\ iter = 1 \end{cases}$$

The optimal path generated by the algorithm is shown in Fig. 5. It can be seen that the UUV formation is in a scattered position after autonomous obstacle avoidance. After the optimization of the IACA, each UUV finds the optimal path that is shown as a straight-line distance in Fig. 5.

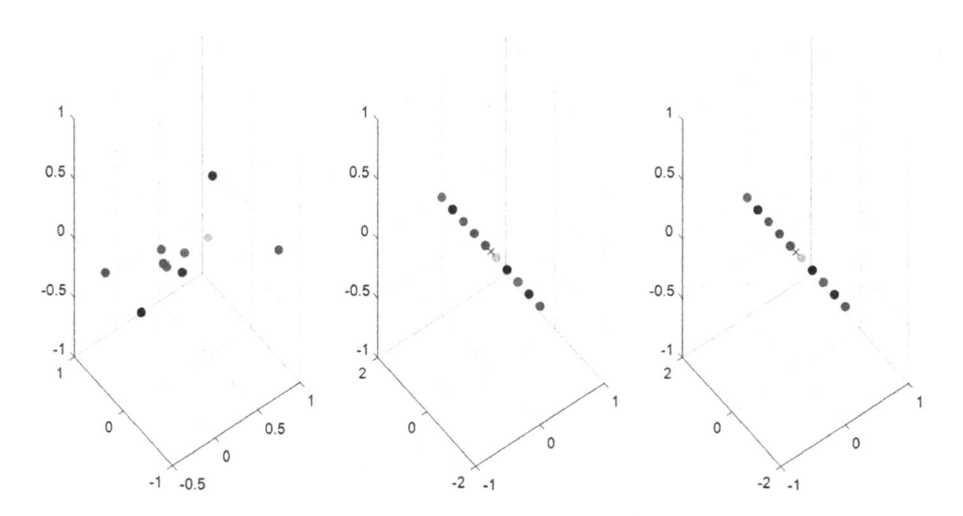

Fig. 3. Formation control and formation maintenance

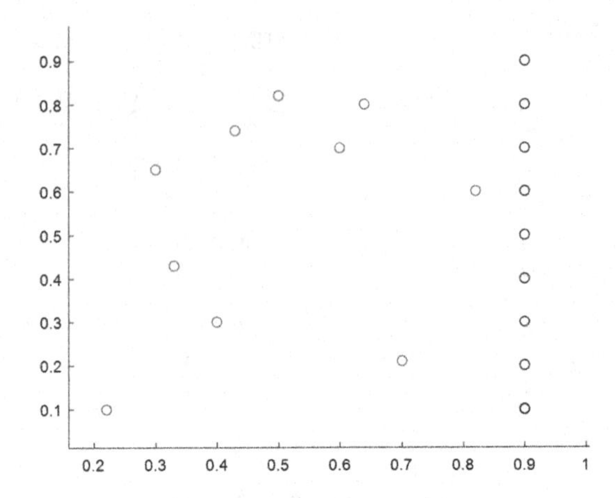

Fig. 4. UUV initial and target points (Color figure online)

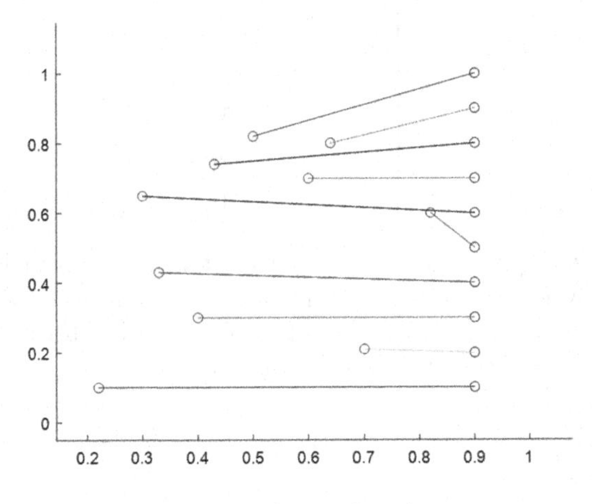

Fig. 5. Optimal path

5 Conclusion

In this paper, a path tracking controller and an IACA have been proposed to solve the problem of multi-UUV formation control and reorganization. A local update method of pheromone has been proposed to speed up the convergence to avoid falling into the local optimum. IACA has been applied for the reorganization of multi-UUV formation. A large number of simulation studies show the effectiveness of the method, and highlight the advantages of IACA.

The next step is to develop studies that include dynamic obstacles to further improve the performance of the IACA. Meanwhile, communication delay and environment disturbance will be taken into account in the future.

References

1. Champion, B.T., Joordens, M.A.: Underwater swarm robotics review. In: 2015 10th System of Systems Engineering Conference (SoSE), pp. 111–116. IEEE (2015)
2. Liu, Y., Ding, P., Wang, T.: Autonomous obstacle avoidance control for multi-UUVs based on multi-beam sonars. In: Global Oceans 2020: Singapore-US Gulf Coast, pp. 1–5. IEEE (2020)
3. Hadi, B., Khosravi, A., Sarhadi, P.: A review of the path planning and formation control for multiple autonomous underwater vehicles. J. Intell. Robot. Syst. **101**(4), 1–26 (2021)
4. Yan, Z., Liu, Y., Yu, C., Zhou, J.: Leader-following coordination of multiple UUVs formation under two independent topologies and time-varying delays. J. Cent. South Univ. **24**(2), 382–393 (2017)
5. Yan, Z., Xu, D., Chen, T., Zhang, W., Liu, Y.: Leader-follower formation control of UUVs with model uncertainties, current disturbances, and unstable communication. Sensors **18**(2), 662 (2018)
6. Ying, Z., Xu, L.: Leader-follower formation control and obstacle avoidance of multi-robot based on artificial potential field. In: The 27th Chinese Control and Decision Conference (2015 CCDC), pp. 4355–4360. IEEE (2015)
7. Zhang, W., Wei, S., Zeng, J., Wang, N.: Multi-UUV path planning based on improved artificial potential field method. Int. J. Robot. Autom. **36**(4), 23–33 (2021)
8. Hao, L., Gu, H., Kang, F., Yang, H., Yang, X.: Improved virtual leader based formation control for nonholonomic multi-UUV. In: Xiao, T., Zhang, L., Ma, S. (eds.) ICSC 2012. CCIS, pp. 172–180. Springer, Heidelberg (2012). https://doi.org/10.1007/978-3-642-34381-0_20
9. Díaz-Pernil, D., Gutiérrez-Naranjo, M.A., Peng, H.: Membrane computing and image processing: a short survey. J. Membr. Comput. **1**(1), 58–73 (2019)
10. Mitrana, V.: Polarization: a new communication protocol in networks of bio-inspired processors. J. Membr. Comput. **1**(2), 127–143 (2019). https://doi.org/10.1007/s41965-018-0001-9
11. Gao, C., Gong, H., Zhen, Z., Zhao, Q., Sun, Y.: Three dimensions formation flight path planning under radar threatening environment. In: Proceedings of the 33rd Chinese Control Conference, pp. 1121–1125. IEEE (2014)
12. Marie, M.J., Mahdi, S.S., Yahia, E.Y.: Hybrid intelligent formation control using PSO_GEN. In: 2019 12th International Conference on Developments in eSystems Engineering (DeSE), pp. 693–698. IEEE (2019)
13. Tian, G., Zhang, L., Bai, X., Wang, B.: Real-time dynamic track planning of multi-UAV formation based on improved artificial bee colony algorithm. In: 2018 37th Chinese Control Conference (CCC), pp. 10055–10060. IEEE (2018)
14. Zeb, A., et al.: Hybridization of simulated annealing with genetic algorithm for cell formation problem. Int. J. Adv. Manuf. Technol. **86**(5), 2243–2254 (2016)
15. Zhou, J., Zhang, Q., Wang, H., Guo, S., Wang, Y.: Multi-UUV formation coordination control based on virtual navigator. In: 2020 39th Chinese Control Conference (CCC), pp. 2090–2095. IEEE (2020)
16. Qu, X., Liang, X., Hou, Y., Li, Y., Zhang, R.: Finite-time sideslip observer-based synchronized path-following control of multiple unmanned underwater vehicles. Ocean Eng. **217**, 107941 (2020)
17. Dorigo, M., Maniezzo, V., Colorni, A.: Ant system: optimization by a colony of cooperating agents. IEEE Trans. Syst. Man Cybern. Part B (Cybern.) **26**(1), 29–41 (1996)

Firefly Algorithm with Opposition-Based Learning

Yanping Qiao[1,2](✉), Feng Li[3,4], Cong Zhang[5], Xiaofeng Li[4], Zhigang Zhou[4], Tao Zhang[3,4], and Quanhua Zhu[4]

[1] School of Power and Energy, Northwestern Polytechnical University, Xi'an 710072, China
qiaoypnwpu@qq.com

[2] Science and Technology on Altitude Simulation Laboratory, Mianyang 621700, China

[3] Southern Marine Science and Engineering Guangdong Laboratory, Guangzhou 511458, China

[4] China Ship Scientific Research Center, Wuxi 214082, China

[5] AECC Sichuan Gas Turbine Establishment, Mianyang 621700, China

Abstract. The firefly algorithm (FA) is a swarm intelligence optimization algorithm based on the firefly's glow and attractive behavior. It possesses a simple design, is easy to implement, and has been applied in many engineering fields. Although FA well solves complex optimization problems, it has flaws regarding premature convergence and overall performance. Its learning style is relatively simple, as it can update individual positions merely by the fluorescence intensity between individuals. This paper proposes the firefly algorithm with opposition-based learning (FA-OL), which enriches learning to improve its capability to jump out of local optima. The opposition-based learning (OL) mechanism can promote individuals to search from the current and opposite points in the search space, and can quickly broaden the solution space. The optimal random parameter settings of FA-OL are obtained through experiments. Comprehensive experiments on nine commonly used test functions compare the proposed algorithm with the latest improved FA and seven other algorithms. The experimental results show that the introduction of OL greatly improved the search capability of FA, and FA-OL achieved better results than other algorithms.

Keywords: Firefly algorithm · Opposition-based learning · Global optimization

---

This work is supported by a project commissioned by AECC Sichuan Gas Turbine Establishment and by Key Special Project for Introduced Talents Team of Southern Marine Science and Engineering Guangdong Laboratory (Guangzhou, No. GML2019ZD0502).

L. Pan et al. (Eds.): BIC-TA 2021, CCIS 1565, pp. 342–352, 2022.
https://doi.org/10.1007/978-981-19-1256-6_26

1 Introduction

Metaheuristic algorithms have been widely used to solve complex and highly nonlinear optimization problems. Among them, swarm intelligence optimization, such as particle swarm optimization, ant colony optimization, and the firefly algorithm (FA), simulates group behavior in nature, and has performed well at solving complex optimization problems. Proposed in 2008 by Dr. Xinshe Yang of the University of Cambridge, FA simulates the group behavior of fireflies. Each firefly positions itself by its own fluorescence, through whose intensity it can attract other fireflies, catch prey, or send out warning signals. It can move individually to complete a search process in a solution space. The algorithm has seen many improvements. An improved FA to learn between mates was proposed, expanding upon the original FA, which has a single-sex firefly, and improving its optimization performance [14]. Wang et al. [23] proposed Ying-Yang FA based on a dimensional Cauchy mutation. A randomized attraction model promoted convergence. To better use FA to solve global continuous optimization problems, Wu et al. designed a logarithmic spiral search method to improve the individual's exploration capability, and a dynamic search method to improve individual exploration capability [25]. Nand et al. [11] proposed an improved FA called FA-Step, and designed a new step search strategy and a step strategy mixed with CMAES. Liu et al. proposed dynamic adaptive FA, which improved the convergence speed and solution accuracy, and stopped the algorithm from suffering from premature convergence [9].

FA has been successfully used in many engineering applications. A hybrid cooperative FA was used to solve the restricted vehicle routing problem, achieving a good balance between intensification and diversity [1]. Cheng et al. proposed a hybrid algorithm based on FA and GA, whose group attraction operator reduced the time complexity. Three combined mutation strategies were introduced, and the algorithm was used for constrained optimization problems. Tao et al. [18] proposed an adaptive strategy in the FA attraction model to design random models and penalty functions. It was used to solve constrained engineering optimization problems, where it achieved rapid convergence and high accuracy. Krishna et al. designed an improved random attraction FA for fuzzy image clustering problems [6]. Tian et al. [20] improved and optimized individual fireflies by combining mutation operations, used them to guide the direction of the group's operation, and applied the algorithm to the design of the SMC-PHD filter. Xie et al. [26] proposed two improved methods based on FA to overcome the local convergence problem in k-means clustering. Ireneus applied FA to the problem of entity selection in machine learning [4]. Wang et al. [24] designed an improved chaotic FA for wireless sensor resource allocation in the IoT environment. Dimitra et al. [22] designed an individual encoding and decoding strategy to use FA for the prize-collecting vehicle routing problem. Cheng et al. [2] improved the FA attraction model to promote convergence at a lower cost, designed three staged adaptive firefly functions to determine its parameters, and used the designed SAFA for the UAV charging planning problem in wireless sensor networks. Dash et al. designed an improved FA-based optimal design of special signal blocking

IIR filters [5]. Rohulla et al. designed a restricted multi-objective optimization model for the privacy protection problem in social networks, and combined FA and fuzzy clustering to solve it [8]. Tian et al. [19] designed an improved FA for particle filtering and used it for multi-target tracking, achieving better accuracy.

This paper addresses the problem of FA easily falling into local optima. Our improved FA based on opposition-based learning (OL), FA-OL, broadens the search space of the algorithm and improves its ability to cover the feasible solution area. The remainder of this article is organized as follows. Section 2 introduces the principle of the standard FA and FA based on the offline decreasing random factor. Section 3 introduces the principle of OL and its application to FA. Section 4 relates the results of experiments comparing our algorithm to seven others, and discusses the randomized parameter-setting method that affects our algorithm. Section 5 discusses our conclusions.

2 Introduction to FA

2.1 FA Concept

The two most basic behaviors of fireflies are attracting fireflies of the opposite sex in the same population and attracting prey, both of which depend on the fluorescence they emit, which is the fundamental principle of FA design. We know from optics that the intensity of light is attenuated because it is absorbed by the air, and the degree of attenuation is directly related to the air absorption and distance of propagation. The intensity of light will also decrease as the distance between the light source and the object increases.

In FA, each feasible solution is expressed as the position of each firefly, and the fluorescence intensities of the fireflies is used as the target function value. In standard FA, the fireflies are all of the same sex, so the attractive relationship between them does not need to be considered. In addition, FA believes that the degree of attraction between fireflies is proportional to their own fluorescence intensity. Individuals with lower fluorescence intensity are attracted to fireflies with higher fluorescence intensity, and will fly to them. The attraction behavior between fireflies is also the convergence behavior of the algorithm, i.e., the fireflies are searching for areas with better target function values.

The individual position update equation of FA has three parts, namely the individual flight inertia; the movement of the i-th firefly attracted by the j-th firefly is equivalent to the convergence behavior of the firefly in the evolutionary process; the randomized search term, which is completely unrelated to fireflies i and j. Through this, random disturbances can be generated to make the fireflies reach a new search position, as follows:

$$\begin{aligned} x_i^k(t+1) = & x_i^k(t) + \beta_{ij}(t) \cdot \left(x_j^k(t) - x_i^k(t)\right) \\ & + \alpha(t) \cdot \left(rand_{ij}^k(t) - 0.5\right) \end{aligned} \tag{1}$$

$$\beta_{ij}(t) = \beta_0 \cdot e^{-\gamma r_{ij}^2(t)} \tag{2}$$

$$r_{ij}(t) = \|x_j(t) - x_i(t)\| = \sqrt{\sum_{k=1}^{D}\left(x_j^k(t) - x_i^k(t)\right)^2} \tag{3}$$

where $x_i^k(t)$ and $x_j^k(t)$ respectively refer to the positions of fireflies i and j in the k-th dimension of the t-th iteration; $r_{ij}(t)$ is the distance between fireflies i and j in the t-th iteration; D is the problem scale; $rand$ is a uniformly distributed random number in $[0, 1]$; α is a random factor; γ is the absorption coefficient; β represents the attraction between two fireflies; and β_0 is the attraction when the distance between two fireflies is zero, i.e., the maximum attraction between them. Regarding the parameters of FA, Yang [27,28] proposed that, when implementing and applying FA, $\beta_0 = 1$, $\gamma \in [0.01, 100]$ is a constant generally set to 1, and $\alpha \in [0, 1]$ is a constant generally set to 0.5.

2.2 Improved FA Based on a Proportionally Decreasing Random Factor

From the position update equation of FA, we know that when the absorption coefficient is zero, the update equation is

$$x_i^k(t+1) = x_j^k(t) + \alpha(t) \cdot \left(rand_{ij}^k(t) - 0.5\right) \tag{4}$$

and when the absorption coefficient tends to infinity, the update equation is

$$x_i^k(t+1) = x_i^k(t) + \alpha(t) \cdot \left(rand_{ij}^k(t) - 0.5\right) \tag{5}$$

It can be seen from formulas (4) and (5) that whether the fluorescence is absorbed or not, the random search term is an important part of the firefly position update, and the random search ability will be affected by the random factor. We carried out research on random factors, and propose control methods based on linear decline, proportional decline, parabolic decline, and fixed constants. Through the experimental and statistical analysis results, it can be known that the improved FA adopting the proportional decreasing method has the best comprehensive optimization performance among the compared algorithms. The proportional decreasing method of random factors has the formula

$$\alpha(t+1) = k \cdot \alpha(t), \alpha(0) = 0.5 \tag{6}$$

where k is the common ratio, which is a constant less than 1. Optimizing the standard test function with the value of k in the interval $[0.97, 0.998]$ can obtain a more reasonable result. Through experiments, it was concluded that the best optimization performance occurs when k is 0.9902 [15]. FA based on a proportionally decreasing random factor is referred to as FA-Prop in this article.

3 Firefly Algorithm with Opposition-Based Learning

3.1 OL Principle

Generally speaking, heuristic algorithms are iterative. Starting from an initial solution, after continuous iterative operations, a heuristic algorithm tries to improve candidate solutions. When the algorithm stop condition is met, the search process is terminated. When the prior information of the solution was lacking, Tizhoosh et al. [21] proposed an OL mechanism, which used a random guess to generate candidate solutions, and whose calculation time was related to the distance between the initial and optimal solutions. Optimization algorithms with superior performance often have a high probability of jumping out of local optima and finding the global optimum, which is closely related to the ability of the particles in the algorithm to explore and open up unknown areas. OL can effectively broaden the search space and cover the feasible solution area.

Definition 1 Opposite number. If $x \in [a, b]$ and $x \in R$ then the opposite number x is $x^* = a + b - x$.

Definition 2 Opposite point. If $P = (x_1, x_2, \cdots, x_D)$ is a point in a D-dimensional space, $x_i \in R$, and $x_i \in [a_i, b_i]\,(i = 1, 2, \cdots, D)$, then P is the corresponding opposite point, where $P^* = (x_1^*, x_2^*, \cdots, x_D^*)$ and $x^* = a + b - x$.

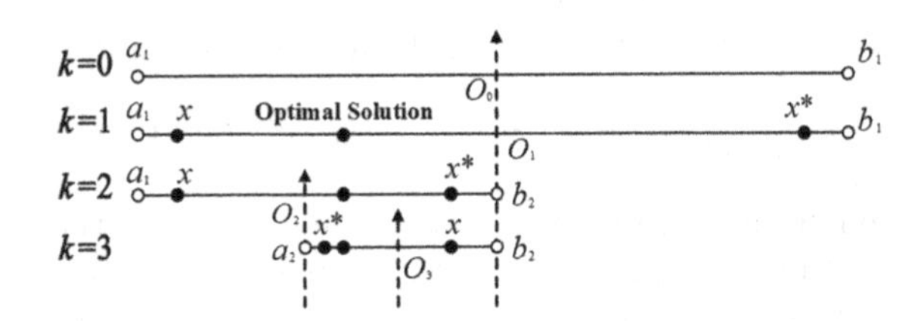

Fig. 1. Opposition-based learning in one dimension [16]

Opposition-based optimization thinking: Assume $P = (x_1, x_2, \cdots, x_D)$ is a point in a D-dimensional space (it can be viewed as a candidate solution), with opposite point $P^* = (x_1^*, x_2^*, \cdots, x_D^*)$. Suppose $f(\cdot)$ is an evaluation function to calculate the fitness of the candidate solution. Calculate $f(P)$ and $f(P^*)$ in each iteration. If $f(P^*) \leq f(P)$, then replace P with P^*, and otherwise still iterate with P. The one-dimensional space OL process is shown in Fig. 1, where O_k is the k-th OL reference point. Its value is the midpoint of the sum of a_i and b_i. It is also called the symmetric OL mechanism. If you take a random position O_k, it is random OL.

Zhou et al. proposed normalized OL [29]. Let $\varphi(x) = \Delta - x$ and $x^* = \Delta - x$, where Δ is a calculated value and the corresponding opposite solution is $x \in [\Delta - b, \Delta - a]$. Compared with the original OL, the difference is the symmetric position of the current and opposite points. The current and opposite points of the original OL are with regard to $(a + b)/2$, while the normal OL is symmetric with regard to $(2\Delta - a - b)/2$. Let $\Delta = \eta(a + b)$, where η is a real number. Then normal OL is

$$x^* = \eta(a + b) - x \tag{7}$$

According to the different values of η, there are four types of OL. When $\eta = 0$, the current and opposition-based solutions are symmetric with regard to the origin, which is called OL based on solution symmetry; when $\eta = 1/2$, the search interval is symmetric about the origin, which is called OL based on interval symmetry; $\eta = 1$ is the special case of the original OL; and when η is a random number, it is called random OL.

In the swarm intelligence algorithm, the OL mechanism is introduced, and the individual can search at the current and opposite points at the same time. If the opposite point has a better fitness value than the current point, then the search area near the opposite point has a better chance to find the optimal value. The individual can jump directly to the opposite point to continue searching, which can quickly expand the solution space and cover the feasible solution area on a larger scale, ultimately increasing the probability of finding the global optimal solution.

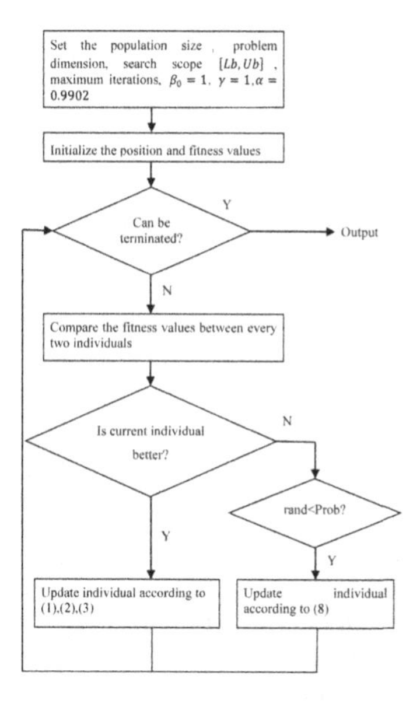

Fig. 2. Flowchart of FA-OL

3.2 Opposition Learning FA

In the standard FA and FA-Prop, when comparing the fluorescence intensity of two individual fireflies, the individual with weaker fluorescence intensity needs to learn from the individual with high fluorescence intensity, i.e., according to formulas (1)–(3), complete the location update of the weaker individuals. In FA-OL,

Table 1. Experimental results of FA-OL on different random parameters

	F1	F2	F3	F4	F5	F6	F7	F8	F9
FA_Prop	2.38E−13	1.70E−03	8.83E+01	−7.97E+03	6.67E+01	1.12E−07	2.37E−03	6.68E−16	1.32E−03
	3.97E−14	8.17E−03	1.40E+02	7.16E+02	2.09E+01	9.20E−09	3.99E−03	1.22E−16	3.64E−03
FA-OL	0	0	2.72E+01	−2.16E+02	0	−8.88E−16	4.75E+01	1.81E−15	4.39E−04
(Prob = 0.1)	0	0	1.33E−01	6.22E+01	0	0	3.92E−01	5.22E−16	2.20E−03
FA-OL	0	0	2.72E+01	−1.62E+02	0	−8.88E−16	5.93E+01	1.82E−15	4.39E−04
(Prob = 0.5)	0	0	1.32E−01	2.57E+01	0	0	1.69E−01	4.16E−16	2.20E−03
FA-OL	0	0	2.73E+01	−6.96E+03	0	−8.88E−16	2.96E−04	1.11E−15	1.32E−03
(Prob = 0.01)	0	0	1.68E−01	7.91E+02	0	0	1.48E−03	3.12E−16	3.64E−03
FA-OL	0	0	2.73E+01	−4.48E+03	0	−8.88E−16	2.96E−04	1.28E−15	1.32E−03
(Prob = 0.02)	0	0	1.47E−01	7.84E+02	0	0	1.48E−03	2.94E−16	3.64E−03
FA-OL	0	0	2.72E+01	−3.32E+03	0	−8.88E−16	5.92E−04	1.41E−15	8.79E−04
(Prob = 0.03)	0	0	1.50E−01	5.81E+02	0	0	2.05E−03	3.27E−16	3.04E−03
FA-OL	0	0	2.73E+01	−2.66E+03	0	−8.88E−16	4.20E−02	1.75E−15	4.39E−04
(Prob = 0.04)	0	0	1.33E−01	3.71E+02	0	0	4.95E−02	3.79E−16	2.20E−03
FA-OL	0	0	2.73E+01	−1.87E+03	0	−8.88E−16	1.57E+00	1.64E−15	1.76E−03
(Prob = 0.05)	0	0	1.54E−01	4.42E+02	0	0	1.30E−01	4.67E−16	4.11E−03

Table 2. Ranking results of FA-OL on different random parameters (ANOVA)

	F1	F2	F3	F4	F5	F6	F7	F8	F9	Total
FA_Prop	8	1	8	1	8	8	4	1	1	40
FA-OL (Prob = 0.1)	1	1	1	7	1	1	7	7	1	27
FA-OL (Prob = 0.5)	1	1	1	8	1	1	8	8	1	30
FA-OL (Prob = 0.01)	1	1	1	2	1	1	1	2	1	**11**
FA-OL (Prob = 0.02)	1	1	1	3	1	1	1	3	1	13
FA-OL (Prob = 0.03)	1	1	1	4	1	1	1	4	1	15
FA-OL (Prob = 0.04)	1	1	1	5	1	1	5	6	1	22
FA-OL (Prob = 0.05)	1	1	1	6	1	1	6	5	1	23

to improve the ability of the algorithm to jump out of local optima, individuals with higher fluorescence intensity expand the search space with probability *Prob* through the OL mechanism, i.e., the position update is completed as

$$x_i^k(t+1) = Ub + Lb - x_i^k(t) + rand(1, D) .* x_i^k(t) \tag{8}$$

where Ub and Lb are the upper and lower bounds, respectively, of the search space, and D is the problem dimension. It can be seen from formula (8) that after an individual performs an opposition search, it can quickly search for the best point in a new search area, which improves the global search capability of the algorithm. The flowchart of FA-OL is shown as Fig. 2.

4 Experiment

4.1 Experiment Setup

We used nine standard test functions [15] to analyze the performance of FA-OL. We also compared it with the latest improved FA, FACL [14]; improved ant colony optimization algorithm NABC [12]; and improved cuckoo algorithm MSSCS [13]; and with classic improved algorithms PSO with inertia weight (PSO-In) [17], PSO with constriction factor (PSO-Co) [3], Gaussian Bare Bone PSO (GBBPSO) [7], and PSO with lbest (PSO-Lb) [10]. The dimension of the test function was $D = 30$, the population size of each algorithm was 30, and the maximum number of evaluations of the target function was 60,000. Other parameters of each algorithm adopted the recommended values in the corresponding literature. Each algorithm independently and randomly optimized each function 25 times, and the average and standard deviation optimal results were recorded.

Table 3. Comparison results of FA-OL with other 7 algorithms

	F1	F2	F3	F4	F5	F6	F7	F8	F9
FA-OL	0	0	2.73E+01	−6.96E+03	0	−8.88E−16	2.96E−04	1.11E−15	1.32E−03
(Prob = 0.01)	0	0	1.68E−01	7.91E+02	0	0.00E+00	1.48E−03	3.12E−16	3.64E−03
FA-CL	1.23E+00	3.06E+01	3.07E+02	−4.38E+03	1.50E+02	5.49E+00	8.91E−02	3.15E+00	3.80E+01
	2.88E−01	1.33E+01	3.45E+02	3.19E+02	1.60E+01	1.60E+00	2.34E−02	1.48E+00	2.48E+01
MSSCS	1.53E−37	1.66E−04	1.84E+01	−1.26E+04	1.95E−04	7.50E−15	1.39E−04	1.15E−12	8.79E−04
	7.23E−37	4.18E−04	1.21E+01	1.24E−05	4.75E−04	2.49E−15	6.97E−04	8.67E−13	3.04E−03
NABC	1.29E−23	1.38E+03	2.05E+00	−1.26E+04	1.56E−14	1.02E−11	2.72E−12	3.31E−24	1.25E−22
	1.19E−23	5.30E+02	2.94E+00	1.44E−05	4.40E−14	3.88E−12	1.31E−11	8.45E−24	1.48E−22
PSO_In	1.94E−09	1.69E+03	1.54E+02	−3.20E+04	4.05E+01	1.90E−05	1.05E−02	5.39E−02	3.52E−03
	3.97E−09	1.41E+03	1.88E+02	4.43E+03	1.21E+01	1.81E−05	1.53E−02	6.77E−02	5.23E−03
PSO_CO	3.52E−03	5.19E−01	4.88E+01	−2.11E+04	6.42E+01	2.19E+00	2.40E−02	1.29E−01	2.84E−02
	5.23E−03	8.71E−01	4.15E+01	3.03E+03	1.92E+01	1.24E+00	2.65E−02	1.67E−01	1.24E−01
GBPSO	2.99E−31	1.49E+04	1.48E+04	−8.92E+03	8.45E+01	1.83E+00	2.51E+00	4.15E−02	3.52E−03
	6.20E−31	6.11E+03	3.35E+04	6.38E+02	2.64E+01	5.37E+00	1.25E+01	6.69E−02	5.23E−03
PSO_LB	1.41E−05	3.63E+03	8.79E+01	−3.17E+04	5.05E+01	1.79E−01	3.84E−03	7.97E−02	4.70E−02
	1.55E−05	1.16E+03	5.26E+01	3.85E+03	9.77E+00	5.22E−01	7.90E−03	1.05E−01	7.25E−02

4.2 Effect Analysis of Random Parameters

In the proposed FA-OL, the execution of OL depends on the random parameter Prob. To clearly understand the impact of its setting on the optimization performance of the algorithm, we set its value to 0.5, 0.1, 0.05, 0.04, 0.03, 0.02, and 0.01. The nine test functions were solved randomly and independently 25 times, with average and standard deviation as displayed in Table 1. The first row of data corresponding to each algorithm is the average value of the optimal solution, and the second row is the standard deviation. It can be seen from Table 1 that the introduction of OL significantly improved the optimization performance of FA on most test functions, especially on the functions of F1, F2, F5, and F6, on which higher solution accuracy was attained.

We used analysis of variance (ANOVA) to test for significant differences between the mean value of the algorithm for each test function at the level of 0.05. The score of each algorithm on each function is shown in Table 2. At the same time, the total score of each algorithm was calculated. The smaller the score the better the comprehensive optimization performance. It can be seen from Table 2 that the comprehensive optimization performance of FA-OL is best when the random parameter value is 0.01.

Table 4. Ranking results of 8 algorithms (ANOVA)

	F1	F2	F3	F4	F5	F6	F7	F8	F9	Total
FA-OL (Prob = 0.01)	1	1	3	7	1	1	1	2	1	18
FA-CL	8	4	7	8	8	8	7	8	8	66
MSSCS	1	1	2	4	3	2	1	3	1	18
NABC	4	5	1	4	1	3	1	1	1	21
PSO_In	5	5	6	1	4	4	4	5	4	38
PSO_CO	7	3	4	3	6	4	6	7	4	44
GBPSO	3	8	8	6	7	4	7	4	4	51
PSO_LB	6	7	5	1	5	4	4	6	4	42

4.3 Comparison with Other Algorithms

Table 3 shows the experimental results of FA-OL and the other seven algorithms, and Table 4 shows their scores on the nine test functions. It can be seen from Table 3 that the proposed algorithm achieves the best optimization results on six functions (F1, F2, F5, F6, F7, and F9), especially in F1, F2, and F5 where the position of the optimal solution was found, showing a strong global search capability. From the statistical scoring results in Table 4, we can see that FA-OL and MSSCS achieve the best statistical results, followed by NABC.

5 Conclusion

We proposed FA-OL, whose OL mechanism can effectively expand the current search area, help individuals jump out of local optima, enhance individuals' search capability, and improve the algorithm's global search capability. The OL mechanism and the attraction mechanism of FA act on two individuals participating in comparative fitness, so that the learning capability of the group is enhanced and the optimization capability is improved. The optimization effect of the proposed algorithm was verified based on nine standard test functions, and compared with seven improved algorithms. The optimization results were analyzed by ANOVA. The experimental results show that FA-OL achieves the best optimization performance and robustness.

References

1. Altabeeb, A.M., Mohsen, A.M., Abualigah, L., Ghallab, A.: Solving capacitated vehicle routing problem using cooperative firefly algorithm. Appl. Soft Comput. **108**, 107403 (2021). https://doi.org/10.1016/j.asoc.2021.107403. https://www.sciencedirect.com/science/article/pii/S1568494621003264
2. Cheng, L., Zhong, L., Zhang, X., Xing, J.: A staged adaptive firefly algorithm for UAV charging planning in wireless sensor networks. Comput. Commun. **161**, 132–141 (2020). https://doi.org/10.1016/j.comcom.2020.07.019. https://www.sciencedirect.com/science/article/pii/S0140366420307763

3. Clerc, M., Kennedy, J.: The particle swarm - explosion, stability, and convergence in a multidimensional complex space. IEEE Trans. Evol. Comput. **6**(1), 58–73 (2002). https://doi.org/10.1109/4235.985692
4. Czarnowski, I.: Firefly algorithm for instance selection. Procedia Comput. Sci. **192**, 2269–2278 (2021). https://doi.org/10.1016/j.procs.2021.08.240. https://www.sciencedirect.com/science/article/pii/S1877050921017373
5. Dash, J., Dam, B., Swain, R.: Improved firefly algorithm based optimal design of special signal blocking IIR filters. Measurement **149**, 106986 (2020). https://doi.org/10.1016/j.measurement.2019.106986. https://www.sciencedirect.com/science/article/pii/S0263224119308528
6. Dhal, K.G., Das, A., Ray, S., Gálvez, J.: Randomly attracted rough firefly algorithm for histogram based fuzzy image clustering. Knowl.-Based Syst. **216**, 106814 (2021). https://doi.org/10.1016/j.knosys.2021.106814. https://www.sciencedirect.com/science/article/pii/S0950705121000770
7. Kennedy, J.: Probability and dynamics in the particle swarm. In: Proceedings of the 2004 Congress on Evolutionary Computation (IEEE Cat. No. 04TH8753), vol. 1, pp. 340–347. https://doi.org/10.1109/CEC.2004.1330877
8. Langari, R.K., Sardar, S., Amin Mousavi, S.A., Radfar, R.: Combined fuzzy clustering and firefly algorithm for privacy preserving in social networks. Expert Syst. Appl. **141**, 112968 (2020). https://doi.org/10.1016/j.eswa.2019.112968. https://www.sciencedirect.com/science/article/pii/S0957417419306864
9. Liu, J., Mao, Y., Liu, X., Li, Y.: A dynamic adaptive firefly algorithm with globally orientation. Math. Comput. Simul. **174**, 76–101 (2020). https://doi.org/10.1016/j.matcom.2020.02.020. https://www.sciencedirect.com/science/article/pii/S0378475420300598
10. Mendes, R., Kennedy, J., Neves, J.: The fully informed particle swarm: simpler, maybe better. IEEE Trans. Evol. Comput. **8**(3), 204–210 (2004). https://doi.org/10.1109/TEVC.2004.826074
11. Nand, R., Sharma, B.N., Chaudhary, K.: Stepping ahead firefly algorithm and hybridization with evolution strategy for global optimization problems. Appl. Soft Comput. **109**, 107517 (2021). https://doi.org/10.1016/j.asoc.2021.107517. https://www.sciencedirect.com/science/article/pii/S1568494621004403
12. Peng, H., Deng, C., Wu, Z.: Best neighbor-guided artificial bee colony algorithm for continuous optimization problems. Soft Comput. **23**(18), 8723–8740 (2019). https://doi.org/10.1007/s00500-018-3473-6. https://doi.org/10.1007/s00500-018-3473-6
13. Peng, H., Zeng, Z., Deng, C., Wu, Z.: Multi-strategy serial cuckoo search algorithm for global optimization. Knowl.-Based Syst. **214**, 106729 (2021). https://doi.org/10.1016/j.knosys.2020.106729. https://www.sciencedirect.com/science/article/pii/S0950705120308583
14. Peng, H., Zhu, W., Deng, C., Wu, Z.: Enhancing firefly algorithm with courtship learning. Inf. Sci. **543**, 18–42 (2021). https://doi.org/10.1016/j.ins.2020.05.111. https://www.sciencedirect.com/science/article/pii/S0020025520305363
15. Qiao, Y., Li, F., Zhang, C., Li, X., Zhou, Z.: Study on the random factor of firefly algorithm. In: Tan, Y., Shi, Y. (eds.) ICSI 2021. LNCS, vol. 12689, pp. 58–71. Springer, Cham (2021). https://doi.org/10.1007/978-3-030-78743-1_6
16. Shao, P., Wu, Z., Zhou, X., Deng, C.: Improved particle swarm optimization algorithm based on opposite learning of refraction. Acta Electron. Sin. **43**(11), 2137–2144 (2015)

17. Shi, Y., Eberhart, R.: A modified particle swarm optimizer. In: 1998 IEEE International Conference on Evolutionary Computation Proceedings. IEEE World Congress on Computational Intelligence (Cat. No. 98TH8360), pp. 69–73 (1998). https://doi.org/10.1109/ICEC.1998.699146
18. rl Tao, R., Meng, Z., Zhou, H.: A self-adaptive strategy based firefly algorithm for constrained engineering design problems. Appl. Soft Comput. **107**, 107417 (2021). https://doi.org/10.1016/j.asoc.2021.107417. https://www.sciencedirect.com/science/article/pii/S1568494621003409
19. Tian, M., Bo, Y., Chen, Z., Wu, P., Yue, C.: Multi-target tracking method based on improved firefly algorithm optimized particle filter. Neurocomputing **359**, 438–448 (2019). https://doi.org/10.1016/j.neucom.2019.06.003. https://www.sciencedirect.com/science/article/pii/S0925231219308240
20. Tian, M., Bo, Y., Chen, Z., Wu, P., Yue, C.: A new improved firefly clustering algorithm for SMC-PHD filter. Appl. Soft Comput. **85**, 105840 (2019). https://doi.org/10.1016/j.asoc.2019.105840. https://www.sciencedirect.com/science/article/pii/S1568494619306210
21. Tizhoosh, H.R.: Opposition-based learning: a new scheme for machine intelligence. In: International Conference on Computational Intelligence for Modelling, Control and Automation and International Conference on Intelligent Agents, Web Technologies and Internet Commerce (CIMCA-IAWTIC 2006), vol. 1, pp. 695–701 (2006). https://doi.org/10.1109/CIMCA.2005.1631345
22. Trachanatzi, D., Rigakis, M., Marinaki, M., Marinakis, Y.: A firefly algorithm for the environmental prize-collecting vehicle routing problem. Swarm Evol. Comput. **57**, 100712 (2020). https://doi.org/10.1016/j.swevo.2020.100712. https://www.sciencedirect.com/science/article/pii/S2210650220303655
23. Wang, W.c., Xu, L., Chau, K.w., Xu, D.m.: Yin-yang firefly algorithm based on dimensionally Cauchy mutation. Expert Syst. Appl. **150**, 113216 (2020). https://doi.org/10.1016/j.eswa.2020.113216. https://www.sciencedirect.com/science/article/pii/S0957417420300427
24. Wang, Z., Liu, D., Jolfaei, A.: Resource allocation solution for sensor networks using improved chaotic firefly algorithm in IoT environment. Comput. Commun. **156**, 91–100 (2020). https://doi.org/10.1016/j.comcom.2020.03.039. https://www.sciencedirect.com/science/article/pii/S0140366420302164
25. Wu, J., Wang, Y.G., Burrage, K., Tian, Y.C., Lawson, B., Ding, Z.: An improved firefly algorithm for global continuous optimization problems. Expert Syst. Appl. **149**, 113340 (2020). https://doi.org/10.1016/j.eswa.2020.113340. https://www.sciencedirect.com/science/article/pii/S0957417420301652
26. Xie, H., Zhang, L., Lim, C.P., Yu, Y., Liu, C., Liu, H., Walters, J.: Improving k-means clustering with enhanced firefly algorithms. Appl. Soft Comput. **84**, 105763 (2019). https://doi.org/10.1016/j.asoc.2019.105763. https://www.sciencedirect.com/science/article/pii/S1568494619305447
27. Yang, X.-S.: Firefly algorithms for multimodal optimization. In: Watanabe, O., Zeugmann, T. (eds.) SAGA 2009. LNCS, vol. 5792, pp. 169–178. Springer, Heidelberg (2009). https://doi.org/10.1007/978-3-642-04944-6_14
28. Yang, X.S.: Firefly algorithm, Levy flights and global optimization. In: Bramer, M., Ellis, R., Petridis, M. (eds.) Research and Development in Intelligent Systems XXVI, pp. 209–218. Springer, London (2010). https://doi.org/10.1007/978-1-84882-983-1_15
29. Zhou, X., Wu, Z., Wang, H.: A differential evolution algorithm using elite opposition-based learning. J. Chin. Comput. Syst. **34**(09), 2129–2134 (2013)

An Optimization Method of Course Scheduling Problem Based on Improved Genetic Algorithm

Yikun Zhang[1] and Jian Huang[1,2](✉)

[1] School of Artificial Intelligence and Automation, Huazhong University of Science and Technology, Wuhan 430074, China
huang_jan@mail.hust.edu.cn
[2] Wuhan National Laboratory for Optoelectronics, Huazhong University of Science and Technology, Wuhan 430074, China

Abstract. Course scheduling is a difficult problem in real school management, which takes a lot of effort of the staff. In this paper, an optimization method based on genetic algorithm is proposed to solve the course scheduling problem in the background of Chinese College Entrance Examination Reform, which means that students of high school should learn 3 compulsory courses and choose 3 from 6 (or 7) optional courses. Through analysis, the scheduling problem can be divided into three subproblems optimized separately, which can reduce the complexity by avoiding some latent conflicts. Besides, we also propose two measures of improvement for genetic algorithm. First, the objective function is adapted to adjust automatically. Then, the crossover and mutation are also improved. In simulation experiment and convergence performance comparison experiment, the statistical results show that the improved genetic algorithm is more efficient than the original one, especially in the speed of searching feasible solution.

Keywords: Genetic algorithm · Course scheduling · Multi-objective optimization · Optional class system

1 Introduction

Course scheduling is a difficult, but important problem in management of educational system. The focus of this problem is the conflicts among educational resources. In mathematical theory, course scheduling problem is regarded as a kind of timetabling problem with multiple elements and constraints [1,2]. In 1963, Gotlieb put forward timetabling problem [3], and in 1975, timetabling problem was proved to be an NP-complete problem by Even et al. [4].

In traditional way, the scheduling problem of whole school is manually solved by educational administrators in short period. Such way of solution wastes time and effort and is inefficient in reducing the conflicts. As the development of computer technology and science, the course scheduling problem can

L. Pan et al. (Eds.): BIC-TA 2021, CCIS 1565, pp. 353–368, 2022.
https://doi.org/10.1007/978-981-19-1256-6_27

be solved by computer programs, which range from mathematical programming to heuristics [5]. Among the meta-heuristic methods, the representative algorithms include genetic algorithm, stimulated annealing algorithm, tabu search, etc. [6], which can produce high quality solutions [7].

Genetic algorithm is an efficient meta-heuristic algorithm, which owns high parallel, random and adaptive global searching characteristics [8]. But it also has many defects. For example, the parameters setting methods is not common [6], and it may experience premature convergence, which may lead to them being trapped into local optima [9].

In this paper, we focus on the improvement on efficiency of course scheduling problem. First, this study sets up the model and objective function of scheduling problem based on Chinese College Entrance Examination Reform, and chooses menu-styled subject selection as basic strategy, which makes dividing the problem into three independent subproblems become possible. And this division benefits reducing the quantity of possible solutions and latent conflicts. Second, genetic algorithm is selected to find the near-optimal solution of this problem. Faced with the problems encountered in the experiment, we propose several methods to improve the drawbacks of the basic genetic algorithm. In experiments, the improved algorithm performs better in convergence speed and shows the ability to jump out of local optimum.

2 Mathematical Model of the Problem

The course scheduling problem in this study is based on Chinese College Entrance Examination Reform, and the major difficulty of it lies in optional class system. The representative solution of scheduling optional courses is the scheme used in universities. However, unlike universities, most of high schools lack educational resources to imitate the scheme. Therefore, many schools turn to adopt the menu-styled subject selection [10], which means schools only provide limit kinds of combination of optional courses for students. And in this condition, the students who choose the same optional courses can form a new kind of class called optional class, which is in the charge of three fixed teachers.

In this way, a student will belong to two classes, i.e. administrative class and optional class. Furthermore, such situation enables the schools to separate the two classes and put their courses into different class hours. Therefore, the course scheduling can be divided into three parts: arranging the time of two classes, scheduling the courses of administrative classes and scheduling the courses of optional classes. Based on this strategy, it can reduce the complexity of the problem and avoid the conflicts between compulsory courses and optional courses. In addition, this strategy is also feasible in real situation.

2.1 Model Hypothesis

1. Students will have three compulsory courses, which are Chinese, English and Math, and choose three from six optional courses, which are Politics, History, Geology, Physics, Chemistry and Biology;

2. The students will be in two classes, i.e. administrative class and optional class;
3. The number of classrooms is just enough for all the classes to use at the same time.

2.2 Constraints Analysis

As introduced above, the scheduling problem can be divided into three parts, and each part has its own hard constraints, which must be satisfied, and soft constraints, which only need fulfilling as far as possible.

Generally Arranging the Time of Administrative and Optional Classes. In this step, the algorithm will arrange compulsory courses, optional courses and self-study courses for all students. First, the number of compulsory courses and optional courses must satisfy the requirement, which means the arrangement must satisfy following inequations:

$$\begin{cases} n_{com} \geq n_{\Sigma com} \\ n_{opt} \geq n_{\Sigma opt} \end{cases} \tag{1}$$

In inequations, n_{com} represents the total number of compulsory courses, n_{opt} represents the total number of optional courses, $n_{\Sigma com}$ and $n_{\Sigma opt}$ represent total required quantity of compulsory and optional class hours respectively.

In the meanwhile, given the difficulties in solving conflicts and the extra tutorials existing in practical education, the total difference value between the number of compulsory courses and optional courses and the required number plus 2 should be small as far as possible. The equations are as follows:

$$d_{com} = |n_{com} - (n_{\Sigma com} + 2)| \tag{2}$$

$$d_{opt} = |n_{opt} - (n_{\Sigma opt} + 2)| \tag{3}$$

The d_{com} and d_{opt} is the difference value of compulsory courses and optional courses respectively.

The courses also need to be uniform as far as possible, which can be described as follows:

$$u_{com} = \frac{1}{n_{com}} \sum_{n=1}^{n_{com}} \left(\tau_{com}(n) - \bar{\tau}_{com}\right)^2 \tag{4}$$

$$u_{opt} = \frac{1}{n_{opt}} \sum_{n=1}^{n_{opt}} \left(\tau_{opt}(n) - \bar{\tau}_{opt}\right)^2 \tag{5}$$

$$ui_{com} = |\bar{\tau}_{com} - n_{\Sigma}/n_{\Sigma\,\mathrm{com}}| \tag{6}$$

$$ui_{opt} = |\bar{\tau}_{opt} - n_{\Sigma}/n_{\Sigma \mathrm{opt}}| \tag{7}$$

Among these, n_{Σ} is the total number of class hours, u_{com} represents the degree of uniformity, n_{com} represents the quantity of compulsory courses, $\bar{\tau}_{com}(n)$ represents the nth interval, i.e. the difference value of two number of class hour,

$\bar{\tau}_{com}$ represents average interval and ui_{com} represents the absolute difference value between $\bar{\tau}_{com}$ and ideal average interval. u_{opt}, n_{opt}, $\tau_{opt}(n)$, $\bar{\tau}_{opt}$ and ui_{opt} are the counterparts in optional courses (Fig. 1 and 2).

According to research [11], students show preference among class hours, the value is listed as followed:

Table 1. Value of course schedule

	Mon.	Tues.	Wed.	Thurs.	Fri.
1st, 2nd lesson	0.94	0.99	0.98	0.95	0.92
3rd, 4th lesson	0.85	0.89	0.87	0.84	0.82
5th, 6th lesson	0.65	0.69	0.67	0.61	0.50
7th, 8th lesson	0.55	0.55	0.55	0.55	0.2
9th, 10th lesson	0.2	0.2	0.2	0.2	0.15

And students also hold different attitude toward importance of different courses. The degree of importance is listed as follows:

Table 2. Importance degree of different courses

	Compulsory course	Optional course	Self-study course
Score	1	0.9	0.1

Therefore, the goodness of course schedule can be expressed by following equation:

$$G = \sum_{k=1}^{k_p} p_k s_k \tag{8}$$

In the equation, k_p represents total class hours, p_k represents the preference of kth class hour, and s_k represents the significance of the course in this class hour.

Scheduling the Courses of Administrative and Optional Classes. In this step, optimization has two constraints. First, the number of three courses must be more than or equal to the required quantity.

$$n_{c,i} \geq n_{i0}, c = 1, 2, \ldots, n_c, i \in I_c \tag{9}$$

$n_{c,i}$ is the quantity of ith course, n_c is total quantity of classes, I_c is the course list of class c and n_{i0} is the corresponding required quantity.

And due to the hypothesis, a classroom can be matched to one class. Meanwhile, teachers who are responsible for the class are fixed. Consequently, there is no conflict caused by classrooms, and the conflicts can only be resulted from classes and teachers and the constraints can be expressed as follows:

$$\sum_{n=1}^{n_{te}} q_{mn}(k) = 1, m = 1, 2, \ldots, n_c, k \in K \tag{10}$$

$$\sum_{m=1}^{n_c} q_{mn}(k) = 1, n = 1, 2, \ldots, n_{te}, k \in K \tag{11}$$

m and n represent the number of class and teacher of three course respectively, n_{te} is the total number of teachers, and k represents the number of class hour. In the schedule, if class m is taught by teacher n during class hour k, $q_{mn}(k) = 1$. Otherwise, $q_{mn}(k) = 0$. K is the class hours belonging to compulsory or optional courses.

And there are also some additional soft constraints. Obviously, the quantity of each course is better to equivalent to the required number.

$$d_{\Sigma} = \sum_{c=1}^{n_c} \sum_{i \epsilon I_c} |n_{c,i} - n_{i0}| \tag{12}$$

d_{Σ} represents the total difference value between the real quantity and required quantity. And the courses are better to be uniform as far as possible.

$$u_{\text{com},i} = \frac{1}{n_{\text{com},i}} \sum_{c=1}^{n_c} \sum_{n=1}^{n_{\text{com},c,i}} \left(\tau_{\text{com,c},i}(n) - \bar{\tau}_{\text{com,c},i}\right)^2, i \in I_c \tag{13}$$

$$u_{opt,\Sigma} = \frac{1}{n_{opt,\Sigma}} \sum_{c=1}^{n_c} \sum_{i \in I_c} \sum_{n=1}^{n_{opt,c,i}} \left(\tau_{opt,c,i}(n) - \bar{\tau}_{opt,c,i}\right)^2 \tag{14}$$

In Eq. (13), $u_{com,i}$ is the uniformity of ith compulsory course. $n_{com,i}$ represents the total quantity of interval of ith compulsory course in all administrative classes, $n_{com,c,i}$ represents the total quantity of interval of ith compulsory course in class c, $\tau_{com,c,i}(n)$ represents the nth interval of course i in class c and $\bar{\tau}_{com,c,i}$ is the average interval.

The variables in Eq. (14) are similar to those in Eq. (13) except that $n_{opt,\Sigma}$ represents total quantity of interval of all optional courses in n_{opt} optional classes.

3 Basic Genetic Algorithm

3.1 Representation of Solution

Generally Arranging the Time of Administrative and Optional Classes. The chromosome of the solution is as follows (Fig. 1):

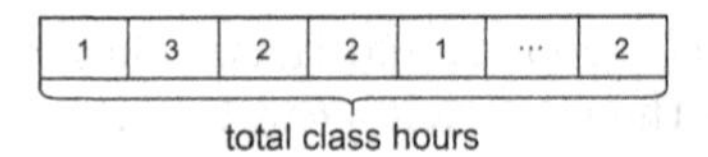

Fig. 1. Chromosome of general arrangement

The chromosome of first step is an array with n_Σ elements, which corresponds to n_Σ class hours. The value of each element is chosen from 1, 2 and 3. 1 and 2 represent the compulsory and optional course respectively, and 3 represents the common self-study course.

Scheduling the Courses of Administrative and Optional Classes. The chromosome is expressed as follows (Fig. 2):

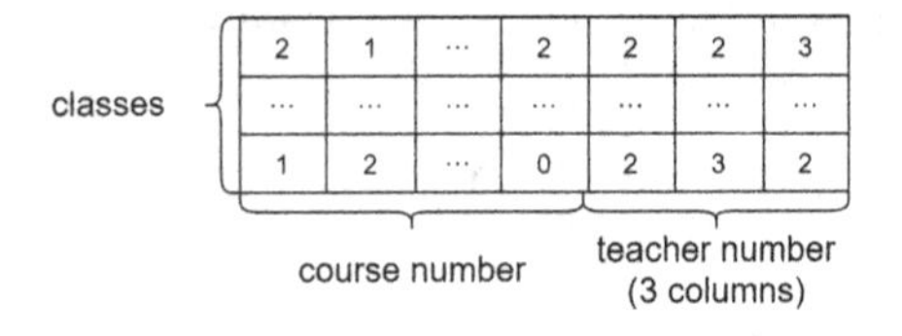

Fig. 2. Chromosome of compulsory and optional course schedule

The chromosome is a matrix with n_c rows in accord with n_c classes. The left side is the part of course. The total number of columns is related to the result of general arrangement. The value of this part is selected from 0, 1, 2 and 3. 1, 2 and 3 represent three courses, while 0 refers to tutorial. What should be noted is that the value in the one of optional courses only refers to the code of courses, while the real course is stored in another $n_{\Sigma opt}$x3 matrix. On the right side is the part of teacher who is responsible for the class. The value of elements refers to the code of teachers, which is an integer ranging from 1 to n_{te}.

Because of the coding pattern, the schedules avoid a kind of conflict that a class of students learn multiple courses at one time, which further simplifies the problem.

In addition, the relation among three chromosomes is showed in Fig. 3.

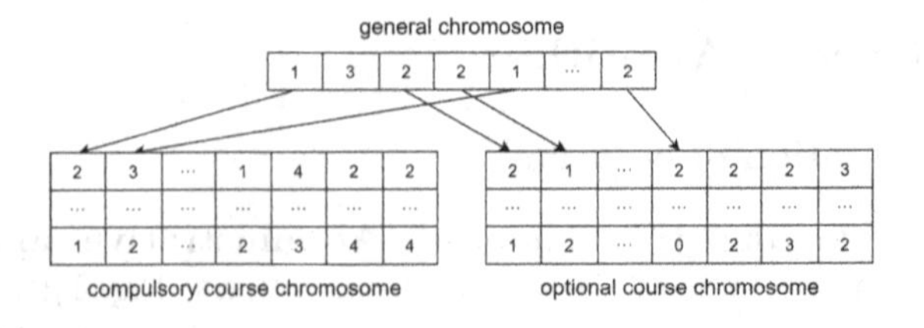

Fig. 3. The mapping relation among three chromosomes

And due to the pattern of coding, the hard constraints Eq. (10) and Eq. (11) will be transformed into following form:

$$\sum_{c_1=1}^{n_c-1}\sum_{c_2=c_1+1}^{n_c} r_{c_1c_2}(k)=0 \tag{15}$$

$$r_{c_1c_2}(k)=\begin{cases}1, c_1 \text{ and } c_2 \text{ has same course and teacher at } k\\ 0 \qquad\qquad\qquad\qquad\qquad\qquad\qquad\qquad\quad, \text{else}\end{cases} \tag{16}$$

3.2 Objective Function

Generally Arranging the Time of Administrative and Optional Classes. The general objective function of this step is as follows:

$$fit=h+s \tag{17}$$

And h is expressed as follows:

$$h=\frac{1}{1+N_v} \tag{18}$$

N_v is the time that individual violates Eq. (1). And s can be describe as:

$$s=0.54*\frac{G}{40}+0.06u+0.14ui+0.08*\left(1-\frac{d_{com}+d_{opt}}{10}\right) \tag{19}$$

Among these

$$h=\frac{u_{com}+u_{opt}}{2} \tag{20}$$

$$h=\frac{ui_{com}+ui_{opt}}{2} \tag{21}$$

Scheduling the Courses of Administrative and Optional Classes. The general objective function of this step is same as Eq. (17), which is:

$$fit=h+s \tag{22}$$

Among these

$$h=\frac{1}{1+N_p/5+N_t/50} \tag{23}$$

$$s=0.4*\left(1-\frac{d_\Sigma}{400}\right)+0.4*\sigma \tag{24}$$

N_p is the time that individual transgresses Eq. (9), and N_t is:

$$N_t=\sum_{k\in K}\sum_{c_1=1}^{n_c-1}\sum_{c_2=c_1+1}^{n_c} r_{c_1c_2}(k) \tag{25}$$

When scheduling administrative classes, σ is described in Eq. (26), and when scheduling optional classes, σ is described in Eq. (27)

$$\sigma = \sum_{i=1}^{3} \frac{1}{1 + u_{com,i}} \tag{26}$$

$$\sigma = 3 * \frac{1}{1 + u_{opt,\Sigma}} \tag{27}$$

3.3 Selection

The aim of selection operation is to reserve the individual with higher fitness. The existing methods of selection includes roulette wheel selection, sequential selection, tournament selection, etc. [12]. In this paper, tournament selection is used in this step. This operator chooses 2 individuals randomly based on their fitness, and then compares the 2 individuals, copying the one with higher fitness to new population.

3.4 Crossover

In crossover operation, the gene in the same slot of two individuals will be exchanged for global search. The crossover operator used in three subproblems will be introduced below.

Generally Arranging the Time of Administrative and Optional Classes. In this step, a kind of two-point crossover is adopted. The operator selects a successive segment with fixed length, and then exchanges it with the counterpart in another individual.

Scheduling the Courses of Administrative and Optional Classes. The crossover used in this study can be depicted as follows:

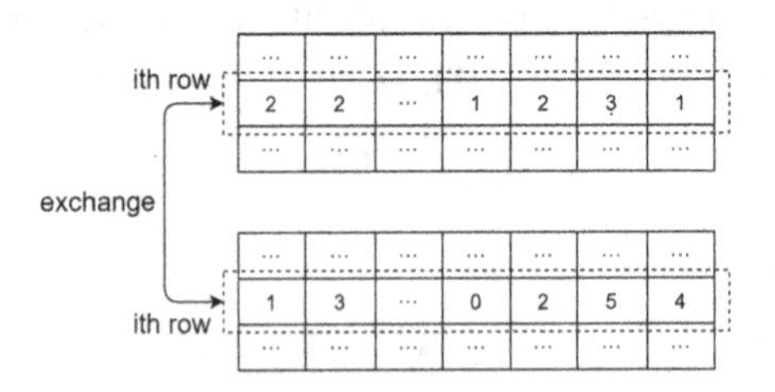

Fig. 4. The process of crossover

Figure 4 illustrates that the crossover only happens between the whole schedule of the same class in two individuals. And this operation can also be understood in another way: the operation tries to recombine the existing schedules to find solutions with higher fitness.

3.5 Mutation

The mutation operation changes several genes in the an individual. In this study, the adapted simple mutation is used in three subproblems.

Generally Arranging the Time of Administrative and Optional Classes. The mutation operator will randomly select one of genes several times. If the generated random number is less than preestablished probability, the operator changes the value according to the proportion of different courses.

Scheduling the Courses of Administrative and Optional Classes. In each class, the operator will select several genes in course part and one gene in teacher part, which will be changed according to their proportion, if the random number is smaller than the preset probability.

3.6 Reservation of Elites from Father Generation

This operation will choose several individual with highest fitness from father generation to replace corresponding number of individuals with lowest fitness in offspring generation. In this way, the best individual in father generation will be preserved.

3.7 Correcting of Schedule

The schedule of administrative classes and optional classes generated by genetic algorithm doesn't always march requirement completely. Therefore, a correcting method should be adopted. If the number of one of 3 courses is more than the requirement, serval courses will be transformed into tutorial. The criterion is the uniformity described as follows:

$$u = \frac{1}{n}\sum_{i=1}^{n}(\tau_i - \bar{\tau})^2 \tag{28}$$

τ_i is the ith interval, and $\bar{\tau}$ is the average interval. And the most uniform case will be selected and reserved.

4 Improvement of Genetic Algorithm

4.1 Dynamic Objective Function

In experiment, the basic genetic algorithm performs not well, especially in speed of finding feasible solution. But when the genetic algorithm only optimizes the hard constraints the algorithm will become easier to find the feasible solution. The contrast is displayed as follows (the total iterations are 300 with a population of 200) (Fig. 5):

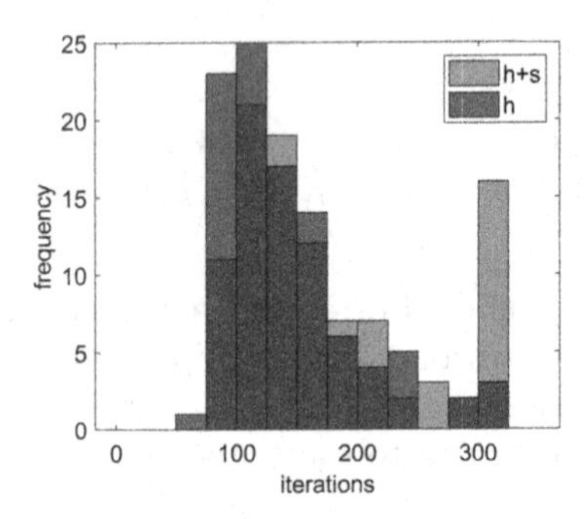

Fig. 5. The comparison of convergence speed in two conditions

In the histogram, it is easy to discover that when focusing on hard constraints, the algorithm is easier to converge. In the most cases, the latter algorithm converges within 200 iterations, while only a few cases cannot converge, which is showed by the rightmost bar.

The reason behind is that the fitness of soft constraints misleads the selection and elites reservation operation, which makes the algorithm choose the individuals with higher s and lower h. Furthermore, the process can be divided into three stages: in stage 1 the h increases rapidly; in stage 2 the increasing of h slows down; in stage 3 the h reaches 1. Based on the division, the serious misleading mainly occurs in stage 2. In addition, in stage 3, the algorithm will sometimes choose the individuals with h lower than 1. If no interference is token, the feasible solution may be abandoned.

Inspired by this phenomenon, we propose a general model of dynamic objective function:

$$fit = h + k(i) * s \tag{29}$$

$k(i)$ is an arbitrary sequence changing through iteration, whose elements range from 0 to 1. Such design enables the function to repress s adaptively. According to analysis, the requirement is that $k(i)$ should repress the value of s at least in stage 2, and in stage 3, when the abandonment occurs, $k(i)$ should also decrease the value of s. And according to Eq. (23), what should be noted is that h is also a variable ranging from 0 to 1, so $k(i)$ can be related to h directly. In this study, the $k(i)$ is set as follows:

$$k(i) = \begin{cases} \underset{1 \le k \le n}{min} \left\{ h_k^4(1) \right\}, i = 1 \\ h_{k_{\text{best}}}^4(i-1), \text{else} \end{cases} \tag{30}$$

Among these, $h_k(i)$ is the h of individual k in the ith iteration, and $h_{k_{best}}(i-1)$ represents the h of best individual in the $(i-1)$th iteration. The fourth power of h can increase the repression degree of soft constraints. In this way, the objective function fulfills the requirement based on the last h. However, the repression is temporary, after the optimization of hard constraints, the fit will go back to the original formation.

4.2 Improvement on Mutation and Crossover

Expect for the improvement on objective function, we also modify the mutation process of scheduling of compulsory courses and optional courses. When mutation occurs, the improved algorithm will try to fix the obvious conflicts in the schedule.

When mutation occurs in compulsory (or optional) course part of chromosome, the algorithm will judge whether the number of this course at this class hour is more than the number of teachers. If the number is greater, the algorithm will try to change the course type, and this process will be operated until the condition is reached or iterations are excessive. When it occurs in teacher part of chromosome, the judgement turns into whether the number of course that the teacher generated by mutation is responsible for is over the threshold.

Besides, according to research [6], when the individuals accumulate, the increment of probability will increase the capacity of local search. Therefore, in addition to the improvement on fixing capacity, the probability of mutation and crossover will increase with iteration. The formulas are displayed below:

$$p_c = p_{c0} * \frac{1}{3 * \sqrt{1 + \left(\frac{250}{3i}\right)^2}} \tag{31}$$

$$p_m = p_{m0} * \frac{1}{3 * \sqrt{1 + \left(\frac{250}{3i}\right)^2}} \tag{32}$$

5 Experiments and Results

This section will carry out the simulation experiment and the convergence performance comparison experiment between the original genetic algorithm and the improved one to analyze the performance of the improved algorithm. In two experiments, MATLAB R2018a is used in the comparative experiment.

5.1 Setup of Experiments

1. The students can be distributed into 9 administrative classes and 13 optional classes;
2. According to hypothesis 3, the school has 13 classrooms, which are just enough for all administrative or optional classes to use at same time;
3. The total number of class hour is 40, i.e. the students learn from Monday to Friday, and each day has 8 courses;
4. The number of each compulsory course in a week is 6, and the number of each optional course in a week is 5;
5. The courses of each optional class are generated randomly.

5.2 Simulation Experiment

Parameters Setting. The parameters are displayed as follows:

Table 3. The parameters of simulation experiment

	General arrangement	Compulsory classes scheduling	Optional classes scheduling
N	100	1000	1000
n	50	200	200
n_{resv}	1	2	2
p_{c0}	0.7	0.5	0.5
n_{cross}	10	–	–
p_{m0}	0.1	0.01	0.01
n_{mutate}	6	4	5

N is the total iterations. n is the number of individuals. n_{resv} is the number of reserved elites. p_{c0} is the initial probability of crossover. n_{cross} is the length of exchanged genes. p_{m0} is the initial probability of mutation, and n_{mutate} is total times of mutation happening in the course's part of chromosome (Fig. 1 and 2).

And through random generation, the list of selected optional course is

Table 4. Selected optional course list

Class 1	Physics	Chemistry	Biology
Class 2	Politics	Chemistry	Biology
Class 3	Politics	Physics	Chemistry
Class 4	History	Physics	Biology
Class 5	Geology	Physics	Chemistry
Class 6	Politics	History	Geology
Class 7	Geology	Physics	Biology
Class 8	Politics	History	Physics
Class 9	Politics	History	Biology
Class 10	Politics	History	Biology
Class 11	Physics	Chemistry	Biology
Class 12	Geology	Physics	Biology
Class 13	History	Geology	Chemistry

Results of Simulation. The variation of best individual's fitness is depicted as follows (Fig. 6, 7 and 8):

In general arrangement, the genetic algorithm converges near 60 iterations.

In scheduling of administrative classes, the increasing rate slows down near 250 iterations. Besides, when it reaches 600 iterations, the hard constrains of best individual temporarily jumps out, and reaches higher fitness in fast rate. This phenomenon illustrates that the improved algorithm has the ability to jump out of local optimum and make the fitness of hard constraints go back to 1 again, finding a better feasible solution.

In scheduling of optional classes, the increasing rate slows down near 400 iterations.

In general, the improved optimized method can solve the course scheduling problem with relatively small population and iterations.

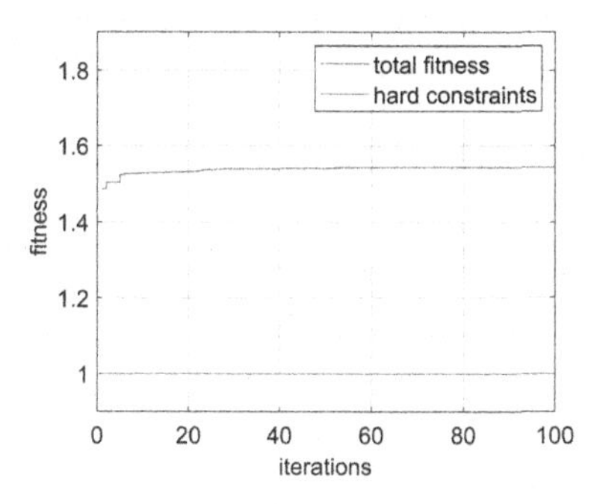

Fig. 6. Fitness variation of general arrangement

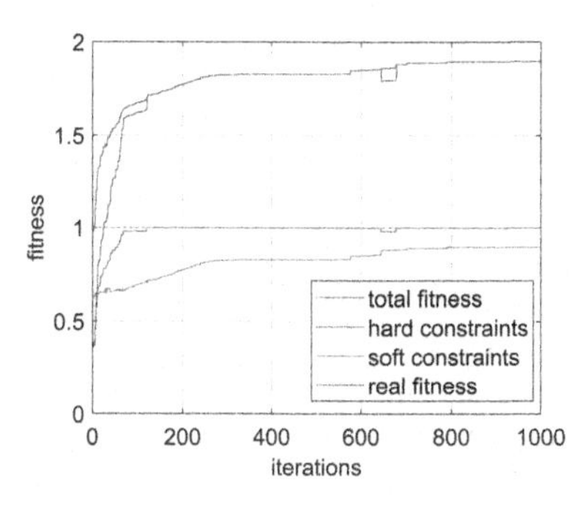

Fig. 7. Fitness variation of scheduling of administrative classes

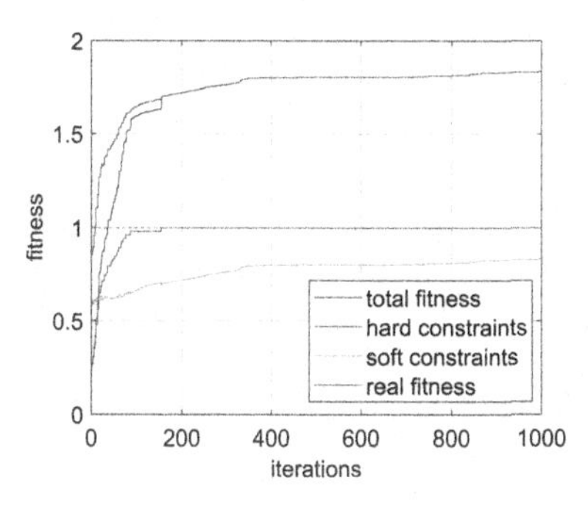

Fig. 8. Fitness variation of scheduling of optional classes

5.3 Convergence Performance Comparison Experiment

Parameters Setting. The parameters of original and improved genetic algorithms are completely the same, which are displayed as follows (Table 5):

Table 5. Parameters of two algorithms used to compare

	Original (compulsory)	Improved (compulsory)	Original (optional)	Improved (optional)
N	300	300	300	300
n	200	200	200	200
n_{resv}	2	2	2	2
p_{c0}	0.5	0.5	0.5	0.5
p_{m0}	0.01	0.01	0.01	0.01
n_{mutate}	4	4	5	5

In addition, we record the highest h of each generation and the historical maximum h in sequences $h_{max}(i)$ and $h_{his}(i)$ respectively, in order to calculate the total difference which is expressed as follows:

$$\Delta_{\max} = \sum_{i=1}^{N} h_{\max}(i) - h_{k_{\text{best}}}(i) \tag{33}$$

$$\Delta_{his} = \sum_{i=1}^{N} h_{his}(i) - h_{k_{\text{best}}}(i) \tag{34}$$

The Δ_{max} and Δ_{his} enable we to analyze the capacity of selecting and reserving the individual with higher h.

The general arrangement is generated by proposed algorithm (the parameters of it are displayed in Table 3), in which the whole schedule has 20 compulsory courses and 18 optional courses. And the number of runs is set as 100.

Results and Analysis. The iterations histograms for comparison are showed as follows:

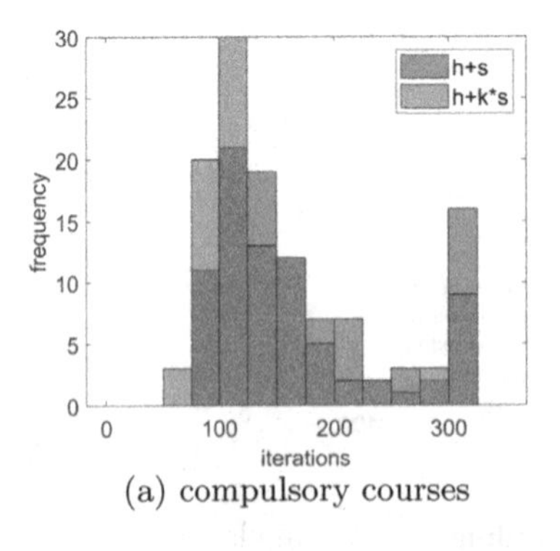

(a) compulsory courses

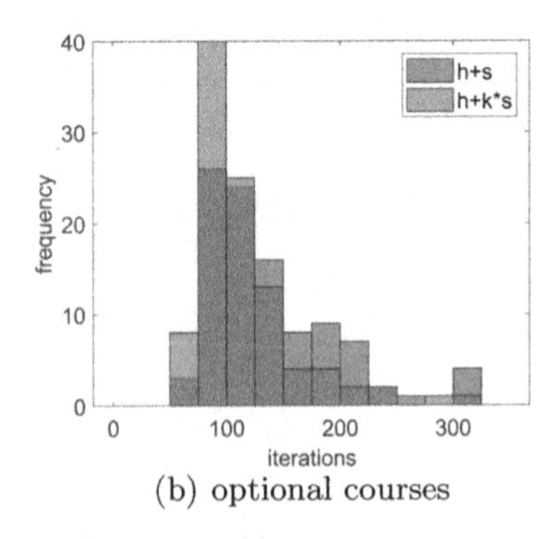

(b) optional courses

Fig. 9. The comparison of convergence speed

And the statistical results are showed as follows:

Table 6. The experimental statistical results of two algorithms

	Original (compulsory)	Improved (compulsory)	Original (optional)	Improved (optional)
Average iterations	149.036	131.846	131.25	111.828
Non-convergent cases	16	9	4	1
Mean Δ_{max}	0.2084	0.0254	0.1685	0.0191
Mean Δ_{his}	0.2801	0.0571	0.2128	0.0613

It should be stated that the non-convergent cases are excluded from the calculation of average iterations in the Table 6.

In Fig. 9a, Fig. 9b and Table 6, it can be found that the improved algorithm performs better than the original one, not only in the number of non-convergent cases but also in the average convergence speed, which is in accord with the analysis in Sect. 4.1. And the lower mean Δ_{max} shows that the improved algorithm tends to choose the individual with highest h in each generation, while the lower mean Δ_{his} the individual with historically highest h is not easy to be abandoned. Through this experiment, the improved algorithm is proved to have higher efficiency.

6 Conclusion

In this paper, a new optimization method of course scheduling problem in College Entrance Examination reformation is proposed. First, the whole problem is divided into three subproblems, which are the general arrangement, the scheduling of administrative and optional classes. And the genetic algorithm is adopted, by which the original nonlinear optimized problem with constraints is transformed into unrestricted problem. Then, two improved measures are proposed. First, a dynamic objective function is proposed, which can repress fitness of soft constraints adaptively. Then the crossover and mutation are also improved. Finally, in the simulation experiment, the problem is solved smoothly with relatively small population and iterations, while the results show that the improved genetic algorithm has the ability to jump out of local optimum. And the convergence performance comparison experiment also illustrates that the improved genetic algorithm shows better performance in finding feasible solutions.

Acknowledgement. This work was supported by the Innovation Fund of WNLO.

References

1. Chen, L., Wang, X.: Solution to course-timetabling problem of mobile learning system based on improved genetic algorithm. Comput. Eng. Appl. **6**, 218–224 (2019)
2. Ni, J., Yang, N.N.: Genetic algorithm and its application in scheduling system. Telkomnika Indonesian J. Electr. Eng. **11**(6), 1934–1939 (2013)
3. Gotlieb, C.C.: The construction of class-teacher time-tables. Commun. ACM **5**(6), 73–77 (1962)
4. Even, S., Itai, A., Shamir, A.: On the complexity of time table and multi-commodity flow problems. In: Proceedings of the 16th Annual Symposium on Foundations of Computer Science, pp. 184–193. IEEE Computer Society, Los Alamitos (1975)
5. MirHassani, S.A., Habibi, F.: Solution approaches to the course timetabling problem. Artif. Intell. Rev. **39**(2), 133–149 (2013)
6. Zhang, J.: Design and implementation of a course scheduling system based on genetic algorithm. University of Electronic Science and Technology of China, Master (2012)
7. Burke, E.K., Petrovice, S.: Recent research directions in automated timetabling. Eur. J. Oper. Res. **140**(2), 166–280 (2002)
8. Yang, Y.M., Gao, W.C., Gao, Y.: 9. Mathematical modeling and system design of timetabling problem based on improved GA. In: Liu, Y., Zhao, L. (eds.) 13th International Conference on Natural Computation, Fuzzy Systems and Knowledge Discovery, pp. 214–220. IEEE, New York (1975)
9. Yang, S.X., Jat, S.N.: Genetic algorithms with guided and local search strategies for university course timetabling. IEEE Trans. Syst. Man Cybern. Part C Appl. Rev. **41**(1), 93–106 (2011)
10. Jiang, L., Chen, P.Z., Shen, X.H., et al.: Research on the allocation of educational resources in the reform of college entrance examination. Math. Model. Appl. **9**(2), 53–64 (2020)
11. Yu, G.L.: Research on Timetable Problem of College Based on Genetic Algorithms. Hebei University of Technology, Master (2007)
12. Kaya, M.: The effects of a new selection operator on the performance of a genetic algorithm. Appl. Math. Comput. **217**(19), 7669–7678 (2011)

DNA and Molecular Computing

Graphene Oxide-triplex Structure Based DNA Nanoswitches as a Programmable Tetracycline-Responsive Fluorescent Biosensor

Luhui Wang[1], Yue Wang[2], Mengyang Hu[2], Sunfan Xi[1], Meng Cheng[1], and Yafei Dong[1,2(✉)]

[1] College of Life Science, Shaanxi Normal University, Xi' an 710119, China
Dongyf@snnu.edu.cn

[2] College of Computer Sciences, Shaanxi Normal University, Xi' an 710119, China

Abstract. The overuse of antibiotics will lead to the emergence of drug resistance so that many common diseases can not be effectively treated. Therefore, based on the special structure of graphene oxide (GO) and DNA triple strand, a programmable DNA nanodevice for quantitative detection of tetracycline (TC) is designed in this paper. The introduction of GO as a quencher can effectively reduce the background fluorescence and whether there is any change in the fluorescence in the auxiliary system; Stabilizing the trigger chain with three chain structure will reduce the error more likely. The feasibility analysis preliminarily confirmed the realizability of the designed model, and the optimal conditions were obtained by condition optimization, which laid the foundation for the subsequent quantitative detection of TC, while the selective experiments in different systems fully showed that the model had excellent specificity.

Keywords: Tetracycline · Triplex structure · Graphene oxide · Quantitative detection

1 Introduction

Antibiotics are one of the most important medical discoveries in the 20th century and have made great contributions to the treatment of bacterial infection and the animal breeding industry [1]. It is well known that the abuse and accumulation of antibiotics will pose a threat to human health and the ecosystem. Tetracycline (TC) is a broad-spectrum antibiotic produced by Streptomyces, which is widely used in aquaculture and veterinary medicine. Introducing such residual compounds into the environment from different sources will lead to serious environmental problems [2]. TC will exist in food and accumulate in the human body through the food chain, causing damage to human health [3]. Therefore, although the efficacy of TC in medical care can not be

L. Wang and Y. Wang—These authors contributed equally to this work.

L. Pan et al. (Eds.): BIC-TA 2021, CCIS 1565, pp. 371–379, 2022.
https://doi.org/10.1007/978-981-19-1256-6_28

ignored, its harm to the environment and health can not be underestimated. Therefore, it is very necessary to design a simple equipment to detect TC quickly and sensitively.

The ever-changing sensing devices add components to the information sensing toolbox of the new generation of Internet of things technology. Biosensor converts the signals to be detected into detectable signals through identification elements and appropriate transducers, which are carried out as required in the biochemical environment after computer software sequence design and experimental element simulation in advance. Among them, biosensors have been widely discussed in the detection of cancer markers [4, 5], harmful and toxic substances [6, 7], metal ions [8–10], and are of great significance to biomedicine, food safety, and environmental detection. The recent research on coronavirus covid-19 [11] is also a new display in this booming scientific field.

The aptamer is a sequence (DNA/RNA) screened from a random oligonu-cleotide library synthesized in vitro by SELEX technology. It not only has the characteristics of high specificity and affinity binding targets comparable to antibodies but also has many advantages as recognition elements of the detection system, such as high stability and miniaturization [12]. The structural flexibility of aptamers and the simplicity of chemical modification provide hope for the development of reagent-free one-step analysis [13]. Using aptamers as functional nucleic acids can be specifically triggered under biochemical conditions to construct target induced DNA logic gates, so that logical operation can be carried out in response to the physiological environment. Aptamer sensors also show excellent performance in detecting diverse analytes [14]. A shortened 8-mer ssDNA aptamer fragment has been used here and has been shown to have a high affinity for TC [15].

Biological logic gate is also an ideal device for intelligent detection, which can be used to properly identify a variety of targets. Although biological logic gates operate logically at the molecular level with a behavior similar to that of computers, their input and output signals cannot be simply defined as the conversion of electrical signals like computer logic. Designers need to indicate the type of input and the Boolean value of output meaning in advance [16]. Fluorescence can be used as one of the best methods to explore the biological mechanism because of its relatively simple imaging method [17]. Compared with expensive, time-consuming, and complex signal transduction methods, fluorescent signal sensors are more attractive in high-performance sensors because of their simple operation, fast sensitivity, and diverse functions.

In recent years, special materials/structures play a more and more important role in biological logic gates. As a new nano material, graphene oxide(GO) has the function of adsorbing DNA single strand. Combined with its fluorescence quenching property, GO has more applications in fluorescent biosensors. DNA usually exists in the form of single strand or double strand. When changing the pH in the system, it may form a special three strand structure to make the binding of intermediate chains more stable and reduce the background noise to a certain extent. Based on the above conditions, we developed a logical sensing model for the quantitative detection of TC. The realization of the model is mainly based on the fact that GO can adsorb DNA single strand but not double strand, and can quench fluorescence by fluorescence common energy transfer to complete the signal transformation and output; In order to reduce the leakage as much as possible, a three chain model is introduced to further stabilize the trigger chain to reduce the possibility of its dissociation in the system. After that, a series of experiments proved that the designed model was feasible and had a good detection limit and specificity.

2 Experiment

2.1 Reagents and Materials

All DNA was purchased from Sangon Biotechnology Co., Ltd. (Shanghai, China) and purified by PAGE and ULTRAPAGE, and all of them were dissolved in ultrapure water as stock solutions (10 μM). TC, amoxicillin and ofloxacin were purchased from Beijing Shiji Aoke Biotechnology Co., Ltd. GO dispersion (0.5 mg/mL) is purchased from Xianfeng nano material technology Co., Ltd. (Nanjing, China).

2.2 Construction of Triple Helix DNA and P1/GO Platform

In order to reduce the leakage of trigger strand, S1 concentration is greater than S2 concentration. The two were mixed in buffers with different pH to prepare for subsequent pH optimization. The triple helix structure is formed stably after the process from high temperature chain breaking to gradual cooling to room temperature.

P1 strand and GO dispersion were mixed in buffer and reacted at room temperature for 1h to obtain fluorescence quenched P1/GO system for subsequent experiments.

2.3 Fluorescent Signal Detection

In this study, we chose to label substrates with the fluorophore FAM. The fluorescence results were obtained for FAM at 492 nm excitation and 518 nm emission using a fluorescence-scanning spectrometer (EnSpire ELISA; PerkinElmer, Waltham, MA, USA).

3 Principle

Acidic conditions can induce DNA strand to form triplex structure. Therefore, triple helix DNA, as one of the substrates of this design, was prepared by S1/S2 in acidic Tris-HCl buffer. Another important substrate in this design is the P1-GO composite structure. In detail, P1 is a single strand, which is easy to be adsorbed by GO, and the fluorescence-labeled by P1 is quenched by GO through fluorescence co energy transfer. In other words, in the absence of other inputs, there are two substrates triple helix DNA and P1-GO in the system, and the total system has a low fluorescence value. When the target TC exists, the situation will change greatly. Specifically, TC can bind to S1 in triple helix DNA, destroy the structure of triple helix DNA and release S2; The free S2 sequence and P1 sequence are highly coincident, which can form S2/P1 double-strand structure, so that the fluorophore is far away from the GO surface, and the fluorescence value in the system is greatly increased (Scheme 1).

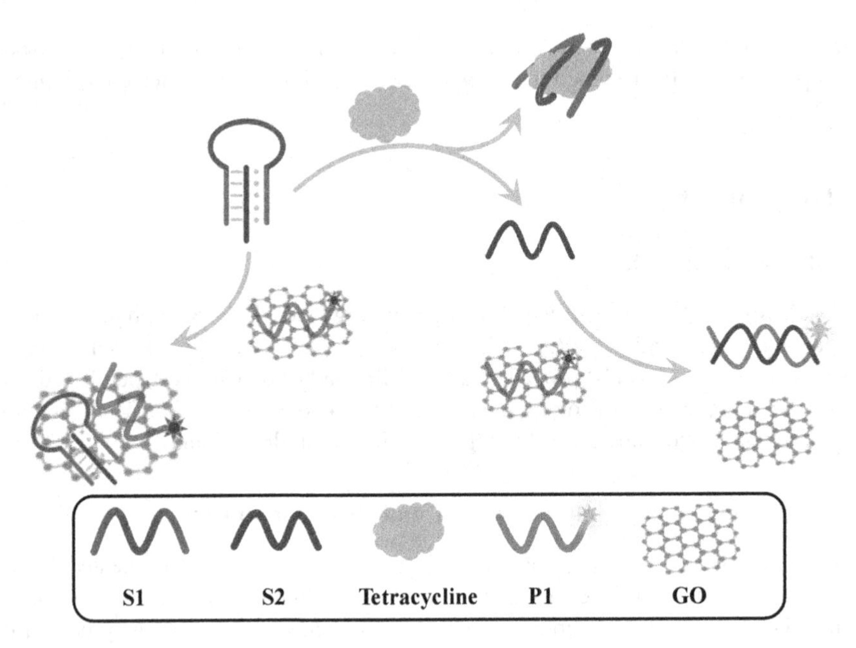

Scheme 1. Principle diagram of TC detection

4 Results and Discussion

4.1 Feasibility Study

The first step of experimental verification is feasibility analysis. The fluorescence value comparison curve shown in Fig. 1 is obtained through fluorescence spectroscopy under different conditions. Without target TC, the fluorescence value is low (black curve), which is due to the formation of triplex and the fluorescence resonance energy transfer between GO and FAM. The fluorescence increased significantly after the addition of target TC, which represents that after TC binds to the aptamer, the S2 chain is free from the binding of Watson Crick and Hoogsteen base pairing, and can be more fully hybridized with P1. FAM also has fluorescence recovery because it is far away from GO (red curve). The two curves with obvious difference can reasonably distinguish whether there is a target or not, which is an effective description of the scheme.

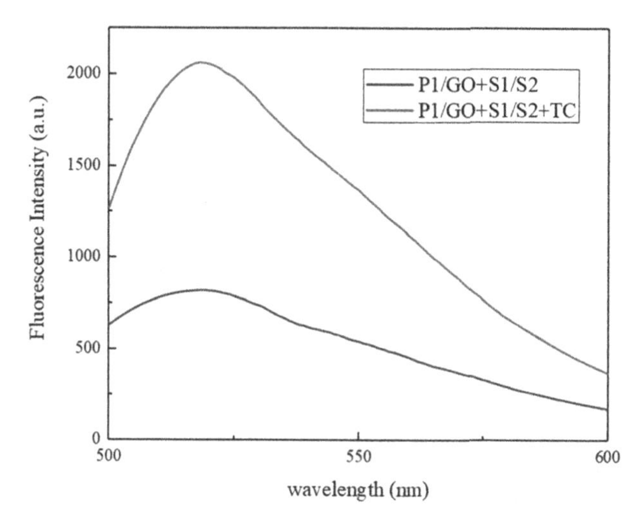

Fig. 1. Fluorescence spectra of the feasibility study for TC detection

4.2 Condition Optimization

By selecting appropriate reaction conditions, the intermolecular reaction is in the best condition and the mismatch ratio between biomolecules is reduced. The four elements are optimized respectively to obtain the signal that can best reflect the biological essence. F represents the fluorescence value in the presence of TC, and F_0 represents the blank control group, that is, in the absence of TC. The results are shown in Fig. 2. The abscissa identifies different conditions, and the ordinate is measured by the difference between F and F_0.

Firstly, the length of the triplex structure part is explored by the three strands S1–8, S1–9 and S1–10. The number after S1 represents the base number complementary between S1 and S2. According to Fig. 2A, the number of bases is 10 bp had the best value, so S1–10 is selected. Then, Mg^{2+} plays an important role in the formation of triplex structures. It can also be seen from Fig. 2B that a better difference can be obtained only when Mg^{2+} reaches a sufficient concentration. The Mg^{2+} concentration in subsequent experiments is set as 10 mM. Secondly, the triplex structures are pH dependent, so pH value is a factor that can not be ignored. According to the response interval of Hoogsteen hydrogen bond, four different values are selected from pH 5 to 7 to achieve the best effect when pH = 6.6 (Fig. 2C). Finally, the fluorescence signal regulator GO plays an indisputable role in the results. The fluorescence output at different concentrations shows a significant difference (Fig. 2D), and it is obvious that the GO concentration is 10μg/ml D-value is the largest, so it is considered became the final choice.

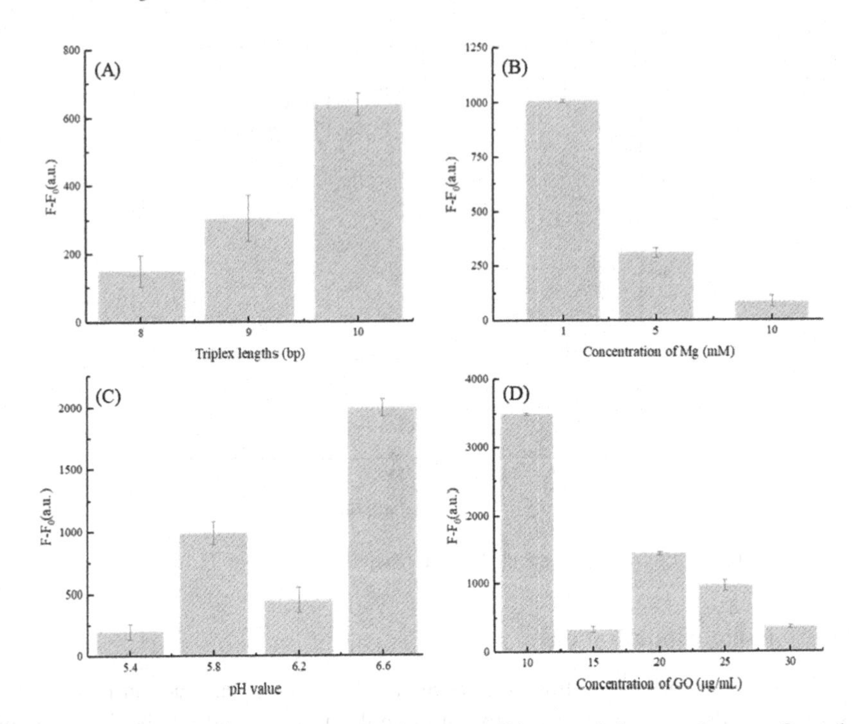

Fig. 2. The optimization diagram under different conditions, and the error bars are from three repeated experiments. (A) Triple-strand length; (B) Mg^{2+} concentration; (C) GO concentration and (D) pH value.

4.3 Performance Analysis

The sensitivity and specificity of the sensor are two important properties. These two indexes were evaluated under the above selected experimental conditions. During the detection of sensitivity, TC solutions with concentrations of 0 nM, 50 nM, 100 nM, 200 nM, 300 nM, 400 nM, 600 nM, 800 nM, and 1000 nM were added respectively for fluorescence characterization under the same conditions.

Figure 3A shows that in the range of 0 –1000 nM, the fluorescence intensity value will increase with the increase of TC solution concentration and maintain a linear growth in the range of 0 –300 nM (Fig. 3B). The linear regression equation is: $y = 6.4356x + 764.37$, and the determination coefficient R^2 is 0.9952. Use formula $3\sigma/S$ (σ Is the standard deviation of the background signal, S represents the slope of the linear equation), and the detection limit of the method is estimated to be 590 pM. Compared with other similar models (Table 1), it reached a lower detection limit without enzyme assistance and cumbersome operation.

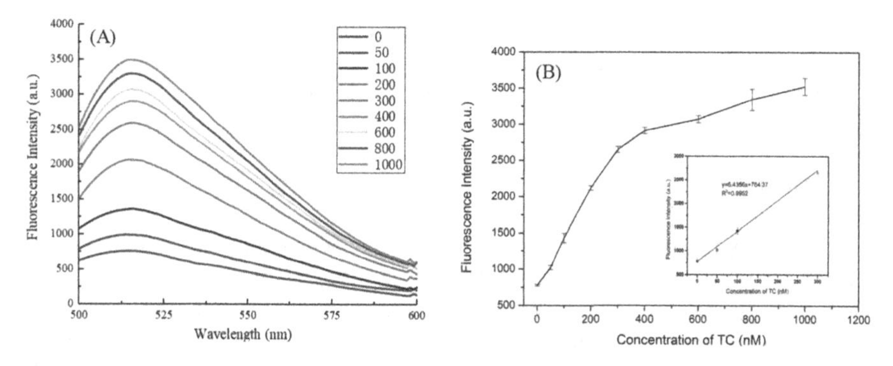

Fig. 3. Fluorescence spectra after adding different concentrations of TC(A) and Fluorescence intensity values of different concentrations of tetracycline(B). The inset shows the linear relationship between TC concentration and fluorescence intensity in the response range of 0 ~ 300nM. Error bars are calculated from three repeated experiments

Table 1. Comparison of the detection method in this article with other research methods

Method	LOD	Reference
Colorimetrc assay	45.8 nM	[18]
Fluorescent assay	970.0 pM	[19]
Fluorescent assay	6.49 nM	[20]
Fluorescent assay	590 pM	This work

In order to explore the selectivity of the sensor for different substances, five groups of controls were selected to detect the other two antibiotics. The first group is the blank control group, the second group is added with amoxicillin, the third group is added with ofloxacin, the fourth group is TC, and the fifth group is the mixture of amoxicillin, ofloxacin and TC, which can verify whether TC can be detected sensitively in complex

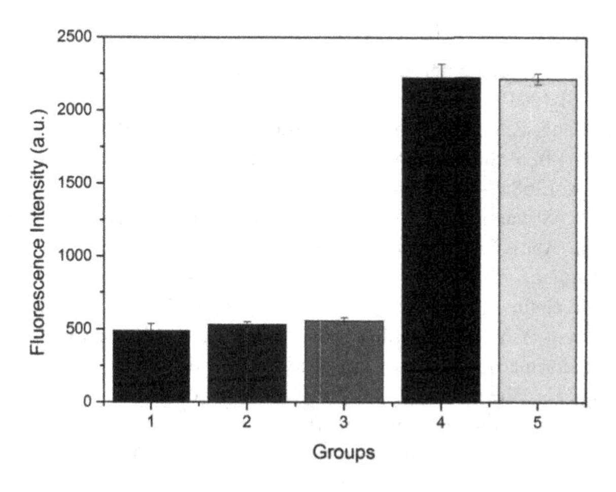

Fig.4. Specificity of the Fluorescence strategy

situations. According to the results in Fig. 4, the fam fluorescence of the non-specific substance measured by the sensor is weak, while the fluorescence of the system with the target is significantly enhanced, indicating that it has good specificity for TC.

5 Conclusion

Based on the logic gate driven by a triple helix nano switch, a fluorescent biosensor for the detection of TC was constructed in this paper. After TC was input into the sensing system as the sample to be tested, after specific recognition, reaction with aptamer, and fluorescence detection, the results were analyzed according to the scanned fluorescence value to prove its feasibility. The parameters affecting the experimental results, such as three chain length, Mg^{2+} concentration, pH value, and GO concentration, were carefully optimized. After comprehensive consideration, the best combination was selected for subsequent experiments. The performance of the sensing system is tested under optimal conditions, and it is concluded that the strategy has good sensitivity and selectivity. This simple and general scheme has a good attraction in the applicability of the sensor. If it is used to detect other substances, it only needs to replace the aptamer sequence.

References

1. Xu, D., Xiao, Y., Pan, H., et al.: Toxic effects of tetracycline and its degradation products on freshwater green algae. Ecotoxicol. Environ. Saf. **174**(6), 43–47 (2019)
2. Daghrir, R., Drogui, P.: Tetracycline antibiotics in the environment: a review. Environ. Chem. Lett. **11**(3), 209–227 (2013)
3. Liu, X.G., Huang, D.L., Lai, C., et al.: Recent advances in sensors for tetracycline antibiotics and their applications. TrAC Trends Anal. Chem. **109**(11), 260–274 (2018)
4. Ma, C., Liu, H.Y., Tian, T., et al.: A simple and rapid detection assay for peptides based on the specific recognition of aptamer and signal amplification of hybridization chain reaction. Biosens. Bioelectron. **83**(11), 15–18 (2016)
5. Sun, Y.L., Fan, J.F., Cui, L.Y., et al.: Fluorometric nanoprobes for simultaneous aptamer-based detection of carcinoembryonic antigen and prostate specific antigen. Microchim. Acta **186**(3), 152–161 (2019)
6. Wang, Y.H., Fang, Z.Y., Ning, G., et al.: G-quadruplex-bridged triple-helix aptamer probe strategy: a label-free chemiluminescence biosensor for ochratoxin A. Sens. Actuators, B Chem. **298**(11), 126867–126874 (2019)
7. Malhotra, B.D., Srivastava, S., Ali, M.A., et al.: Nanomaterial-based biosensors for food toxin detection. Appl. Biochem. Biotechnol. **174**(3), 880–896 (2014)
8. Liu, C., Wang, J., Zhang, X., et al.: Mercury ion fluorescence biosensor based on oligonucleotide chain. Anal. Chem. **45**(2), 164–168 (2017)
9. Chen, J.L., Zhang, Y.Y., Cheng, M.P., et al.: Highly active G-quadruplex/hemin DNAzyme for sensitive colorimetric determination of lead (II). Microchim. Acta **186**(12), 1–8 (2019)
10. Lin, T., Zeng, G., Shen, G., et al.: Study on a novel glucose oxidase sensor based on inhibition for the determination of mercury ions in environmental pollutants (2005)
11. Santiago, I.: Trends and innovations in biosensors for COVID-19 mass testing. ChemBioChem **21**(20), 2880–2889 (2020)

12. Cho, E.J., Lee, J.W., Ellington, A.D.: Applications of aptamers as sensors. Annu. Rev. Anal. Chem. **2**(1), 241–264 (2009)
13. Rajendran, M., Ellington, A.D.: Selection of fluorescent aptamer beacons that light up in the presence of zinc. Anal. Bioanal. Chem. **390**(4), 1067–1075 (2008)
14. Kim, J., Jang, D., Park, H., et al.: Functional-DNA-driven dynamic nanoconstructs for biomolecule capture and drug delivery. Adv. Mater. **30**(45), 1707351–1707382 (2018)
15. Kwon, Y.S., Raston, N.H.A., Gu, M.B.: An ultra-sensitive colorimetric detection of tetracyclines using the shortest aptamer with highly enhanced affinity. Chem. Commun. **50** (1), 40–42 (2014)
16. Barnoy, E.A., Popovtzer, R., Fixler, D.: Fluorescence for biological logic gates. J. Biophotonics **13**(9), 158–172 (2020)
17. Wang, Y.F., Zhang, Y.X., Wang, J.J., et al.: Aggregation-induced emission (AIE) fluorophores as imaging tools to trace the biological fate of nano-based drug delivery systems. Adv. Drug Delivery Rev. **143**(3), 161–176 (2019)
18. He, L., Luo, Y.F., Zhi, W.T., et al.: Colorimetric sensing of tetracyclines in milk based on the assembly of cationic conjugated polymer-aggregated gold nanoparticles. Food Anal. Meth. **6**(6), 1704–1711 (2013)
19. Chen, T.X., Ning, F., Liu, H.S., et al.: Label-free fluorescent strategy for sensitive detection of tetracycline based on triple-helix molecular switch and G-quadruplex. Chin. Chem. Lett. **28**(7), 1380–1384 (2017)
20. Yang, P., Zhu, Z.Q., Chen, M.M., et al.: Microwave-assisted synthesis of xylan-derived carbon quantum dots for tetracycline sensing. Opt. Mater. **85**(11), 329–336 (2018)

Construction of Complex Logic Circuit Based on DNA Logic Gate AND and OR

Mengyang Hu[1], Luhui Wang[2], Sunfan Xi[2], Rong Liu[1], and Yafei Dong[1,2](✉)

[1] Department of Computer Science, Shaanxi Normal University, Xi' an 710119, China
dongyf@snnu.edu.cn
[2] Department of Life Science, Shaanxi Normal University, Xi' an 710119, China

Abstract. DNA, as one of the most promising research objects for molecular computing, is often used to construct the basic modules of computing. A simple and multiplexable DNA logic computing platform comprising a basic two input logic gate AND and OR, is designed by using G-quadruplex combined with NMM to generate fluorescence mechanism as output signal and DNA strand as input signal. Designed DNA strand displacement and catalytic hairpin self-assembly, achieving enzyme-free and label-free, simplified the design and reduced the cost. Strand displacement reactions not only enable signal amplification, but also provide methods to implement multi-layer cascading and solve complex computational problems by utilizing the basic logic gate AND and OR. In summary, this logic platform can realize the design of complex logic circuit by the constructed simple underlying logic gate AND and OR, which provides the potential to solve complex computational problems.

Keywords: Molecule logic gate · DNA strand displacement · G-quadruplex · Cascade circuits

1 Introduction

Biomolecules possess excellent characteristics of building blocks, such as naturally available amounts, good biocompatibility, structural diversity, a variety of possible coordination modes, and price adjustability [1]. In recent years, biochemical based research on logic components has flourished, which also directly promoted the construction of complex logic computing devices with biomolecules as the basic materials [2–4]. Among various biomolecules, deoxyribonucleic acid is a highly programmable biomolecule because of its precise base pairing properties, which makes it possible to construct complex DNA circuits [5]. In the design and construction of logic elements and molecular circuits, DNA is also rated as a high-performance nanomaterial because of its outstanding features, such as huge data storage space, massively parallel processing capacity, functional predictability, and excellent molecular recognition capability. At this stage, there are many kinds of molecular logic gates, computational circuits and molecular robots constructed based on DNA molecules, and many research teams have developed a series of simple logic gates and high-level logic circuits [6, 7],

L. Pan et al. (Eds.): BIC-TA 2021, CCIS 1565, pp. 380–389, 2022.
https://doi.org/10.1007/978-981-19-1256-6_29

but there are still some problems to be solved in the current research. On the one hand, constructing complex logic gates is accompanied by problems of high chain complexity, complicated design and operation; On the other hand, most logic designs involve enzymatic reactions or modification of fluorescent groups, quenching groups on the DNA strand ends, leading to excessive costs because of the generally high price of enzymes and the price of modified DNA is dozens of times that of ordinary DNA. Therefore, seeking materials that can achieve enzyme free labeling is also key to advancing the study of DNA logic computing.

G-quadruplex is a special structure of deoxyribonucleic acid. The 4 guanines joined by Hoogsteen hydrogen bonds form a cyclic plane through which two or more tetrads pass π-π stacking to form a quadruplex [8]. The G-quadruplex structure was confirmed to produce strong fluorescence in combination with organic dyes, such as N-methylmesoporphyrin IX(NMM) [9], and was widely applied in biological computation and as a signal output.

DNA strand displacement (DSD) is an enzyme free molecular technique and an effective programming language for network behavior [10]. As a programming framework, DSD can be rationally designed to deal with extraordinary computational problems because of its unique features including predictability and programmability [11]. In many DNA devices, thermodynamic deviations generated during DNA strand hybridization alter the equilibrium state of the system, using the difference in energy to prompt nanoparts made of DNA in turn to displace the DNA strand along the thermodynamically stable direction. The extra base pairs generated by hybridization generate enthalpies, coupled with the increased entropy upon release of the strands [12], thereby driving the strand displacement system to form an energetically more favorable product, key to creating programmed DNA reactions. The catalytic hairpin self-assembly amplification (CHA) technique, also used as an enzyme free signal amplification technique, has attracted much attention due to its unique features such as precise molecular recognition and huge design space [13]. CHA has higher catalytic efficiency, lower background signal, and more stable reaction system [14], which is convenient for constructing enzyme free and label free DNA logic circuits with high sensitivity.

In 2020, Chang et al. developed molecular and logic gates for triggered signal amplification based on the CRISPR/cas9n system [15]; In 2020, Wang et al. developed a logic gate based on an Eu^{3+}-functionalized GA MOF turn-on fluorescent probe [16]; In 2020, Chen et al. developed a dual anchored proximity aptamer DNA logic gate for accurate identification of circulating tumor cells [17]. Most logical operations here are used as biorecognition and disease diagnosis, and involve complex chemical materials, which have the advantages of precise identification and high efficiency [18, 19], but have some disadvantages, such as longer experimental time, complicated operation and cost. Simple and convenient logic gate strategies are urgently needed to provide methods for the development of novel logic computing.

Inspired by previous studies [18, 20], we constructed an enzyme free, tag free logic operational platform based on DSD and CHA techniques, developed a two input basic logic gate (AND, OR) by using whether the hidden G-rich sequence in the hairpin stem is released, and verified it by biological experiments to demonstrate the operability of the logic gate. In addition, a three-layer cascade circuit is constructed based on this platform, which can realize the corresponding logic operation and provide ideas for solving complex logic calculation problems. Most importantly, the constructed logic assembly can be realized under any enzyme free condition without any modification of fluorescent groups or quenching groups, making the logic platform much simpler and economical. In summary, this simple and robust mechanism has great potential for applications in many fields, such as digital computing [21], logical reasoning [22], classifiers [23], as well as biological imaging [24], which shed new light on future molecular computing.

2 Experimental principle

DNA strand reaction based on strand displacement reaction and catalytic hairpin self-assembly design as shown in Figure 1, the first trigger strand TP1 (purple) is able to specifically recognize the naked part of the strand auxiliary strand AS (blue) in the duplex, and binding AS will displace the trigger strand TP2 out, followed by the trigger hairpin self-assembly reaction such that HP1 and HP2 open, and combine to form the duplex. The naked part of the duplex, rich in base G, is able to form a G-quadruplex under buffer environment and thus is able to combine with NMM to produce a fluorescence signal as output. The sequences of the DNA strands involved in the experiments are shown in Table1.

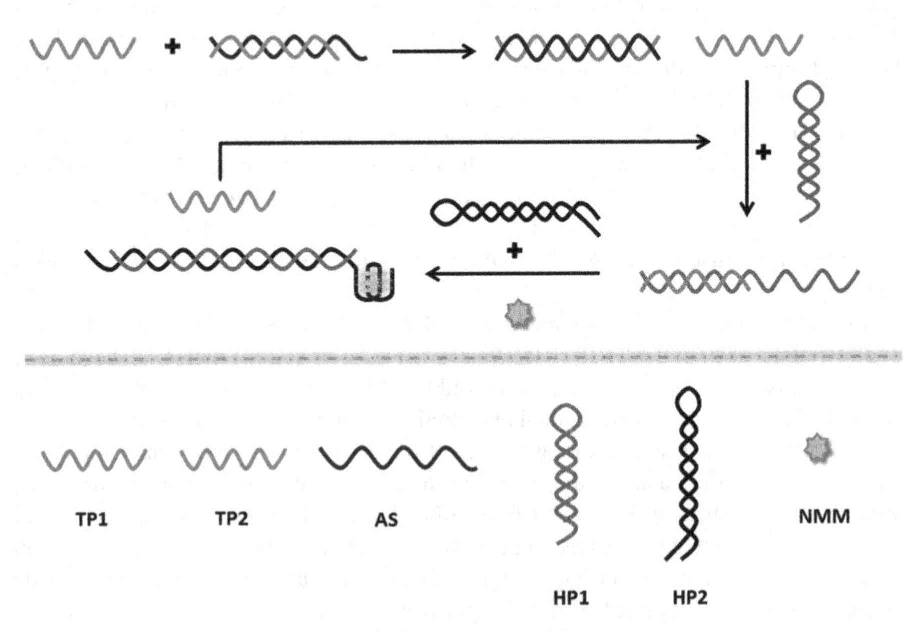

Fig. 1. Principle model diagram of logical gate construction

Table 1. DNA sequences

Name	Sequence
TP1	TCTTCATAATATGGTGTCAG
TP2	AGTCAGTGTGGATATGGTGTCAG
HP1	CTGACACCATATCCACACTGACTCGGGTAGGGCGAGTCAGTGTGG
HP2	CTGACTCGCCCTACCCGAGTCAGTGTGGCGGGTAGGGCGGGTTGGG
AS	CTGACACCATATTATGAAGA

3 Experimental Materials and Methods

3.1 Chemicals and Materials

The ssDNA used in the experiments were synthesized and further purified by JKchemicalCo., Ltd. (Beijing, China). N-methyl-mesoporphyrin IX (NMM) were bought from Sangon Biotech Co., Ltd. (Shanghai, China) and stored at −20oC in the dark before use. The ultra-pure water used in our experiment was purified by a Milli-Q system (18 MΩ cm). 96-Well microplates were purchased from Lingyi Biotech Co., Ltd. (Shanghai, China).

3.2 Instrumentation

The fluorescence spectrum of NMM was measured using a fluorescence scanning spectrometer with excitation at 399 nm and emission at 610 nm through EnSpire ELIASA from PerkinElmer USA company (Shanghai, China).

3.3 Fluorescence Verification of Functional Logic Gates

Prior to experiments, the diluted solution with HP1 (1 μM) and HP2 (1 μM) were heated at 95 °C by PCR for 5 min, and cooled to room temperature at a slow rate to form hairpin structure. And mixing TP2 (1 μM) and AS (1 μM) to form a double-stranded structure.

For the fluorescence verification of the OR logic gate, four different experimental groups were set: no TP1 and TP2, only TP1 (100 nM), only TP2 (100 nM), TP1 (100 nM) and TP2 (100 nM). Then mixed with TP2/AS (100 nM), Hp1 (300 nM) and Hp2 (300 nM) were mixed with the working buffer and incubated for 75 min at 35 °C. After that, NMM (2 μM) was added and further incubated for 25 min at 25 °C. Finally, a part of the solution was removed to a 96-well plate, and the fluorescence intensity was detected and recorded by the instrument mentioned above.

For the fluorescence verification of the AND logic gate, four different experimental groups were set: no HP1 and HP2, only HP1 (300 nM), only HP2 (300 nM), HP1 (300 nM) and HP2 (300 nM). Then mixed with TP1 (100 nM), TP2/as (100 nM) and were mixed with the working buffer and incubated for 75 min at 35 °C. After that, NMM (2 μM) was added and further incubated for 25 min at 25oC. Finally, a part of the solution was removed to a 96-well plate, and the fluorescence intensity was detected and recorded by the instrument mentioned above.

Several representative input states are set for the cascade circuit for verification. According to whether to add or not given in the truth table, the specific steps are the same as above, and will not be repeated here one by one.

4 Results and Discussion

4.1 OR Logic Gate

In discrete mathematics, a proposition is a statement that is true or false, but it cannot be both true and false. A conjunction is a short name for a logical or propositional conjunction [25].Disjunctive is one of the most commonly used logical conjunctions, which means "or". If one of its two variables is true, the result is true. Both variables are false, and the result is false. In the model design based on SDR and CHA, this logical operation is implemented by building "OR" logic gates, whose true values are shown in Fig. 2B. We took the designed TP1 and TP2 chains as two inputs and defined them as input truth as "true" or "false" based on their presence or absence. The remaining desired DNA strands and fluorescent dyes were included in the logic gate system, after the reaction ended, the fluorescence intensity at 610 nm of G-quadruplex/NMM detected in different cases was taken as the corresponding output signal. As shown in Fig. 2A, whether TP1 (purple) or TP2 (orange) is present triggers the subsequent reaction with a fluorescence signal, i.e., output as "true". When both are absent i.e. the input is "false", the subsequent reaction cannot be triggered and there is no production of a fluorescent signal i.e. the output is "false". The fluorescence signals obtained through biological experiments after normalization are shown in Fig. 2C. Of note, we defined above threshold 0.4. Notably, we defined an excess of 0.4 above the threshold as "true" and a fall below the threshold as "false". To sum up, the fluorescence detection results in different input situations and their counterparts are in accordance with the truth table characteristics of "OR" logic gate, which enables disjunctive operations.

4.2 AND Logic Gate

Joint is also one of the most commonly used logical conjunctions, which means "and". If one of its two variables has a true value of "false", the result is "false"; Both variables are true and the result is true [19].In this model design based on SDR and CHA, this logical operation is implemented through the firmware AND logic gate, and its true value table is shown in Fig. 3B.We use HP1 (green) and HP2 (red) hairclip as two inputs, which contain trigger chain TP1, double chain TP2/AS and fluorescent dye. As shown in Fig. 3A, the final fluorescence signal was generated because, in the presence of trigger chains, the bare part of the double-chain structure formed by self-assembly of HP1 and HP2 formed a G-quadruplex. Therefore, when both HP1 and HP2 exist, that is, when the input is true, the fluorescence signal will be generated, that is, when the output is "true". When any hairpin is not present and the input is "false", no fluorescent signal will be generated, that is, the output is "false". The results of standardized

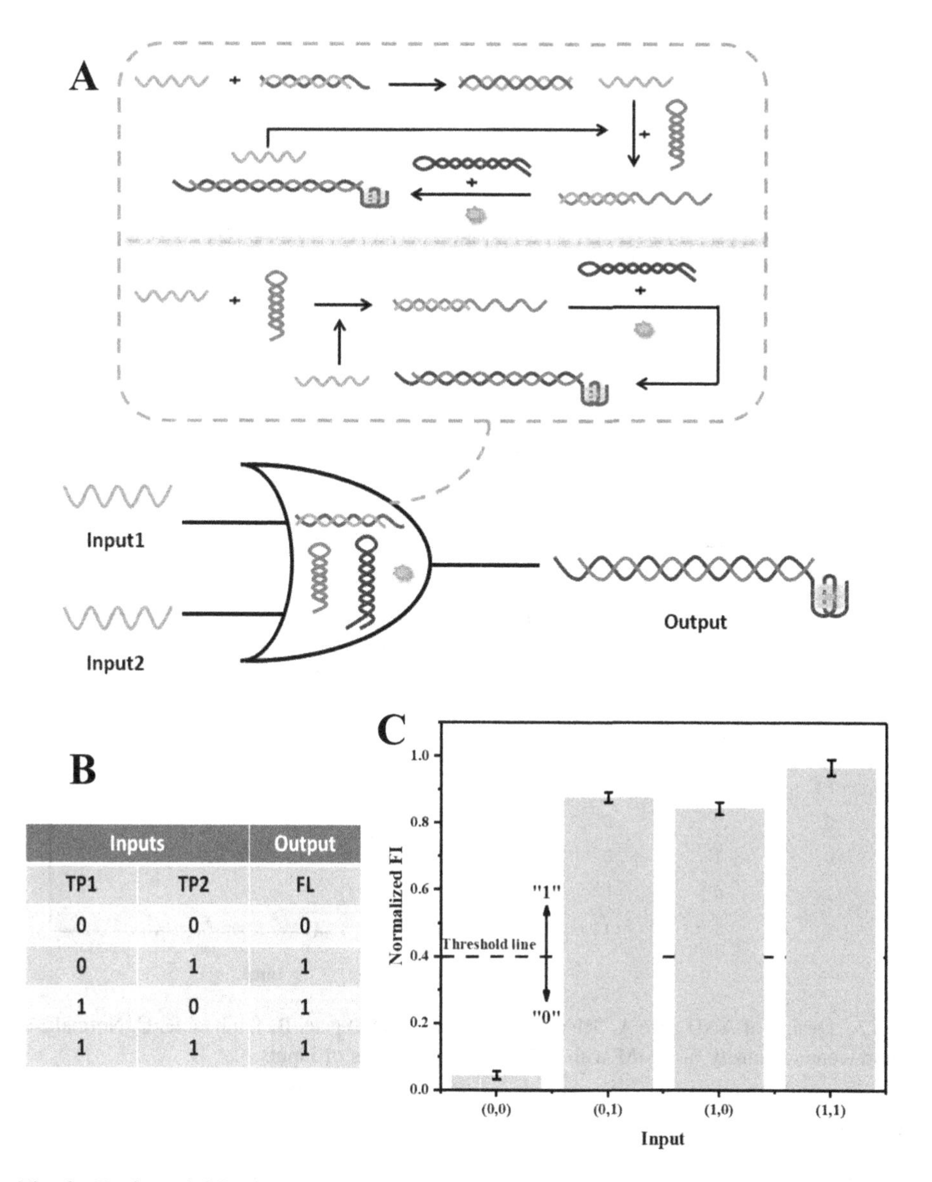

Inputs		Output
TP1	TP2	FL
0	0	0
0	1	1
1	0	1
1	1	1

Fig. 2. Design of OR gate A. Schematic diagram of OR gate. **B.** Truth table. **C.** Normalized fluorescence intensity of NMM with different combinations of inputs.

fluorescence signals obtained from biological experiments are shown in Fig. 3C. Similarly we define 0.4 above the threshold as "true" and below the threshold as "false". In summary, the results of fluorescence detection under different input conditions conform to the characteristics of the true value table of the AND logic gate, and the combination operation can be implemented.

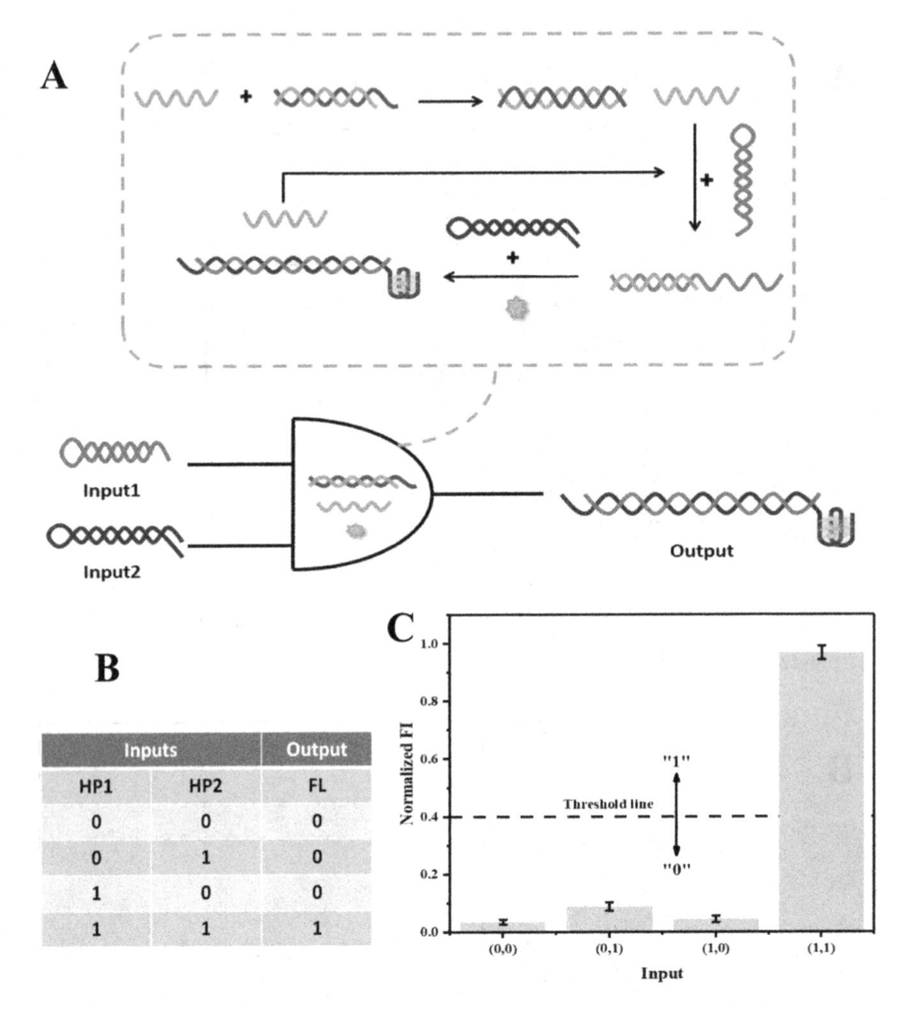

Inputs		Output
HP1	HP2	FL
0	0	0
0	1	0
1	0	0
1	1	1

Fig. 3. Design of AND gate A. Schematic diagram of AND gate. **B.** Truth table. **C.** Normalized fluorescence intensity of NMM with different combinations of inputs.

4.3 Cascade Circuit

Based on this system, the cascade circuit can also be constructed. For example, TP1, TP2, HP1, HP2 and NMM can be used as input at the same time, and the final fluorescence signal can be used as output, and a three-layer cascade OR-AND combined logic gate can be constructed as shown in Fig. 4A. It is worth noting that the test of the circuit is to demonstrate the basic functions of the combined logic gate by displaying the output of the upper logic gate as the input of the lower logic gate in different layouts. Therefore, only several representative inputs are selected to verify the feasibility of the cascade circuit. The true value table is shown in Fig. 4B. The results of the standardized treatment of the fluorescent signal are shown in Fig. 4C. Based on the idea of chain replacement reaction and self-assembly of catalytic hairpin and the diversity of base arrangement of trigger chain sequence, more layers of cascade circuits can be constructed, and more complex logic operations can be realized.

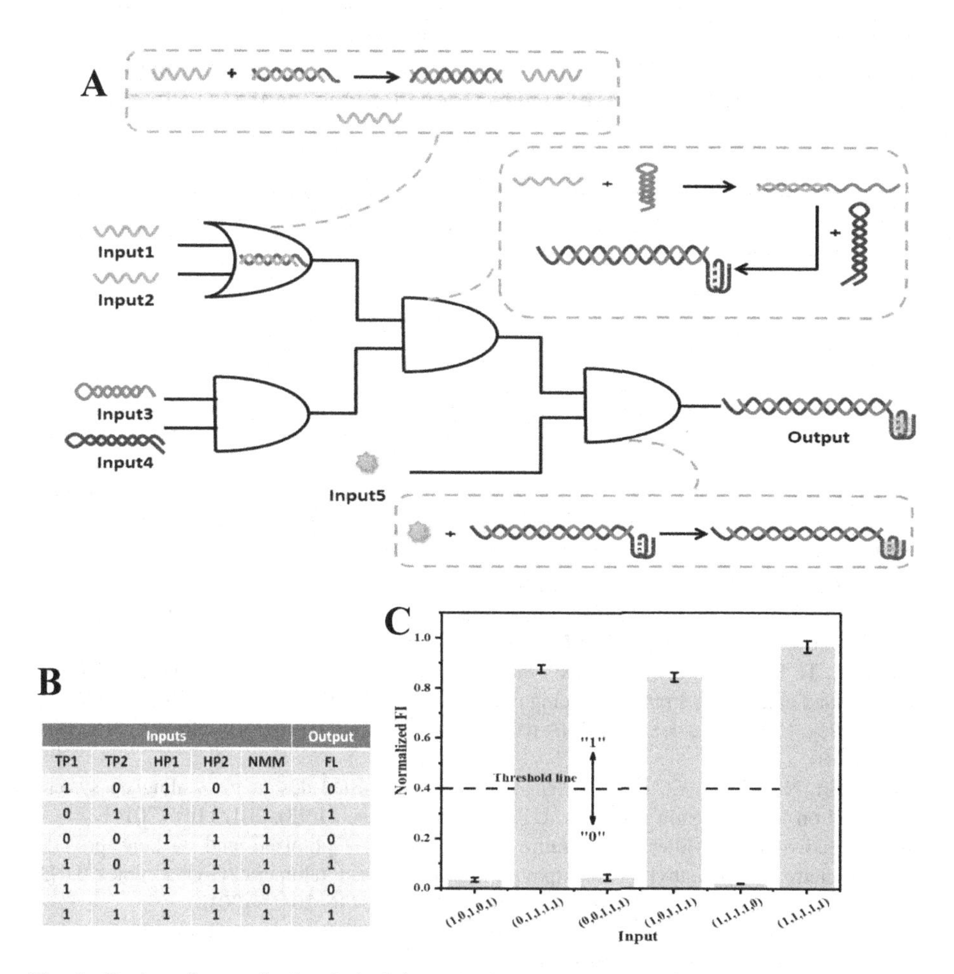

Inputs					Output
TP1	TP2	HP1	HP2	NMM	FL
1	0	1	0	1	0
0	1	1	1	1	1
0	0	1	1	1	0
1	0	1	1	1	1
1	1	1	1	0	0
1	1	1	1	1	1

Fig. 4. Design of cascade circuit A. Schematic diagram of cascade circuit. **B.** Partial truth table. **C.** Partial normalized fluorescence intensity of NMM with different combinations of inputs.

5 Conclusion

In a word, we have constructed a simple, economical, enzyme-free and label free logic computing platform based on DSD and CHA, and confirmed its feasibility by biological experiments with fluorescence detection. It is worth noting that we adopt a unified threshold of 0.4, and there is a certain space between high expression signal and low expression signal, so that the system has good fault tolerance. The realization of basic logic gates AND and OR provides the base support for the construction of complex logic circuits. The idea of DNA strand replacement and catalytic hairpin self-assembly provides a technical method for the realization of multi-layer cascade circuits. Practical computational problems can be solved through the assembly of logical components. Because our system does not contain not gate, the problem circuit can be transformed into dual track logic form when solving problems, which provides a new way to solve various complex logic calculation problems. For the future application of DNA logic circuits, the choice of architecture will depend on the actual needs of the application.

Acknowledgement. This work is supported by the National Natural Science Foundation of China (No. 62073207) and the basic natural science research program of Shaanxi Province (No. 2020JM-298).

References

1. Wang, S.C., et al.: Visual multiple recognition of protein biomarkers based on an array of aptamer modified gold nanoparticles in biocomputing to strip biosensor logic operations (2016)
2. Chatterjee, G., Dalchau, N., Muscat, R.A., et al.: A spatially localized architecture for fast and modular DNA computing. Nat. Nanotechnol. **12**, 920–927 (2017)
3. Srinivas, N., Parkin, J., Seelig, G., et al.: Enzyme-free nucleic acid dynamical systems. Science **358**(6369), 1401–1401 (2017)
4. Cherry, K.M., Qian, L.: Scaling up molecular pattern recognition with DNA-based winner-take-all neural networks. Nature **559**(7714), 370–376 (2018)
5. Song, T., Eshra, A., Shah, S., et al.: Fast and compact DNA logic circuits based on single-stranded gates using strand-displacing polymerase. Nat. Nanotechnol. **14**, 1075–1081 (2019)
6. Bradley, C.A.: Drug delivery: DNA nanorobots—seek and destroy. Nat. Rev. Cancer **18**(4) (2018)
7. Aubert, N., Mosca, C., Fujii, T., et al.: Computer-assisted design for scaling up systems based on DNA reaction networks. J. Roy. Soc. Interface **11**(93), 20131167 (2014)
8. Nicoludis, J.M., Miller, S.T., Jeffrey, P.D., et al.: Optimized end-stacking provides specificity of N-Methyl Mesoporphyrin IX for Human telomeric G-Quadruplex DNA. Journal of the American Chemical Society **134**(50), 20446–20456 (2012)
9. Guo, L., Nie, D., Qiu, C., et al.: A G-quadruplex based label-free fluorescent biosensor for lead ion. Biosensors Bioelectron. **35**(1), 123–127 (2012)
10. Chen, Y.J., Dalchau, N., Srinivas, N., et al.: Programmable chemical controllers made from DNA. Nat. Nanotechnol. **8**, 755–762 (2013)
11. Lakin, M.R., Stefanovic, D.: Supervised learning in adaptive DNA strand displacement networks. ACS Synt. Biol. **5**, 885 (2016)
12. Zhang, D.Y., Turberfield, A.J., Yurke, B., et al.: Engineering entropy-driven reactions and networks catalyzed by DNA. Science **318**(5853), 1121–1125 (2007)
13. Si, Y., Xu, L., Deng, T., et al.: Catalytic hairpin self-assembly-based SERS sensor array for the simultaneous measurement of multiple cancer-associated miRNAs. ACS Sens. **5**(12), 4009–4016 (2020)
14. Wang, K., Fan, D., Liu, Y., et al.: Cascaded multiple amplification strategy for ultrasensitive detection of HIV/HCV virus DNA. Biosensors Bioelectron. **87**, 116–121 (2017)
15. Wca, B., Wl, A., Hs, A., et al.: Molecular AND logic gate for multiple single-nucleotide mutations detection based on CRISPR/Cas9n system-trigged signal amplification - ScienceDirect. Analytica Chimica Acta **1112**, 46–53 (2020)

16. Bw, B., Bing, Y.: A turn-on fluorescence probe Eu 3+ functionalized Ga-MOF integrated with logic gate operation for detecting ppm-level ciprofloxacin (CIP) in urine. Talanta **208**, 120438 (2020)
17. Chen, T., et al.: A DNA logic gate with dual-anchored proximity aptamers for the accurate identification of circulating tumor cells. Chem. Commun. **56**, 6961–6964 (2020)
18. Barnoy, E.A., Popovtzer, R., Fixler, D.: Fluorescence for biological logic gates. J. Biophoton. (2020)
19. Wang, Y., Yuan, G., Sun, J.: Four-input multi-layer majority logic circuit based on DNA strand displacement computing. IEEE Access **8**, 3076–3086 (2019)
20. Wang, Y., Ji, H., Wang, Y., et al.: Stablity based on PI control of three-dimensional chaotic oscillatory system via DNA chemical reaction networks. IEEE Trans. NanoBiosci. **20**, 311–322 (2021)
21. Wang, F., Lv, H., Li, Q., et al.: Implementing digital computing with DNA-based switching circuits. Nature Commun. **11**(1), 121 (2020)
22. Ordóñez-Guillén, N.E., Martínez-Pérez, I.M.: Catalytic DNA strand displacement cascades applied to logic programming. IEEE Access **7**, 100428–100441 (2019)
23. Wibowo, A., Adhy, S., Kusumaningrum, R., et al.: Parallel rules based classifier using DNA strand displacement for multiple molecular markers detection. In: International Workshop on Big Data & Information Security. IEEE (2017)
24. Donati, M., Abbozzo, G.: Optical imaging of individual biomolecules in densely packed clusters. Riv. Clin. Pediatr. **11**(9), 798–807 (2016)
25. Rose, K.H.: Discrete Mathematics and its Application. China Machine Press, Beijing (2007)

Tetracycline Intelligent Target-Inducing Logic Gate Based on Triple-Stranded DNA Nanoswitch

Sunfan Xi[1], Yue Wang[2], Mengyang Hu[2], Luhui Wang[1], Meng Cheng[1], and Yafei Dong[1,2](✉)

[1] College of Life Science, Shaanxi Normal University, Xi'an 710119, Shaanxi, China
dongyf@snnu.edu.cn
[2] School of Computer Science, Shaanxi Normal University, Xi'an 710119, Shaanxi, China

Abstract. A previously unreported three-strand DNA (ts-DNA) non-metal structure with hairpin structure was designed based on a strategy of logic switching and tetracycline (TC) signaling molecule switching. Will target and TC-binding aptamer (TBA) is the combination of body, and its relation with three functional chain DNA conformation, with programmable DNA sensor signal by FAM input molecule fluorescence emission monitoring. After the combination, the input signal is converted to a fluorescent signal to perform a logical operation. By introducing a novel nanomaterial, graphene oxide (GO), as a signal regulator, the fluorescence response changed by intermolecular interactions is the key to achieving logical calculation. The harm of tetracycline can generally cause kidney damage, ear damage, gastrointestinal adverse reactions, skin allergies, etc.Using tetracycline as a molecular trigger to induce conformational changes in DNA, a logical operation target detection method was constructed. In addition, nanoscale devices can be reconfigured and optimized to accommodate additional logical calculations and target detection.

Keywords: DNA triplexes · Target recognition · Tetracycline · DNA logic gate

1 Introduction

The programmability of DNA ensures the elasticity of DNA computation during the elucidation of nanodevice self-assembly [1]. DNA is particularly well suited to regard as the foundation of autonomous molecular computers taking advantage of its outstanding properties. DNA recognition technology facilitates the development of a small integrated, economical and rapid molecular diagnostic platform [2]. The technique raises some new perspectives for solving logic computing problems, which may eventually help to provide new means for biomedical medicine. Boolean logic gate,

S. Xi and Y. Wang—These authors contributed equally to this work.

L. Pan et al. (Eds.): BIC-TA 2021, CCIS 1565, pp. 390–401, 2022.
https://doi.org/10.1007/978-981-19-1256-6_30

which can process data and information triggered by biomolecules, has a broad prospect for early disease diagnosis and personalized treatment at the molecular level [3, 4]. Current studies had broad impacts in various areas ranging from theories to applications. Biological logic gate is also an ideal device for intelligent detection with the ability to identify multiple targets appropriately [5]. Among them, biosensors have been widely discussed in detecting cancer markers [6, 7], harmful and toxic substances [8, 9] and metal ions [10–12], which were of great significance to biomedical, food safety and environmental detection. The current research on the diagnosis and discovery of coronavirus disease (COVID-19) also demonstrates the wide applicability of such nanoscale devices [13].

Polymorphism in DNA structures has far-reaching research value and its key role in biological processes has aroused a great enthusiasm for the exploration of specific structural sequences [14]. Triple-stranded DNA is formed by high-pyrimidine oligonucleotides that occupy the major groove in DNA double-stranded bodies through Hoogsteen interactions when pyrimidine or purine bases are paired undergoing classical Waston-Crick complementary principle [15]. A large number of studies have revealed the potential biological significance of triplets in vitro and in vivo, including triple helices conformation influence on gene expression regulation [16], fast and efficient biological sensing platform [17–19], as well as nanoswitch for biological analysis and treatment [20], widen the intelligent detection and treatment of medicine. It is worth mentioning that many controlled DNA nanomachines have been made based on the unique PH responsiveness of the triple helices [21–23].

Aptamers are single-stranded ligand-binding nucleic acids differ from traditional nucleic acids, which are isolated from combinatorial oligonucleotide libraries chemically synthesized in vitro selection [24]. The selection and binding ability of aptamers with high-affinity and high-specificity to their target can be on a par with those of antibodies. Regarding aptamers as functional nucleic acid that can be specifically triggered in biochemical conditions to construct target-induced DNA logic gates, making logical operation conceivable in responding to physiological environments. Herein, tetracycline was chosen as the target because of the threat to human health and the environment posed by the abuse and accumulation of antibiotics [25]. A shortened 8-mer ssDNA aptamer fragments with a highly enhanced affinity for tetracycline was used here [26].

Nanoparticles and nanomaterials can intelligently respond to related biological stimuli, paving the way for nanodevices implanted into living organisms. The tiny size of nanomaterials does not hide its powerful function. Among the new products in the nanomaterials industry in recent years, graphene oxide is the thinnest and hardest two-dimensional nano porous material [27]. Graphene oxide has unusual thermodynamic and optical properties, first of all, the active oxygenated body on the GO surface makes it good water dispersion in our reaction solution to fully engage in the interaction with DNA strands [28]. Secondly,graphene oxide (GO) is also a powerful fluorescence quenching agent thanks to its large surface area and wide absorption spectrum [29], which acts as an energy donor through Fluorescence resonance energy transfer (FRET) [30]. Thirdly, GO has the characteristic of priority binding for single-stranded DNA, but not so strong adsorption for double-stranded, three-stranded or G-quadruplex bodies [31]. This specific selection capacity of GO allows the single strand with fluorophore to be anchored on the GO surface. In addition, GO can be easily prepared in large quantities, thus eliminating the marking of fluorescent quenching groups. Given these virtues of performance and economy, the value of GO deserves to be further applied and developed.

Although the biological gate acts logically at the molecular level using computer-like behavior, its inputs and outputs cannot be simply defined as the transformation of electrical signals, thus designers need to pre-indicate the type of inputs and the Boolean value of outputs [32]. Of particular interest is fluorescence labeling technique. Fluorescence can be one of the first choices for exploring biological mechanisms due to its relatively simple imaging methods [33]. Compared with expensive, time-consuming and complex signal transduction methods, fluorescent signal sensors are more attractive in high-performance sensors with good reason—their simple operation, quick sensitivity, and convenient modification.

Heretofore, many attempts have been devoted to elucidating molecular mechanisms and controlling biological systems. Nevertheless, the exploration of newfangled structures in DNA/RNA nanotechnology remains extensive attention. Our reasonably designed logical strategy based on triple helix molecular switch could selectively identify specific targets and program the PH response system. In this paper, a new functional hairpin triple helices structure was devised as a nano-switch, then the logic device was constructed from the perspective of the allosteric triplexes effect caused by special conditions. The FAM with high fluorescence emission was used as an energy donor. If integrated with GO, it is possible to transform input signals into the switch of fluorescence to spark off subsequent logic-driven target detection.

2 Experimental

2.1 Materials and Reagents

The signal probe (SP) along with the fluorophore FAM which denotes 5(6)-Carboxyfluorescein was purified by HPLC (High Performance Liquid Chromatography). The sequences of three-strand DNA (ts-DNA) flanked at the 5'- and 3'- terminals with two limb segments and trigger probe (TP) were ULTRAPAGE-purified. All oligonucleotides in this work were synthesized and purchased from Sangon Biological Engineering Technology & Co. Ltd. (Shanghai, China). All of the DNA strands were diluted in ultrapure water to give the stock solution of 10 μM and stored in a fridge at 4 °C for later use. Graphene Oxide was purchased from Nanjing XFNANO Materials Tech Co. Ltd. (Nanjing, China) and suspend in water via sonication. Tetracycline was obtained from Beijing Century Aoko Biotechnology Co. Ltd. (Beijing, China). Tris was bought from Xi'an JingBo Bio-Technique.

2.2 Synthesis of Triple-Stranded DNA (Ts-DNA)

To form a triple helices structure, aptamer (TBA) (1.2 μM final concentration) was mixed with TP (400 nM final concentration) into reaction buffer (10 mM Tris-HCl, containing 100 mM NaCl, 5 mM $MgCl^2$, 5 mM KCl and 0.5 mM EDTA, pH 6.2). The mixture solution was annealed using a polymerase-chain-reaction (PCR) machine at the reaction of 95 °C for 5 min, 75 °C for 15 min, 55 °C for 15 min, 37 °C for 15 min, 25 °C for 30 min, and finally kept at 25 °C, resulting in DNA hybridization and stable ts-DNA formation.

2.3 Procedure for Logic Gate Operation and Fluorescent Detection

Playing the role of signal conditioning, GO dispersion (75 μg mL-1) and SP (100 nM final concentration) was dispersed into Tris-HCl buffer to obtain GO/SP work platform. At 37 °C temperature, a succession of concentrations was mixed with ts-DNA solution, and the mixture was allowed to incubate for 1h to release TP. Then, GO/SP was added dropwise into the above reaction mixture. After incubating for about 30 min at room temperature, the final solution was subsequently used for fluorescence detection. All fluorescence measurements were carried out on fluorescence-scanning spectrometer (EnSpire ELISA; PerkinElmer, Waltham, MA, USA) with a step of 1 nm at the excitation of 492 nm and emission of 518 nm.

2.4 Sequences Design

Pertinent DNA sequences design is one of the essential conditions for controlling the assembly process. There are three basic DNA strands involved in our system. In order to precisely monitor the reaction, 5` end of the signal probe is labeled with the single fluorescein. Since Förster resonant energy transfer depends on the resonance condition between the Rydberg and molecular transitions, an electric field can be used to shift the Rydberg transition in and out of resonance, thereby changing the energy transfer rate. TP and SP are fully complementary paired to form a stable two-stranded structure that separates the SP/TP complex from GO, upon which the green fluorescence of FAM is restored.

As a result of tetracycline as an identification element, TBA contains a loop that can bind to tetracycline (TC) specifically, with two stems complementary to part of the TP on both sides, which folds the linear conformation into a triple conformation when TP presented under appropriate conditions. Special attention is paid to the sequences of the triplexes. On the one hand, if the T•A * T and C•G * C+ parts are long, the triplexes will be too stable to deconstruct. In this case, TP may not be released effectively, and fluorescent signal output is not obvious. Conversely, if this part matches too few bases, it will be difficult to form a stable triplex, which will cause circuit leakage and excessive background fluorescence. In order to achieve better recognition effect, the triplexes under metastable state is more neutral and more flexible [34]. On the other hand, the content of TAT should be taken into account, which was inspired by the previous research [35]. A matching 60% TAT was designed according to the expected space in response to the PH value in our experiment.

3 Results and Discussion

3.1 Model Principle

The mechanism and its functional model are portrayed in Fig. 1. FAM-modified SP adsorbs onto the basal plane of GO via the π–π stacking force, fluorescence quenching of the FAM took place because of FRET. In this condition, that is in the absence of inputs, the fluorescence signal was shut off. The middle portion of TBA (i.e. the loop part after the formation of the ts-DNA) possesses a high affinity to TC. On the introduction of TC, the specific binding of TC and its aptamer can catalyze the decomposition of triplex, during which TP is liberated and the SP/TP duplex is formed. Due to the weak interaction between GO and double-stranded DNA, SP/TP is easily detached from GO, causing the fluorescent signal to turn on. With the goals of proving the feasibility, the successful progress was clearly demonstrated by fluorescence intensity.

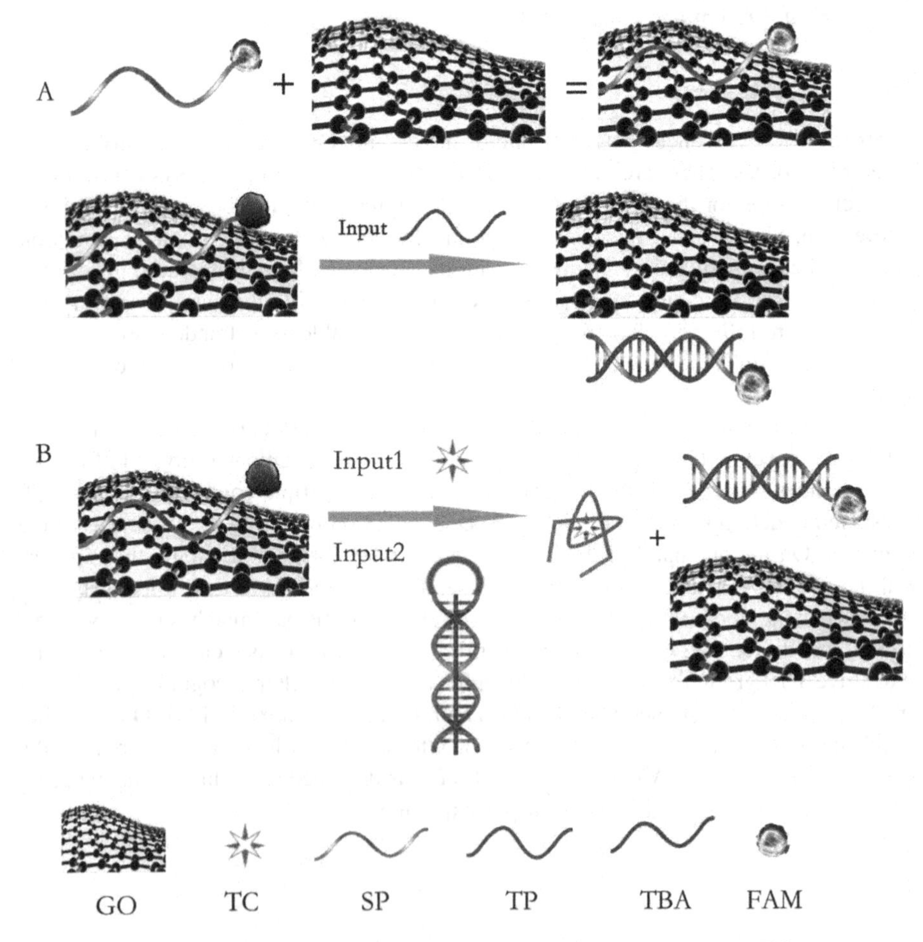

Fig. 1. Schematic diagram of logic gate (A) "YES" gate and (B) "AND" gate.

The aforementioned basic reaction mechanism was rationally thought to construct the logic gate. The SP/GO substrates via noncovalent assembly could be viewed as an initial computational platform. The Boolean values "1" and "0" are assigned with the presence or absence of inputs. The normalized fluorescence intensities of FAM measured at excitation/emission wavelengths of 492/518 nm above or below the threshold correspond to output "1"/"0".

The signal probe appended with FAM stably anchored on the GO. Then TP/SP dsDNA products were repelled by GO and produced strong fluorescence intensity when TP was added to the solution. Thus, a YES gate was established which took TP as the input (see Fig. 2).

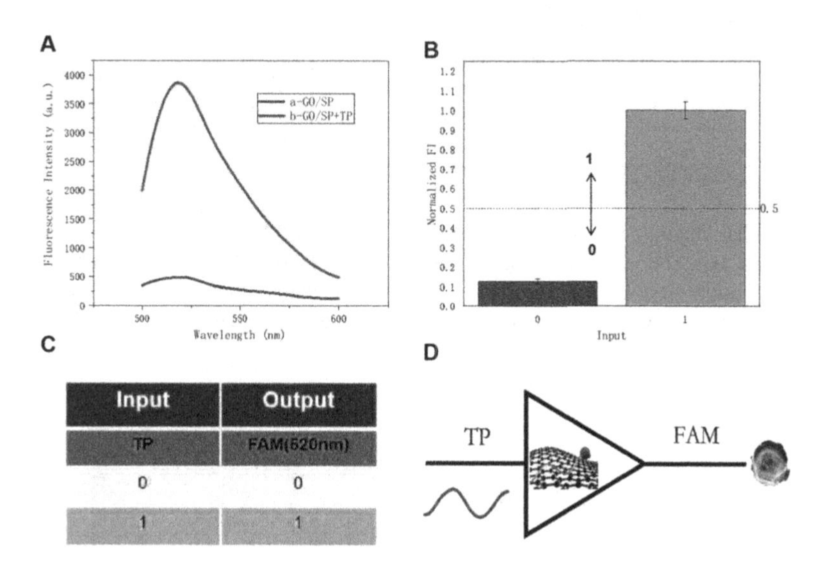

Fig. 2. (A) The fluorescence spectra of FAM for the "YES" logic gate against the input combinations; (B) The different output fluorescence values of FAM have different output at 520 nm excitation wavelengths. Error bars show the standard deviation of three experiments; (C) the corresponding truth table of the "YES" logic gate; (D) diagram of the "YES" logic gate.

On this basis, the DNA nanomachine was used to construct an AND logic gate, where TC and ts-DNA were defined as input1 and input2, respectively. In the presence of any one of input1 or input2, there was no significant signal change, due to FAM were adsorbed onto the surface of GO. When two inputs were present together, the TP was released because TC bound with TBA, and this leads to the restoration of fluorescence due to the removal of SP/TP from GO (see Fig. 3).

Figure 4 shows the activation of another AND logic gate based on the basic work platform, which gives an output of 1 only if both of the two inputs (OH- and ts-DNA) are held at "1". In this logic gate, the destruction of the Hoogsteen bond under neutral or basic conditions will lead to the deconstruction of the triple configuration. In presence of ts-DNA and OH-, that is to say, there is no target but under alkaline

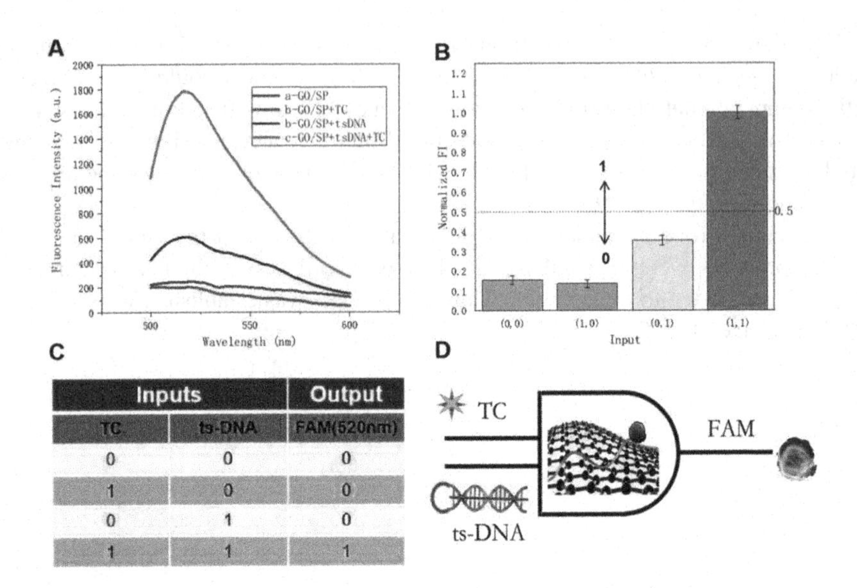

Fig. 3. (A) The fluorescence spectra of FAM for the "AND" logic gate against various input combinations; (B) The different output fluorescence values of FAM have different output at 520 nm excitation wavelengths. Error bars show the standard deviation of three experiments; (C) truth table of the "AND" logic gate; (D) the corresponding diagram of the "AND" logic gate.

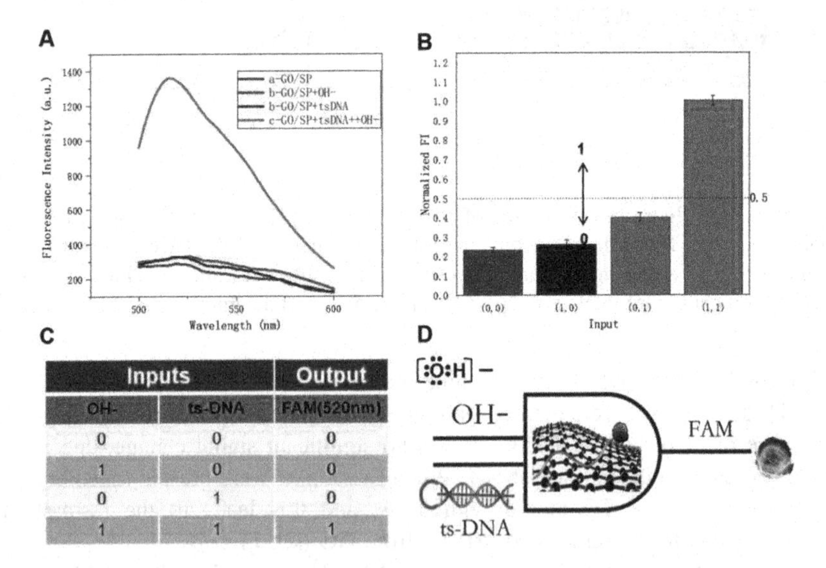

Fig. 4. (A) The fluorescence spectra of FAM for the "AND" logic gate against various input combinations; (B) The different output fluorescence values of FAM have different output at 520 nm excitation wavelengths. Error bars show the standard deviation of three experiments; (C) truth table of the "AND" logic gate; (D) the corresponding diagram of the "AND" logic gate.

condition, the Hoogsteen interactions in the ts-DNA are destabilized, resulting in the accessibility of TP to hybridize with SP. The formation of a stable duplex enabled FAM to be desorption to GO. The results showed enhanced fluorescence, suggesting that the strategy has the ability to trigger and control self-assembly of DNA nanostructures through simple PH alteration.

3.2 Sensor System Algorithm Model

The model principle of the sensor system is shown in Fig. 5, which is also built based on the SP/TP reaction platform of the logic gate. It is mainly divided into three basic steps: recognition, reaction and scanning. The algorithm is as follows:

3.3 Optimization of Assay Conditions

In order to achieve the best performance of the designed fluorescence sensor, the following four aspects were optimized.The main factors affecting the fluorescence

检测流程算法描述

输入:TC

输出: fluorescence

```
For i = 1,⋯,N do
    ts-DNA←Annealing(TBA, TP)
     SP/GO←Adsorption(SP, GO)
     TP, TC/TBA←Hybridization (TC, ts-DNA)
    GO, TP/SP←Hybridization (TP, SP/GO)
    FAM(i)←Signal (GO,TP/SP)
End for
     mFAM←mean(FAM)
If  mFAM > threshold  then
   fluorescence ← true
Else
   fluorescence ← false
End if
End
```

Fig. 5. Pseudo-code of sensor model

response of tetracycline were studied. Firstly, the effects of the length of the three chains, Mg2 +, pH and GO concentration on the fluorescence intensity were investigated (Fig. 6).

According to Fig. 6-A, the time difference and ratio of base number of 10bP were the best, so TbA10 was selected. Then, Mg2+ played an important role in the formation of the three chains. As can be seen from Fig. 6-B, only when Mg2+ reaches enough concentration can better difference be achieved. The concentration of Mg2+ was set at 10nM in subsequent experiments. Secondly, the three chains are pH-dependent, so the pH value is a factor that cannot be ignored. According to the response interval of Hoogsteen hydrogen bond, four different values were selected for measurement within the pH range of 5–6, and the best effect could be achieved when pH = 6.6 (Fig. 6-C) after comprehensive consideration. Finally, the fluorescence modulator GO played an unquestionable role in the results, and the fluorescence output at different concentrations showed significant differences (Fig. 6-D). Although the GO concentration was 10 μg/mL, the time difference was the largest, but the fluorescence quenching effect was poor at this time, leading to extremely high background fluorescence. Therefore, 25 μg/mL became the best choice. And then the ratio is at its maximum.

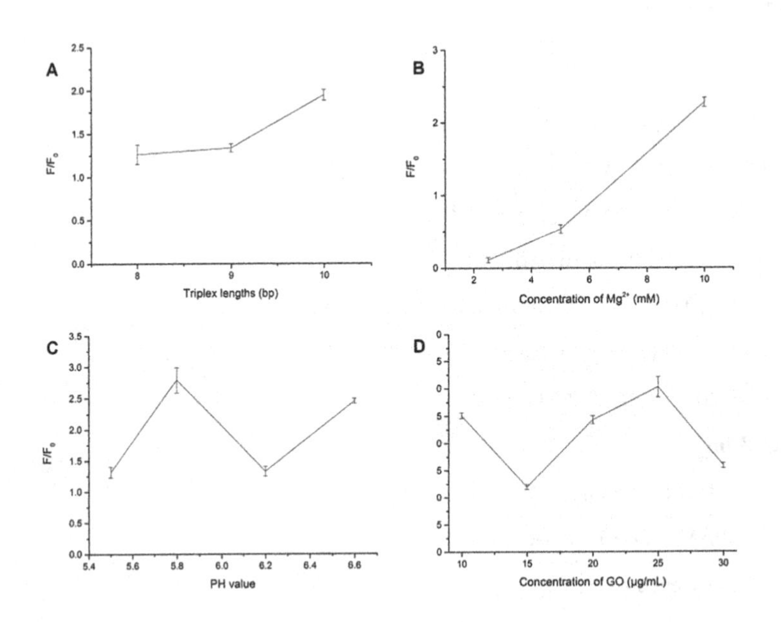

Fig. 6. The optimization diagram underdifferent conditions. (A)Triple-stand length. (B) The influence of Mg^{2+} on experimental environment. (C)Effect of PH on experimental environment. (D)The optimization of GO concentration.

4 Conclusion

The proposed target-induced structural switching is viable to dynamically and isothermally manipulate triple DNA conformational interconversion. This model has conspicuous advantages including easy synthesis, enzyme-free and moderate reaction conditions, which contribute to its expansion. The distinct properties of DNA triplexes with unique PH response characteristics and design flexibility, provide a friendly user design experience. Boolean logic gates could recognize the user-appointed target to make a pondered decision. It is possible to rely on flexibility of DNA sequences design to incorporate multiple factors subject to errant logical expressions. The reasoning systems can also lend themselves universal to expediently reconfiguring by matching aptamer for a range of target quantifications. In view of the above, our conception has broadened inspiration in this highly interdisciplinary field, with a focus on DNA computation by going in quest of delicately structural and computational DNA nanotechnology.

Acknowledgment. This research was supported by National Natural Science Foundation of China (no. 61572302), Natural Science Basic Research Plan in Shaanxi Province of China (no. 2020JM-298). The authors are very grateful to the anonymous reviewers for their valuable comments and suggestions for improving the quality of the paper.

Data Availability. The data used to support the findings of this study are available from the corresponding author upon request.

Conflicts Interest. The authors declare that they have no competing interests.

References

1. Douglas, S.M., Bachelet, I., Church, G.M.: A logic-gated nanorobot for targeted transport of molecular payloads. Science **335**(6070), 831–834 (2012)
2. Petralia, S., Forte, G., Zimbone, M., Conoci, S.: The cooperative interaction of triplex forming oligonucleotides on DNA-triplex formation at electrode surface: Molecular dynamics studies and experimental evidences. Coll. Surf. B **187**, 110648 (2020)
3. Deng, J.K., Tao, Z.H., Liu, Y.Q., Lin, X.D., Qian, P.C., Lyu, Y.L., et al.: A target-induced logically reversible logic gate for intelligent and rapid detection of pathogenic bacterial genes. Chem. Commun. **54**(25), 3110–3113 (2018)
4. Li, J., Green, A.A., Yan, H., Fan, C.H.: Engineering nucleic acid structures for programmable molecular circuitry and intracellular biocomputation. Nat. Chem. **9**(11), 1056–1067 (2017)
5. Wang, K.M., He, Q., Zhai, F.H., Wang, J., He, R.H., Yu, Y.L.: Biosens. Bioelectron. **105**, 159–165 (2018)
6. Ma, C., Liu, H.Y., Tian, T., Song, X.R., Yu, J.H., Yan, M.: A simple and rapid detection assay for peptides based on the specific recognition of aptamer and signal amplification of hybridization chain reaction. Biosens. Bioelectron. **83**, 15–18 (2016)
7. Sun, Y.L., Fan, J.F., Cui, L.Y., Ke, W., Zheng, F.J., Zhao, Y.: Fluorometric nanoprobes for simultaneous aptamer-based detection of carcinoembryonic antigen and prostate specific antigen. Microchim. Acta. **186**(3), 1–10 (2019)

8. Wang, Y., et al.: G-quadruplex-bridged triple-helix aptamer probe strategy: a label-free chemiluminescence biosensor for ochratoxin A. Sens. Actuat. B Chem. **298**, 126867 (2019). https://doi.org/10.1016/j.snb.2019.126867
9. Malhotra, B.D., Srivastava, S., Ali, M.A., Singh, C.: Nanomaterial-based biosensors for food toxin detection. Appl. Biochem. Biotechnol. **174**(3), 880–896 (2014). https://doi.org/10.1007/s12010-014-0993-0
10. Zhou, D.H., Wu, W., Li, Q., Pan, J.F., Chen, J.H.: A label-free and enzyme-free aptasensor for visual Cd2+ detection based on split DNAzyme fragments. Anal. Methods-UK **11**(28), 3546–3551 (2019)
11. Chen, J., et al.: Highly active G-quadruplex/hemin DNAzyme for sensitive colorimetric determination of lead(II). Microchim. Acta **186**(12), 1–8 (2019). https://doi.org/10.1007/s00604-019-3950-3
12. Xiong, Z.W., Liu, Q.L., Yun, W., Hu, Y., Wang, X.M., Yang, L.Z.: Dual-entropy-driven catalytic amplification reaction for ultra-sensitive and visible detection of Hg2+ in water based on thymine-Hg2+-thymine coordination chemistry. Analyst **144**(17), 5143–5149 (2019)
13. Santiago, I.: Trends and innovations in biosensors for COVID-19 mass testing. ChemBioChem. **21**(20), 2880–2889 (2020). https://doi.org/10.1002/cbic.202000250
14. Kaushik, S., Kukreti, S.: Formation of a DNA triple helical structure at BOLF1 gene of human herpesvirus 4 (HH4) genome. J. Biomol. Struct. Dyn. **39**, 3324–3335 (2020)
15. Frank-Kamenetskiiun, M.D., Mirkin, S.M.: Triplex DNA structures. Annu. Rev. Biochem. **64**, 65–95 (1995)
16. Dervan, P., Doss, R., Marques, M.: Programmable DNA binding oligomers for control of transcription. Curr. Med. Chem.-Anti-Cancer Agents **5**(4), 373–387 (2005). https://doi.org/10.2174/1568011054222346
17. Zhao, Y.H., Wang, Y., Liu, S., Wang, C.L., Liang, J.X., Li, S.S., et al.: Triple-helix molecular-switch-actuated exponential rolling circula amplification for ultrasensitive fluorescence detection of miRNAs. Analyst **144**(17), 5245–5253 (2019)
18. Zeng, P., Hou, P., Jing, C.J., Huang, C.Z.: Highly sensitive detection of hepatitis C virus DNA by using a one-donor-four-acceptors FRET probe. Talanta **185**, 118–122 (2018)
19. Han, Y.P., Zhang, F., Gong, H., Cai, C.Q.: Functional three helix molecular beacon fluorescent "turn on" probe for simple and sensitive simultaneous detection of two HIV DNAs. Sens. Actuat. B-Chem. **281**, 303–310 (2019)
20. Chen, X.X., Chen, T.S., Ren, L.J., Chen, G.F., Gao, X.H., et al.: Triplex DNA Nanoswitch for pH-sensitive release of multiple cancer drugs. ACS Nano **13**(6), 7333–7344 (2019)
21. Amodio, A., Adedeji, A.F., Castronovo, M., Franco, E., Ricci, F.: pH-controlled assembly of DNA tiles. J. Am. Chem. Soc. **138**(39), 12735–12738 (2016)
22. Amodio, A., et al.: Rational design of pH-controlled DNA strand displacement. J. Am. Chem. Soc. **136**(47), 16469–16472 (2014)
23. Yao, D.B., Li, H., Guo, Y.J., Zhou, X., Xiao, S.Y., Liang, H.J.: A pH-responsive DNA nanomachine-controlled catalytic assembly of gold nanoparticles. Chem. Commun. **52**(48), 7556–7559 (2016)
24. Zhou, J.H., Rossi, J.: Aptamers as targeted therapeutics: current potential and challenges. Nat. Rev. Drug. Discov. **16**(3), 181–202 (2017)
25. Liu, X.G., Huang, D.L., Lai, C., Zeng, G.M., Qin, L., Zhang, C., et al.: Recent advances in sensors for tetracycline antibiotics and their applications. Trac-Trend. Anal. Chem. **109**, 260–274 (2018)
26. Kwon, Y.S., Raston, N.H.A., Gu, M.B.: An ultra-sensitive colorimetric detection of tetracyclines using the shortest aptamer with highly enhanced affinity. Chem. Commun. **50**(1), 40–42 (2014)

27. Geim, A.K., Novoselov, K.S.: The rise of graphene. Nat. Mater. **6**(3), 183–191 (2007)
28. Allen, M.J., Tung, V.C., Kaner, R.B.: Honeycomb carbon: a review of graphene. Chem. Rev. **110**(1), 132–145 (2010)
29. Sharma, R., Baik, J.H., Perera, C.J., Strano, M.S.: Anomalously large reactivity of single graphene layers and edges toward electron transfer chemistries. Nano. Lett. **10**(2), 398–405 (2010)
30. He, S.J., Song, B., Li, D., Zhu, C.F., Qi, W.P., Wen, Y.Q., et al.: A graphene nanoprobe for rapid, sensitive, and multicolor fluorescent DNA analysis. Adv. Funct. Mater. **20**(3), 453–459 (2010)
31. Wu, M., Kempaiah, R., Huang, P.J., Maheshwari, V., Liu, J.W.: Adsorption and desorption of DNA on graphene oxide studied by fluorescently labeled oligonucleotides. Langmuir **27** (6), 2731–2738 (2011)
32. Barnoy, E., Popovtzer, R., Fixler, D.: Fluorescence for biological logic gates. J. Biophoton. **13**(9), e202000158 (2020). https://doi.org/10.1002/jbio.202000158
33. Wang, Y.F., Zhang, Y.X., Wang, J.J., Liang, X.J.: Aggregation-induced emission (AIE) fluorophores as imaging tools to trace the biological fate of nano-based drug delivery systems. Adv. Drug. Deliver. Rev. **143**, 161–176 (2019)
34. Tian, J., Chu, H., Zhu, L., Xu, W.: Duplex-specific nuclease-resistant triple-helix DNA nanoswitch for single-base differentiation of miRNA in lung cancer cells. Anal. Bioanal. Chem. **412**(19), 4477–4482 (2020). https://doi.org/10.1007/s00216-020-02713-6
35. Idili, A., Vallee-Belisle, A., Ricci, F.: Programmable pH-triggered DNA nanoswitches. J. Am. Chem. Soc. **136**(16), 5836–5839 (2014)

Application of Chain P Systems with Promoters in Power Coordinated Control of Multi-microgrid

Wenping Yu[1], Fuwen Chen[1], Jieping Wu[1,2](✉), Xiangquan Xiao[1], and Hongping Pu[3]

[1] School of Automation and Electrical Engineering, Chengdu Technological University, Chengdu 611730, People's Republic of China
densitywelldone@hotmail.com

[2] College of Electronics and Information Engineering, Sichuan University, Chengdu 610065, People's Republic of China

[3] Artificial Intelligence Key Laboratory of Sichuan Province, Zigong 643000, People's Republic of China

Abstract. In view of the mobile energy storage characteristic and random dispersion of electric vehicles, the coordination and optimization of electric vehicles combined with distributed generation units are conducive to achieve the economical and reliable operation of multi-microgrid. In this paper, a multi-microgrid system with electric vehicles is established, and its power coordinated control is researched. Based on the characteristics of electric vehicles and multi-microgrid, a chain P system with promoters (PCPS) is proposed to describe the complex structure and a large amount of uncertain information of multi-microgrid. At the same time, its internal rules are set. Finally, a multi-microgrid system with two sub-microgrids is considered for reasoning test. The simulation results show that the proposed power coordinated control method based on PCPS are feasible and effective.

Keywords: Chain P systems · Promoters · Multi-microgrid · Coordinated control · Electric vehicles

1 Introduction

As a new branch of natural computing, membrane computing has become a research focus, because of its strong information processing capacity and maximum parallelism [1,2]. At present, there are three main membrane computing models, namely cell-like P systems [3,4], tissue-like P systems [5,6] and neural-like P systems [7,8]. Cell-like P system is considered to have a tree-like hierarchical structure. Tissue-like P system emphasizes the bi-directional communication between cells, and can be described as a unidirectional graph. Neural-like P system has a network diagram structure. In order to explore new structures of P systems, a membrane system with chain structure, also called Chain P system,

L. Pan et al. (Eds.): BIC-TA 2021, CCIS 1565, pp. 402–414, 2022.
https://doi.org/10.1007/978-981-19-1256-6_31

is proposed in [9]. This system combines discrete Morse theory with membrane computing, and its advantages and computational efficiency are proved by solving arithmetic algorithms. In [10], the Chain P systems are applied to cluster analysis, and better results are obtained.

In recent years, membrane computing has been studied in the field of microgrid. In [11], cell-like P systems are used to handle the coordinated control and economic operation of microgrid, and excellent control and optimization results are obtained. In [12], combined with fuzzy theory and cell-like P systems, the language fuzzy cell-like P systems are applied to the coordinated control of microgrid, which achieves reasonable power distribution and coordinated control requirements. In [13], distributed fuzzy P systems with promoters are proposed for energy control of multi-microgrid, and good results are obtained. Therefore, membrane computing has a good application prospect in the field of control and optimization of microgrids.

Microgrid is an effective way for renewable energies to be connected to the power grid, and multiple adjacent microgrids are interconnected to form a multi-microgrid system in a certain area [14]. Renewable energies are the main force to achieve the goal of carbon peak and carbon neutrality in the field of energy and power. However, high proportion of renewable energy generations has strong intermittency and randomness, which has a considerable impact on the stability of multi-microgrid. Due to the duality of electric vehicles [15], good regulation advantages can be realized after being incorporated into microgrid. Moreover, relying on clean energies as their power, electric vehicles have broad development prospects. However, the addition of large-scale renewable energies generation units and electric vehicles will cause the information of multi-microgrid to be more complex and uncertain. Therefore, a reasonable and effective operation control strategy plays an important part for coordinating the output power of each unit in multi-microgrid and ensuring the operation stability of multi-microgrid.

P systems have a variety of characteristics (such as distributed, parallel, non-deterministic and easy to achieve communication, etc.) to make themselves suitable for solving all kinds of application problems. Thus, combined with promoters and chain P systems, a chain P system with promoters (PCPS) is proposed in this paper. And the proposed PCPS is applied to power coordination control of multi-microgrid, which can make the proposed coordinated control method not only has powerful information processing ability of the conventional P system, but also has clear and strict control logic of chain P systems. Eventually, the randomness and impact on the power grid caused by the access of many renewable energies and electric vehicles can be improved.

The rest of this paper is organized as follows. In Sect. 2, multi-microgrid structure is introduced. In Sect. 3, the specific definition and rules setting of PCPS are proposed. In addition, the correctness and feasibility of the proposed method are preliminarily verified by modeling the structure of multi-microgrid based on PCPS. In Sect. 4, the simulation process and results are presented. In Sect. 5, the conclusion is discussed.

2 Multi-microgrid

In recent years, multi-microgrid (MMG) mainly has two basic structures, series and parallel. In this paper, the parallel multi-microgrid is adopted, and its structure is shown in Fig. 1. The MMG contains two sub-microgrids, namely MG1 and MG2. In MG1, there are wind power generation (WT), fuel cell (FC), energy storage battery (ESB1) and flexible load (load1). In MG2, there are photovoltaic power generation (PV), electric vehicles (EV), another energy storage battery (ESB2) and important load (load2). In addition, the controllable objects of the multi-microgrid are ESB1, ESB2, FC, load1 and EV.

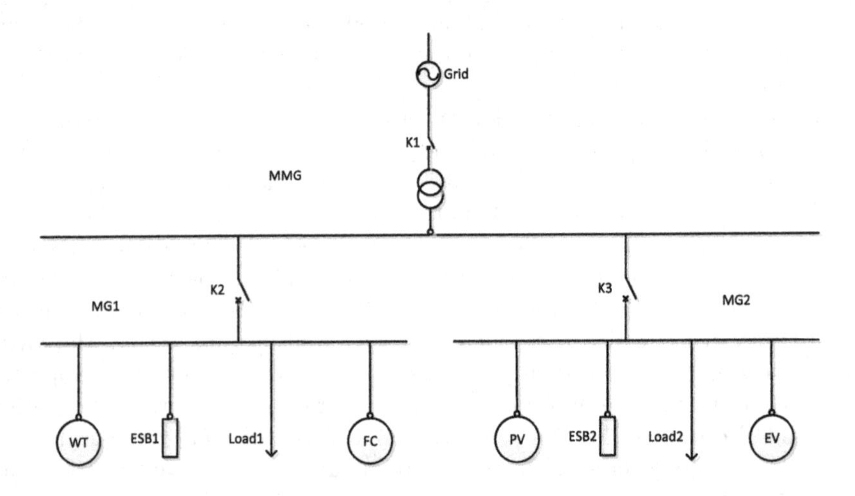

Fig. 1. Schematic diagram of multi-microgrid in parallel structure.

When the power of multi-microgrid is unbalanced, the strategy can give priority to the power interaction between sub-microgrids. If the MMG cannot restore the balance, the internal regulation of each sub-microgrid should be considered. Firstly, the internal regulation of each sub-microgrid shall give priority to the use of energy storage battery. Secondly, if sub-microgrids cannot restore the balance, the regulation of adjustable micro-sources and EV should be considered. Thirdly, if the balance still cannot be satisfied, the regulation of flexible load should be considered. Finally, if the MMG still has power shortages or surpluses after a series of regulation, the power difference of the MMG should be calculated, and then MMG applies to the large power grid for power interaction to achieve the power balance.

3 Application of PCPS to Power Coordinated Control in Multi-microgrid

3.1 Chain P Systems with Promoters

The position and value of catalysts in the membrane can not be changed before or after participating in the chemical reaction. When the multi-microgrid model based on PCPS is established, the power difference is used as the control quantity, and its value can be changed with the execution of rules. However, the value of the catalysts can not be changed before or after the reaction, it is impossible to use catalyst to characterize the change of power difference value. Therefore, the standard chain P systems are difficult to solve this problem. In [13] and [16,18], a chemical called promoter is mentioned. Compared with the catalysts, the value of the promoters can be changed before or after the chemical reaction. Moreover, promoters can also evolve independently, and their evolution process can be carried out simultaneously with the biochemical reaction promoted. So, promoters are introduced into chain P systems to improve their performance in this paper.

Definition 1. *A PCPS with degree $m (m \geq 1)$ is as follows.*

$$\Delta = (O, T, Z, C, u, \sigma_1, \sigma_2, \ldots, \sigma_m, syn, in, out, P) \tag{1}$$

where:

(1) O is an alphabet. Its elements are called objects.
(2) T is an output alphabet, and $T \in O$.
(3) Z is a natural number set, and P represents external environment.
(4) C is the set of promoters, $C = \{C_i^*, i \in (1, 2, \ldots, m)\}$. In this paper, the position of promoters does not be changed in the evolution process, but their value can evolve with the implementation of rules.
(5) μ is the set of membrane structure, each membrane and its enclosed area are represented by the label set H, $H = \{1, 2, \ldots, m\}$.
(6) $\sigma_i, i \in (1, 2, \ldots, m)$ represents one skin membrane unit σ_i and m chain membrane units $\sigma_{i1}, \ldots, \sigma_{im}$ in the system. The form of σ_i is $\sigma_i = (n_i, R_i), 1 \leq i \leq m$, where:
   a) $n_i \geq 0$ represents the number of objects in σ_i at the beginning of calculation.
   b) R_i represents a finite set of all rules in σ_i. There are three forms of R_i:
      i) Rewriting rules, the form is $u \rightarrow v$. u is an object representing a multivariate set in character set O. v is a string in set $O^*\{here, in, out\}$, the form is (a, tar), where a is an object from O, tar represents here, in or out.
      ii) Communication rules, it also can be divided into symport rules and antiport rules. The form of symport rule is $(u, (tar1, tar2))$, the form of antiport rule is $(u, (tar1, tar2); v, (tar1, tar2))$. Among them, u and v are objects representing multivariate sets in character set O. In

addition, tar1 = {here, in, out}, tar2 = {pre, sub}, tar1 is used to represent the direction of the control objects in and out of the membrane, and tar2 represents which membrane unit the object is ultimately located in. For example, $(u, (out, sub); v, (in, pre))$ represents that the object u in the membrane where the rule is located enters the subsequent membrane of this membrane, while the object v from the subsequent membrane of this membrane enters this membrane.

iii) Forgetting rules, the form is $u \to \lambda$. u is an object representing a multivariate set in character set O, λ is null string. Those rules are used to eliminate the execution of rules, which can reduce the number of some objects in the membrane units where the rule is located.

(7) $syn = \{1, 2, \ldots, m\}$, represents the connection relationship between the membrane units in chain P systems. For each $0 \leq i \leq m$, there is $\sigma_0 \to \ldots \to \sigma_m$.

(8) $in, out \in (1, 2, \ldots, m)$ respectively represent input and output units.

3.2 Modeling Process

Literature [9] proposed two chain membrane structure forms, namely graphic membrane and tree-like membrane structure. Based on the improved chain P systems, the graphic membrane structure and the control strategy of the multi-microgrid proposed in this paper, the membrane calculation model of the multi-microgrid is constructed as shown in Fig. 2. The calculation model has two types of membranes, namely outer chain membranes and inner chain membranes. The outer chain membrane is $\sigma_0 \to \sigma_1 \to \sigma_2$, the three inner chain membranes are $\sigma_{01} \to \sigma_{02}, \sigma_{21} \to \sigma_{22}$ and $\sigma_{11} \to \sigma_{12} \to \sigma_{13}$. Besides, only membranes with a precursor or successor relationship between chain membrane units can communicate. For example, $\sigma_{11} \to \sigma_{12} \to \sigma_{13}$, where σ_{11} is the precursor membrane of σ_{12} or σ_{12} is the successor membrane of σ_{11}, they can communicate with each other. However, there is no predecessor or successor relationship between σ_{11} and σ_{13}, so they cannot communicate directly. Moreover, the chain membrane unit and the skin membrane unit can communicate with each other. For example, σ_1 is the skin membrane unit of $\sigma_{11} \to \sigma_{12} \to \sigma_{13}$, it can communicate with any of $\sigma_{11}, \sigma_{12}, \sigma_{13}$. σ_{11} and σ_{13} cannot communicate directly, but they can communicate indirectly through the skin membrane σ_1.

Figure 3 is the computation model of multi-microgrid based on PCPS. In this model, the skin membrane σ_0 and its objects and rules represent the parameters of the multi-microgrid level and its control strategy. C_0^*, C_1^*, C_2^*represent the promoters in this model, where $C_0^* = \Delta P_1 * \Delta P_2, C_1^* = \Delta P_1, C_2^* = \Delta P_2$,$\Delta P_i = P_{i.give} - P_{i.load}$. a, b, d and e are the objects in P system, representing whether ΔP_i exist and its concrete number value. m_i, n_j, x_i, y_j are the natural numbers corresponding to ΔP_i, representing the number of objects. At the same time, a negative sign in front of an object indicates an operation to reduce the number of objects, B indicates that the balance has been reached, α means that multi-microgrid buy power from the large power grid, β means that multi-microgrid sell power to the large power grid. Figure 4 is the model of MG1 based on

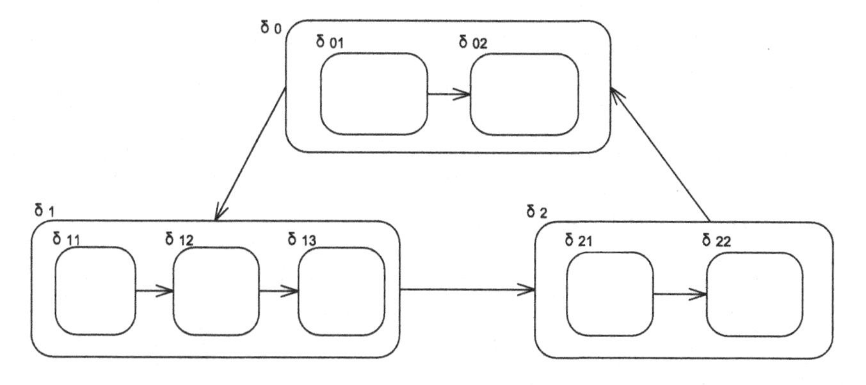

Fig. 2. The computation model of multi-microgrid based on PCPS.

PCPS. $C^*_{soc1}, C^*_{FC}, C^*_{L1}$ are the promoters in this model, their value are the adjustable amounts of each adjustable unit at the corresponding sampling time. x_i represents the number of objects a and d, where $x_3 = x_1$ or x_2, $x_5 = x_3$ or x_4. Figure 5 is the model of MG2 based on PCPS. $C^*_{soc2}, C^*_{EV \cdot S}, C^*_{EV \cdot L}$ are the promoters in this model, their value are the adjustable amounts of each adjustable unit at the corresponding sampling time. y_i represents the number of objects b and e, where $y_3 = y_1$ or y_2.

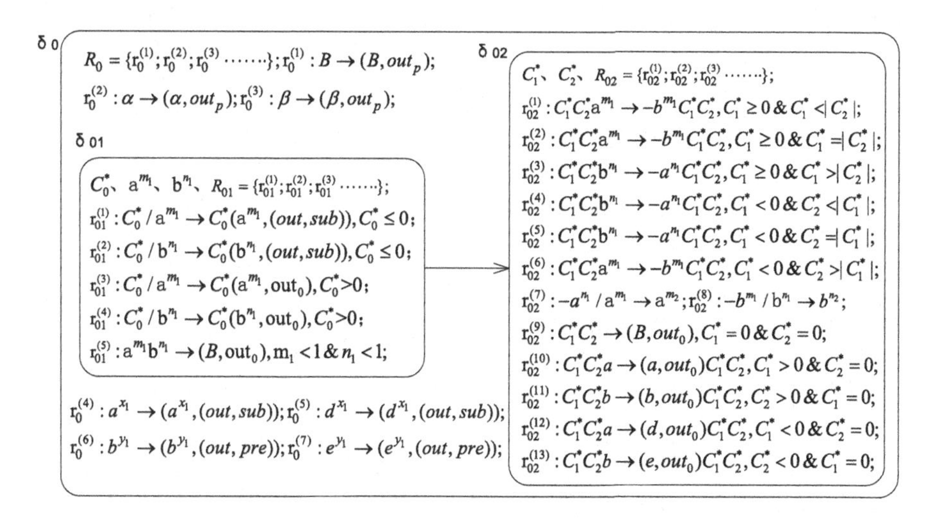

Fig. 3. The computation model of multi-microgrid based on PCPS.

In each calculation model, each membrane contains objects, promoters and corresponding rules. Rules are mainly divided into rewriting rules and communication rules. These two kinds of rules are used to change the type and number of

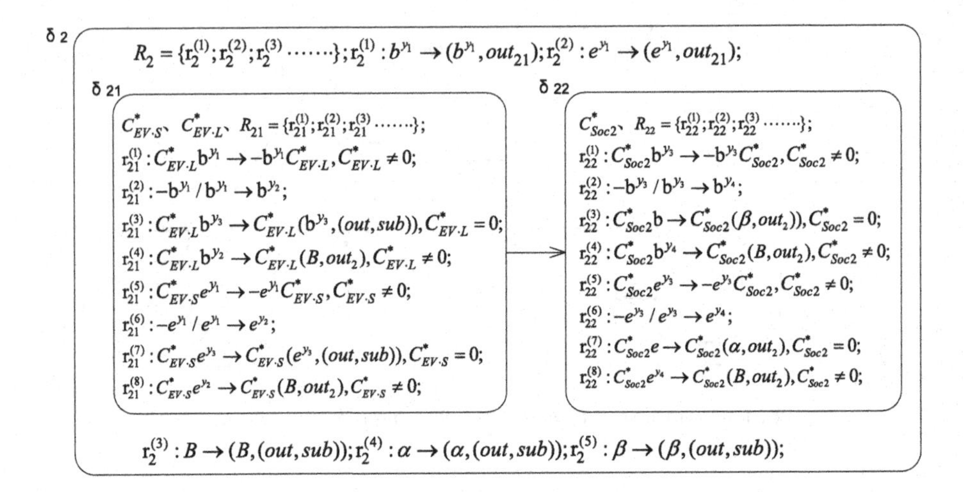

Fig. 4. The computation model of MG1 based on PCPS.

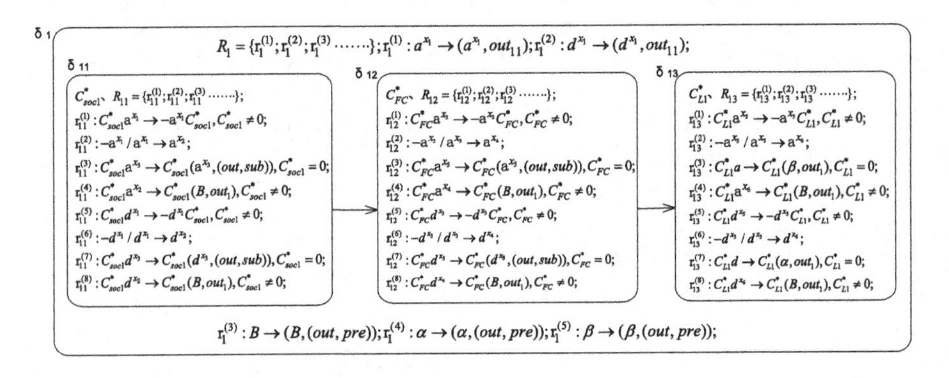

Fig. 5. The computation model of MG2 based on PCPS.

objects in the membrane and complete the communication between membranes, to finally achieve the purpose of control.

(1) Rewriting rules

$$C_1^* C_2^* a^{m1} \rightarrow -b^{m1} C_1^* C_2^*, C_1^* \geq 0 \& C_1^* < |C_2^*| \tag{2}$$

when there are objects a and b in membrane and promoters, C_1^* and C_2^* meet the corresponding conditions, the rule is fired to change the type of objects and reduce the corresponding number of objects. Currently, the promoters evolve with the execution of the rule.

(2) Communication rules

$$C_1^* C_2^* \rightarrow (B, out_0), C_1^* = 0 \& C_2^* = 0 \tag{3}$$

when promoters C_1^* and C_2^* meet $C_1^* = 0 \& C_2^* = 0$, the rule is fired to generate an object B and send it into the skin membrane σ_0.

3.3 Reasoning Test

The sampling data of multi-microgrid at a certain moment is selected as the input data to preliminarily verify the correctness of the calculation model. In addition, the power unit is kW, and 1 kW power is set to a corresponding object.

(1) Input data
  a) Input value of each unit
  $P_{WT} = 14, P_{WT}^{min} = 10, P_{WT}^{max} = 20, P_{PV} = 8, P_{PV}^{min} = 0, P_{PV}^{max} = 18, P_{L2} = 10, SOC1 = 10, SOC1^{min} = 2, SOC1^{max} = 20, SOC2 = 12, SOC2^{min} = 2, SOC2^{max} = 20, P_{FC} = 10, P_{FC}^{min} = 0, P_{FC}^{max} = 14, P_{L1} = 14, P_{L1}^{min} = 10, P_{L1}^{max} = 27, P_{EV} = 14, P_{EV}^{min} = 2, P_{EV}^{max} = 16$.
  b) Initialization of power balance and corresponding objects
  $\Delta P_1 = P_{WT} + P_{FC} - P_{L1}, \Delta P_2 = P_{PV} - P_{L2}$. So, $\Delta P_1 = 10, \Delta P_2 = -2$. Since $1kW$ power corresponds to one object, it can be concluded that ΔP_1 corresponds to 10 objects a (i.e., a^{10}), and ΔP_2 corresponds to 2 objects b (i.e., b^2).
  c) Initialization of promoters
  $C_0^* = \Delta P_1 * \Delta P_2 = -20, C_1^* = \Delta P_1 = 10, C_2^* = \Delta P_2 = -2$, $C_{soc1}^* = |SOC1 - SOC1^{max}| = 10$ (when it needs to charge), $C_{soc1}^* = |SOC1 - SOC1^{min}| = 8$ (when it needs to discharge), $C_{soc2}^* = |SOC2 - SOC2^{max}| = 8$ (when it needs to charge), $C_{soc2}^* = |SOC2 - SOC2^{min}| = 10$ (when it needs to discharge), $C_{FC}^* = |P_{FC} - P_{FC}^{max}| = 4$ (when it needs to increase output), $C_{FC}^* = |P_{FC} - P_{FC}^{min}| = 10$ (when it needs to decrease output), $C_{L1}^* = |P_{L1} - P_{L1}^{max}| = 13$ (when it needs to increase load), $C_{L1}^* = |P_{L1} - P_{L1}^{min}| = 4$ (when it needs to decrease load), $C_{EV \cdot L}^* = |P_{EV} - P_{EV}^{max}| = 2$ (when it needs to charge as a load), $C_{EV \cdot S}^* = |P_{EV} - P_{EV}^{min}| = 2$ (when it needs to discharge as a source).

(2) Regulation process of multi-microgrid
  a) Based on the above input data, there are C_0^*, a^{10}, b^2 in membrane σ_{01}. It meets the firing conditions of the rules $r_{01}^{(1)}$ and $r_{01}^{(2)}$, the objects a^{10} and b^2 are sent into σ_{02} which is the successor membrane of σ_{01}.
  b) Firstly, there are $C_1^*, C_2^*, a^{10}, b^2$ in σ_{02}. the firing condition of $r_{02}^{(3)}$ is met, the rule is executed. Secondly, the firing condition of $r_{02}^{(8)}$ is met, $r_{02}^{(8)}$ is executed. After that, the number of object b becomes 0, i.e., $\Delta P_2 = 0$, the number of object a becomes 8, and $C_1^* = 8, C_2^* = 0$. Thirdly, the firing condition of $r_{02}^{(10)}$ is met, and the object a^8 is sent into the skin membrane σ_0. Finally, there is a^8 in the skin membrane σ_0, the firing condition of $r_0^{(4)}$ is met, and the object a^8 is sent into the σ_1. At this moment, the firing condition of $r_1^{(1)}$ in σ_1 is met, and the object a^8 is sent into σ_{11}.
  c) There are a^8, C_{soc1}^* in σ_{11}, the firing conditions of $r_{11}^{(1)}$ and $r_{11}^{(2)}$ are met, $C_{soc1}^* = 2$, and the number of object a decreases to 0, i.e., $\Delta P_1 = 0$. And

then the firing conditions of $r_{11}^{(3)}$ is met, one object B is generated and sent into σ_1. After that, the firing conditions of $r_1^{(3)}$ in σ_1 is met, and the object B is sent into σ_0 which is the precursor membrane of σ_1. Finally, the firing conditions of $r_0^{(1)}$ in σ_0 is met, the object B is sent out to the environment. That means the multi-microgrid system has reached power balance through the coordinated control of itself.

(3) Output data after the coordinated control $P_{WT} = 14, P_{WT}^{min} = 10, P_{WT}^{max} = 20, P_{PV} = 8, P_{PV}^{min} = 0, P_{PV}^{max} = 18, P_{L2} = 10, SOC1 = 18, SOC1^{min} = 2, SOC1^{max} = 20, SOC2 = 12, SOC2^{min} = 2, SOC2^{max} = 20, P_{FC} = 10, P_{FC}^{min} = 0, P_{FC}^{max} = 14, P_{L1} = 14, P_{L1}^{min} = 10, P_{L1}^{max} = 27, P_{EV} = 14, P_{EV}^{min} = 2, P_{EV}^{max} = 16, \Delta P_1 = 0, \Delta P_2 = 0, C_0^* = 0, C_1^* = 0, C_2^* = 0$, $C_{soc1}^* = 2$ (when it needs to charge), $C_{soc1}^* = 16$ (when it needs to discharge), $C_{soc2}^* = 8$ (when it needs to charge), $C_{soc2}^* = 10$ (when it needs to discharge), $C_{FC}^* = 4$ (when it needs to increase output), $C_{FC}^* = 10$ (when it needs to decrease output), $C_{L1}^* = 13$ (when it needs to increase load), $C_{L1}^* = 4$ (when it needs to decrease load), $C_{EV \cdot L}^* = 2$ (when it needs to charge as a load), $C_{EV \cdot S}^* = 2$ (when it needs to discharge as a source).

The reasoning test results show that the power balance of the multi-microgrid system is achieved through the PCPS-based coordinated control method of the multi-microgrid, which initially verifies the correctness and feasibility of the proposed method.

4 Simulation Analysis

In this section, simulation experiment is carried out by sampling 24 h data from multi-microgrid to further verify the correctness and effectiveness of the proposed PCPS-based coordinated control method for multi-microgrid. In the experiment, the charging and discharging of energy storage are restricted, charging cannot exceed 90% state-of-charge (SOC), and discharging cannot be lower than 10% SOC. Besides, Electric vehicles can be charged as a load or discharged as a power source, and the remaining adjustable units cannot exceed the upper and lower limits of their own power constraints. The simulation experimental data and simulation results are shown in Figs. 6, 7, 8, 9, 10, 11 and 12. Figure 6 is the 24-h power distribution curve of each unit in multi-microgrid. Figure 7 is the 24-h SOC curve of each storage in sub-microgrids. Figure 8 is the 24-h SOC curve of electric vehicles in multi-microgrid. Figure 9 is the power fluctuation curve of the two sub-microgrids before coordinated control. Figure 10 is the power interaction curve between two sub-microgrids. Figure 11 is the frequency fluctuation curve of the multi-microgrid system. The power interaction curve between multi-microgrid and the large power grid is shown in Fig. 12.

As is shown in Fig. 6 and Fig. 7 that the batteries of energy storages and electric vehicles can not be in the full charge state and zero charge state, to avoid over charging and discharging and affect the service life of the batteries. It can be seen from Fig. 8, Fig. 11, the power of each sub-microgrid fluctuates greatly before the coordinated control. After the coordinated control of the

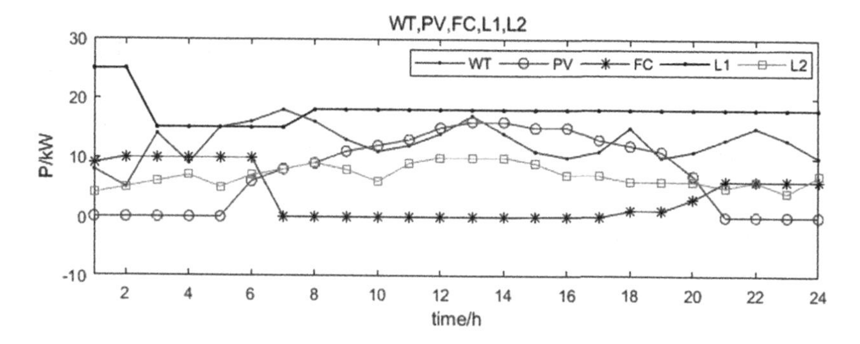

Fig. 6. The 24-h power distribution curve of each unit in multi-microgrid.

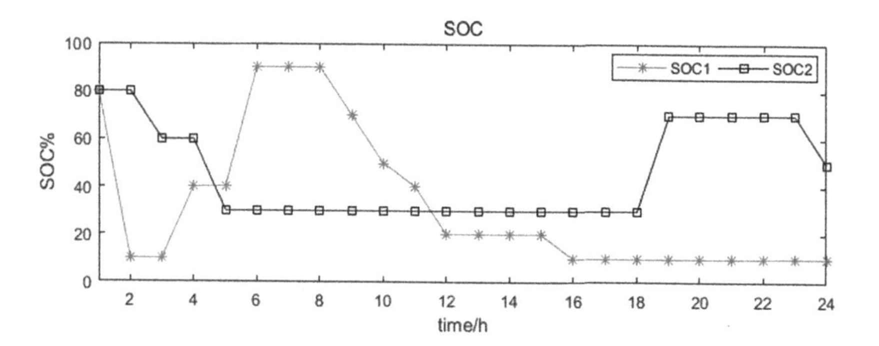

Fig. 7. The 24-h SOC curve of each storage in sub-microgrids.

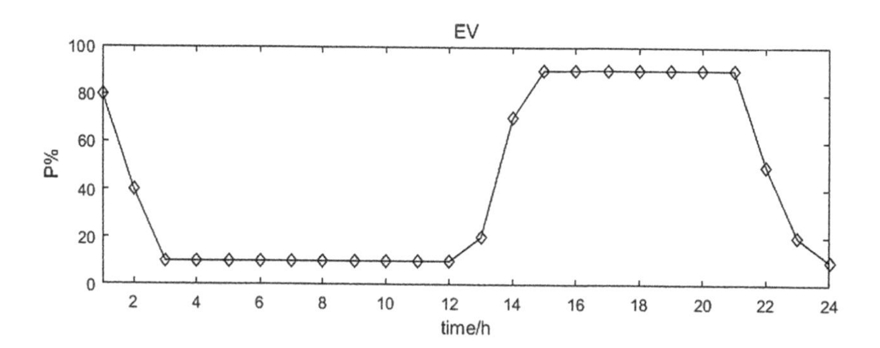

Fig. 8. The 24-h SOC curve of electric vehicles in multi-microgrid.

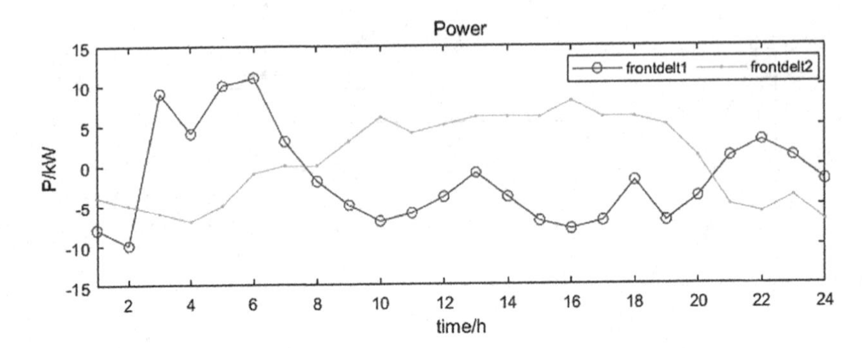

Fig. 9. The power fluctuation curve of the two sub-microgrids before coordinated control.

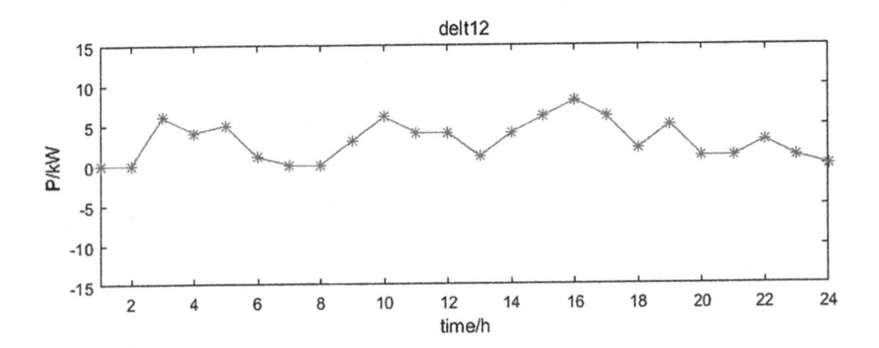

Fig. 10. The power interaction curve between two sub-microgrids.

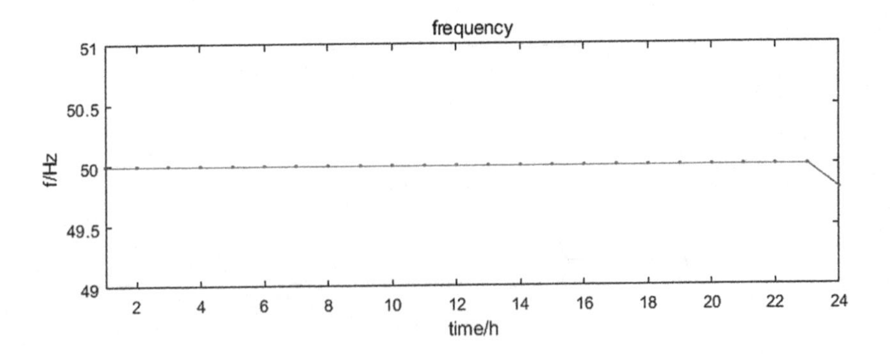

Fig. 11. The frequency fluctuation curve of the multi-microgrid system.

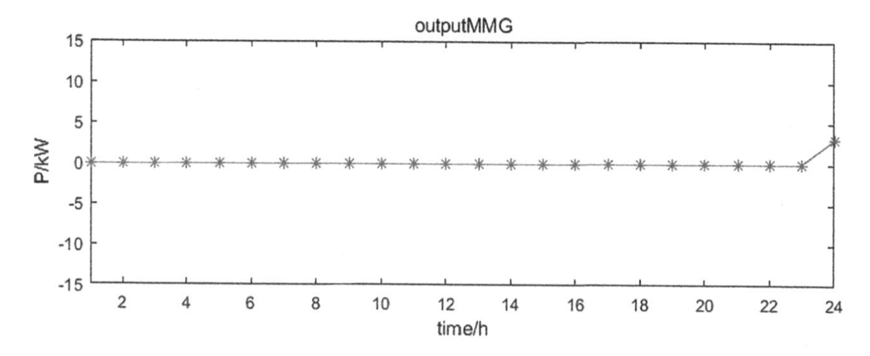

Fig. 12. The power interaction curve between multi-microgrid and the large power grid.

multi-microgrid itself and the reasonable and effective interaction between the sub-microgrids, the power and frequency fluctuations of the multi-microgrid are very small at 23 o'clock. As is shown in Fig. 12, After PCPS-based coordinated control, the multi-microgrid only needs to sell electricity to the large power grid at 23 o'clock to maintain the balance.

In summary, by adopting the coordinated control method based on PCPS, the power interaction between the sub-microgrids and the adjustment of each adjustable units are performed to achieve the power balance of the MMG for 23 hours. There is only one hour of power imbalance, so it must sell power to the large power grid at 23 o'clock, it greatly improves the utilization of distributed generation, and the stability of the grid can not be affected due to little interaction. Thus, the simulation results show that the proposed coordinated control method of multi-microgrid based on PCPS is reasonable and effective.

5 Conclusion

Combined promoters and chain P systems, an improved chain P system is proposed in this paper. Compared with other coordinated control models in [12] and [13], PCPS have many advantages, such as directional, predecessor and successor properties, which make the control of multi-microgrids clear and logical, more organized, more reliable, and more flexible. At the same time, because the entire control is chained, it is easier to program. For the multi-microgrid structure constructed in this paper, a multi-microgrid coordinated control model based on PCPS is built, and the reasoning test is carried out to verify the feasibility of the model. Finally, MATLAB is used to simulate the proposed method to verify its correctness and effectiveness. The simulation results show that the multi-microgrid not only realizes the safe and stable operation of the system, but also conducts reasonable and effective energy interaction with the external grid. In the future, based on this research method, the improved membrane optimization algorithm will be proposed and applied to the energy optimization of the multi-microgrid to make the operation of the multi-microgrid more economical and efficient.

Acknowledgment. This work was partially supported by the National College Student Innovation and Entrepreneurship Training Program Project (No. 202111116014 and No. 202111116022) and the Sichuan Key Laboratory Project of artificial intelligence (No. 2021RYY05).

References

1. Păun, G.: Computing with membranes. J. Comput. Syst. Sci. **61**(1), 108-C143 (2000)
2. Zhang, G., et al.: Membrane Computing Models: Implementations. Springer, Singapore (2021). https://doi.org/10.1007/978-981-16-1566-5
3. Zhao, Y., Zhang, W., Sun, M., et al.: An improved consensus clustering algorithm based on cell-like p systems with multi-catalysts. IEEE Access **99**(1), 106934 (2020)
4. Song, B., Li, K., Orellana-Martín, D., et al.: Cell-like p systems with evolutional symport/Antiport rules and membrane creation. Inf. Comput. **275**, 104542 (2020)
5. Peng, H., Wang, J., Shi, P., et al.: Fault diagnosis of power systems using fuzzy tissue-like p systems. Integr. Comput. Aided Eng. **24**(4), 401–411, 104542 (2017)
6. Luo, Y., Guo, P., Jiang, Y., et al.: Timed homeostasis tissue-like p systems with evolutional symport/Antiport rules. IEEE Access **8**(1–1), 104542 (2020)
7. Wu, T., Zhang, L., Pan, L.: Spiking neural p systems with target indications. Theor. Comput. Sci. **862**, 250–261, 104542 (2020)
8. Wang, J., Peng, H., Yu, W., et al.: Interval-valued fuzzy spiking neural p systems for fault diagnosis of power transmission networks. Eng. Appl. Artif. Intell. **82**(6), 102–109, 104542 (2019)
9. Luan, J., Liu, X.: A variant of p system with a chain structure. J. Inf. Comput. Sci. **10**(10), 3189–3197, 104542 (2013)
10. Jie, X., Liu, X.: Communication p system with oriented chain membrane structures and applications in graph clustering. J. Comput. Theor. Nanosci. **13**(7), 4198–4210, 104542 (2016)
11. Luo, J., Wang, J., Shi, P., et al.: Micro-grid economic operation using genetic algorithm based on p systems. ICIC Exp. Lett. **9**(2), 609–618, 104542 (2015)
12. Wang, J., Chen, K., Li, M., et al.: Cell-like fuzzy p system and its application in energy management of micro-grid. J. Comput. Theor. Nanosci. **13**(6), 3643–3651, 104542 (2016)
13. Yu, W., Wang, J., Tao, W., et al.: Distributed fuzzy p systems with promoters and their application in power balance of multi-microgrids. In: International Conference on Bio-inspired Computing: Theories and Applications, pp. 329–342 (2017)
14. Wang, C., Zhang, G., Chen, S., et al.: Bilevel energy optimization for grid-connected AC multimicrogrids. Int. J. Electr. Power Energy Syst. **130**(1), 106934 (2021)
15. Tang, W., Qin, H.: Multi-Microgrid Optimal Scheduling Considering Electric Vehicle Participation. IEEE (2020)
16. Ping, G., Xie, J.: AP system for k-medoids-based clustering. Int. J. Adv. Comput. Sci. Appl. **9**(10), 41–48, 106934 (2018)
17. Song, B., Pan, L.: The computational power of tissue-like p systems with promoters. Theor. Comput. Sci. **641**, 43–52, 106934 (2016)
18. Yan, S., Xue, J., Liu, X.: An improved quicksort algorithm based on tissue-like p systems with promoters. In: Hinze, T., Rozenberg, G., Salomaa, A., Zandron, C. (eds.) CMC 2018. LNCS, vol. 11399, pp. 258–274. Springer, Cham (2019). https://doi.org/10.1007/978-3-030-12797-8_18

Solution to Satisfiability Problem Based on Molecular Beacon Microfluidic Chip Computing Model

Jing Yang[1], Zhixiang Yin[2(✉)], Zhen Tang[1], Jianzhong Cui[1], and Congcong Liu[1]

[1] School of Mathematics and Big Data, Anhui University of Science and Technology, Huainan 232001, China
[2] School of Mathematics, Physics and Statistics, Shanghai University of Engineering Science, Shanghai 201620, China
zxyin66@163.com

Abstract. The satisfiability problem is a well-known NPC problem in mathematics. In this paper, we report construction of a computational model based on molecular beacon microfluidic chip, to solve the satisfiability problem by using advantages such as; rapid, simple and economical measurement of molecular beacon probe. Taking 3-satisfiability problem as an example, we encode the variable and construct molecular beacon probe by combining the complement of corresponding DNA strands with variable whose clause is 1. They are then fixed on the chip and placed in the microfluidic chip. It takes DNA strands corresponding to all combinations of values as the initial data pool, and utilize the microfluidic chip to detect. If there are DNA strands in the final data pool, it will be sequenced, and we call that this problem can be satisfied. If there are no DNA strands, the problem is not satisfied. The complexity of the model is $O(m) + O(n)$, where m is the number of clauses in conjunctive normal form, and n is the number of variables contained in the conjunctive normal form. This model represents great parallelism, massive storage capacity and high throughput, integration and multi-function of microfluidic chips.

Keywords: Satisfiability problem · Molecular beacon · Microfluidic chip · DNA computing

1 Introduction

The emergence of DNA computing has opened up a new field of computational science. Based on the background of DNA computing, DNA computing not only has the same computing power as electronic computers, but also has great potential and functions that traditional electronic computers cannot match. This is mainly since DNA computer has four outstanding advantages: (1) high degree of parallelism; (2) massive storage capacity; (3) low energy consumption; (4) abundant resources. Its calculation principle is to use the DNA molecule with massive storage capacity and biochemical reaction of huge parallelism and other characteristics to calculate, which is not for electronic computers. There are four bases in DNA: adenine (A), cytosine (C), guanine

L. Pan et al. (Eds.): BIC-TA 2021, CCIS 1565, pp. 415–425, 2022.
https://doi.org/10.1007/978-981-19-1256-6_32

(G) and thymine (T). Electronic computer units are coded in 0 and 1 binaries, while DNA as an operation unit is encoded in terms of A, C, G, and T. The molecular structures involved in DNA computing are divided into single - stranded, double - stranded, triple - stranded, single - double - stranded mixture and molecular beacon. DNA computing is a brand new computing mode and an emerging thinking mode combining information science and biological science [1, 2]. The basic idea is to use the information processing ability of biological organic molecules to replace digital physical switching components, that is, to use the principle of base complementary pairing between DNA double chains to encode the problem to be dealt with as the specific DNA molecular chain as the input chain, through the specific recognition and pairing between sequences. Using biochemical reactions, strands of DNA produced according to the principle of non-pairing are output. The reaction may take place in solution or on other carriers. The reaction may take place in solution or on other carriers. In 2009, IBM announced the development of the next generation of micro-processing chips using DNA and nanotechnology, ushering in a new era of DNA computing. Therefore, the biological chip based on DNA computing will be the commanding point of the core technology of computer chips in the future, and its national strategic significance is self-evident.

Among stem-loop oligonucleotides, molecular beacon probes are most commonly used for bimolecular recognition. Molecular beacon probes are fast, simple and economical for determination, and can monitor nucleic acid reactions in vivo and in vitro in real time. The different design of molecular beacon structure can make it to have different applications in different fields. The special 'hairpin' structure of molecular beacon also determines its special superiority in DNA calculation. For example, Japanese scholars Sakatomo et al. [3] established DNA calculation model to solve the satisfiability problem by using this special hairpin structure. Yin Zhixiang et al. [4] used the molecular beacon model to solve the 0–1 integer programming problem. Molecular beacons generally consist of two parts: (1) ring region, which is composed of 15–30 base sequence and can specifically bind to target sequence, and is generally called ring recognition region; (2) stem region, consisting of complementary sequences of about 8 bases in length. The hybridization specificity of molecular beacons is significantly higher than that of conventional linear probes according to the thermodynamic equilibrium relationship between the double-strand structure of molecular beacons with target sequences and double-strand structure in stem region [5–8].

Microfluidic chip, also known as microfluidic laboratory on a chip or laboratory on a chip, refers to a miniaturized, integrated and automated chemical and biological experiment platform built on a chip of a few square centimeters or smaller [9–13]. Microfluidic chips manipulate fluids at the micron or nanometer scale to make substances interact with each other, to achieve the transfer of matter, energy and information to achieve the required physical, chemical or biological effects. Manz and Harrison published the first paper on capillary electrophoresis separation on a microfluidic chip in 1992 [14]. After that, many laboratories conducted research on capillary electrophoresis microfluidic chip and realized multi-quantity high-speed DNA sequencing, laying a foundation for practical application and industrialization of microfluidic chip in gene sequencing. In 2001, a professional journal of microfluidic chips, Lab on a Chip, was founded. It soon became the authoritative journal in the field of microfluidic chips, and promoted the rapid development of microfluidic chip research worldwide. In 2002, the Stephen Quake research group [15] developed the microfluidic chip that integrated thousands of microvalves and reaction pools, allowing thousands of nanoupgrades to be performed simultaneously in a very short time, which marked an important step towards high-throughput, integrated, and multifunctional microfluidic chips. The size of a microfluidic chip is generally around a few square centimeters, and the size of each functional area is usually in the micron scale. Therefore, compared with the fluid at the macro scale, its physical and chemical properties were both consistent and unique [16–19]. Compared with traditional analysis and detection methods, microfluidic chips not only greatly improve the speed and efficiency of analysis and reduce the consumption of reagents and samples due to their miniaturization and integration, but also can be combined with a variety of detection methods to greatly improve the detection sensitivity [20, 21].

In this study, the molecular beacon microfluidic chip is prepared by using the structural advantages of molecular beacon combined with microfluidic technology to extract the combination with each clause having a value of 1 in the satisfiability problem. We then get the combination that makes the whole conjunctive paradigm assignment true. Finally, we construct the satisfiability problem computing model based on molecular beacon microfluidic chip.

2 Satisfiability Problem

The satisfiability problem is a well-known NP problem in mathematics and the first NPC problem to be proved. The algorithms for solving SAT problems have been sought by scholars, and some algorithms of acceptable complexity have been presented. Since Professor Aldeman used DNA molecular computation to solve the Hamiltonian road with seven vertices, some scholars have tried to use the huge parallelism of DNA to solve SAT problems, and made certain achievements, which greatly reduced the computational complexity. In 2000, Liu [22] et al. presented a solution based on surface to SAT problem. In 2002, Braich [23] et al. solved the satisfiability problem of 20 variables by using DNA calculation. In 2004, Kristiane et al. [24] solved the satisfiability problem with 4 variables and 4 statements using single-stranded molecular hybridization detection technology. In 2007, Lin et al. [25] solved the satisfiability problem by using DNA chips. In the same year, Yin Zhixiang et al. [26] presented a plasmid-based DNA computing model to satisfiability problem. In 2008, Brun [27] proposed a tile self-assembly model to solute satisfiability problem with a fixed number of tiles. In the same year, Yin Zhixiang et al. gave the molecular beacon model to satisfiability problem. In 2013, Chen Mei et al. [28] presented a computational model of bio brick flip cell for satisfiability problem. In 2021, Yang Jing et al. [29] constructed molecular beacon tile by using molecular beacon, and provided the solution for 3 SAT problem by using self-assembly model [30]. In 2020, Yin et al. applied the DNA hybridization chain reaction to solve the 3-SAT problem [31]. The HCR was used to verify whether each set of values of variables could make the clause result 1. Finally, fluorescence spectrometer was used to detect and read the solution.

Definition: Given n Boolean arguments $x_1, x_2, \cdots, x_n$, and m clauses $C_1, C_2, \cdots, C_M$ $(C_j = x_{j1} \vee x_{j2} \vee \cdots \vee x_{jk}, j = 1, 2, \cdots, m)$, asking whether there is a set of 0, 1 assignments of $x_1, x_2, \cdots, x_n$ that make $C_1 \wedge C_2 \wedge \cdots \wedge C_m$ of true is called an SAT problem.

If the maximum value of k does not exceed 3, it is called the 3-SAT problem. This problem seems less complex than the SAT problem, but it's still an NPC problem. Because of its harsh conditions, it is easier to be reduced to other problems, so 3-SAT is more widely used in NPC problems than SAT problems.

Basic algorithm:

1. Generate all combinations of variables;
2. Use the microfluidic chip to find the combination satisfying each clause one by one;
3. The last remaining combination satisfies that each clause is 1, thus making the conjunction paradigm assignment true.

3 Establishment of Computing Model of Microfluidic Chip

Taking conjunctive normal form $\phi = (x_1 \vee x_2 \vee x_3) \wedge (\bar{x}_1 \vee x_2 \vee x_3) \wedge (\bar{x}_1 \vee x_2 \vee \bar{x}_3)$ as example, the computing model of microfluidic chip is established as follows; Variables $x_1, x_2, x_3, \bar{x}_1, \bar{x}_2, \bar{x}_3$ value as 0 or 1. $\bar{x}_i$ is the negative of x_i.All combinations of the 3 variables are $(0,0,0)$, $(0,0,1)$, $(0,1,0)$, $(1,0,0)$, $(0,1,1)$, $(1,0,1)$, $(1,1,0)$, $(1,1,1)$.We then determine whether there are combinations of assignments with ϕ equal to 1 (true). If ϕ is required to be true, then each clause must be 1. The problem is to find a combination of values in which every clause is 1. If so, the problem is satisfiable. If not, the problem is unsatisfiable.

3.1 Biological Algorithm

Step 1: In the above problems, the value of each variable is 0 or 1. There are two encoding methods for each variable, one is 1 and the other is 0. $\bar{x}_i$ is the negative of x_i, when $\bar{x}_i$ is 1, x_i is 0. So, the encoding of the variable when $\bar{x}_i$ is 1 is the same as when x_i is 0. So we only need is 2^n types of DNA codes. The values of the variables are encoded and linked into 2^3 DNA fragments, which are then amplified by PCR for use.

Step 2: According to the combination of true value for each clause, the molecular beacon probe is made by using its complement strand, and it is fixed on the chip.

Step 3: Then, using the parallel reaction of the microfluidic chip, the DNA fragment that satisfies each clause can be screened out.

Step 4: Repeat steps 2 and 3 to the last clause.

Step 5: The data pool that remains after each clause has been validated is checked. If there are DNA strands, they are combinations that satisfies the whole conjunctive paradigm assignment to be true, to determine whether the conjunctive normal form is satisfiable. If no DNA strands are present, then this conjunctive paradigm is unsatisfiable.

3.2 Example Analysis

The variables x_1, x_2, x_3 are represented by short DNA strands of 10 bp each, which allow the ligase to link the three pieces together. $x_1 \rightarrow x_2 \rightarrow x_3$ is a combination of variables values. Since $\overline{x}_i$ is the negative form of x_i, if the value of x_i is represented by 10 bp DNA fragment, then $\overline{x}_i$ can be represented by DNA fragment that is base complementary to x_i. So, there are 2^3 combinations of all the values of these three variables x_1, x_2, x_3 (Fig. 1). Then we connect the $x_1^* \rightarrow x_2^* \rightarrow x_3^*$ strands to two complementary 8 bp DNA fragments to make molecular beacon probe (Fig. 1). So, there are 2^3 groups of probes.

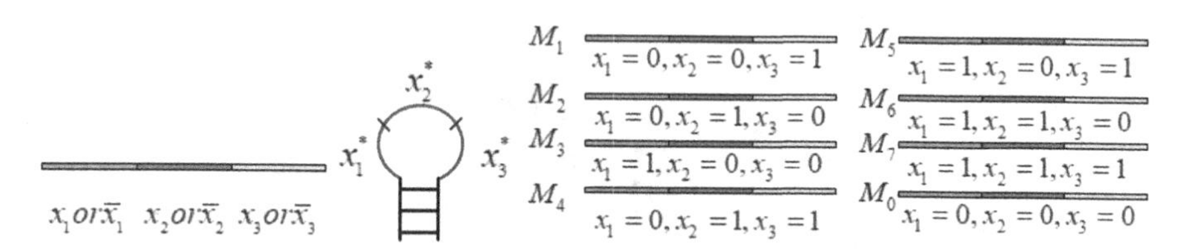

Fig. 1. Shows the DNA strand of variable values

The extension of molecular beacon's stem fixes it to the chip. Mercapto modified glass can be selected as the chip. The general method is as follows: take ordinary slide into 95% concentrated sulfuric acid and 5% potassium dichromate solution soak overnight. Then wash the slides with distilled water and immerse them in 25% ammonia overnight. Take out the slides, wash them with pure water and let them dry. After rinsing with anhydrous ethanol, the slides are immersed in MEA (1% 3-mercaptopropyl trimethoxysilane, 95% ethanol, 16 mM acetic acid) solution for 1 h, followed then by immersion in 95% ethanol and 16 mM acetic acid (pH4.5) solution for 5 min, and quickly placed in dry nitrogen at room temperature overnight. So the coded molecular beacon probes are fixed on the chip, and the self-assembly method is adopted. The chip is processed and the gold nanoparticles are self-assembled on the chip. This not only increases the effective surface area for the chip, but also increases the amount of DNA fragment (probe) and hybridization. At the same time, the special chemical reactivity of gold nanoparticles also increases the sensitivity of the probe. The DNA is mercaptized to form Au-S bonds on the processed chip, allowing the molecular beacon probe to self-assemble on the chip. The chip surface is highly ordered and stable. High hybridization efficiency is achieved for complementary DNA sequences at appropriate fixed densities. When the solution to be examined contains the target sequence, the hairpin structure of the molecular beacon changes to open (Fig. 2).

The entire experiment can be built on a substrate and microchannels can be lithographed. The micropump can be used to input the coded variable value into the microchannel according to a certain flow rate. Under certain conditions, self-assembly reaction occurs (Fig. 3). After a certain time, the chip is removed and a new chip is inserted. After the chip is removed, the output is then read. The resulting DNA

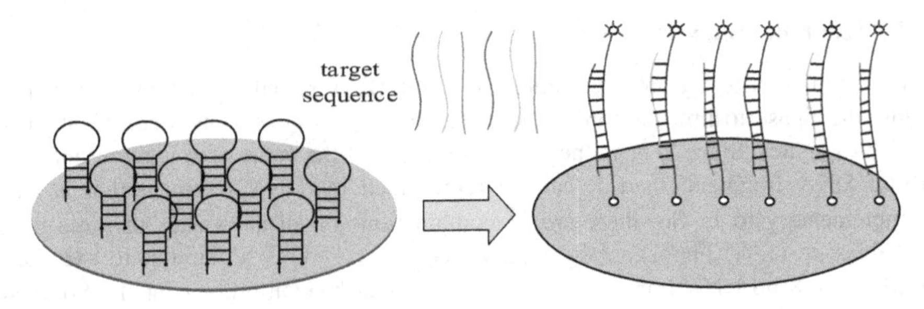

Fig. 2. Preparation and reaction principle of the chip

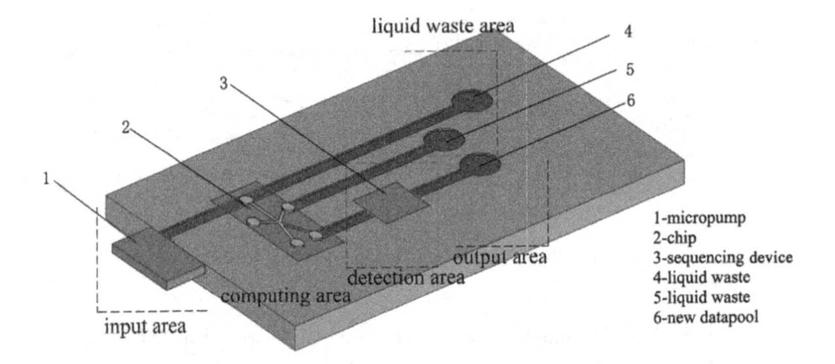

Fig. 3. Microchannel chip

fragment is reusable for the chip designed here. Due to the rapid biochemical reaction, multiple chips can be designed and different combinations of values satisfying the sub-formula can be extracted circularly. Therefore, the microfluidic chip is divided into five areas: input area, computing area, detection area, output area and waste area. The first clause $(x_1 \vee x_2 \vee x_3)$ contains three variables x_1, x_2, x_3. For the clause to have a value of 1, at least one variable must have value of 1. We encode the values of variables as shown in Fig. 4. The complement stands of DNA strands $M_1, M_2, M_3, M_4, M_5, M_6, M_7$ corresponding to the value combination representing the first clause are made into molecular beacon probe and fixed on the substrate (Fig. 4).

The DNA strands $M_0, M_1, M_2, M_3, M_4, M_5, M_6, M_7$ representing the combination of all values of the variable are then amplified by PCR and put into the micropump. Slowly, the single stranded PCR complexes are injected into the main channel according to certain flow rate and enter the detection area. In the detection region, the DNA strands satisfying the value of the first clause are hybridized with the molecular beacon probe. They are fixed to the chip, while the unsatisfied strands of DNA flow into waste pool in the solution. The hybridized chip then come to the detection area. Here the denaturation reaction is heated and the chip is flushed. The opened molecular beacon again returns to the hairpin structure (Fig. 5). The DNA strand that satisfies the first clause goes to the new data pool as the initial data pool for the second clause to be tested.

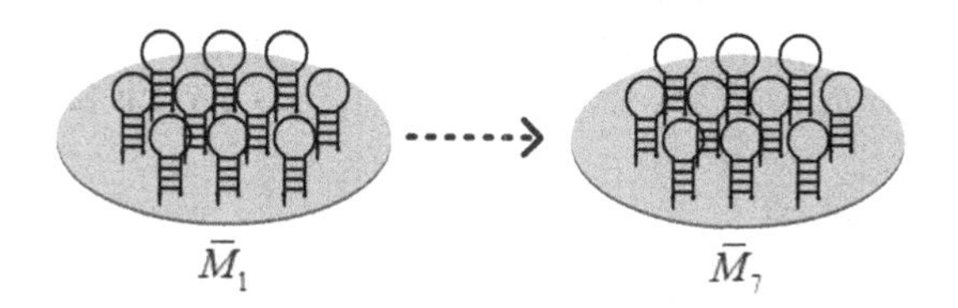

Fig. 4. Chip with molecular beacon probe

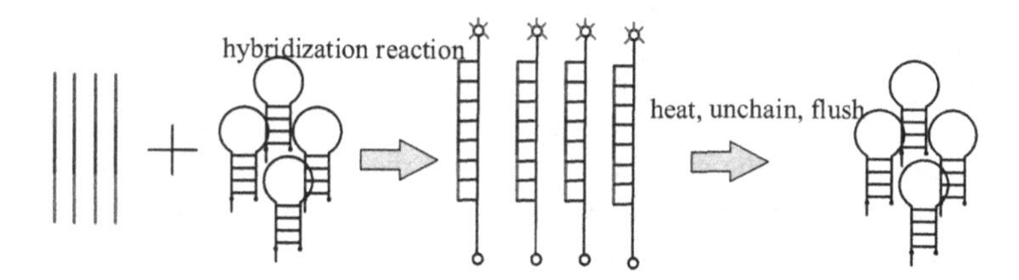

Fig. 5. Schematic diagram of reaction between variables and molecular beacon probe

The second clause $(\bar{x}_1 \vee x_2 \vee x_3)$ contains the variables $\bar{x}_1, x_2, x_3$. In the same way, at least one of the variables takes 1 to give the value of the second clause 1. And because $\bar{x}_1$ is the negative of x_1, the DNA strands corresponding to the combination of values satisfying the second clause are $M_0, M_1, M_2, M_4, M_5, M_6, M_7$.The molecular beacon probes are made from the strands complement and then fixed on the chip by self-assembly method. The data pool obtained from the first biochemical reaction is put into the micro pump, and injected into the main channel slowly according to certain flow rate, and enters the detection area. In the detection region, the DNA strands $M_1, M_2, M_4, M_5, M_6, M_7$ satisfying the value of 1 in the second clause are hybridized with the molecular beacon probe. Then they are fixed to the chip. Unsatisfied strands of DNA flow into a waste pool in the solution and hybridized chip then come to the detection area. Here the denaturation reaction is heated and the chip is flushed. The opened molecular beacons again return to the hairpin structure. The combination of values satisfying the first and second clauses lead to the new data pool, which serves as the initial data pool for testing the third clause.

The third clause $(\bar{x}_1 \vee x_2 \vee \bar{x}_3)$ contains variables $\bar{x}_1, x_2, \bar{x}_3$. Variables $\bar{x}_1, \bar{x}_3$ are negative values for x_1, x_3.Therefore, the combination of values satisfying the third clause are $M_0, M_1, M_2, M_3, M_4, M_6, M_7$. The molecular beacon probes are made from the complementary strands and then fixed on the chip by self-assembly method. In the same way as above, we can obtain the value combination M_1, M_2, M_4, M_6, M_7 which satisfies the first clause, second clause and third clause simultaneously. Again, they act as the next reaction to the new data pool.

The fourth clause $(\bar{x}_2 \vee \bar{x}_3)$ contains only two variables $\bar{x}_2, \bar{x}_3$, and $\bar{x}_2, \bar{x}_3$ are the negative of x_2, x_3. The variable x_1 is missing in the clause, so the value of the variable x_1 does not affect the value of the clause. Thus, the combination of values satisfying the fourth clause are $M_0, M_1, M_2, M_3, M_5, M_6$. The molecular beacon probes are made from the complementary strands and then fixed on the chip by self-assembly method. In the same way as above, we can obtain the value combination M_1, M_2, M_6 which satisfies the four clauses simultaneously, which are $(0,0,1)$,$(0,1,0)$,$(1,1,0)$. So $\phi = (x_1 \vee x_2 \vee x_3) \wedge (\overline{x}_1 \vee x_2 \vee x_3) \wedge (\overline{x}_1 \vee x_2 \vee \overline{x}_3) \wedge (\overline{x}_2 \vee \overline{x}_3)$ is satisfiable.

3.3 Simulated Analysis

Visual DSD is an automatic simulation of all possible strand displacement reactions between various DNA strands. Here we take the M7 and M2 molecular beacon probes as examples, and use Visual DSD to simulate the binding between the probe and target DNA strand. The simulation results are shown in the figure below. (a) and (c) in Fig. 6 are the time series diagrams of probe binding to the target DNA strand. (c) and (d) in Fig. 6 are sequence structures of probes, target DNA, and end products. The initial concentration of probe and target DNA strand are set to 20 nM, and reaction time is set to 6000 s. According to the simulated time series diagrams (a) and (c): when targeted DNA strands input, they can be united with the probe. Then the initial concentration goes down as green curve, representing target DNA strands, and blue curve representing the molecular beacon probe. The final product, DNA strand and probe hybridization of DNA complexes (red curve) concentration increase gradually until completion of the initial concentration, and the reaction is over. (e) and (f) in Fig. 6 are histograms showing the concentration of probe after sufficient reaction with target DNA strands. The histogram of concentration after the full reaction show that no erroneous products are present except for the expected DNA complexes. It shows that the binding efficiency between the target DNA strand and molecular beacon probe is high, the reaction is sufficient, and design of the model in this paper is reasonable. Figure 6, is worth noting that the simulations (a) and (b) of the M7 probe and target combination are essentially the same as simulations (e) and (f) of the M2 probe and target combination. The reason for this is that the reactions are essentially the same between the designed probe and target DNA strand. But the difference is that the DNA base sequence for the probe is different (the sequence layout in Figure (c) and (d) is different). In conclusion, the molecular beacon probes designed in this study have high binding efficiency, simple reaction conditions and accurate product. In addition, compared with previous DNA computing models, the molecular beacon probes designed in this study combined with microfluidic chip technology are easier to realize the chip for the computing model, which is an advantage that other theoretical computing models do not have.

This is 3-SAT problem in which there are no more than three variables in each clause. The solution for the whole model is based on the microfluidic chip, which makes full use of high throughput microfluidic chip and the great parallelism of DNA computation. The detection of each clause can be completed at one time. The complexity of the model is therefore only related to the number of clauses contained in the conjunctive normal form to be determined, that is $O(m)$, where m is the number of clause in conjunctive normal form normal. For the general 3-SAT problem, as the number of variables in the problem increases, the number of variables in each clause does not exceed 3. In calculation principle, the number of detection probes in each clause does not change much, but the types of probes increase with number of variables. The complexity of the model should be $O(m) + O(n)$, where m is the number of

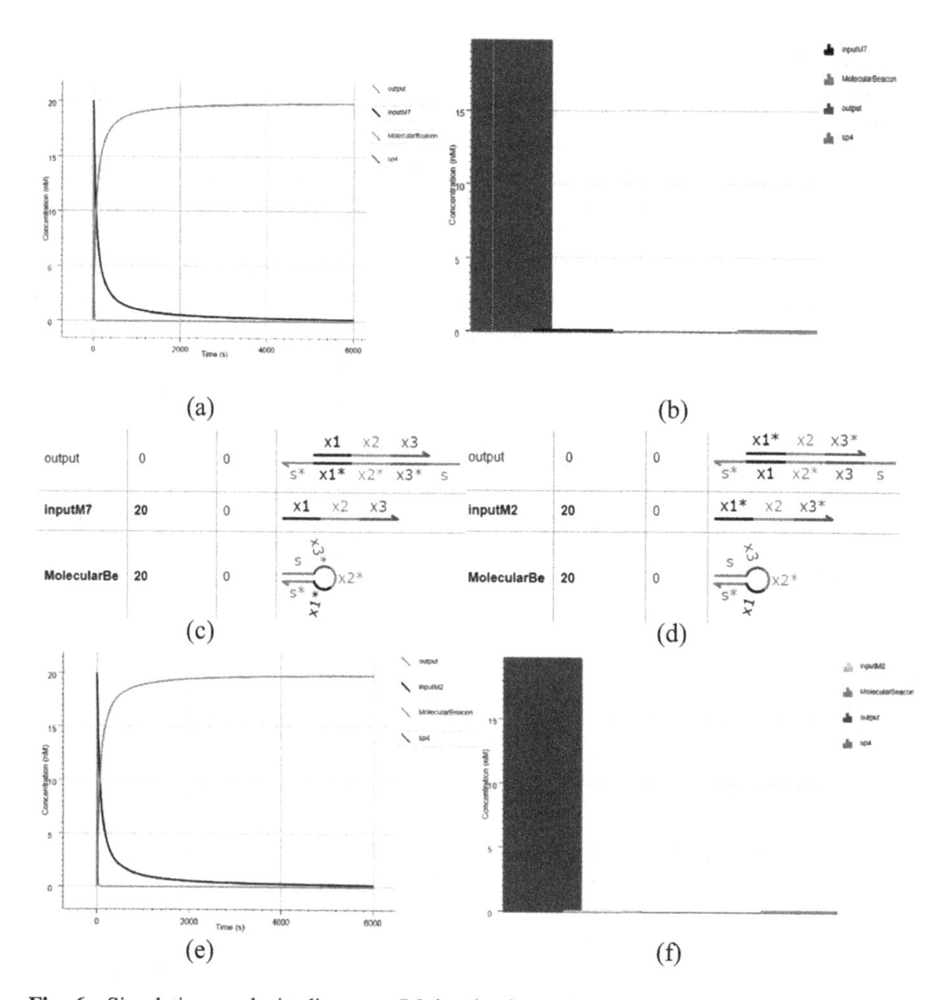

Fig. 6. Simulation analysis diagram. (Molecular beacon represents a molecular beacon that represents variables in a simulated reaction. Input represents a probe that complements the variable as input. The products of the reaction between the probe and the variable represented by output are shown in figures c and d. Figure a and e show the progress of the reaction time at any time, the concentration of the initializer begins to decrease and the concentration of the product begins to increase. To a certain time after the reaction is sufficient, the concentration of the initial and the product does not change. Figure e and f represent the concentrations of initiators and products in the solution after the reaction.)

clause in conjunctive normal form normal, and n is the number of variables contained in the conjunctive normal form. For the general satisfiability problem, the complexity of the model is still $O(m) + O(n)$.

4 Conclusion

In this paper, we report construction of a DNA chip using molecular beacon probe, and build the solution for satisfying problem on microfluidic chip. In the process of calculation, the great parallelism of DNA calculation and the miniaturization and reusable characteristics of microfluidic chip can effectively reduce the computational complexity and save the computational cost. By making full use of specificity and sensitivity of molecular beacons, and also high throughput of microfluidic chips, the complexity of the computational model for satisfiability problem is reduced to $O(m) + O(n)$, where m is the number of clause in conjunctive normal form normal. n is the number of variables contained in the conjunctive normal form. Molecular beacon microfluidic chip combines the advantages of molecular beacon and microfluidic chip, making full use of the huge parallelism and massive storage capacity of DNA computing, and providing a solution method for solving NP problems.

Acknowledgements. The project is supported by National Natural Science Foundation of China (No. 61702008, No. 62072296), Natural Science Foundation of Anhui Province (No. 1808085MF193) and Anhui Province postdoctoral fund (No. 2019B331).

References

1. Hameed, K.: DNA computation based approach for enhanced computing power. Int. J. Emerg. Sci. **1**(1), 23–30 (2011)
2. Kumar, S.N.: A proper approach on DNA based computer. Am. J. Nanomater. **3**(1), 1–4 (2015)
3. Sakakibara, Y., Suyama, A.: Intelligent DNA chips: Logical operation of gene expression profiles on DNA computers. Geneme Inf. **11**, 33–42 (2002)
4. Yin, Z.X., Zhang, F.Y., Xu, J.: 0–1 programming problem based on DNA computing. J. Electron. Inf. Technol. **15**(1), 1–5 (2003)
5. Yang, J., Yin, Z.X., Tang, Z., et al.: Visual solution to minimum spanning tree problem based on DNA origami. Mater. Express **11**(10), 1700–1706 (2021)
6. Yang, X.M., Yang, J., Yin, Z.X., Tang, Z., et al.: DNA tetrahedron walker calculation model for 0–1 integer programming problem. J. Fuyang Normal Univ. (Nat. Sci.) **37**(02), 93–98 (2020)
7. Yang, J., Yin, Z.X., Tang, Z., et al.: DNA computing model for process problem based on hybridization chain reaction. J. Guangzhou Univ. (Nat. Sci. Ed.) **19**(1), 14–21 (2020)
8. Yang, J., Yin, Z.X., Tang, Z., et al.: Search computing model for the knapsack problem based on DNA origami. Mater. Express **9**(6), 535–544 (2019)
9. Whitesides, G.M.: The origins and the future of microfluidics. Nature **442**(7101), 368–373 (2006)
10. Chow, A.W.: Lab-on-a-chip: opportunities for chemical engineering. AIChE J. **48**(8), 1590–1595 (2002)
11. Sohila, Z., Francoise, R., Raphael, L.: Microfluidic chip with molecular beacons detects miRNAs in Human CSF to reliably characterize CNS-specific disorders. RNA Dis. **3**, 2375–2567 (2016)

12. Guo, Q., Bai, Z., Liu, Y., et al.: A molecular beacon microarray based on a quantum dot label for detecting single nucleotide polymorphisms. Biosens. Bioelectron. **77**, 107–110 (2016)
13. Lin, B.C.: Research and industrialization of microfluidic chip. Chin. J. Anal. Chem. **44**(4), 491–499 (2016)
14. Harrison, D.J., Manz, A., Fan, Z., et al.: Capillary electrophoresis and sample injection systems integratedon a planar glass chip. Anal. Chem. **64**(17), 1926–1932 (1992)
15. Thorsen, T., Maerkl, S.J., Quake, S.R.: Microfluidic large-scale integration. Science **298** (5593), 580–584 (2002)
16. Yager, P., Edwards, T., Fu, E., et al.: Microfluidic diagnostic technologies for global public health. Nature **442**(7101), 412–418 (2006)
17. Nge, P.N., Rogers, C.I., Woolley, A.T.: Advances in microfluidic materials, functions, integration, and applications. Chem. Rev. **113**(4), 2550–2583 (2013)
18. Amini, H., Lee, W., Di, C.D.: Inertial microfluidic physics. Lab Chip **14**(15), 2739–2761 (2014)
19. Yetisen, A.K., Akram, M.S., Lowe, C.R.: Paper-based microfluidic point-of-care diagnostic devices. Labon a Chip **13**(12), 2210–2251 (2013)
20. Chen, G.: Fabrication and Application of Surface-Enhanced Raman Scattering (SERS) Substrates in Microfluidic Channels. Jilin University, Changchun (2015)
21. Zhang, C.X.: Research on DNA Biosensor and its Application in Microfluidic Laboratory on a Chip. Beijing Institute of Technology, Beijing (2015)
22. Liu, Q.H., Wang, L.M., Frutos, A.G., et al.: DNA computing on surfaces. Nature **403**(13), 175–179 (2000)
23. Braich, R.S., Chelyapov, N., Johnson, C., et al.: Solution of a 20 variable 3-SAT problem on a DNA computer. Science **296**(5567), 499–502 (2002)
24. Kristiane, A., Christiaan, V.H., Grzegorz, R., et al.: DNA computing using single-molecule hybridization detection. Nucleic Acids Res. **32**(17), 4962–4968 (2004)
25. Lin, C.H., Cheng, H.P., Yang, C.B., et al.: Solving satisfiability problems using a novel microarray-based DNA compute. Biosystems **90**(1), 242–252 (2007)
26. Yin, Z.X., Cui, J.Z., Liu, W.B., et al.: Plasmid resolving the satisfiability problem with DNA computing models. J. Comput. Theor. Nano-Sci. **4**, 1243–1248 (2007)
27. Brun, Y.: Solving satisfiability in the tile assemblymodelwith a constant-size tileset. J. Algorithms Cogn. Inf. Logic **63**(4), 151–166 (2008)
28. Chen, M., Chen, X.Q., Zhang, L., Xu, J.: A biobrick inversion cellular computing model for satisfiability problem. Chin. J. Comput. **3**(12), 2537–2544 (2014)
29. Yang, J., Yin, Z.X., Cui, J.Z.: Satisfiability problem based on self-assembly model of molecular beacons. J. Bionanosci. **9**(3), 197–202 (2015)
30. Yin, Z., Yang, J., Zhang, Q., Tang, Z., Wang, G., Zheng, Z.: DNA computing model for satisfiability problem based on hybridization chain reaction. Int. J. Pattern Recogn. Artif. Intell. **35**(03), 2159010,1-2159010,16 (2021). https://doi.org/10.1142/S0218001421590102

Construction of Four-Variable Chaotic System Based on DNA Strand Displacement

Haoping Ji[1,2], Yanfeng Wang[1,2], and Junwei Sun[1,2](✉)

[1] Henan Key Lab of Information-Based Electrical Appliances, Zhengzhou University of Light Industry, Zhenzhou 450002, China
junweisun@yeah.net

[2] School of Electrical and Information Engineering, Zhengzhou University of Light Industry, Zhenzhou 450002, China

Abstract. As a typical DNA nanotechnology, DNA strand displacement reaction has attracted many scholars to investigate. In this paper, a four-variable chaotic system is proposed by multiplication, degradation, and adjustment DNA strand displacement modules, and the dynamic behaviors of designed chaotic system are analysed by the software of Matlab and Visual DSD. The simulation results show that our strand displacement chaotic system can realize the production of chaotic signals. Our work uses biological reactions to achieve the construction of chaotic systems, which can provide a reference for the future researches of chaotic system.

Keywords: DNA strand displacement reactions · Chaotic systems · Dynamic behaviors

1 Introduction

With the increasing demand of computing power, silicon-based integrated circuits can not meet the demands of people [1,2]. Many experts predict that 5 nm silicon technology node will be the end of Moore's Law, since quantum tunnelling will occur when the transistors smaller than 7 nm scale silicon-based integrated [3–5]. Researchers are constantly looking for new ways to replace silicon-based integrated circuits to promote the continued development of computing technology [6,7]. Due to its fast computing speed and strong parallel computing capabilities, DNA computing has attracted many researchers and has done a lot of works. DNA strand displacement reactions can be used for DNA computing. In addition, DNA strand displacement technology using the bonding technology between the single chain and double chain of the DNA molecules for releasing a single chain [8]. DNA strand displacement reactions have the advantages of sensitivity, precision and controllability. This reactions can be cascaded to realize complex chemical reaction networks (CRNs) [9]. Complex circuit systems

L. Pan et al. (Eds.): BIC-TA 2021, CCIS 1565, pp. 426–436, 2022.
https://doi.org/10.1007/978-981-19-1256-6_33

are designed by integration, multiplication, degradation and annihilation DNA strand displacement modules in work [10]. Catalysis, integration, and sum DNA modules are proposed to realize feedback control circuit in work [11]. QSM controller are gained by catalysis, degradation and annihilation DNA modules to achieve nonlinear feedback control [12].

Chaos is a random phenomenon produced by a fixed system, its movement is uncertain and trajectory cannot be predicted [13–15]. when disturbing small disturbances for parameters in control system, the dynamic behaviors will have huge changes. However, the dynamic behaviors for the chaotic system are different with the dynamic behaviors of real random system, since the dynamic behaviors for chaotic systems are reproducible [16]. Since the chaotic system has the characters of irregularity and repeatability, the chaotic systems have attracted many scholars to investigate this research field. They want to understand the general laws of chatic systems. The fields of information transmission, image encryption weather prediction and so on [17–20] have attracted more attention. Classical three-variable Lotka-Cvolterra chaotic system is designed via strand displacement reactions by work [21–27].

In this work, we have designed three strand displacment modules that multiplication modules, degradation modules, and adjustment DNA modules. In addition, these DNA modules can be studied to achieve four-variable chaotic system. Compared with previous works, we using Visual DSD for designing strand displacement circuit and realizing four-variable chaotic system.

2 Modules Construction

2.1 Principle of DNA Strand Displacement

As shown in Fig. 1, the DNA strand displacement reaction follows the principle of base complementary pairing. From Fig. 1(a), the toehold "*TCTG*" of input strand X binds with the toehold "*AGAC*" of auxiliary strand A to promote the branch migration of auxiliary A, generating a new stable double-stranded $sp2$ structure and output strand Y. In order to simplify the expression of DNA strand displacement reaction, the representations of the bases are hidden, and Fig. 1(b) is proposed to simplify the Fig. 1(a).

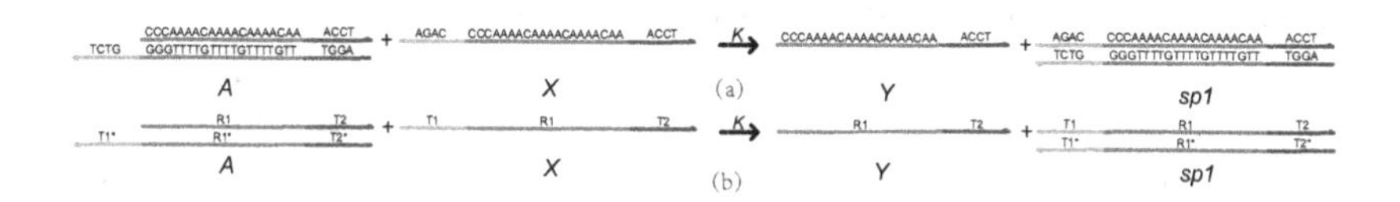

Fig. 1. DNA strand displacement reaction. (a) full representation of DNA strand displacement reaction, (b) simple representation of DNA strand displacement reaction.

2.2 Modules of Four-Variable Chaotic System

In order to achieve the ideal system, multiplication, degradation, and adjustment three DNA modules are designed. The cascade of these modules can realize a four-variable chaotic system and generate chaotic oscillation signals. In this part, we analyze the construction of each module and their dynamic behavior. The simulation experiments prove that the modules can achieve the ideal function, providing a basis for us to build a chaotic system.

To implement the ideal multiplication module (1), strand displacement reactions (2) and (3) are proposed that Y is the input strand. $A1$ and $A2$ are the auxiliary strands. Similarly, $sp2$ and $sp4$ are waste strands. $sp3$ is the middle strand to cascade these two reactions. Based on the law of mass action, $\frac{d[Y]_t}{dt} = -q_i C_m[Y_t]$ and $\frac{d[Y]_t}{dt} = q_m C_m[sp3_t]$ are obtained by DNA strand displacement reactions (2) and (3), respectively. When the initial concentration of the input strand Y is set as 10 nM and auxiliary strands $A1$ and $A2$ are defined to 10000 nM. Displacement rates q_i and q_m are $10^{-6}\,\mathrm{nMs}^{-1}$ and $q_i = 10^{-3}\,\mathrm{nMs}^{-1}$, respectively. The changes in the concentration of Y are shown in Fig. 2. It can be seen from the Fig. 2 that the DNA reactions can achieve the purpose of multiplication. Due to $q_i \ll q_m$, the rate k_a of multiplication module (1) can be obtained by the product of q_i and C_m.

The ideal multiplication module is:

$$Y \xrightarrow{k_a} 2Y \tag{1}$$

The DNA strand displacement reactions are:

$$A1 + Y \xrightarrow{q_i} sp2 + sp3 \tag{2}$$

$$A2 + sp3 \xrightarrow{q_m} sp4 + 2Y \tag{3}$$

To implement the ideal adjustment module (4), strand displacement reactions (5), (6), (7) and (8) are proposed that Y and Z are the input strands. $A3$, $A4$, $A5$ and $A6$ are the auxiliary strands. Similarly, $sp2$, $sp6$, $sp7$ and $sp9$ are waste strands. $sp5$ and $sp8$ are the middle strands to cascade these four reactions. Based on the law of mass action, $\frac{d[Y]_t}{dt} = -q_i C_m[Y_t]$ and $\frac{d[Z]_t}{dt} = q_m C_m[sp5_t]$ are obtained by DNA strand displacement reactions (5) and (6). Similarly, $\frac{d[Z]_t}{dt} = -q_s C_m[Z_t]$ and $\frac{d[Z]_t}{dt} = q_s C_m[sp8_t]$ are obtained based on DNA strand displacement reactions (7) and (8). When the initial concentrations of the input strand Y and Z are set as 10 nM and auxiliary strands $A3$, $A4$, $A5$ and $A6$ are defined to 10000 nM. Displacement rates q_i, q_m and q_s are $10^{-6}\,\mathrm{nMs}^{-1}$, $10^{-3}\,\mathrm{nMs}^{-1}$ and $10\,\mathrm{nMs}^{-1}$, respectively. The changes in the concentrations of Y and Z are shown in Fig. 3. It can be seen from the Fig. 3 that the DNA reactions can achieve the purpose of adjustment. Due to $q_i \ll q_m \ll q_s$ the rate k_b of adjustment module (4) can be obtained by the product of q_i and C_m.

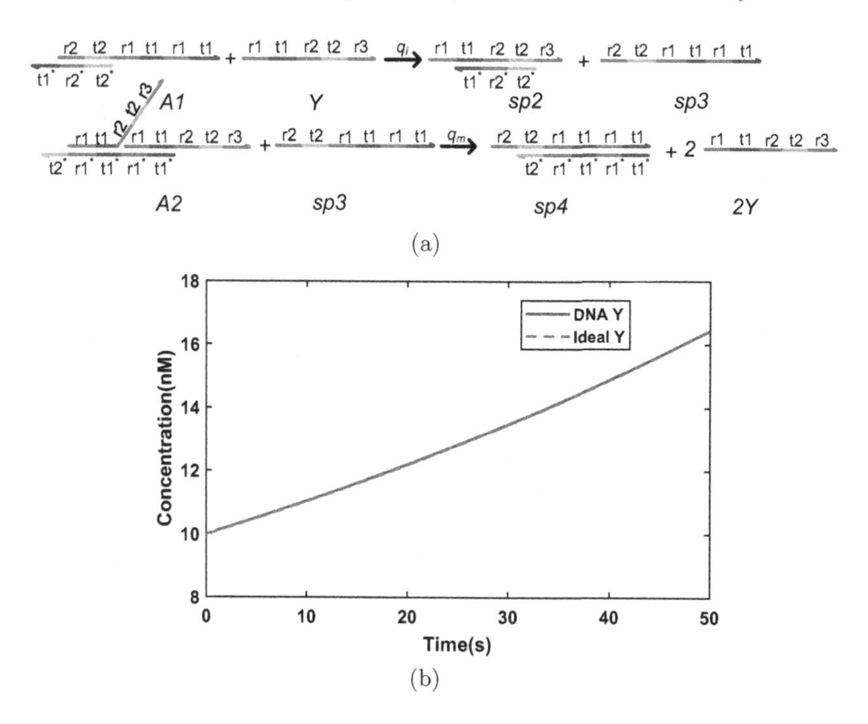

Fig. 2. Multiplication module. (a) DNA strand displacement reactions, (b) the changes in the concentration of Y.

The ideal multiplication module is:

$$Y + Z \xrightarrow{k_b} 2Z \tag{4}$$

The DNA strand displacement reactions are:

$$A3 + Y \xrightarrow{q_i} sp2 + sp5 \tag{5}$$

$$A4 + sp5 \xrightarrow{q_m} sp6 + Z \tag{6}$$

$$A5 + Z \xrightarrow{q_s} sp7 + sp8 \tag{7}$$

$$A6 + sp8 \xrightarrow{q_s} sp9 + Z \tag{8}$$

To implement the ideal elimination module (9), strand displacement reactions (10) and (11) are proposed that W is the input strand. $A6$ is the auxiliary strand. Similarly, $sp11$ and $sp12$ are waste strands. $sp10$ is the middle strand to cascade these two reactions. Based on the law of mass action, $\frac{d[W]_t}{dt} = -(q_i - q_j)C_m[W_t]$ and $\frac{d[sp10]_t}{dt} = (q_s - q_j)[sp10_t]$ are obtained by DNA strand displacement reactions (10) and (11), respectively. When the initial concentration of the input strand W is set as 6 nM and auxiliary strand $A6$ is defined to 10000 nM. Displacement

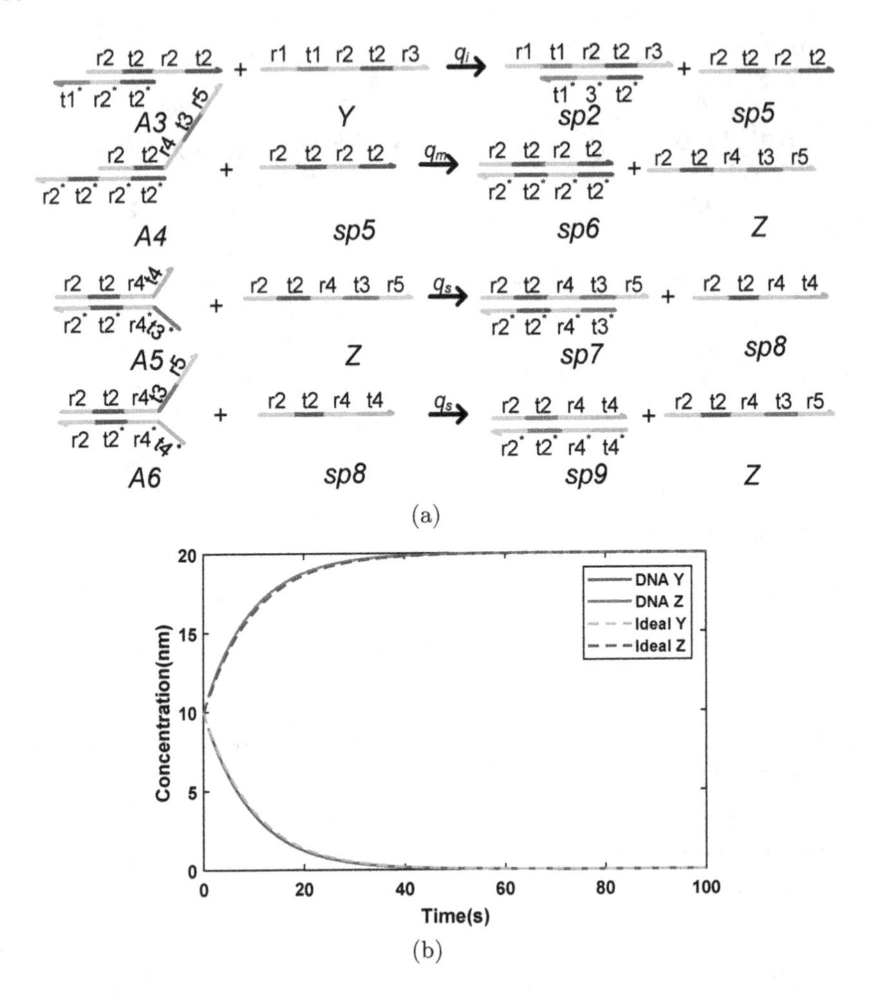

Fig. 3. Adjustment module. (a) DNA strand displacement reactions, (b) the changes in the concentration of Y and Z.

rates q_i, q_s and q_j are $10^{-5}\,\mathrm{nMs}^{-1}$, $10\,\mathrm{nMs}^{-1}$ and $10^{-10}\,\mathrm{nMs}^{-1}$, respectively. The changes in the concentration of W are shown in Fig. 4. It can be seen from the Fig. 4 that the DNA reactions can achieve the purpose of elimination. Due to $q_j \ll q_i \ll q_s$, the reverse rate can be neglected and the rate k_c of elimination module (9) can be obtained by the product of q_i and C_m.

The ideal multiplication module is:

$$W \xrightarrow{k_c} \phi \tag{9}$$

The DNA strand displacement reactions are:

$$A6 + W \underset{q_j}{\overset{q_i}{\rightleftharpoons}} sp10 \tag{10}$$

$$sp10 \underset{q_j}{\overset{q_s}{\rightleftharpoons}} sp11 + sp12 \tag{11}$$

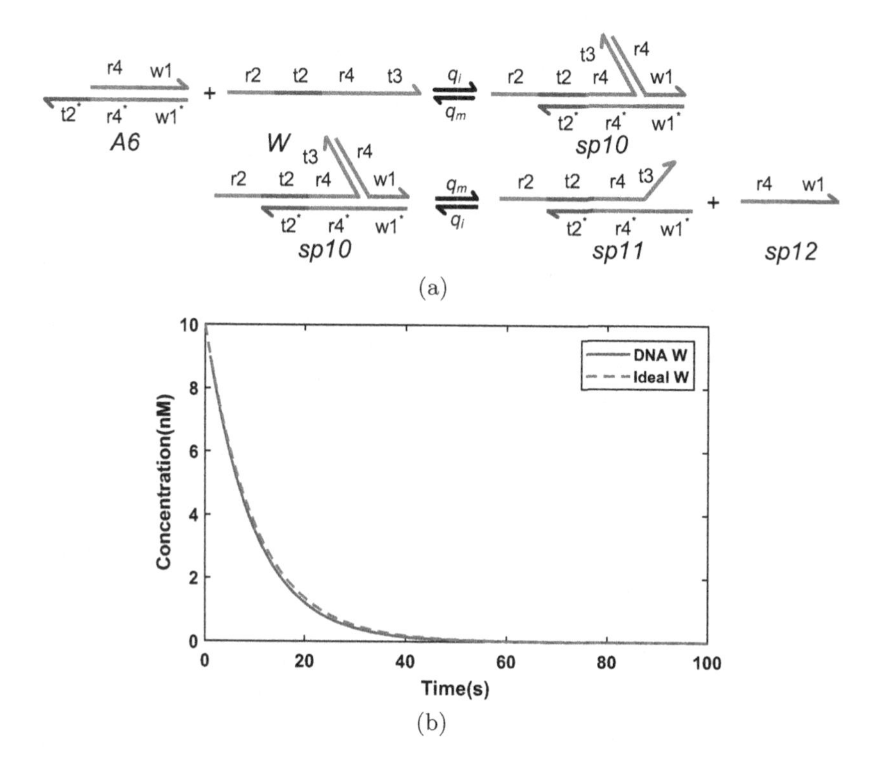

Fig. 4. Elimination module. (a) DNA strand displacement reactions, (b) the changes in the concentration of W.

3 Construction and Simulation of Chaotic System

A 4-variable chaotic system can be described by the following set of ordinary differential equations (ODEs)

$$\begin{cases} \frac{d[x]_1}{dt} = -2x_1 + x_1 y_1 \\ \frac{d[y]_1}{dt} = 2y_1 - y_1 x_1 + z_1 y_1 \\ \frac{d[z]_1}{dt} = -2z_1 - z_1 y_1 + z_1 w_1 \\ \frac{d[w]_1}{dt} = 2w_1 - w_1 z_1 \end{cases} \tag{12}$$

CRNs for (12) are given as

$$\begin{cases} X_1 \xrightarrow{2} \phi \\ X_1 + Y_1 \xrightarrow{1} 2X_1 \\ Y_1 \xrightarrow{2} 2Y_1 \\ Y_1 + Z_1 \xrightarrow{1} 2Y_1 \\ Z_1 \xrightarrow{2} \phi \\ Z_1 + W_1 \xrightarrow{1} 2Z_1 \\ W_1 \xrightarrow{2} 2W_1 \end{cases} \tag{13}$$

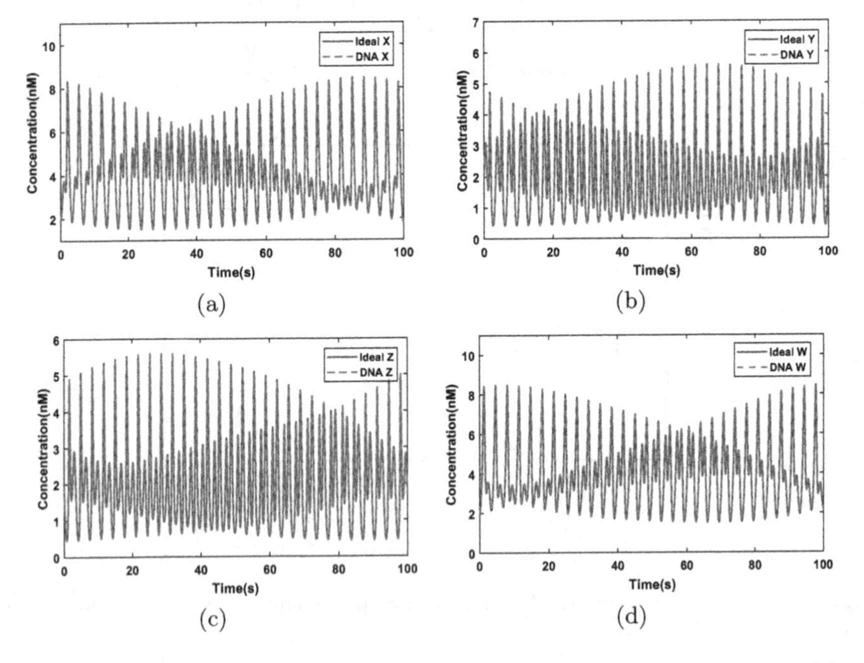

Fig. 5. Changes of concentration. (a) the changes in the concentrations of X_1, (b) the changes in the concentrations of Y_1, (c) the changes in the concentrations of Z_1, (d) the changes in the concentrations of W_1.

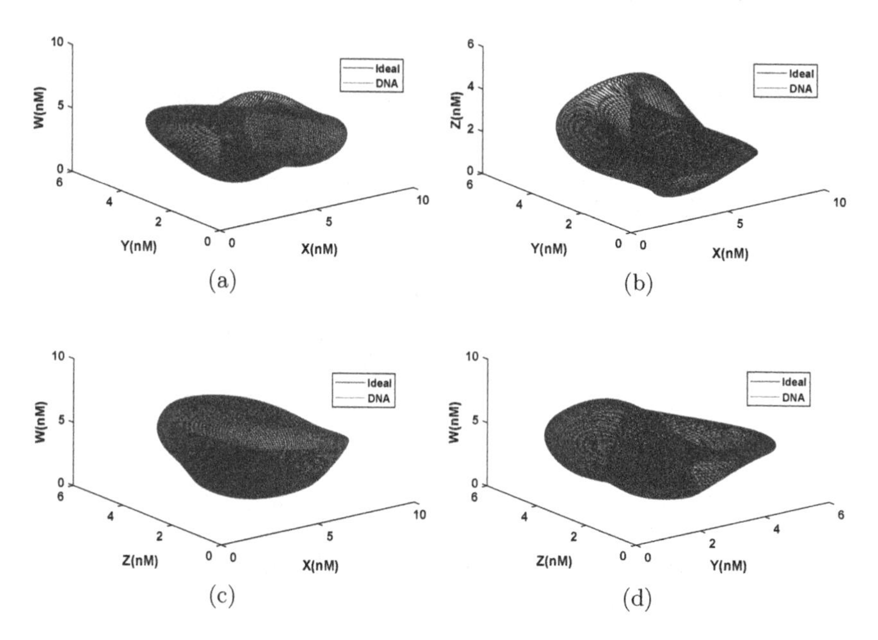

Fig. 6. Chaotic attractors of 3-variable. (a) the chaotic attractor of XYW, (b) the chaotic attractor of XYZ, (c) the chaotic attractor of XZW, (d) the chaotic attractor of YZW.

Based on the CRNs (13) above, we have the following conclusions and obviously, (12) and (14) are the same, the CRNs are correct. When the initial concentrations of X_1, X_2, X_3 and X_4 are set to 2 nM, the changes in the concentrations of X_1, X_2, X_3 and X_4 are shown in Fig. 5 and chaotic attractors are as Fig. 6. Through the simulation results, we can know that the chaotic CRNs we designed is right and the generated time-domain waveform is oscillating (Fig. 7).

$$\begin{cases} \frac{d[X]_1}{dt} = -2X_1 + X_1Y_1 \\ \frac{d[Y]_1}{dt} = 2Y_1 - Y_1X_1 + Z_1Y_1 \\ \frac{d[Z]_1}{dt} = -2Z_1 - Z_1Y_1 + Z_1W_1 \\ \frac{d[W]_1}{dt} = 2W_1 - W_1Z_1 \end{cases} \tag{14}$$

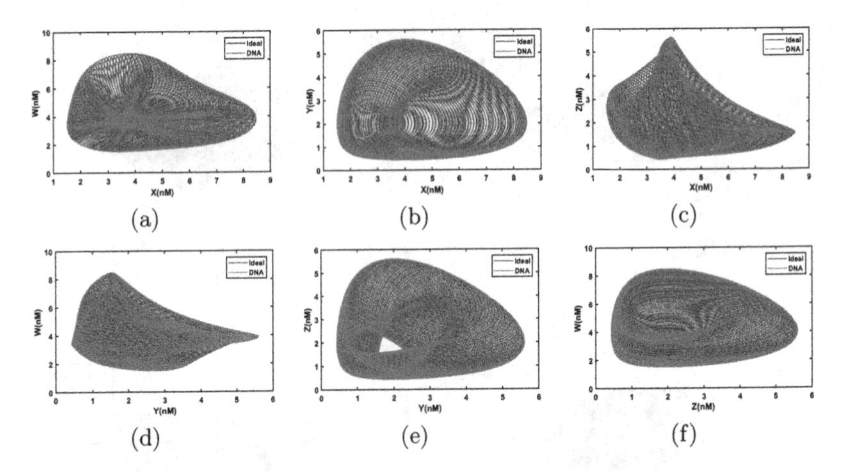

Fig. 7. Chaotic attractors of 2-variable. (a) the chaotic attractor of XW, (b) the chaotic attractor of XY, (c) the chaotic attractor of XZ, (d) the chaotic attractor of YW, (e) the chaotic attractor of YZ, (f) the chaotic attractor of ZW.

4 Conclusions

In this work, a method to realize the designing of four-variable chaotic oscillatory system. Furthermore, these strand displacement modules can be caseded to reaizing complex circuit. Our design can be proved reliable by Visual DSD. In addition, our strand displacement reactions will provide reference for chaotic system investigating. Our future works are realizing PI control for our designed chaotic system.

Acknowledgements. This work was supported in part by the National Key R and D Program of China for International S and T Cooperation Projects (2017YFE0103900), in part by the Joint Funds of the National Natural Science Foundation of China (U1804262), in part by the State Key Program of National Natural Science of China under Grant 61632002, in part by the National Natural Science of China under Grant 61603348, Grant 61775198, Grant 61603347, and Grant 61572446, in part by the Foundation of Young Key Teachers from University of Henan Province (2018GGJS092), and in part by the Youth Talent Lifting Project of Henan Province and Henan Province University Science and Technology Innovation Talent Support Plan under Grant 20HASTIT027.

References

1. Wunsch, B.H., et al.: Gel-on-a-chip: continuous, velocity-dependent DNA separation using nanoscale lateral displacement. Lab Chip **19**(9), 1567–1578 (2019)
2. Cui, Y.X., Feng, X.N., Wang, Y.X., Pan, H.Y., Pan, H., Kong, D.M.: An integrated-molecular-beacon based multiple exponential strand displacement amplification strategy for ultrasensitive detection of dna methyltransferase activity. Chem. Sci. **10**(8), 2290–2297 (2019)
3. He, J.L., et al.: Hybridization chain reaction based dnazyme fluorescent sensor for l-histidine assay. Anal. Methods **11**(16), 2204–2210 (2019)
4. Liu, N., Xu, K., Liu, L., Chen, X., Zou, Y., Xiao, X.: A star-shaped DNA probe based on strand displacement for universal and multiplexed fluorometric detection of genetic variations. Microchimica Acta **185**(9), 413 (2018)
5. Xie, W., Zhou, C., Lv, H., Zhang, Q.: Logic operation model of the complementer based on two-domain DNA strand displacement. Fundamenta Informaticae **164**(2–3), 277–288 (2019)
6. Zou, C., Wei, X., Zhang, Q., Liu, C., Zhou, C., Liu, Y.: Four-analog computation based on DNA strand displacement. ACS Omega **2**(8), 4143–4160 (2017)
7. Sun, J., Li, X., Cui, G., Wang, Y.: One-bit half adder-half subtractor logical operation based on the DNA strand displacement. J. Nanoelectron. Optoelectron. **12**(4), 375–380 (2017)
8. Li, W., Zhang, F., Yan, H., Liu, Y.: DNA based arithmetic function: a half adder based on DNA strand displacement. Nanoscale **8**(6), 3775–3784 (2016)
9. Song, T., Garg, S., Mokhtar, R., Bui, H., Reif, J.: Analog computation by DNA strand displacement circuits. ACS Synth. Biol. **5**(8), 898–912 (2016)
10. Cui, G., Zhang, J., Cui, Y., Zhao, T., Wang, Y.: DNA strand-displacement digital logic circuit with fluorescence resonance energy transfer detection. J. Comput. Theor. Nanosci. **12**(9), 2095–2100 (2015)
11. Engelen, W., Wijnands, S.P., Merkx, M.: Accelerating DNA-based computing on a supramolecular polymer. J. Am. Chem. Soc. **140**(30), 9758–9767 (2018)
12. Ito, K., Murayama, Y., Takahashi, M., Iwasaki, H.: Two three-strand intermediates are processed during rad51-driven DNA strand exchange. Nat. Struct. Molec. Biol. **25**(1), 29 (2018)
13. Guo, Y., et al.: DNA and DNA computation based on toehold-mediated strand displacement reactions. Int. J. Mod. Phys. B **32**(18), 1840014 (2018)
14. Sawlekar, R., Montefusco, F., Kulkarni, V.V., Bates, D.G.: Implementing nonlinear feedback controllers using DNA strand displacement reactions. IEEE Trans. Nanobiosci. **15**(5), 443–454 (2016)
15. Zhang, Z., Fan, T.W., Hsing, I.M.: Integrating DNA strand displacement circuitry to the nonlinear hybridization chain reaction. Nanoscale **9**(8), 2748–2754 (2017)
16. Barati, K., Jafari, S., Sprott, J.C., Pham, V.T.: Simple chaotic flows with a curve of equilibria. Int. J. Bifurcation Chaos **26**(12), 1630034 (2016)
17. Li, C., Sprott, J.C., Xing, H.: Constructing chaotic systems with conditional symmetry. Nonlinear Dyn. **87**(2), 1351–1358 (2016). https://doi.org/10.1007/s11071-016-3118-1
18. Zou, C., Wei, X., Zhang, Q.: Visual synchronization of two 3-variable lotka-volterra oscillators based on DNA strand displacement. RSC Adv. **8**(37), 20941–20951 (2018)
19. Li, Y., Kuang, Y.: Periodic solutions of periodic delay lotka-volterra equations and systems. J. Math. Anal. Appl. **255**(1), 260–280 (2001)

20. Dunbar, S.R.: Traveling wave solutions of diffusive lotka-volterra equations: a heteroclinic connection in r4. Trans. Am. Math. Society **286**, 557–594 (1984)
21. Zou, C., Wei, X., Zhang, Q., Liu, Y.: Synchronization of chemical reaction networks based on DNA strand displacement circuits. IEEE Access **6**, 20584–20595 (2018)
22. Apraiz, A., Mitxelena, J., Zubiaga, A.: Studying cell cycle-regulated gene expression by two complementary cell synchronization protocols. JoVE (J. Vis. Exp.) (124), e55745 (2017)
23. Fries, P., Reynolds, J.H., Rorie, A.E., Desimone, R.: Modulation of oscillatory neuronal synchronization by selective visual attention. Science **291**(5508), 1560–1563 (2001)
24. Zhang, Q., Wang, X., Wang, X., Zhou, C.: Solving probability reasoning based on DNA strand displacement and probability modules. Comput. Biol. Chem. **71**, 274–279 (2017)
25. Olson, X., Kotani, S., Yurke, B., Graugnard, E., Hughes, W.L.: Kinetics of DNA strand displacement systems with locked nucleic acids. J. Phys. Chem. B **121**(12), 2594–2602 (2017)
26. Lakin, M.R., Stefanovic, D.: Supervised learning in adaptive DNA strand displacement networks. ACS Synth. Biol. **5**(8), 885–897 (2016)
27. Lakin, M.R., Youssef, S., Polo, F., Emmott, S., Phillips, A.: Visual DSD: a design and analysis tool for DNA strand displacement systems. Bioinformatics **27**(22), 3211–3213 (2011)

Synchronization of Chaos with a Single Driving Variable Feedback Control Based on DNA Strand Displacement

Zijie Meng, Xiaoyu An, and Junwei Sun(✉)

Zhengzhou University of Light Industry, Zhengzhou 450002, China
junweisun@yeah.net

Abstract. DNA molecular reaction is one of methods to build next generation operational circuit. It has demonstrated the capacity to design control systems. In this paper, a PID controll based on DNA strand displacement (DSD) is presented to realize synchronization of chaotic system. Based on the theory of synchronization and design principle of control, the proportion terms, integration terms, and differentiation terms are added to chaotic oscillatory systems for implementing synchronization by DSD. The two four-variable chaotic oscillatory systems are tended to be synchronization by PID control. There are four chemical reaction modules which are proposed to realize the systems. The modeling and simulation results demonstrate the validity of chemical reaction terms and control systems by visual DSD and matelab. Achieved the ideal results when it is applied to the chaotic system synchronization.

Keywords: DNA strand displacement · PID control · Chaos synchronization

1 Introduction

Recently, DNA computing development has advanced significantly due to the advantages of parallelism and low energy consumption. DNA strand displacement (DSD) has been deeply development in the area of molecular computing, nanomachines, disease diagnosis and treatment [3,6,8,15]. The communicating facility in nanonetworks has emerged, and it is used in fields like biomedicine, environmental monitoring and manufacturing [7,12].

DNA molecules are widely used in information processing of biological networks because of their advantages such as miniaturization, large storage capacity and high parallelism, and a DNA reaction network for information processing is designed [2,5,17]. CRNs (Chemical reaction networks) has been used as a formal language for modeling molecular systems with three elements: reactants, products and reaction rates, the reaction networks are modeled by using the law of mass action kinetics. Each reaction network has a corresponding ordinary differential equation system through mass action [4,13,16]. Thus chemical reaction

L. Pan et al. (Eds.): BIC-TA 2021, CCIS 1565, pp. 437–446, 2022.
https://doi.org/10.1007/978-981-19-1256-6_34

networks which consisting by DSD reactions can approximate arbitrary polynomial ordinary differential equations. But chaos appeared in the construction of reaction networks.

Chaos is one of the hot spot in nonlinear science fields, due to the great value and broad application prospect in engineering technology. Chaos is being combined with many disciplines to produce new marginal disciplines, such as chaos image processing, chaos medicine, chaos control, chaos prediction, chaos art and *etc* [9,10,14,18]. The United States Naval Laboratory creatively proposed and realized the synchronization of chaotic systems [11]. The chaotic synchronization system is developed so that the signal processing can be realized more effectively and then it can be applied to communication.

To date, the coupling synchronization between the PI controller of chaotic system based on DSD and chaotic system has been realized [1,14]. Four-variable chaotic system is implemented and PID controller is realized and used to achieve the synchronization in this paper based on DSD. Firstly, the CRNs of the whole system is constructed, and then the DSD reaction module of CRNs is designed and implemented. The dynamic characteristics of the design system can be approached by cascading these modules. According to the results of simulation, the chemical reaction modules which have been designed to show that DSD implement CRNs can imitate the dynamical behavior of the target CRN. In this experiment, the DNA strands are built and the feasibility of the CRNs is tested by DSD by visual DSD software.

In this paper, the PID controller is introduced into the synchronization control of chaotic system based on DSD reaction for the first time, and other three controllers are realized by adjusting the controller parameters, and the advantages of each controller are compared and analyzed. That makes our work more meaningful.

2 Construction of Chaotic System

2.1 DNA Strand Displacement Modules

Several DNA strand displacement (DSD) reaction modules are proposed in this section such as annihilation, double, and catalysis reaction modules.

Double reaction module of ideal CRN is shown as Eq. (1), the DSD double reaction module is described by Eq. (2) and it is demonstrated in Fig. 1,

$$x \xrightarrow{k_i} 2x \tag{1}$$

$$A + X \xrightarrow{q_i} B + 2x \tag{2}$$

There are DSD reaction element x and auxiliary species A. $[X]_0$ is the initial strand concentration of species X, initial strand concentration of the auxiliary species A is set to C_m, where $[X]_0 \ll C_m$. The rate of DSD CRNs is $\frac{d[X]_t}{dt} =$

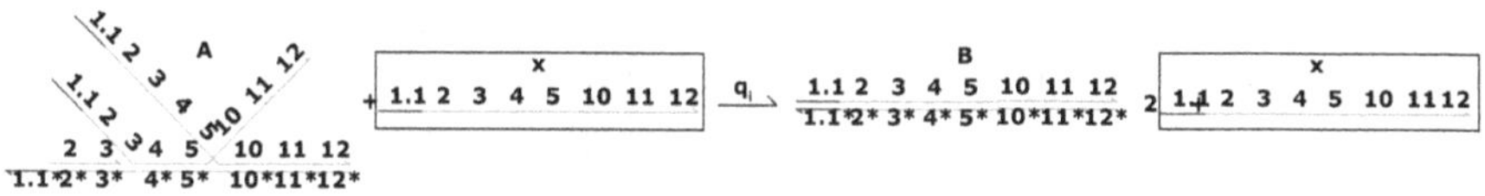

Fig. 1. DSD double reaction module.

$q_iC_m[X]_t$. At the time of DSD chemical reaction is in equilibrium, the expression $q_iC_m = k_1$ can be derived from the law of mass action.

Catalysis reaction module 1 of ideal CRN is shown as Eq. (3), the DSD catalysis reaction module 1 is described by Eq. (4) and it is demonstrated in Fig. 2,

$$x + y \xrightarrow{k_i} x + y + z \tag{3}$$

$$\begin{cases} A + X \underset{q_i}{\overset{q_i}{\rightleftharpoons}} B & B \underset{q_k}{\overset{q_m}{\rightleftharpoons}} C + D \\ C + Y \underset{q_m}{\overset{q_k}{\rightleftharpoons}} E & E \underset{q_k}{\overset{q_m}{\rightleftharpoons}} F + G \\ H + G \xrightarrow{q_k} I + x + y + z \end{cases} \tag{4}$$

There are DSD reaction element x and auxiliary species H and the concentration of the auxiliary species A and H are set to C_m, $C_m = 10^6 nM$, where $[X]_0 \ll C_m$.

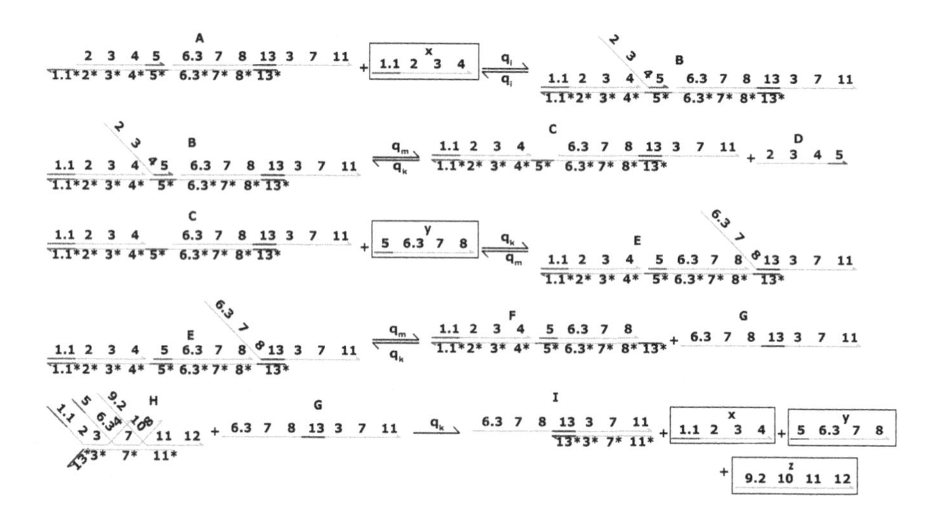

Fig. 2. DSD catalysis reaction module 1.

Catalysis reaction module 2 of ideal CRN is shown as Eq. (5), the DSD catalysis reaction module 2 is described by Eq. (6) and it is demonstrated in Fig. 3,

$$x + y + z \xrightarrow{k_i} x + y + z + w \tag{5}$$

$$\begin{cases} A + X \underset{q_i}{\overset{q_i}{\rightleftharpoons}} B & B \underset{q_k}{\overset{q_m}{\rightleftharpoons}} C + D \\ C + Y \underset{q_m}{\overset{q_k}{\rightleftharpoons}} E & E \underset{q_k}{\overset{q_m}{\rightleftharpoons}} E + Waste \\ F \underset{q_k}{\overset{q_m}{\rightleftharpoons}} F + G & H + I \xrightarrow{q_k} I + x + y + z + W \end{cases} \tag{6}$$

The Catalysis reaction module 2, there are chemical reaction element x and auxiliary species A and I. As mentioned earlier $[x]_0$ is the initial strand concentration of species x, initial strand concentrations of the auxiliary species AB are set to C_m, where $\frac{d[X]_t}{dt} = q_k qs[X]0Cm$. Auxiliary species A reacts with x to form intermediate B. B unbinds single-strand DNA D and intermediate product C by reversible reaction. C reacts with y to form species E, E through reversible reaction to unbind a single-strand DNA and species F. F reacts with z to form species H. H reacts with auxiliary material I to form x, y, z, w.

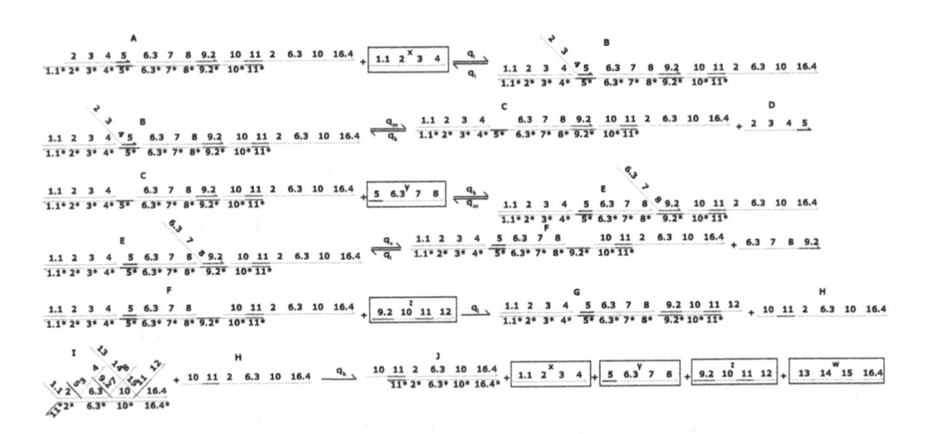

Fig. 3. DSD catalysis reaction module 2.

As we know, the systems signals can take positive or negative value, however the DNA strand concentrations can only take the non-negative values. For solving this problem we use the difference between two DNA strands concentrations to represent a signal like Eq. (7) and the value of x is $x = x^+ - x^-$. It is ca be represented by annihilation reaction module, the DSD reaction module is represented by Eq. (8), $q_i = 10^6 nMs^{-1}$. A is an auxiliary species, and the concentration is set to $C_m = 10(6)$. $[X]_0 \ll C_m$, $q_i C_m = k_i$.

$$x + y \xrightarrow{k_i} \phi \tag{7}$$

$$\begin{cases} A + X \underset{q_m}{\overset{q_i}{\rightleftharpoons}} B \qquad C + Y \xrightarrow{q_k} E + Waste \\ B \underset{q_k}{\overset{q_k}{\rightleftharpoons}} E + C \end{cases} \tag{8}$$

There is an implication that the value of x can be represented by many combinations of x^+ and x. The value of combination $x^+ = 1nM$ and $x^+ = 0nM$ can be represented by the combination $x^+ = 11nM$ and $x^+ = 10nM$. So the rule is used to get an unique representation of value x which $x^+ = 0nM$ or $x^+ = 0$ nM as the minimal representation of x. This rule can be more efficient in an actual operation. This rule is implemented by the annihilation reaction module, where $k_i C_m = k_4$.

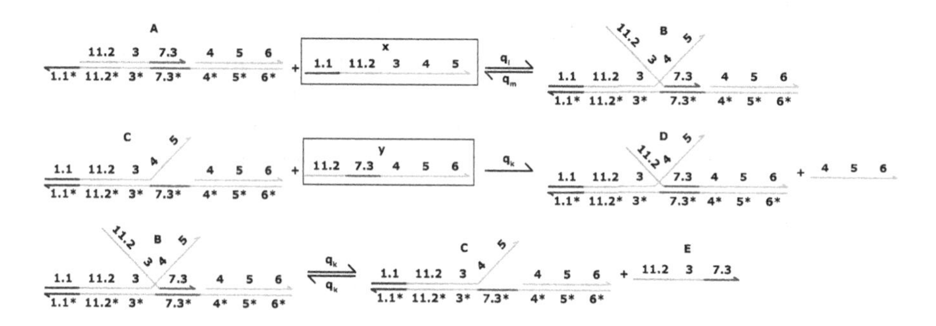

Fig. 4. DSD annihilation reaction module.

2.2 Chaotic System Implementation

The four-variables chaotic system can be described by ODEs (9)

$$\dot{x} = AF(x) + f(x) \tag{9}$$

where $x = [x_1, x_2, x_3, x_4]^T$. The matrix A and $f(x)$ is shown as: $A = \begin{bmatrix} 0 & 0 & 0 & 0 \\ 0 & 1 & 0 & 0 \\ 0 & 0 & -1 & 0 \\ 0 & 0 & 0 & 1 \end{bmatrix}$,

$f(x) = \begin{bmatrix} x_1^2 x_2 \\ -x_1^2 x_1 + x_2 x_3 \\ -x_2 x_3 + x_3 x_4 \\ -x_3 x_4 \end{bmatrix}$. The initial value is $x_1 = x_2 = x_3 = 1$ and $x_4 = 2$.

Its CRNs is obtained according to the law of mass action. For example, $\dot{X} = -a\lambda X^a Y^b, \dot{Y} = -b\lambda X^a Y^b, \dot{Z} = -c\lambda X^a Y^b$ are obtained among equation $aX + bY \xrightarrow{\lambda} cZ$ by the law of mass action. So the CRNs of system (9) is shown as

$$\begin{cases} 2X_1^{\pm} \xrightarrow{r_1} X_1^{\pm} & 2X_1^{\pm} + X_2^{\pm} \xrightarrow{r_2} 3X_1^{\pm} \\ X_2^{\pm} \xrightarrow{r_3} 2X_2^{\pm} & X_2^{\pm} + X_3^{\pm} \xrightarrow{r_4} 2X_2^{\pm} \\ X_3^{\pm} + X_4^{\pm} \xrightarrow{r_5} 2X_2^{\pm} & X_3^{\pm} \xrightarrow{r_6} \phi \\ X_4^{\pm} \xrightarrow{r_7} 2X_4^{\pm} \end{cases} \tag{10}$$

By adding the double rail representation, we can get the Eq. (11)

$$\begin{cases} \dot{x}_1^+ = r_1 x_1^+ x_1^- + r_2 (x_1^+)^2 x_2^+ + r_3 (x_1^-)^2 x_2^- + r_4 x_1^+ x_1^- x_2^- - \eta x_1^+ x_1^- \\ \dot{x}_1^- = r_2 (x_1^+)^2 + r_3 (x_1^-)^2 + + r_2 (x_1^+)^2 x_2^- + r_3 (x_1^-)^2 x_2^+ + r_4 x_1^+ x_1^- x_2^+ - \eta x_1^+ x_1^- \\ \dot{x}_2^+ = r_5 x_2^+ + r_2 (x_1^+)^2 x_2^- + r_3 (x_1^-)^2 x_2^+ + r_4 x_1^+ x_1^- x_2^+ + r_6 x_2^+ x_3^+ + r_6 x_2^- x_3^- - \eta x_2^+ x_2^- \\ \dot{x}_2^- = r_5 x_2^- + r_2 (x_1^+)^2 x_2^+ + r_3 (x_1^-)^2 x_2^- + r_4 x_1^+ x_1^- x_2^- + r_6 x_2^+ x_3^- + r_6 x_2^- x_3^+ - \eta x_2^+ x_2^- \\ \dot{x}_3^+ = r_7 x_3^- + r_6 x_2^+ x_3^- + r_6 x_2^- x_3^+ + r_6 x_3^+ x_4^+ + r_6 x_3^- x_4^- - \eta x_3^+ x_3^- \\ \dot{x}_3^- = r_7 x_3^+ + r_6 x_2^+ x_3^+ + r_6 x_2^- x_3^- + r_6 x_3^+ x_4^- + r_6 x_3^- x_4^+ - \eta x_3^+ x_3^- \\ \dot{x}_4^+ = r_8 x_4^+ + r_6 x_3^+ x_4^- + r_6 x_3^- x_4^+ - \eta x_4^+ x_4^- \\ \dot{x}_4^- = r_8 x_4^- + r_6 x_3^+ x_4^+ + r_6 x_3^- x_4^- - \eta x_4^+ x_4^- \end{cases} \tag{11}$$

The CRNs corresponding to Eq. (11) can be obtained according to the law of mass action. The cascade of DSD reaction modules can approach the kinetic characteristics of CRNs. The chaotic system (9) based on DSD modules can be obtained. By visual DSD, the simulation results are shown in Fig. 7.

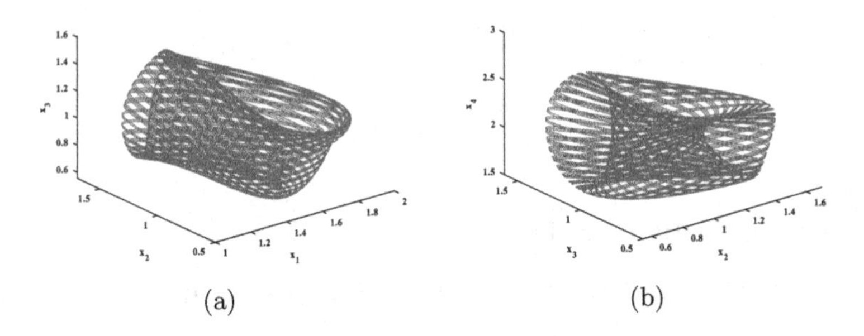

Fig. 5. Four-Variable chaotic oscillatory based on DSD (a) Attractor of $x_1 - x_2 - x_3$ (b) Attractor of $x_2 - x_3 - x_4$.

3 Synchronization of Chaotic System by PID Control

This paper uses a single driving variable to drive (signal) master-slaver synchronization. Master-slaver synchronization is that in chaotic synchronization, there is a master system or drive system, "a slaver system or response system", the slaver system is controlled and the master system is not controlled.

The difference between a signal of the master system and the signal corresponding to the slave system is input to the PID controller to form a drive signal (control signal), and the drive signal is returned to the slave system, so as to achieve the synchronization of the two systems. The flow of its work is shown in Fig. 6 In this paper, two chaotic systems with different initial values are used as control objects. The initial concentration of master system is $x_1 = x_2 = x_3 = 1$ and $x_4 = 2$. The initial concentration of slaver system is $x_1 = x_2 = x_3 == x_4 = 1.5$. The simulation results of synchronization of chaotic systems with different initial values are shown in Fig. 7 and Fig. 8. The synchronization of chaotic system controlled by PID can be easily converted to P control, PI control and PD control by adjusting the control parameters. The comparison is shown in Fig. 9.

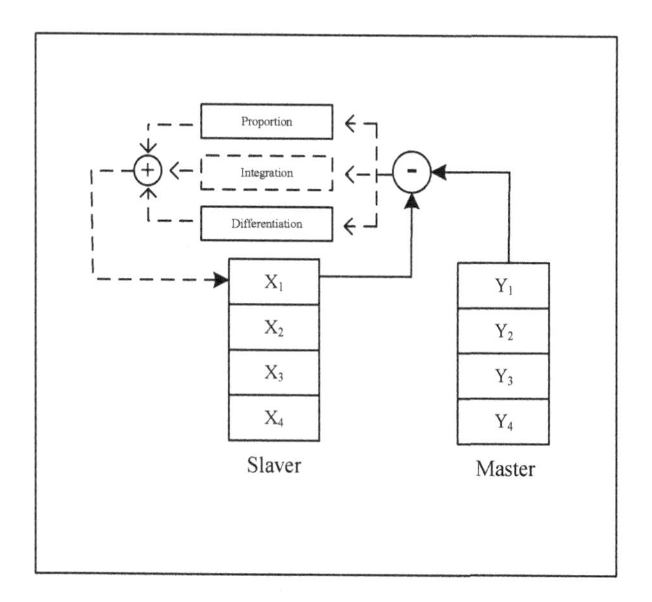

Fig. 6. Synchronization flow chart of chaotic system.

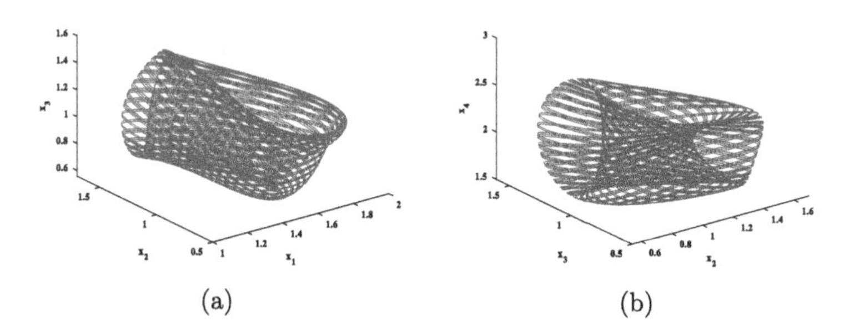

Fig. 7. Slaver system phase diagram (a) Attractor of $x_1 - x_2 - x_3$ (b) Attractor of $x_2 - x_3 - x_4$.

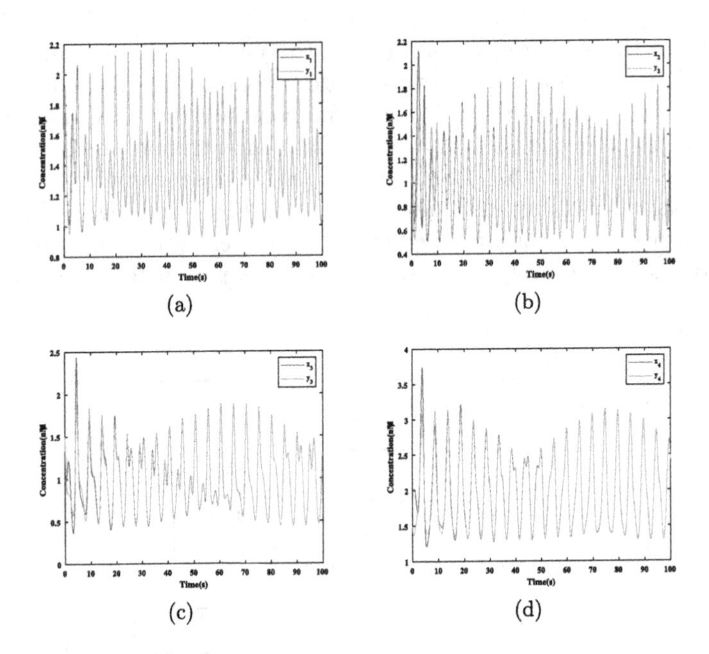

Fig. 8. Time domain waveform of chaotic system synchronization. (a) comparison of master system y_1 and slaver system X_1, (b) comparison of master system y_2 and slaver system X_2, (c) comparison of master system y_3 and slaver system X_3, (d) comparison of master system y_4 and slaver system X_4.

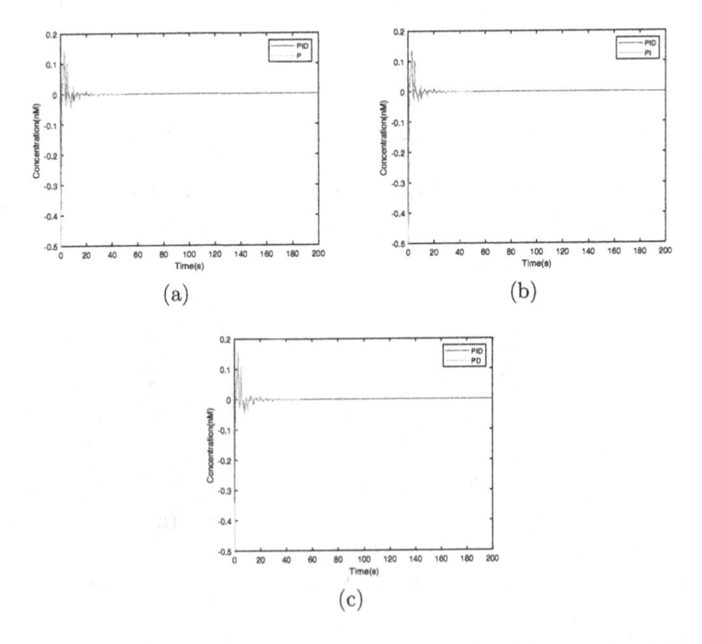

Fig. 9. Synchronization with PID controller, P controller, PI controller, PD controller

4 Conclusion

In this paper, PID control is used to realize the synchronization of four-variable chaotic oscillation system based on DNA chain displacement. Four DSD reaction modules of double, catalysis and annihilation were designed. According to the law of mass action, the CRNs construction of chaotic oscillation system and synchronous PID controller is completed, and it is realized by the cascade of DSD reaction module. It is proved that it is feasible to use DNA chain displacement to represent the synchronization of chaotic oscillation system and control system. Through the DNA chain permutation technology, the synchronization of chaotic systems with different initial values is realized by using PID control items. Through the simulation and comparison of P control, PI control, PD control and PID control, the results show that PID control achieves higher regulation quality.

References

1. An, X., Meng, Z., Wang, Y., Sun, J.: Proportional-integral-derivative control of four-variable chaotic oscillatory circuit based on DNA strand displacement. J. Nanoelectron. Optoelectron. **16**(4), 612–623 (2021)
2. Apraiz, A., Mitxelena, J., Zubiaga, A.: Studying cell cycle-regulated gene expression by two complementary cell synchronization protocols. JoVE (J. Visualized Exp.) (124), e55745 (2017)
3. Cardelli, L., Tribastone, M., Tschaikowski, M.: From electric circuits to chemical networks. Nat. Comput. **19**(1), 237–248 (2019). https://doi.org/10.1007/s11047-019-09761-7
4. Foo, M., Kim, J., Sawlekar, R., Bates, D.G.: Design of an embedded inverse-feedforward biomolecular tracking controller for enzymatic reaction processes. Comput. Chem. Eng. **99**, 145–157 (2017)
5. Fries, P., Reynolds, J.H., Rorie, A.E., Desimone, R.: Modulation of oscillatory neuronal synchronization by selective visual attention. Science **291**(5508), 1560–1563 (2001)
6. Huang, C., Han, Y.J., Sun, J.W., Zhu, W.J., Wang, Y.F., Zhou, Q.L.: The design and application of exclusive or logical computation based on DNA 3-Arm sub-tile self-assembly. J. Nanoelectron. Optoelectron. **15**(11), 1327–1334 (2020)
7. Li, J.X., Wang, Y.F., Sun, J.W.: Odd judgment circuit of four inputs based on DNA strand displacement. J. Nanoelectron. Optoelectron. **15**(3), 415–424 (2020)
8. Liu, X., Parhi, K.K.: Molecular and DNA artificial neural networks via fractional coding. IEEE Trans. Biomed. Circ. Syst. **14**(3), 490–503 (2020)
9. Oishi, K., Klavins, E.: Biomolecular implementation of linear I/O systems. IET Syst. Biol. **5**(4), 252–260 (2011)
10. Paulino, N.M.G., Foo, M., Kim, J., Bates, D.G.: On the stability of nucleic acid feedback control systems. Automatica **119**, 109103 (2020)
11. Pecora, L.M., Carroll, T.L.: Synchronization in chaotic systems. Phys. Rev. Lett. **64**(8), 821 (1990)
12. Qian, L., Winfree, E.: A simple DNA gate motif for synthesizing large-scale circuits. J. Royal Soc. Interface **8**(62), 1281–1297 (2011)

13. Sawlekar, R., Montefusco, F., Kulkarni, V.V., Bates, D.G.: Implementing nonlinear feedback controllers using DNA strand displacement reactions. IEEE Trans. Nanobiosci. **15**(5), 443–454 (2016)
14. Wang, Y., Li, Z., Sun, J.: Three-variable chaotic oscillatory system based on DNA strand displacement and its coupling combination synchronization. IEEE Trans. Nanobiosci. **19**(3), 434–445 (2020)
15. Wang, Y., Wang, P., Huang, C., Sun, J.: Five-input cube-root logical operation based on DNA strand displacement. J. Nanoelectron. Optoelectron. **13**(6), 831–838 (2018)
16. Yordanov, B., Kim, J., Petersen, R.L., Shudy, A., Kulkarni, V.V., Phillips, A.: Computational design of nucleic acid feedback control circuits. ACS Synth. Biol. **3**(8), 600–616 (2014)
17. Zou, C., Wei, X., Zhang, Q., Liu, Y.: Synchronization of chemical reaction networks based on DNA strand displacement circuits. IEEE Access **6**, 20584–20595 (2018)
18. Zou, C., Zhang, Q., Wei, X.: Compilation of a coupled hyper-chaotic Lorenz system based on DNA strand displacement reaction network. IEEE Trans. NanoBiosci. **20**(1), 92–104 (2020)

Sequential Spiking Neural P Systems with Polarizations Based on Minimum Spike Number Working in the Accepting Mode

Li Liu[1,2] and Keqin Jiang[1,2](✉)

[1] The University Key Laboratory of Intelligent Perception and Computing of Anhui Province, Anqing Normal University, Anqing 246133, Anhui, China
lliu150@yeah.net, jiangkq0519@163.com

[2] School of Computer and Information, Anqing Normal University, Anqing 246133, Anhui, China

Abstract. Inspired by the information transmission method of electrical signals in the biological impulse nervous system, a new variant of the spiking neural P systems, called spiking neural P systems with polarizations (PSN P systems for short), is proposed. In this research, the excitation conditions of PSN P systems are mainly determined by the polarization and the number of spikes together. Based on the fact that spiking neural P systems can be used as a different working device, the computational power of the sequential spiking neural P systems with polarization induced by the number of spikes can work in the receptive mode. Specifying the computational result as the temporal distance between the first two spike moments of input neuron reception proves that the PSN P systems are Turing universal as number accepting devices.

Keywords: Membrane computing · Spiking neural P system · Polarization · Accepting mode

1 Introduction

Since membrane computing was first proposed by European academician G. Păun in 1998, scholars from various countries have started to study the membrane computing boom. The distributed and parallel computing capabilities in membrane computing not only brought a new computing model to the computer community but also provided a better-applied research method to the fields of biology and linguistics [1]. According to the in-depth study of different biological backgrounds, the branch of membrane computing is divided into three categories: cell-like, tissue-like, and neural-like membrane computing models [2–4].

With the continuous development of research in each branch of membrane computing, in 2006, M. Ionescu et al. put forward spiking neural P system (SN P system for short), a biological phenomenon based on the transmission and

L. Pan et al. (Eds.): BIC-TA 2021, CCIS 1565, pp. 447–458, 2022.
https://doi.org/10.1007/978-981-19-1256-6_35

processing of signals between neurons. The spiking behavior in SN P systems is described using a formal language generated by spiking rules, introducing the concept of time as the encoding of information [5,6]. In terms of model characteristics, SN P systems also use impulse coding and are therefore called third generation neural network model [7]. To this day, spiking neural P systems are still the concentration of research for scholars at home and abroad, and there are countless research results in this field, such as: various variant models of spiking neural P systems [8–15], optimization algorithms and language generation [16,17], practical application problems [18–20], etc.

Spiking neural P systems with polarizations (PSN P systems for short) [21], one of the variants of spiking neural P systems, which were inspired by the potential changes during the transmission of information by neurons by Tingfang Wu et al. Based on the charged nature of neurons (−, 0, and + charges), a new rule-triggering mechanism was proposed to simplify the regular expressions, i.e., rules triggered by both charge and spiking, so as to achieve a free application of rules in neurons. Subsequently, the computational power and universality of different variants of models under PSN P systems were also studied by Tingfang Wu et al. [22,23], and the problem of rules on synapse was investigated by Jiang Suxia et al. [24]. So far, there are no more research results of spiking neural P systems with polarizations. So, more scholars need to continue to study more thoroughly.

Usually, there are four kinds of research devices for spiking neural P systems: the number generating devices, the number accepting devices, language generators and function computation devices, and similarly spiking neural P systems with polarizations have such a mode of operation. In the generating mode, Tingfang Wu et al. investigated the sequential PSN P systems caused by the minimum spikes number [25], i.e., if more than one neuron can be fired in each step of the system computation, then it is stipulated that one of the neurons possessing the minimum number of spikes is fired first. In this work, we study the computational universality of the sequential spiking neural P systems with polarizations caused by the minimum spikes number operating in the accepting mode, which is composed of an input module, a deterministic addition module and a subtraction module. The time interval between the reception of the first two spikes by the input neuron in the systems are taken as the final result of the computation. If the system eventually stops, then it indicates that the number is accepted by the system.

The main innovation of this study: reconstructing the system module and demonstrating the Turing universality of sequential PSN P systems based on the minimum number of spikes as the accepting device. In addition, as number accepting devices, the research uses fewer neurons than the module constructed in the literature [25].

The research results of this work are described in detail in the following sections: the next Sect. 2 presents a formal definition of PSN P systems and a brief introduction to the acceptance model. Section 3 investigates Turing universality of PSN P systems based on minimum spike number working in the accepting mode. Finally, in Sect. 4, conclusions and some derived questions are given.

2 Spiking Neural P Systems with Polarizations

This section briefly introduces the definition of the rules of spiking neural P systems with polarizations and the concept of related symbols, and how the execution of the rules and the definition of the results are implemented when the sequential spiking neural P systems with polarizations induced by the number of spikes is constructed.

A PSN P system of degree m ($m \geq 1$) has the following structure

$$\Pi = (O, \sigma_1, \sigma_2, \ldots, \sigma_m, syn, in),$$

where

(1) $O = \{a\}$ refers to the singleton alphabet, where a stands for the spike;
(2) $\sigma_i = (\alpha_i, n_i, R_i)$, $1 \leq i \leq m$, $(m \geq 1)$, σ_i denotes the i-th neuron and each neuron has the following element:
  a) $\alpha_i \in \{+, 0, -\}$ denotes the initial polarity (charge) in the neuron σ_i;
  b) $n_i \geq 0$ is the initial number of spikes in neuron σ_i;
  c) R_i is the set consisting of the following two forms:
    ① spiking rules: $\alpha/a^c \rightarrow a; \beta$, for α, $\beta \in \{+, 0, -\}$, $c \geq 1$;
    ② forgetting rules: $\alpha'/a^s \rightarrow \lambda; \beta'$, for α', $\beta' \in \{+, 0, -\}$, $s \geq 1$;
    If both ① and ② are in R_i, note the restriction: $\alpha' \neq \alpha$.
(3) $syn \subseteq \{1, 2, \ldots, m\} \times \{1, 2, \ldots, m\}$ represents the synaptic connections between neurons, where $i \neq j$ for each $(i, j) \in syn$, $1 \leq i, j \leq m$;
(4) $in \in \{1, 2, \ldots, m\}$ represents the input neuron.

In the spiking rule of ①, if the neuron σ_i contains spiking rules, once the polarity in that neuron is α ($\alpha \in \{+, 0, -\}$) and possesses c spikes ($c \geq 1$), the spiking rule is triggered immediately, i.e., meaning that the process consumes c spikes from neuron σ_i and immediately sends a spike with β charge to all neighboring neurons that are synaptically connected to it.

In the forgetting rule of ②, if the neuron σ_i contains forgetting rules, once the polarity in that neuron is α' ($\alpha' \in \{+, 0, -\}$) and possesses s spikes ($s \geq 1$), the forgetting rule is triggered immediately, i.e., meaning that the process neuron σ_i consumes s spikes by itself and immediately sends a β' charge to all adjacent neurons connected by the synapse. However, note that if there are both spiking and forgetting rules in a neuron, the polarity satisfaction conditions of both rules cannot be the same ($\alpha' \neq \alpha$), and only one rule can be used for application when the rule is enabled (there are multiple rules in the neuron, and when one rule is enabled by satisfying the condition, none of the other rules is available).

In (4) the neuron σ_{in} denotes the input neuron. When spiking neural P systems with polarizations operate in the receptive mode, the system is computed as the time interval between the first two spikes received by the input neuron ($t_2 - t_1 - 1$, where t_1, t_2 are the moments of the first two spikes received, respectively), that is, the number n is encoded as spiking string, e.g., $10^{2n}1$. The moment when the input neuron σ_{in} receives a spike with charge is marked as 1, and input neuron σ_{in} does not accept spike with charges are marked as 0. Thus, the spiking string in the environment is fed to a specific register in the system through the input neuron σ_{in}. If the system computation stops, it means that the number n is accepted by the system Π, and the set of all numbers accepted by the system is noted as $N_{acc}(\Pi)$.

Spiking neural P systems with polarizations use a regular trigger mechanism in which not only the number of spikes but also the charge condition is met, so that when the presynaptic neuron sends a spike with a charge to the postsynaptic neuron, but due to the reaction principle that comes with the charge (0, +, and −), the postsynaptic neuron receives the charge and undergoes a polarization reaction (here the charge reaction is considered as not consuming any time). The polarization reactions are as follows:

(i) when some positive (negative) charges meet (priority in the polarization reaction) or when a positive (negative) charge meets a neutral charge (0), the polarization reaction results in a positive (negative) charge;
(ii) when a positive charge meets a negative charge, the polarization reaction results in a neutral charge.

In the sequential PSN P system Π, each step of systems operation, there are multiple active neurons, only one of the active neurons with the minimum number of spikes can be fired first, i.e., the min-sequentiality-strategy—one of the excitable neurons containing the minimum number of spikes is selected to start firing.

When the system Π operates as number accepting devices, the input neuron is used to receive spikes from the environment, and when the system Π eventually stops, the number is said to be received by the system Π, and the cluster of the set of all numbers received by the system Π corresponds to the set of Turing computable numbers (denoted by NRE). $N_{2,acc}PSNP(polar_p)$ is used to denote the set of numbers accepted by the PSN P system using the min-sequentiality-strategy, where the subscript 2 indicates that the computed number is obtained from the time interval between the first two spikes received, p stands for polarity ($p \leq 3$).

3 Turing Universality of PSN P Systems Based on Minimum Spike Number Working in the Accepting Mode

This section proves PSN P systems using the min-sequentiality strategy as number accepting devices are Turing universality by simulating deterministic register machines, i.e., it is able to describe NRE (Turing computable number set clusters).

A deterministic register machine is a five-tuple $M = (m, H, l_0, l_h, I)$, where m represents the number of registers, H stands for the set of instruction labels. l_0 and l_h denote the starting and halt labels (corresponding to the instruction HALT), and I is the set of instructions, of which there are three types.

- deterministic ADD instructions: l_i : (ADD(r), l_j): Make the value in register r incremented by 1, then choose to execute the instruction l_j.
- SUB instructions: l_i : (SUB(r), l_j, l_k): If the value in register r is not zero, its stored value is subtracted by 1 and the generated spiking is released from instruction l_j; if the value in register r is zero, its stored value remains unchanged and the spiking is released from instruction l_k.
- the halt instruction: l_h: terminate instruction denotes HALT.

In the accepting mode, the computation starts by storing a natural number n in a particular register r (usually register 1), and all other registers are empty. At the end of the input module (end of stored data), the start instruction l_0 is first executed and then runs as described above for the use of instructions. If the termination instruction l_h is reached, which means that the computation halts, the natural number n is accepted by the system Π. The time interval between the first two spikes received by the input neuron is used as the result of the system's computation, i.e., $t_2 - t_1 - 1$.

Theorem 1. $N_{2,acc}PSNP(polar_3) = NRE$.

Proof. Here we only need to prove the inclusion relationship $NRE \subseteq N_{2,acc}PSNP(polar_3)$. The converse inclusion is straightforward but cumbersome (for similar technical details, please refer to Section 8.1 in [1]). In the acceptance mode, the deterministic register M is simulated by constructing PSN P systems adopting the min-sequentiality strategy, where system Π consists of input, deterministic addition, and subtraction modules.

Module INPUT: simulating a input module

The input module, shown in Fig. 1, first reads into a particular register through the neuron σ_{in} the number n to be computed (accepted) encoding a spike string shaped as $10^{2n}1$. If the read symbol is 1, the input neuron receives a spike; if the read symbol is 0, the input neuron does not receive a spike.

Suppose that at step t the neuron σ_{in} receives the first spike from the environment and begins to fire, sending a spike with a positive charge to each of the neurons σ_{d_1} and σ_{d_2}. At this point, neuron σ_{d_1} will remain inactive and wait for the arrival of the second spike. While neuron σ_{d_2} fires at step $t+1$, neuron σ_{d_3} is in the excited state due to the receipt of a spike with negative charge, and a spike with negative charge will be added to neuron σ_1.

At step $t+2$, both neurons σ_{d_2} and σ_{d_3} are in the excited state, and according to the min-sequentiality strategy, neuron σ_{d_3} fires preferentially, sending a spike with a positive charge to neurons σ_{d_2} and σ_1. Thus, neurons σ_{d_2} and σ_1 have three spikes and two spikes (neuron σ_1 is polarized as neutral and unexcitable), respectively, so that neurons σ_{d_2} and σ_{d_3} will fire alternately next, increasing the number of spikes in neuron σ_1 by two every two steps.

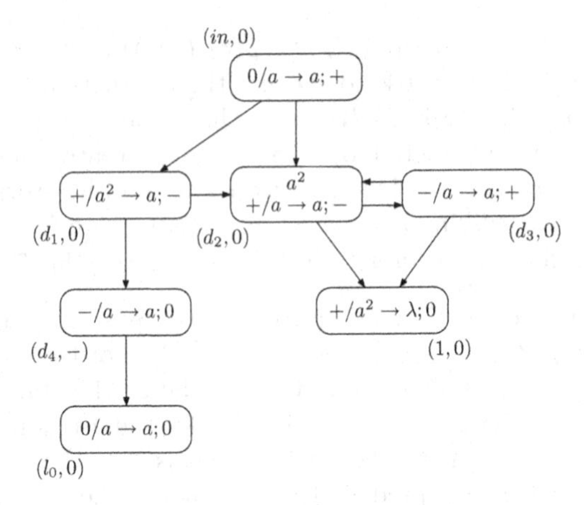

Fig. 1. Input module of of Π.

Table 1. The polarization evolution and the spike number in the neurons of the Input module.

Step	Neuron						
	in	d_1	d_2	d_3	d_4	σ_1	l_0
t	(0,1)	(0,0)	(0,2)	(0,0)	(–,0)	(0,0)	(0,0)
$t+1$	(0,0)	(+,1)	(+,3)	(0,0)	(–,0)	(0,0)	(0,0)
$t+2$	(0,0)	(+,1)	(+,2)	(–,1)	(–,0)	(–,1)	(0,0)
$t+3$	(0,0)	(+,1)	(+,3)	(–,0)	(–,0)	(0,2)	(0,0)
$t+4$	(0,0)	(+,1)	(+,2)	(-,1)	(–,0)	(–,3)	(0,0)
...	...	...	...	...	...	...	...
$t+2n+1$	(0,1)	(+,1)	(+,3)	(–,0)	(–,0)	(0,2n)	(0,0)
$t+2n+2$	(0,0)	(+,2)	(+,4)	(–,0)	(–,0)	(0,2n)	(0,0)
$t+2n+3$	(0,0)	(+,0)	(0,5)	(–,0)	(–,1)	(0,2n)	(0,0)
$t+2n+4$	(0,0)	(+,0)	(0,5)	(–,0)	(–,0)	(0,2n)	(0,1)

If by step $t+2n+1$, neuron σ_{in} receives a second spike from the environment, and the state of each neuron in the system at this time, neurons σ_{d_2} and σ_1 have three spikes (positive polarity, excitable state), $2n$ spikes (neutral polarity, unexcitable, indicating that the number stored in register 1 is n), according to the min-sequentiality strategy, neuron σ_{in} fires first, sending a second spike to neurons σ_{d_1} and σ_{d_2}. At step $t+2n+2$, both system neurons σ_{d_1} and σ_{d_2} are excitable (neuron σ_{d_1} has two spikes with positive polarity and neuron σ_{d_2} has four spikes with positive polarity), and again based on the min-sequentiality strategy, neuron σ_{d_1} fires first. At step $t+2n+3$, the neuron σ_{d_2} is in a non-excitable state because it receives a spike with a negative charge and the positive and negative charges cancel each other and the polarity becomes neutral. In contrast, neuron σ_{d_4} receives a spike transmitted by neuron σ_{d_1} and is excitable. In the next step, neuron σ_{l_0} fires and system Π begins to simulate the starting instruction σ_{l_0} in register M.

The specific evolution of the polarization and spike number can be referred to Table 1.

Module ADD: simulating a deterministic addition $l_i : (\mathtt{ADD}(r), l_j)$.

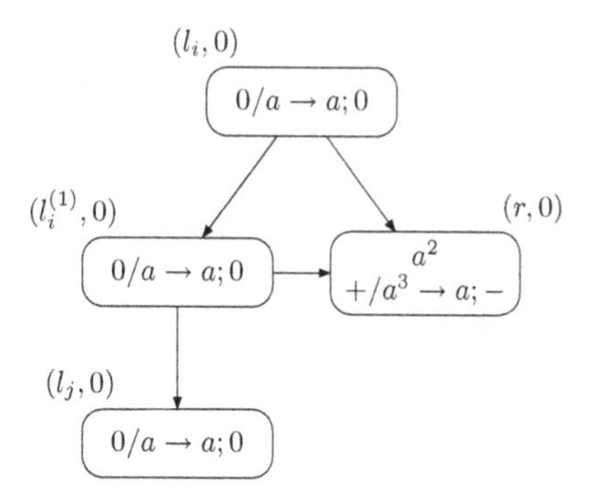

Fig. 2. ADD module of Π to simulate instruction $l_i : (\mathtt{ADD}(r), l_j)$.

The deterministic addition module is shown in Fig. 2. Neuron σ_{l_i} receives a spike and fires it, sending a spike and a neutral charge to neuron $\sigma_{l_i^{(1)}}$ and σ_r, so that neurons $\sigma_{l_i^{(1)}}$ and σ_r have one and $2n+3$ spikes (both neutral in polarity), respectively. Next, neuron $\sigma_{l_i^{(1)}}$ fires and transmits a spike with neutral charge to neurons σ_{l_j} and σ_r. Finally, the neuron σ_{l_j} fires and begins simulating the l_j instruction in the register M, while the number of spikes in the neuron σ_r is $2n+4$ (i.e., it simulates the number stored in the register r plus 1).

The addition module runs a relatively simple process, so the table description is omitted here.

Module SUB: simulating a subtraction instruction $l_i : (\mathtt{SUB}(r), l_j, l_k)$.

Since in the subtraction module, the computation in register 1 in register machine M is only incremental, we construct the subtraction module with $r \neq 1$ as shown in Fig. 3.

Suppose that at step t, the system Π begins to simulate the instruction $l_i : (\mathtt{SUB}(r), l_j, l_k)$. The neurons σ_{l_i}, $\sigma_{l_i^{(1)}}$, and $\sigma_{l_i^{(2)}}$ fire sequentially and transmit both the spikes with charge generated by each to the neuron $\sigma_{l_i^{(4)}}$. At step $t+3$, the neurons σ_r and $\sigma_{l_i^{(4)}}$ have $2n+3$ $(n \geq 0)$, four spikes (both are positive polarizations and in the excitable state) respectively. Thus, depending on the value of n encoded by the number of spikes in neuron σ_r, they can be classified as follows.

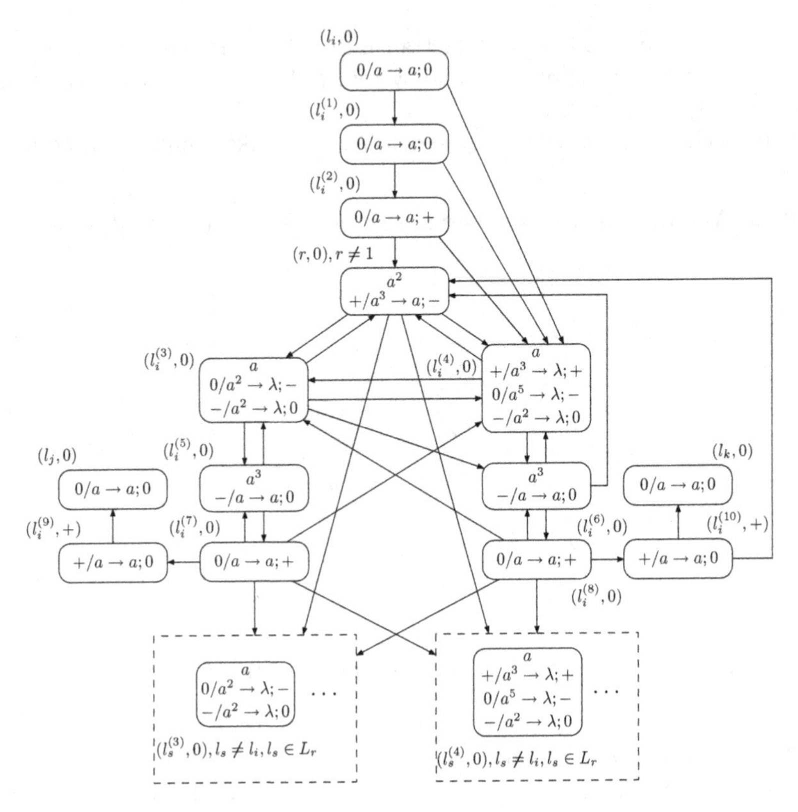

Fig. 3. SUB module of Π to simulate instruction l_i : $(\texttt{SUB}(r), l_j, l_k)$, $r \neq 1$, where $L_r = \{l \mid l$ is a label of a SUB instruction operating on register $r\}$.

(1) If the number of spikes in neuron σ_r are $2n + 3$ ($n > 0$, the corresponding register r turns out to be non-empty), then at step $t + 3$ neuron σ_r has at least five spikes. According to the min-sequentiality strategy, neuron $\sigma_{l_i^{(4)}}$ fires preferentially, using the spiking rule $+/a^3 \to a; -$, itself consuming three spikes, sending a positive charge to each of the neurons σ_r, $\sigma_{l_i^{(3)}}$ and $\sigma_{l_i^{(6)}}$. At step t + 4, neuron σ_r fires, consumes three spikes of its own ($2n$ spikes remaining, which corresponds to minus 1 in register r), and sends a spike with negative charge to each of neurons $\sigma_{l_i^{(3)}}$ and $\sigma_{l_i^{(4)}}$. At step $t + 5$, the only excitable neuron $\sigma_{l_i^{(3)}}$ in the system, sends a negative charge to the adjacent neuron using the forgetting rule $0/a^2 \to \lambda; -$, and according to the polarization response rule, both neurons σ_r and $\sigma_{l_i^{(6)}}$ restore the initial polarity state of the system. At step t+6, neurons $\sigma_{l_i^{(4)}}$ and $\sigma_{l_i^{(5)}}$ are both excitable and have two and three spikes, respectively. According to the min-sequentiality strategy, neuron $\sigma_{l_i^{(4)}}$ fires preferentially and consumes two spikes each by itself. Next, neurons $\sigma_{l_i^{(5)}}$, $\sigma_{l_i^{(7)}}$ and $\sigma_{l_i^{(9)}}$ fire sequentially,

and the charges and spikes generated during this process are transmitted to the neighboring neurons connected to the synapse, so that neurons $\sigma_{l_i^{(3)}}$, $\sigma_{l_i^{(4)}}$ and $\sigma_{l_i^{(5)}}$ are restored to the initial values of each neuron. Finally, the neuron σ_{l_j} fires and the system Π begins to simulate the instruction l_j in the register M.

Table 2. The polarization evolution and the spike number in neurons of the SUB module when the number encoded by spikes in neuron σ_r exceeds zero.

Step	Neuron										
	σ_{l_i}	σ_r	$\sigma_{l_i^{(1)}}$	$\sigma_{l_i^{(2)}}$	$\sigma_{l_i^{(3)}}$	$\sigma_{l_i^{(4)}}$	$\sigma_{l_i^{(5)}}$	$\sigma_{l_i^{(6)}}$	$\sigma_{l_i^{(7)}}$	$\sigma_{l_i^{(9)}}$	σ_{l_j}
t	(0,1)	(0, $2n+2$)	(0,0)	(0,0)	(0,1)	(0,1)	(0,3)	(0,3)	(0,0)	(+,0)	(0,0)
$t+1$	(0,0)	(0, $2n+2$)	(0,1)	(0,0)	(0,1)	(0,2)	(0,3)	(0,3)	(0,0)	(+,0)	(0,0)
$t+2$	(0,0)	(0, $2n+2$)	(0,0)	(0,1)	(0,1)	(0,3)	(0,3)	(0,3)	(0,0)	(+,0)	(0,0)
$t+3$	(0,0)	(+, $2n+3$)	(0,0)	(0,0)	(0,1)	(+,4)	(0,3)	(0,3)	(0,0)	(+,0)	(0,0)
$t+4$	(0,0)	(+, $2n+3$)	(0,0)	(0,0)	(+,1)	(+,1)	(0,3)	(+,3)	(0,0)	(+,0)	(0,0)
$t+5$	(0,0)	(+, $2n$)	(0,0)	(0,0)	(0,2)	(0,2)	(0,3)	(+,3)	(0,0)	(+,0)	(0,0)
$t+6$	(0,0)	(0, $2n$)	(0,0)	(0,0)	(0,0)	(–,2)	(–,3)	(0,3)	(0,0)	(+,0)	(0,0)
$t+7$	(0,0)	(0, $2n$)	(0,0)	(0,0)	(0,0)	(–,0)	(–,3)	(0,3)	(0,0)	(+,0)	(0,0)
$t+8$	(0,0)	(0, $2n$)	(0,0)	(0,0)	(0,1)	(–,0)	(–,2)	(0,3)	(0,1)	(+,0)	(0,0)
$t+9$	(0,0)	(0, $2n$)	(0,0)	(0,0)	(0,1)	(0,1)	(0,3)	(0,3)	(0,0)	(+,1)	(0,0)
$t+10$	(0,0)	(0, $2n$)	(0,0)	(0,0)	(0,1)	(0,1)	(0,3)	(0,3)	(0,0)	(+,0)	(0,1)

(2) If the number of spikes in neuron σ_r are three ($n = 0$, the corresponding register r turns out to be empty), then at the step $t + 3$, the neuron σ_r fires preferentially according to the min-sequentiality strategy, neuron σ_r consumes three spikes and sends a spike with negative charge to each of neurons $\sigma_{l_i^{(3)}}$ and $\sigma_{l_i^{(4)}}$. At step $t + 4$, neuron $\sigma_{l_i^{(3)}}$ and $\sigma_{l_i^{(4)}}$ have two and five spikes in them and both are in the excitable state. Again, according to the min-sequentiality strategy, neuron $\sigma_{l_i^{(3)}}$ fires preferentially, using the forgetting rule $-/a^2 \rightarrow \lambda; 0$, consuming two spikes itself, and the neutral charge generated by this forgetting rule has no effect on the polarization state of the other adjacent neurons. At step $t+5$, neuron $\sigma_{l_i^{(4)}}$ firesusing the forgetting rule $0/a^5 \rightarrow \lambda; -$, and sends a negative charge to neurons σ_r, $\sigma_{l_i^{(3)}}$ and $\sigma_{l_i^{(6)}}$, respectively, which changes the polarity of neuron σ_r to neutral according to the polarization response rule. At step $t+6$, neuron $\sigma_{l_i^{(6)}}$ fires, consumes a spike and sends a spike with a neutral charge to neurons σ_r, $\sigma_{l_i^{(4)}}$ and $\sigma_{l_i^{(8)}}$. Next, neurons $\sigma_{l_i^{(8)}}$ and $\sigma_{l_i^{(10)}}$ fire sequentially, and the charges and spikes generated during this process are transmitted to the adjacent neurons connected by the synapse, causing neurons σ_r, $\sigma_{l_i^{(3)}}$ and $\sigma_{l_i^{(6)}}$ to all return to the initial values of the respective neurons. At step $t+8$, the neuron σ_{l_k}

Table 3. The polarization evolution and the spike number in the neurons of the SUB module when the number encoded by spikes in neuron σ_r equals to zero.

Step	Neuron									
	σ_{l_i}	σ_r	$\sigma_{l_i^{(1)}}$	$\sigma_{l_i^{(2)}}$	$\sigma_{l_i^{(3)}}$	$\sigma_{l_i^{(4)}}$	$\sigma_{l_i^{(6)}}$	$\sigma_{l_i^{(8)}}$	$\sigma_{l_i^{(10)}}$	σ_{l_k}
t	(0,1)	(0,2)	(0,0)	(0,0)	(0,1)	(0,1)	(0,3)	(0,0)	(+,0)	(0,0)
$t+1$	(0,0)	(0,2)	(0,1)	(0,0)	(0,1)	(0,2)	(0,3)	(0,0)	(+,0)	(0,0)
$t+2$	(0,0)	(0,2)	(0,0)	(0,1)	(0,1)	(0,3)	(0,3)	(0,0)	(+,0)	(0,0)
$t+3$	(0,0)	(+,3)	(0,0)	(0,0)	(0,1)	(+,4)	(0,3)	(0,0)	(+,0)	(0,0)
$t+4$	(0,0)	(+,0)	(0,0)	(0,0)	(–,2)	(0,5)	(0,3)	(0,0)	(+,0)	(0,0)
$t+5$	(0,0)	(+,0)	(0,0)	(0,0)	(–,0)	(0,5)	(0,3)	(0,0)	(+,0)	(0,0)
$t+6$	(0,0)	(0,0)	(0,0)	(0,0)	(–,0)	(0,0)	(–,3)	(0,0)	(+,0)	(0,0)
$t+7$	(0,0)	(0,1)	(0,0)	(0,0)	(–,0)	(0,1)	(–,2)	(0,1)	(+,0)	(0,0)
$t+8$	(0,0)	(0,1)	(0,0)	(0,0)	(0,1)	(0,1)	(0,3)	(0,0)	(+,1)	(0,0)
$t+9$	(0,0)	(0,2)	(0,0)	(0,0)	(0,1)	(0,1)	(0,3)	(0,0)	(+,0)	(0,1)

fires, which indicates that the system Π begins to simulate the instruction l_k in the register M.

The specific evolution of the polarization and spike number can be referred to Tables 2 and 3.

When constructing the subtraction module, two auxiliary neurons $\sigma_{l_s^{(3)}}$ and $\sigma_{l_s^{(4)}}$ are added so that there is no interaction between the subtraction modules, and the details are not repeated here [25].

Based on the above description of the input, deterministic addition, and subtraction modules, we obtain that the system Π correctly simulates the register M under the min-sequentiality strategy, i.e., Theorem 1 is proved.

4 Conclusions and Discussions

In this paper, spiking neural P systems with polarizations using the min-sequentiality strategy as number accepting devices are proved to be Turing universal by constructing an input module, a deterministic addition module and a subtraction module. This work and literature [25] integrate a study on the computational universality of sequential spiking neural P systems with polarization induced by minimum spike number. Therefore, this research also applies to the Turing universality of the PSN P systems using min-pseudo-sequentiality.

In the number generating and number accepting modes, PSN P systems working in the min-sequentiality strategy are computationally universality. As it is well known that spiking neural P systems can also be used as language generators and function computation devices, it is discussed whether the sequential PSN P systems induced by the number of spikes has language generation capability when used as language generator, and whether it can inscribe a series of language generation problems such as recursively enumerable languages. And if it is also possible to study small universalized PSN P systems adopting the min-sequentiality strategy as a function of computational type? These questions are open for further research.

Acknowledgments. This work was supported by Anhui Provincial Natural Science Foundation (No. 1808085MF173), Natural Science Foundation of Colleges and Universities in Anhui Province of China (No. KJ2021A0640).

References

1. Păun, G.: Membrane Computing - An Introduction. Springer-Verlag, Berlin (2002)
2. Păun, G.: Computing with membranes. Journal of Computer and System Sciences. **61**(1), 108–143 (2000)
3. Păun, G., Rozenberg, G., Salomaa, A. (eds.): Handbook of Membrane Computing. Oxford University Press, Cambridge (2010)
4. Zhang, G., Pan, L.: A survey of membrane computing as a new branch of natural computing. Chinese Journal of Computers **33**(2), 208–214 (2010)
5. Ionescu, M., Păun, G., Yokomori, T.: Spiking neural P systems. Fundamenta Informaticae 71(2, 3), 279–308 (2006)
6. Ibarra, O.H., Woodworth, S.: Characterizations of some classes of spiking neural P systems. Natural Computing **7**(4), 499–517 (2008)
7. Tan, C., Šarlija, M., Kasabov, N.: Spiking neural networks: background, recent development and the NeuCube architecture. Neural Processing Letters **52**(2), 1675–1701 (2020)
8. Ibarra, O.H., Păun, A., Rodríguez-Patón, A.: Sequential SNP systems based on min/max spike number. Theoretical Computer Science **410**, 2982–2991 (2009)
9. Pan, L., Păun, G.: Spiking neural P systems with anti-spikes. International Journal of Computers, Communications and Control **4**(3), 273–282 (2009)
10. Song, T., Pan, L., Păun, G.: Asynchronous spiking neural P systems with local synchronization. Information Sciences **219**, 197–207 (2013)
11. Song, T., Pan, L., Păun, G.: Spiking neural P systems with rules on synapses. Theoretical Computer Science **529**, 82–95 (2014)
12. Wu, T., Zhang, Z., Păun, G., Pan, L.: Cell-like spiking neural P systems. Theoretical Computer Science **623**, 180–189 (2016)
13. Peng, H., Yang, J., Wang, J., et al.: Spiking neural P systems with multiple channels. Neural Networks **95**, 66–71 (2017)
14. Wu, T., Wang, Y., Jiang, S., et al.: Spiking neural P systems with rules on synapses and anti-spikes. Theoretical Computer Science **724**, 13–27 (2018)
15. Peng, H., Lv, Z., Li, B., et al.: Nonlinear spiking neural P systems. International Journal of Neural Systems **30**(10), 2050008 (2020)
16. Zhang, G., Rong, H., Neri, F., et al.: An optimization spiking neural P system for approximately solving combinatorial optimization problems. International Journal of Neural Systems **24**(05), 1440006 (2014)
17. Huang, Y., Li, B., Lv, Z., et al.: On string languages generated by spiking neural P systems with multiple channels. International Journal of Unconventional Computing **14**, 243–266 (2019)
18. Zhang, G., Pérez-Jiménez, M.J., Gheorghe, M.: Real-life applications with membrane computing. Springer International Publishing, Berlin (2017)

19. Fan, S., Paul, P., Wu, T., et al.: On applications of spiking neural P systems. Applied Sciences **10**(20), 7011 (2020)
20. Song, T., Pan, L., Wu, T., et al.: Spiking neural P systems with learning functions. IEEE Transactions on Nanobioscience **18**(2), 176–190 (2019)
21. Wu, T., Păun, A., Zhang, Z., et al.: Spiking neural P systems with polarizations. IEEE Transactions on Neural Networks and Learning Systems **29**(8), 3349–3360 (2018)
22. Wu, T., Pan, L., Alhazov, A.: Computation power of asynchronous spiking neural P systems with polarizations. Theoretical Computer Science **777**, 474–489 (2019)
23. Wu, T., Zhang, T., Xu, F.: Simplified and yet Turing universal spiking neural P systems with polarizations optimized by anti-spikes. Neurocomputing **414**, 255–266 (2020)
24. Jiang, S., Fan, J., Liu, Y., et al.: Spiking neural P systems with polarizations and rules on synapses. Complexity **2020**, 1–12 (2020)
25. Wu, T., Pan, L.: The computation power of spiking neural P systems with polarizations adopting sequential mode induced by minimum spike number. Neurocomputing **401**, 392–404 (2020)

Author Index

Printed in the United States
by Baker & Taylor Publisher Services